Lecture Notes in Computer Science 7947

Commenced Publication in 1973
Founding and Former Series Editors:
Gerhard Goos, Juris Hartmanis, and Jan van Leeuwen

Andreas Holzinger Gabriella Pasi (Eds.)

Human-Computer Interaction and Knowledge Discovery in Complex, Unstructured, Big Data

Third International Workshop, HCI-KDD 2013
Held at SouthCHI 2013
Maribor, Slovenia, July 1-3, 2013
Proceedings

Volume Editors

Andreas Holzinger
Medical University of Graz (MUG)
Institute for Medical Informatics, Statistics and Documentation (IMI)
Auenbruggerplatz 2/V, 8036, Graz, Austria
E-mail: andreas.holzinger@medunigraz.at

Gabriella Pasi
University of Milano-Bicocca
Department of Informatics, Systems and Communication
Viale Sarca 336, 20126 Milano, Italy
E-mail: pasi@disco.unimib.it

ISSN 0302-9743 e-ISSN 1611-3349
ISBN 978-3-642-39145-3 e-ISBN 978-3-642-39146-0
DOI 10.1007/978-3-642-39146-0
Springer Heidelberg Dordrecht London New York

Library of Congress Control Number: 2013941040

CR Subject Classification (1998): H.5, H.4, H.3, I.2.6, J.1, C.2, K.4

LNCS Sublibrary: SL 3 – Information Systems and Application, incl. Internet/Web and HCI

Typesetting: Camera-ready by author, data conversion by Scientific Publishing Services, Chennai, India

Printed on acid-free paper

Springer is part of Springer Science+Business Media (www.springer.com)

Preface

One of the grand challenges in our networked digital world includes the increasingly large amounts of data sets. Most of these complex "big data" are weakly structured or even unstructured. To gain insight into these data, to make sense of these data, to gain new scientific knowledge out of them, we must proceed to work on and aim to define efficient and user-friendly solutions to handle and to analyze such data.

Pioneering into a previously unknown territory, there is a vast area of undiscovered land to explore. Seeing the world in data is awesome, and it allows the discovery of novel areas, unexpected facets, and intriguing insights of and into the fascinating science of data, information, and knowledge. One success factor is a cross-disciplinary approach – integrating and appraising different fields, diverse subjects, different perspectives, and different opinions, so as to have a fresh look at novel ideas and at finding the methodologies on how to put these ideas into business.

The task force HCI-KDD pursues the ultimate goal of combining the best of two worlds: Knowledge Discovery from Data (KDD), dealing with computational intelligence and Human–Computer Interaction (HCI), dealing with human intelligence.

Some of the most pressing questions of nature are interwoven in these areas: What is information? What is life? What is computable? How can we build systems to facilitate understanding and sensemaking?

The first international special session of this network of excellence was organized in Graz, Austria, in the context of USAB 2011 "Information Quality in e-Health."

The second session was organized in the context of the World Intelligence Congress 2012 in Macau, China, in light of the Alan Turing centenary celebration.

This third special session was organized in the context of SouthCHI 2013 in Maribor, Slovenia.

The mission is to bring together researchers from diverse areas in a cross-disciplinary way, to stimulate new ideas, and to encourage multi-disciplinary research. Working in an interdisciplinary area requires the ability to communicate with professionals in other disciplines and the willingness to accept and incorporate their particular points of view.

The topics covered by this special session focused on aspects related to interactive analytics of weakly structured or unstructured data, including understanding of human behavior, designing novel quasi-intelligent systems to support interactive analytics, and applications involving novel interactions of humans with a data mining or machine learning system. We emphasize the importance of the human-in-the-loop, as we are of the opinion that the "real intelligence" is in the head of the people, but supporting this human intelligence with machine intelligence can bring real benefits.

The special session received a total of 68 submissions. We followed a careful and rigorous two-level, double-blind review, assigning each paper to a minimum of three and maximum of six reviewers from our international scientific board. On the basis of the reviewers' results, only 20 papers were accepted for the session HCI-KDD, resulting in an acceptance rate of 29 %.

On the basis of what we have seen from current scientific trends, we identified five main topics for future research:

1) Interactive content analytics from "unstructured" (weakly structured) data (e.g., interactive topological mining of big data)
2) Swarm intelligence (collective intelligence, crowd sourcing) and collaborative knowledge discovery/data mining/decision making
3) Intelligent, interactive, semi-automatic, multivariate information visualization and visual analytics
4) Novel search user interaction techniques (supporting human intelligence with computational intelligence)
5) Modeling human search behavior and understanding human information needs

Additionally to the papers of the special session HCI-KDD, this volume contains the papers from the workshop Knowledge Discovery and Smart Homes and the Special Session on Smart Learning Environments (SLE) organized by Kinshuk and Ronghuai Huang.

The organizers see these special sessions as a bridge between the two worlds dealing with "human intelligence" and with "machine intelligence" – all participants who synergistically worked together to organize, participate, and attend these special sessions showed great enthusiasm and dedication.

We cordially thank each and every person who contributed toward making this session a success, for their participation and commitment: the authors, reviewers, partners, organizations, supporters, and all the volunteers, in particular the marvelous host of this session—the team of the University of Maribor. Thank you!

July 2013

Andreas Holzinger
Gabriella Pasi

Organization

Special Session HCI-KDD

Chairs

Andreas Holzinger	Medical University of Graz, Austria
Gabriella Pasi	Università di Milano, Bicocca, Italy

International Scientific Committee

Amin Anjomshoaa	Vienna University of Technology, Austria
Joel Arrais	University of Coimbra, Portugal
Mounir Ben Ayed	Ecole Nationale d'Ingenieurs de Sfax (ENIS) and Université de Sfax, Tunisia
Matt-Mouley Bouamrane	University of Glasgow, UK
Polo Chau	Carnegie Mellon University, USA
Chaomei Chen	Drexel University, USA
Nilesh V. Chawla	University of Notre Dame, USA
Tomasz Donarowicz	Wroclaw University of Technology, Poland
Achim Ebert	Technical University Kaiserslautern, Germany
Max J. Egenhofer	University of Maine, USA
Kapetanios Epaminondas	University of Westminster, London, UK
Massimo Ferri	University of Bologna, Italy
Alexandru Floares	Oncological Institute Cluj-Napoca, Romania
Ana Fred	IST – Technical University of Lisbon, Portugal
Adinda Freudenthal	Technical University Delft, The Netherlands
Wolfgang Gaissmaier	Max Planck Institute of Human Development, Adaptive Behaviour and Cognition, Berlin, Germany
Hugo Gamboa	Universidade Nova de Lisboa, Portugal
Venu Govindaraju	University of Buffalo State New York, USA
Michael Granitzer	University Passau, Germany
Dimitrios Gunopulos	University of Athens, Greece
Helwig Hauser	University of Bergen, Norway
Jun Luke Huan	University of Kansas, Lawrence, USA
Anthony Hunter	UCL University College London, UK
Alfred Inselberg	Tel Aviv University, Israel
Kalervo Jaervelin	University of Tampere, Finland
Igor Jurisica	IBM Life Sciences Discovery Centre and University of Toronto, Canada
Jiri- Klema	Czech Technical University, Prague, Czech Republic

Lubos Klucar	Slovak Academy of Sciences, Bratislava, Slovakia
David Koslicki	Pennsylvania State University, USA
Patti Kostkova	City University London, UK
Damjan Krstajic	Research Centre for Cheminformatics, Belgrade, Serbia
Natsuhiko Kumasaka	Center for Genomic Medicine (CGM), Tokyo, Japan
Nada Lavrac	Joszef Stefan Institute, Ljubljana, Slovenia
Pei Ling Lai	Southern Taiwan University, Taiwan
Alexander Lex	Harvard University, Cambridge (MA), USA
Chunping Li	Tsinghua University, Canada
Luca Longo	Trinity College Dublin, Ireland
Lenka Lhotska	Czech Technical University Prague, Czech Republic
Andras Lukacs	Hungarian Academy of Sciences and Eoetvos University, Budapest, Hungary
Avi Ma' Ayan	The Mount Sinai Medical Center, New York, USA
Ljiljana Majnaric-Trtica	Josip Juraj Strossmayer University, Osijek, Croatia
Martin Middendorf	University of Leipzig, Germany
Silvia Miksch	Vienna University of Technology, Vienna, Austria
Antonio Moreno-Ribas	Universitat Rovira i Virgili, Tarragona, Spain
Marian Mrozek	Jagiellonian University, Krakow, Poland
Daniel E. O'Leary	University of Southern California, USA
Ant Ozok	UMBC, Baltimore, USA
Vasile Palade	University of Oxford, UK
Jan Paralic	Technical University of Kosice, Slovakia
Valerio Pascucci	University of Utah, USA
Gabriella Pasi	Università di Milano Bicocca, Italy
Armando J. Pinho	Universidade the Aveiro, Portugal
Gerald Petz	Upper Austria University of Applied Sciences, Austria
Margit Pohl	Vienna University of Technology, Vienna, Austria
Paul Rabadan	Columbia University College of Physicians and Surgeons, New York, USA
Heri Ramampiaro	Norwegian University of Science and Technology, Norway
Dietrich Rebholz	European Bioinformatics Institute, Cambridge, UK
Renè Riedl	Johannes Kepler University of Linz, Austria
Gerhard Rigoll	Munich University of Technology, Germany
Lior Rokach	Ben-Gurion University of the Negev, Israel

Carsten Roecker	RWTH Aachen University, Germany
Giuseppe Santucci	La Sapienza, University of Rome, Italy
Reinhold Scherer	Graz BCI Lab, Graz University of Technology, Austria
Paola Sebastiani	Boston University, USA
Christin Seifert	University Passau, Germany
Tanja Schultz	Karlsruhe Institute of Technology, Germany
Klaus-Martin Simonic	Medical University Graz, Austria
Andrzej Skowron	University of Warszaw, Poland
Neil R. Smalheiser	University of Illinois at Chicago, USA
Alexander Stocker	Joanneum Research Graz, Austria
Marc Streit	Johannes-Kepler University Linz, Austria
A Min Tjoa	Vienna University of Technology, Austria
Olof Torgersson	Chalmers University of Technology and University of Gothenburg, Goeteburg, Sweden
Patricia Ordonez-Rozo	University of Maryland, Baltimore County, Baltimore, USA
Jianhua Ruan	University of Texas at San Antonio, USA
Pak Chung Wong	Pacific Northwest Laboratory, Washington, USA
William Wong	Middlesex University London, UK
Kai Xu	Middlesex University London, UK
Pinar Yildirim	Okan University, Istanbul, Turkey
Martina Ziefle	RWTH Aachen University, Germany
Minlu Zhang	University of Cincinnati, USA
Ning Zhong	Maebashi Institute of Technology, Japan
Xuezhong Zhou	Beijing Jiaotong University, China

International Industrial Committee

Peter Bak	IBM Haifa Research Lab, Mount Carmel, Israel
Alberto Brabenetz	IBM Vienna Austria, Austria
Anni R. Coden	IBM T.J. Watson Research Center New York, USA
Stefan Jaschke	IBM Vienna Austria, Austria
Homa Javahery	IBM Centers for Solution Innovation, Canada
Jie Lu	IBM Thomas J. Watson Research Center, Hawthorne, New York, USA
Laxmi Parida	IBM Thomas J. Watson Research Center, Yorktown Heights, New York, USA
Hugo Silva	PLUX Wireless Biosensors, Lisbon, Portugal

International Student's Committee

Andre Calero-Valdez	RWTH Aachen, Germany
Pavel Dlotko	Jagiellonian University, Poland
Emanuele Panzeri	University of Milano-Bicocca, Italy
Igor Pernek	University of Maribor, Slovenia
Markus Fassold	Research Unit HCI Graz, Austria
Christof Stocker	Research Unit HCI Graz, Austria
Hubert Wagner	Jagiellonian University, Poland

Special Session on Smart Learning Environments (SLE)

Chairs

Kinshuk	Athabasca University, Canada
Ronghuai Huang	Beijing Normal University, China

International Scientific Committee

Arif Altun	Hacettepe University, Turkey
Gautam Biswas	Vanderbilt University, USA
Maiga Chang	Athabasca University, Canada
Guang Chen	Beijing Normal University, China
Nian-Shing Chen	National Sun Yat-sen University, Taiwan
Christophe Choquet	Universiti du Maine, France
Grainne Conole	University of Leicester, UK
Moushir El-Bishouty	Athabasca University, Canada
Baltasar Fernández-Manjón	Complutense University of Madrid, Spain
Sabine Graf	Athabasca University, Canada
Tsukasa Hirashima	Hiroshima University, Japan
Andreas Holzinger	Medical University Graz, Austria
Gwo-jen Hwang	National Taiwan University of Science and Technology, Taiwan
Mohamed Jemni	University of Tunis, Tunisia
Sandhya Kode	International Institute of Information Technology Hyderabad, India
Siu-Cheung Kong	The Hong Kong Institute of Education, Hong Kong
Yanyan Li	Beijing Normal University, China
Chen-Chung Liu	National Central University, Taiwan
Alke Martens	PH Schwäbisch Gmünd, Germany
Miguel Nussbaum	Catholic University of Chile, Chile
Hiroaki Ogata	Tokushima University, Japan
Elvira Popescu	University of Craiova, Romania
Srinivas Ramani	International Institute of Information Technology, India
Junfeng Yang	Beijing Normal University, China

Stephen J.H. Yang	National Central University, Taiwan
Demetrios G. Sampson	University of Piraeus, Greece
Marcus Specht	Open University of the Netherlands, The Netherlands
Vincent Tam	University of Hong Kong, Hong Kong
Setsuo Tsuruta	Tokyo Denki University, Japan
Gustavo Zurita	University of Chile, Chile

Table of Contents

Human-Computer Interaction and Knowledge Discovery

Knowledge Discovery and Smart Homes

Smart Learning Environments

Visualization and Data Analytics

Hypothesis Generation by Interactive Visual Exploration of Heterogeneous Medical Data

Cagatay Turkay[1], Arvid Lundervold[2], Astri Johansen Lundervold[3], and Helwig Hauser[1]

[1] Department of Informatics, University of Bergen, Norway
{Cagatay.Turkay,Helwig.Hauser}@uib.no
[2] Department of Biomedicine, University of Bergen, Norway
Arvid.Lundervold@biomed.uib.no
[3] Department of Biological and Medical Psychology, University of Bergen, Norway
Astri.Lundervold@psybp.uib.no

Abstract. High dimensional, heterogeneous datasets are challenging for domain experts to analyze. A very large number of dimensions often pose problems when visual and computational analysis tools are considered. Analysts tend to limit their attention to subsets of the data and lose potential insight in relation to the rest of the data. Generating new hypotheses is becoming problematic due to these limitations. In this paper, we discuss how interactive analysis methods can help analysts to cope with these challenges and aid them in building new hypotheses. Here, we report on the details of an analysis of data recorded in a comprehensive study of cognitive aging. We performed the analysis as a team of visualization researchers and domain experts. We discuss a number of lessons learned related to the usefulness of interactive methods in generating hypotheses.

Keywords: interactive visual analysis, high dimensional medical data.

1 Introduction

As in many other domains, experts in medical research are striving to make sense out of data which is collected and computed through several different sources. Along with new imaging methodologies and computational analysis tools, there is a boom in the amount of information that can be produced per sample (usually an individual in the case of medical research). This increasingly often leads to heterogeneous datasets with very large number of dimensions (variables), up to hundreds or even thousands. This already is a challenging situation since most of the common analysis methods, such as regression analysis or support vector machines [1], for example, do not scale well to such a high dimensionality. Consider for instance applying factor analysis to understand the dominant variations within a 500-dimensional dataset. It is a great challenge to correctly interpret the resulting factors even for the most skilled analyst.

On top of this challenge, the number of samples is usually very low in medical research due to a number of factors such as the availability of participants in

A. Holzinger and G. Pasi (Eds.): HCI-KDD 2013, LNCS 7947, pp. 1–12, 2013.

a study or high operational costs. This results in datasets with small number of observations (small n) but a very high number of variables (large p). Since most of the statistical methods need sufficiently large number of observations to provide reliable estimates, such "long" data matrices lead to problematic computations [2]. Both the high dimensionality of the datasets and the "$p \gg n$ problem", pose big challenges for the analyst and the computational tools. These challenges lead to the fact that the experts tend to limit their analyses to a subset of the data based on a priori information, e.g., already published related work. Limiting the analysis to a subset of the data dimensions hides relations in the data that can potentially lead to new, unexpected hypotheses.

At this stage, the field of visual analytics can offer solutions to analysts to overcome these limitations [3] [4]. The visual analysis methods enable analysts to quickly build new hypotheses through interaction with the data. The user also gets immediate feedback on whether or not these hypotheses call for a further investigation. Moreover, the interactive tools enable analysts to check for known hypotheses and relationships that have been already studied and reported in the related literature.

In this application paper, we discuss how interactive visual analysis methods facilitate the hypothesis generation process in the context of heterogeneous medical data. We discuss how we utilize the *dual analysis* of items and dimensions [5] in the interactive visual analysis of high dimensional data. We report on the analysis of data related to a longitudinal study of cognitive aging [6] [7]. We demonstrate how our explorative methods lead to findings that are used in the formulation of new research hypotheses in the related study. We additionally showcase observations that are in line with earlier studies in the literature. We then comment on a number of lessons learned as a result of the analysis sessions that we performed as a team of visualization researchers and domain experts.

2 Interactive Visual Analysis Environment

The analysis of the cognitive aging study data is performed through a coordinated multiple view system [8], that primarily makes use of scatterplots. The user is able to make selections in any of the views and combine these selections through Boolean operators, i.e., $\cup, \cap, \neg$. In order to indicate the selections and achieve the focus+context mechanism, we employ a coloring strategy, i.e, the selected points are in a reddish color and the rest is visualized in gray with a low transparency (see Fig. 1-b) to aid the visual prominence of the selection. One additional note here is that we use a density based coloring such that overlapping points lead to a more saturated red color. We use Principal Component Analysis (PCA) – on demand – to reduce the dimensionality of the data when needed. Additionally, we use Multidimensional Scaling (MDS) directly on the dimensions similar to the *VAR display* by Yang et al. [9]. In this visualization approach, the authors represent a single dimension by a glyph that demonstrates the distribution of the items in the dimension. Later authors apply MDS on the dimensions to lay them out on a $2D$-display. Similarly in this work, we feed the correlations

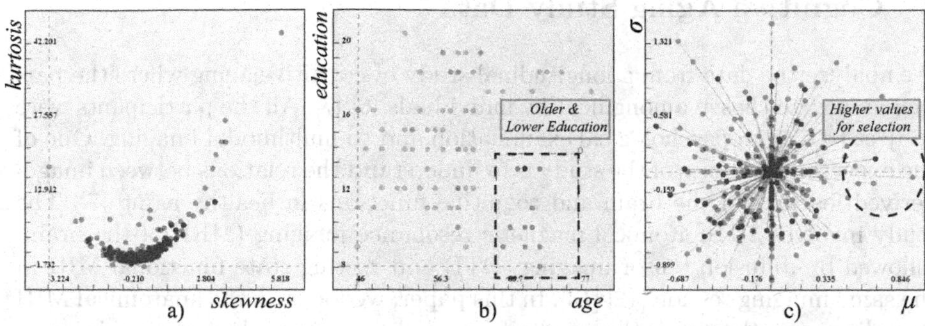

Fig. 1. Dual analysis framework where visualizations of items have a blue and those of dimensions a yellow background. a) We employ a visualization of the dimensions over their *skewness* and *kurtosis* values, where each dot represents a single dimension b) We select a group of participants who are older and have a lower education. c) The deviation plot shows how the μ and σ values change when the selection in (b) is made.

between the dimensions as a distance metric to MDS and as a result, it places the highly inter-correlated groups close to each other. These computational analysis tools are available through the integration of the statistical computation package R [10].

The analysis approach employed in this paper is based on the dual analysis method by Turkay et al. [5]. In this model, the visualization of data items is accompanied by visualizations of dimensions. In order to construct visualizations where dimensions are represented by visual entities, a number of statistics, such as mean (μ), standard deviation (σ), *median*, inter-quartile-range (*IQR*), *skewness*, and, *kurtosis* are computed for each dimension (i.e., column of the data). These computed statistics are then used as the axes of a visualization of dimensions. In Fig. 1-a, the dimensions are visualized with respect to their skewness and kurtosis, where each dot here represents a dimension.

An additional mechanism we employ is the *deviation plot*, which enables us to see the changes in the statistical computations for dimensions in response to a subset selection of items [11]. In Fig. 1-b, we select a sub-group of participants (from the study data) who are older and have a lower education. We now compute the μ and σ values for each dimension twice, once with using all the items (participants) and once with using only the selected subset. We then show the difference between the two sets of computations in a deviation plot (Fig. 1-c). The dashed circle shows the dimensions that have larger values for the selected subset of items, i.e., for the elderly with lower education. Such a visualization shows the relation between the selection and the dimensions in the data and provides a quick mechanism to check for correlations. Throughout the paper, the views that show items have blue background and those that visualize the dimensions have a yellow background. Further details on the methods could be found in the related references [5] [11].

3 Cognitive Aging Study Data

We analyze the data from a longitudinal study of cognitive aging where the participants were chosen among healthy individuals [6] [7]. All the participants were subject to a neuropsychological examination and to multimodal imaging. One of the expected outcomes of the study is to understand the relations between image-derived features of the brain and cognitive functions in healthy aging [7]. The study involves 3D anatomical magnetic resonance imaging (MRI) of the brain, followed by diffusion tensor imaging (DTI) and resting state functional MRI in the same imaging session [12] [13]. In this paper, we focus on the anatomical MRI recordings together with the results from the neuropsychological examination. The examination included tests related to intellectual function (IQ), memory function, and attention/executive function. IQ was estimated from two sub tests from the Wechsler Abbreviated Scale of Intelligence [14]. The total learning score across the five learning trials of list A (learning), the free short and long delayed recall and the total hits on the Recognition scores from the California Verbal Learning Test (CVLT) II [15] were together with the subtest Coding from Wechsler Adult Intelligence Scale-III [16] used to assess memory function. The Color Word Interference Test from the Delis-Kaplan Executive Function System [17] and the Trail Making Test A and B from the Halstead-Reitan Test Battery [18] were used to assess attention/executive function.

The resulting dataset from the study contains information on 82 healthy individuals who took part in the first wave of the study in 2004/2005. T1-weighted MRI images were segmented into 45 anatomical regions. For each segmented brain region, seven features were derived automatically, namely: *number of voxels, volume* and *mean, standard deviation, minimum, maximum and range of the intensity values in the regions.* All these automated computations were done in the FreeSurfer software suite [19]. This automated process creates $45 \times 7 = 315$ dimensions per individual. Additional information on the participants, such as age and sex, and, the results of two neuropsychological tests are added to the data. With this addition, the resulting dataset has 373 dimensions, i.e., the resulting table's size is 82×373. Moreover, meta-data on the dimensions is also incorporated. This meta-data contains whether each dimension is a test score or a brain segment statistic, which brain regions that dimension is related to, and, which statistical feature (e.g., volume or mean intensity) is encoded.

4 Analysis of Cognitive Aging Study Data

In this study, our analysis goal is to determine the relations between age, sex, neuropsychological test scores, and the statistics for the segmented brain regions. The conventional routine to analyze this dataset is to physically limit the analysis to a subset of the dimensions and perform time-consuming, advanced statistical analysis computations on this subset, e.g., loading only the data on specific brain regions and training a neural network with this data. In this setting, if the same analysis needs to be applied on a slightly different subset (which is often the case), all the operations need to be redone from the beginning – a considerably long time to build/evaluate a single hypothesis. On the contrary, in our interactive methods, the whole data is available throughout the analysis and analysts switch the current focus quickly through interactive brushes.

In order to direct the analysis, we treat age, sex, and the test scores as the dependent variables and try to investigate how they relate to the imaging based variables. Moreover, we investigate the relations within the brain segments. In each sub-analysis, we derive a number of observations purely exploratively. We then discuss these findings as an interdisciplinary team of visualization researchers, experts in neuroinformatics and neuropsychology. We comment on the observations using a priori information and suggest explanations/hypotheses around these new findings. These hypotheses, however, needs to be confirmed/rejected through more robust statistical and/or clinical tests to be considered for further studies. Our aim here is to enable analysts to generate new hypotheses that could potentially lead to significant findings when careful studies are carried out.

Prior to our analysis we handle the missing values and perform normalization on the data. To treat missing values, we apply one of the methods known as statistical imputation and replace the missing values with the mean (or mode) of each column [20]. We continue with a normalization step where different normalization schemes are employed for different data types. Here, dimensions related to the imaging of the brain are z-standardized and the rest of the columns are scaled to the unit interval.

Inter-relations in Test Results. We start our analysis by looking at the relations between the test scores. We first focus our attention on the results related to IQ & Memory function and attention/executive functions related tests and apply a correlation-based-MDS on the 15 dimensions. The rest of the dimensions are not used in the computation and are placed in the middle of the view and colored in gray in Fig. 2-a. Here, we choose to focus on the two large groups, that are to the left and to the right of the view. For a micro analysis, one can focus on the sub-groupings that are visible in both of the clusters. The first group relates to test results assessing IQ and memory function (Group-1). The second group relates to test scores assessing attention and executive function (Group-2). This grouping is in line with the interpretation of these scores and we investigate these two sub-groups separately in the rest of the analysis. We interactively select these sub-groups and locally apply PCA on them. We then use the resulting principal components (PC) to represent these two groups of test scores. We observed that for both of the groups much of the variance is captured by a single PC, so we decide to use only the first PC for each group.

Hypothesis 1: There are two dominant factors within the test results, *IQ & memory* and *attention & executive function.*

Findings Based on Sex. As a continuation of our analysis, we now focus on available meta-data on patients, such as age and sex, to derive interesting

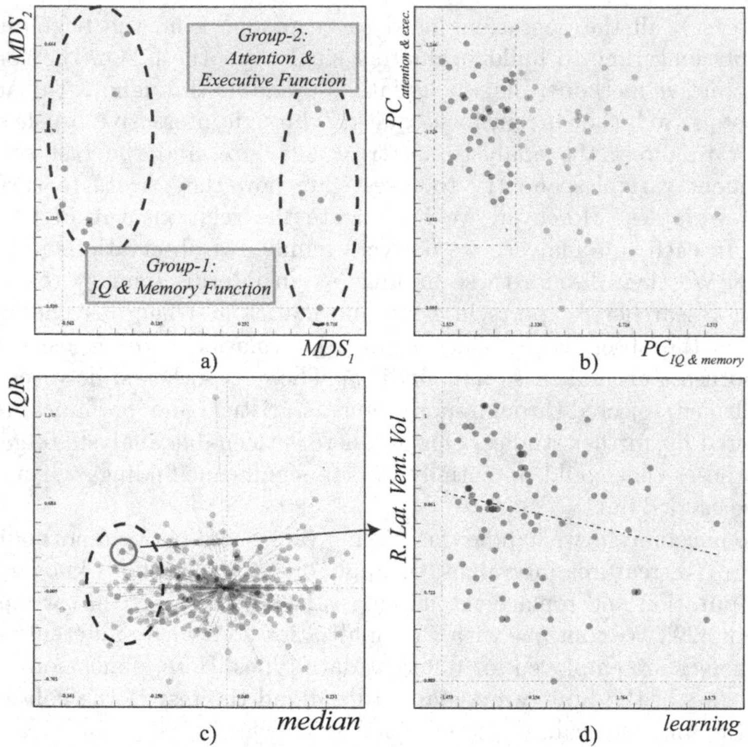

Fig. 2. a) MDS is applied on the *test score* dimensions, where related dimensions are placed close to each other. Two groups for the test scores (Group-1: IQ and memory related, Group-2: attention) show up in the results. b) Each group is represented through an application of PCA and the resulting first principal components are mapped to the axes of the scatterplot. A group of participants, who are better in learning and attentive function is selected. c) Some brain regions are smaller for this subgroup, i.e., have smaller *median* value. d) We select one of the dimensions that shrink the most, *right lateral ventricle volume* (red circle), and visualize these values against the learning scores from CVLT. We notice that there is indeed a negative correlation with the learning score from the CVLT.

relations. We begin by a visualization of *age* vs. *sex* and select the male participants (Fig. 3-a) with a brush and observe how the test scores change in the linked deviation view (Fig. 3-b). The visualization shows that the male participants performed worse in *IQ & memory function* related tasks. In tests related to attention and executive function, however, there were no significant changes between sexes. This is a known finding that has been already observed throughout the study. Another observation that is also confirmed by prior information is the differences in brain volumes between sexes. An immediate reading in Fig. 3-c is that *male participants have larger brains (on average) compared to women*, which is a known fact. We analyze further by selecting one of the regions that changed the most, *Thalamus volume*, and look at its relation with sex (Fig. 3-d).

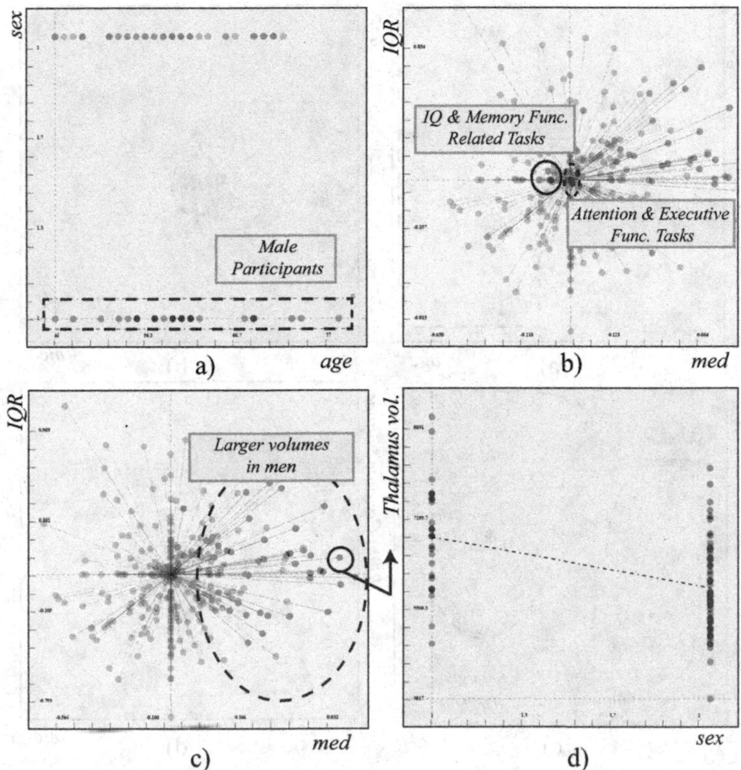

Fig. 3. Male participants are selected (a) and the deviation plot shows that for IQ & memory related tasks, males generally perform worse. However, for attentive and executive function related tests, there is no visible difference (b). When the changes in volume for the brain segments are observed, it is clearly seen that males have larger brains (c). When the volume of one of the segments, thalamus, is visualized with a linear regression line, the sex based difference is found to be significant.

We see that there is a significant change, however, this apparent sex difference in thalamic volume has shown to be negligible when the intracranial volume (ICV) difference between sexes are taken into account [21]. This finding could probably be further explored by normalizing segmented brain volumes with the subject's ICV (if this measure is available).

Hypothesis 2: Males perform worse in *IQ & memory* related tests but not in those related to *attention & executive function*.

Findings Based on Age. We continue our investigation by limiting our interest to the elderly patients to understand the effects of aging on the brain and the test results. We select the patients over the age of 60 (Fig. 4-a) and visualize how brain volumes and test scores change. We observed no significant difference in IQ

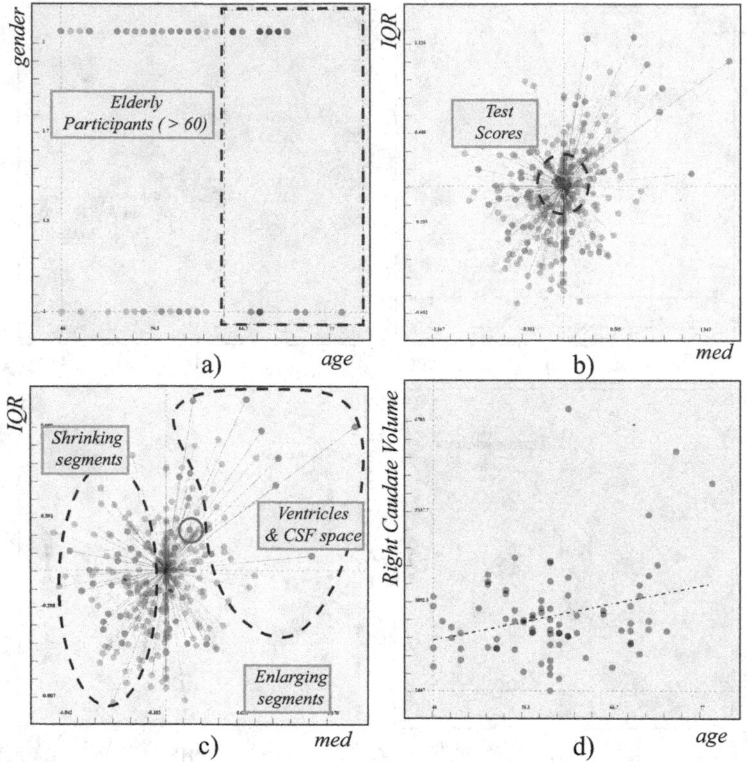

Fig. 4. Elderly patients (> 60 years old) are selected (a). No significant relation is observed in the test scores (b). When we focus on the volumes of the segments, we see most of the regions are shrinking with age, but some, especially the ventricles, are enlarging (c). Apart from the expected enlargement of the ventricles, *the right caudate* is also found to enlarge with age (d).

& memory and attentive functions for the elderly patients (Fig. 4-b). However, when we observe the change in brain volumes, we observe that there is an overall *shrinkage in most of the brain segments with age*. This is clearly seen in Fig. 4-c, where most of the dimensions have smaller *median* values (i.e., to the left of the center line). Although most of the brain regions are known to shrink with age [22], some regions are reported to enlarge with age. When the dimensions that have a larger *median* value due to the selection (i.e., enlargement due to aging) are observed, they are found to be the *ventricles* (not the *4th ventricle*) and the *CSF space*. Since this is a known fact [22], we focused on the regions that shows smaller enlargements and decide to look at *the right caudate* more closely. When *the right caudate* is visualized against age, a significant correlation is observed (Fig. 4-d). This is an unexpected finding that needs to be investigated further.

Hypothesis 3: There is no significant relation between age and performance in IQ & memory and attentive & executive functions for individuals undergoing a healthy aging. Moreover, in contrast to the most of the brain regions, there is a significant enlargement in *the right caudate* in healthy aging individuals.

IQ & Memory Function vs. Brain Segment Volumes. We oppose the first principal components for the two groups of test scores (Fig. 2-a) and select the participants that show better IQ & memory function performance (Fig. 2-b). A linked deviation plot shows the change in *median* and *IQR* values where we observe the change in the imaging related variables (Fig. 2-c). We limit our interest to the variables that are the *volumes of the brain segments* by selecting the volume category through a histogram that displays the related meta-data (not shown in the image). In the deviation plot, we see a sub-group of segments (dashed circle) that have lower volumes for the selected participants (i.e., those that showed better performance). Among those segments are the lateral ventricles that show a significant change. Lateral ventricles are filled with cerebrospinal fluid and have no known function in learning and IQ. We use the integrated linear regression computation on a scatterplot showing *learning* vs. *right lateral ventricle volume* and observe that there is in fact a negative correlation. This could be explained such that, when the ventricles have larger sizes, it indicates less gray matter volume in the brain parenchyma responsible in cognitive function, and is thus associated with reduced performance in IQ & memory function. However, although ventricles tend to grow with age, we observed no significant relation between aging and the performance (See Hypothesis 3). These are now two related observations that leads to an interesting hypothesis.

Hypothesis 4: Regardless of age, the larger sizes of the ventricles are associated with low performance. However, the (expected) enlargement of the ventricles with aging does not directly influence the overall performance.

Relations within Brain Segments. We continue by delimiting the feature set for the brain regions to their *volume* and apply MDS on the 45 dimensions (one for each segment) using the correlation between the dimensions as the distance metric. We identify a group of dimensions that are highly correlated in the MDS plot (Fig. 5-a). This group consists of the volumes for different *ventricles* (lateral, inferior) and *non-white matter hypointensities.* We investigate this finding closely by looking at the relations between *left lateral ventricle* and *non-WM-hypointensities* and found a positive correlation relation (Fig. 5-b) due to a sub-group of patients that have outlying values. This is an interesting finding since non-white matter hypointensities (as segmented by FreeSurfer) might represent local lesions in gray matter such as vascular abnormalities that have a predilection for involving the thalamus and the basal ganglia. Such vascular abnormalities in deeper brain structure could then lead to substance loss and enlarged lateral ventricles. One might further expect that this pathophysiological process would be increasingly frequent with age, but such relationship between age and non-white matter hypointensities was observed to be insignificant in our analysis.

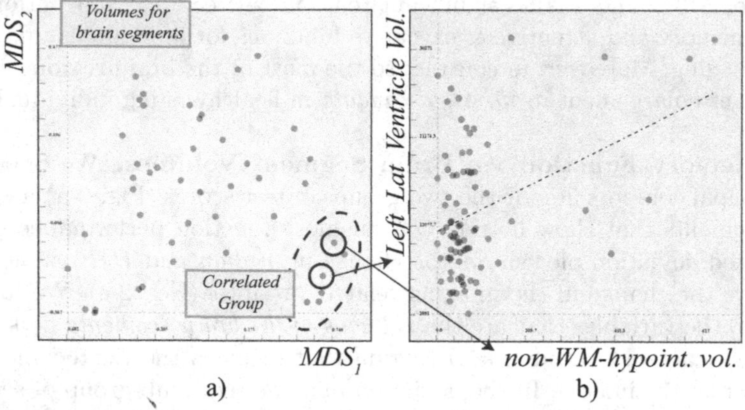

Fig. 5. After MDS is applied on the volume dimensions for brain segments, a correlated group of brain segments is observed (a). Although most of these dimensions are related to the volume of different parts of *the ventricles* (which is expected), *non white matter hypointensities* (scars on the white matter) is also related. This is an interesting finding which led to an hypothesis on the relation between the enlargement of the scars on the white matter and the ventricles.

Hypothesis 5: There is a positive relation between lesions on brain tissue and the volume of the ventricles. However, no significant relation with such lesions and age has been detected, this is likely due to the fact that the study involves only participants going through healthy aging.

5 Discussions, Lessons Learned and Conclusions

In a typical analysis of this data, domain experts usually utilize complex machine learning methods, such as neural networks [1], to analyze the data and confirm hypotheses. With such methods however, the process is not transparent and the results can be hard to interpret.

Explorative methods, such as this one presented here, offers new opportunities in building hypotheses. However, the hypotheses built in such systems may suffer from over-fitting to the data, i.e., the finding could be a great fit for a specific selection but harder to generalize [23]. In order to provide feedback on this problem of over-fitting, interactive systems could include cross-validation (or bootstrapping) functionalities to report on the sensibility of the results [24]. In these methods, the hypotheses are tested for several subsets of the data to check the validity of the findings [24]. Another important feature that needs to be present in such interactive systems is the immediate use of more robust and solid statistical verification methods. In our current framework, we employ linear regression to check for the statistical significance of certain relations (see Fig. 4-d). Such functionalities, and even more advanced inferential statistics, are feasible to incorporate through the embedding of R. Such extensions are desirable for domain experts and can increase the reliability of the results considerably in interactive frameworks.

In this work, we only employed scatterplots and the deviation plot. One can easily extend the selection of visualizations using more advanced methods discussed in the literature. The changes can be encoded by flow-based scatterplots [25] and the comparison of groups can be enhanced by using clustered parallel coordinates [26].

In a significantly short analysis session, we were able to build 5 hypotheses from the healthy aging data. Building this many potential hypotheses using the conventional analysis process would require a considerable amount of time. Throughout the analysis, we discovered relations that lead to novel hypotheses for the healthy aging domain. In addition, we came up with a number of findings that have been already confirmed in the related literature.

Acknowledgments. We would like to thank Peter Filzmoser for the valuable insights on the statistical foundations of this work. The study on cognitive aging was supported by grants from the Western Norway Regional Health Authority (# 911397 and #911687 to AJL and 911593 to AL).

References

1. Mitchell, T.M.: Machine learning. McGraw Hill series in computer science. McGraw-Hill (1997)
2. Chen, J., Chen, Z.: Extended bic for small-n-large-p sparse glm. Statistica Sinica 22(2), 555 (2012)
3. Keim, D.A., Mansmann, F., Schneidewind, J., Thomas, J., Ziegler, H.: Visual analytics: Scope and challenges. In: Simoff, S.J., Böhlen, M.H., Mazeika, A. (eds.) Visual Data Mining. LNCS, vol. 4404, pp. 76–90. Springer, Heidelberg (2008)
4. Kehrer, J., Ladstädter, F., Muigg, P., Doleisch, H., Steiner, A., Hauser, H.: Hypothesis generation in climate research with interactive visual data exploration. IEEE Transactions on Visualization and Computer Graphics (IEEE TVCG) 14(6), 1579–1586 (2008)
5. Turkay, C., Filzmoser, P., Hauser, H.: Brushing dimensions – a dual visual analysis model for high-dimensional data. IEEE Transactions on Visualization and Computer Graphics 17(12), 2591–2599 (2011)
6. Andersson, M., Ystad, M., Lundervold, A., Lundervold, A.: Correlations between measures of executive attention and cortical thickness of left posterior middle frontal gyrus - a dichotic listening study. Behavioral and Brain Functions 5(41) (2009)
7. Ystad, M., Lundervold, A., Wehling, E., Espeseth, T., Rootwelt, H., Westlye, L., Andersson, M., Adolfsdottir, S., Geitung, J., Fjell, A., et al.: Hippocampal volumes are important predictors for memory function in elderly women. BMC Medical Imaging 9(1), 17 (2009)
8. Ward, M.O.: Xmdvtool: integrating multiple methods for visualizing multivariate data. In: Proceedings of the Conference on Visualization, VIS 1994, pp. 326–333. IEEE Computer Society Press (1994)

9. Yang, J., Hubball, D., Ward, M., Rundensteiner, E., Ribarsky, W.: Value and relation display: Interactive visual exploration of large data sets with hundreds of dimensions. IEEE Transactions on Visualization and Computer Graphics 13(3), 494–507 (2007)
10. Team, R.D.C.: R: A Language and Environment for Statistical Computing. R Foundation for Statistical Computing (2009), `http://www.R-project.org`
11. Turkay, C., Parulek, J., Hauser, H.: Dual analysis of DNA microarrays. In: Proceedings of the 12th International Conference on Knowledge Management and Knowledge Technologies, i-KNOW 2012, pp. 26:1–26:8 (2012)
12. Hodneland, E., Ystad, M., Haasz, J., Munthe-Kaas, A., Lundervold, A.: Automated approaches for analysis of multimodal mri acquisitions in a study of cognitive aging. Comput. Methods Prog. Biomed. 106(3), 328–341 (2012)
13. Ystad, M., Eichele, T., Lundervold, A.J., Lundervold, A.: Subcortical functional connectivity and verbal episodic memory in healthy elderly – a resting state fmri study. NeuroImage 52(1), 379–388 (2010)
14. Wechsler, D.: Wechsler abbreviated scale of intelligence. Psychological Corporation (1999)
15. Delis, D.C.: California verbal learning test. Psychological Corporation (2000)
16. Wechsler, D.: Wechsler adult iintelligence scale-iii (wais-iii). Psychological Corporation, San Antonio (1997)
17. Delis, D.C., Kaplan, E., Kramer, J.H.: Delis-Kaplan executive function system. Psychological Corporation (2001)
18. Reitan, R., Davison, L.: Clinical neuropsychology: current status and applications. Series in Clinical and Community Psychology. Winston (1974)
19. FreeSurfer (2012), `http://surfer.nmr.mgh.harvard.edu`
20. Scheffer, J.: Dealing with missing data. Research Letters in the Information and Mathematical Sciences 3(1), 153–160 (2002)
21. Sullivan, E.V., Rosenbloom, M., Serventi, K.L., Pfefferbaum, A.: Effects of age and sex on volumes of the thalamus, pons, and cortex. Neurobiology of Aging 25(2), 185–192 (2004)
22. Walhovd, K.B., Fjell, A.M., Reinvang, I., Lundervold, A., Dale, A.M., Eilertsen, D.E., Quinn, B.T., Salat, D., Makris, N., Fischl, B.: Effects of age on volumes of cortex, white matter and subcortical structures. Neurobiology of Aging 26(9), 1261–1270 (2005)
23. Hawkins, D.M., et al.: The problem of overfitting. Journal of Chemical Information and Computer Sciences 44(1), 1–12 (2004)
24. Kohavi, R., et al.: A study of cross-validation and bootstrap for accuracy estimation and model selection. In: International Joint Conference on Artificial Intelligence, vol. 14, pp. 1137–1145. Lawrence Erlbaum Associates Ltd. (1995)
25. Chan, Y.H., Correa, C.D., Ma, K.L.: Flow-based scatterplots for sensitivity analysis. In: 2010 IEEE Symposium on Visual Analytics Science and Technology (VAST), pp. 43–50. IEEE (2010)
26. Johansson, J., Ljung, P., Jern, M., Cooper, M.: Revealing structure within clustered parallel coordinates displays. In: IEEE Symposium on Information Visualization, INFOVIS 2005, pp. 125–132 (2005)

Combining HCI, Natural Language Processing, and Knowledge Discovery - Potential of IBM Content Analytics as an Assistive Technology in the Biomedical Field

Andreas Holzinger[1], Christof Stocker[1], Bernhard Ofner[1], Gottfried Prohaska[2], Alberto Brabenetz[2], and Rainer Hofmann-Wellenhof[3]

[1] Medical University Graz, A-8036 Graz, Austria
Institute for Medical Informatics, Statistics & Documentation,
Research Unit HCI4MED, Auenbruggerplatz 2/V, A-8036 Graz, Austria
{a.holzinger,c.stocker,b.ofner}@hci4all.at

[2] IBM Austria, Obere Donaustraße 95, A-1020 Vienna, Austria
{gottfried_prohaska,a.brabenetz}@at.ibm.com

[3] LKH-University Hospital Graz
Department for Dermatology, Auenbruggerplatz 22/V, A-8036 Graz, Austria
rainer.hofmann@medunigraz.at

Abstract. Medical professionals are confronted with a flood of big data most of it containing unstructured information. Such unstructured information is the subset of information, where the information itself describes parts of what constitutes as significant within it, or in other words - structure and information are not completely separable. The best example for such unstructured information is text. For many years, text mining has been an essential area of medical informatics. Although text can easily be created by medical professionals, the support of automatic analyses for knowledge discovery is extremely difficult. We follow the definition that knowledge consists of a set of hypotheses, and knowledge discovery is the process of finding or generating new hypotheses by medical professionals with the aim of getting insight into the data. In this paper we present some lessons learned of ICA for dermatological knowledge discovery, for the first time. We follow the HCI-KDD approach, i.e. with the human expert in the loop matching the best of two worlds: human intelligence with computational intelligence.

Keywords: Knowledge discovery, data mining, human-computer interaction, medical informatics, Unstructured Information Management, Content Analytics.

1 Introduction and Motivation for Research

Electronic patient records (EPR) contain increasingly large portions of data which has been entered in non-standardized format, which is often and not quite correctly called *free text* [1, 2]. Consequently, for many years, text mining was

A. Holzinger and G. Pasi (Eds.): HCI-KDD 2013, LNCS 7947, pp. 13–24, 2013.

and is an essential area of medical informatics, where researchers worked on statistical and linguistic procedures in order to dig out (mine) information from plain text, with the primary aim of gaining information from data. Although text can easily be *created* by medical professionals, the support of (semi-) automatic analyses is extremely difficult and has challenged researchers for many years [3–5]. The next big challenge is in Knowledge Discovery from this data. Contrary to the classical text mining, or information retrieval approach, where the goal is to find information, hence the medical professional knows what he wants, in knowledge discovery we want to discover novel insights, get new knowledge which was previously unknown. To reach this goal, approaches from pure computer science alone are insufficient, due to the fact that the "real" intelligence is in the brains of the professionals; the next step consists of making the information both usable and useful. Interaction, communication and sensemaking are still missing within the pure computational approaches [6].

Consequently, a novel approach is to combine HCI & KDD [7] in order to enhance human intelligence by computational intelligence. The main contribution of HCI-KDD is to *enable* end users to *find and recognize* previously unknown and potentially useful, usable, and interesting information. Yet, what is interesting is a matter of research [8]. HCI-KDD may be defined as the process of identifying novel valid, and potentially useful data patterns, with the goal to understand these patterns [9]. This approach is based on the assumption that the domain expert possesses explicit domain knowledge and by enabling him to interactively look at his data sets, he may be able to identify, extract and understand useful information, as to gain new - previously unknown - knowledge [10].

1.1 The Challenges of Text

Text, seen as transcription of natural language, poses a lot of challenges for computational analysis. Natural language *understanding* is regarded as an AI-complete problem [11]. In analogy to NP-completeness from complexity theory this means that the difficulty of the computational problem is equivalent to designing a computer which is as intelligent as a human being [12], and which brings us back to the very roots of the computational sciences [13].

It became evident over the past decades that the understanding of human language requires extensive knowledge, not only about the language itself, but also about the surrounding real world, because language is more than words, and meaning depends on context, and "understanding" requires a vast body of knowledge about this real world context [11] - we call this context-awareness [14]. Consequently, natural language processing (NLP) is a term that does not necessarily target a total understanding of language per se [15].

2 Theory and Background

In this section we define and describe the basic notions we use throughout this paper. Our understanding of terms such as "data" is a little uncommon, but based on what we think are good reasons. One being that natural language understanding is an AI-complete problem [11]. It makes more sense to ground the semantics of those terms closer to human understanding rather than "traditional" computer models.

2.1 Unstructured Information Management

There is still no clear definition given so far and the few definitions are very ambiguous (see a recent work as a typical example: [16]).

First, let us define our understanding of certain basic terms; namely data, information, and knowledge. These terms are interpreted quite differently throughout the scientific literature [17].

We ground our definitions on Boisot & Canals (2003) [18] and their sources. They describe **data** as originating in discernible differences in physical states-of-the-world, registered through stimuli. Theses states are describable in terms of space, time, and energy. Significant regularities in this data - whatever one qualifies as significant - then constitutes **information**. This implies that the information gained from data, depends on the agent extracting it - more precisely: his expectations, or *hypotheses*. This set of hypotheses held by an agent can then be referred to as **knowledge** and is constantly modified by the arrival of information.

Definition 1. *If knowledge consists of a set of hypotheses, then* ***knowledge discovery*** *is the process of finding or generating new hypotheses out of information.*

Since what qualifies as significant depends on the agents individual disposition, information can only appear to be *objective*, if what constitutes as significant regularities is established through convention [18].

It might be interesting to note, that based on these definitions, the commonly used term "*unstructured data*" refers to complete randomness, or noise.

Unstructured information, on the other hand, often refers to natural language, be it in the form of written documents, speech, audio, images or video. This implicit definition makes sense, as it is used to split information into two easily understood classes: databases content and everything else. The reason for this is mostly business motivated, as the term "unstructured" is then used to convey the message of computational inaccessibility through information retrieval methods to the "stored" information, and hence a necessity for action.

Let us state a more precise definition:

Definition 2. ***Unstructured Information*** *is the subset of information, where the information itself describes parts of what constitutes as significant regularity.*

What this essentially means, is that *information* and its *structure* are not completely separable. The best example for unstructured information is in text. The meaning of the text - its nouns, verbs, markers and so fourth - partly depends on the text itself - on the context and discourse. Even for humans it can be difficult. Sometimes sentences have to be re-read to be understood, or are misunderstood completely. While processing text, our knowledge-base is constantly being updated by the text itself, and a combination of our previous knowledge and updated knowledge is used to overcome and interpret uncertainties.

2.2 Model Structure through Annotation

In linguistics, the term annotation most commonly refers to meta data used to describe words, sentences and so fourth. The process of annotation is often described as *tagging*, which is the automatic assignment of descriptors to input tokens [19]. One prominent example is part-of-speech (POS) tagging, which maps a natural language, such as english, to a meta-language made up of word classes, such as nouns, verbs, adverbs and so fourth. We will describe POS tagging in more detail later.

The idea behind tagging is to model human knowledge, be it about language or another topic, in a computationally understandable way, in order to help the computer process unstructured information. This is usually not a trivial task, and its complexity strongly depends on the language, domain and the quality of the text. An universal automatic solution would be desirable, but for now seems out of reach.

However, it is our opinion, that by efficiently including the human in the loop, the research progress of computational techniques can vastly be improved. On a business perspective, a good user interface for the developers significantly speeds up domain specific solutions for unstructured information management.

2.3 On the Origins of IBM Content Analytics

For many years, IBM research groups from various countries are working on the development of systems for text analysis, and text-mining methods to support problem solving in life science. The best known system today is called Biological Text Knowledge Services and integrates research technologies from multiple IBM research labs. BioTeKS is the first major application of the so-called Unstructured Information Management Architecture (UIMA) initiative [20]. These attempts go back to a text mining technology called TAKMI (Text Analysis and Knowledge MIning), which has been developed to acquire useful knowledge from large amounts of textual data - not necessarily focused on medical texts [21].

The subsystem of UIMA is the Common Analysis System (CAS), which handles data exchanges between the various UIMA components, including analysis engines and unstructured information management applications. CAS supports data modeling via a type system and is independent of any programming language. It provides data access through a powerful indexing mechanism, hence provides support for creating annotations on text data [22].

BioTeKS was originally intended to analyze biomedical text from MEDLINE abstracts, where the text is analyzed by automatically identifying terms or names corresponding to key biomedical entities (e.g., proteins, drugs, etc.) and concepts or facts related to them [23]. MEDLINE has been often used for testing text analytics approaches and meanwhile a large number of Web-based tools are available for searching MEDLINE. However, the non-standardized nature of text is still a big issue, and there is much work left for improvement. A big issue is in end-user centred visualisation and visual analytics of the results, required for the support of the sensemaking processes amongst medical professionals [24,25].

3 Related Work

Many solutions for data analytics are available either as commercial or open-source software, ranging from programming languages and environments providing data analysis functionality to statistical software packages to advanced business analytics and business intelligence suites.

Prominent tools focusing on statistical analysis are IBM SPSS, SAS Analytics as well as the open-source R project for statistical computations. Each of the aforementioned tools provides additional packages for text analysis, namely IBM SPSS Modeler, a data mining and text analytics workbench, SAS Text Analytics and the tm package for text mining in R.

Software focusing on text mining and text analysis like the Apache UIMA project or GATE (General architecture for text engineering) are aimed at facilitating the analysis of unstructured content. Several projects based on the UIMA framework provide additional components and wrappers for 3rd-party tools, with the purpose of information extraction in the biomedical and the healthcare domain, including Apache cTAKES (clinical Text Analysis and Knowledge Extraction System) and the BioNLP UIMA Component Repository.

Other solutions for knowledge analysis utilize machine learning algorithms and techniques, with the most prominent frameworks Weka (Waikato Environment for Knowledge Analysis) and RapidMiner.

To our knowledge there are only a few publications concerning the integration of UIMA into clinical routine:

Garvin et al. (2012) [26] built a natural language processing system to extract information on left ventricular ejection fraction, which is a key component of heart failure, from "free text" echocardiogram reports to automate measurement reporting and to validate the accuracy of the system using a comparison reference standard developed through human review. For this purpose they created a set of regular expressions and rules to capture "ejection fraction" using a random sample of 765 echocardiograms. The authors assigned the documents randomly on two sets: a set of 275 used for training and a second set of 490 used for testing and validation. To establish a reference standard, two independent experts annotated all documents in both sets; a third expert resolved any incongruities. The test results for documentlevel classification of EF of $< 40\%$ had a sensitivity (recall) of 98.41%, a specificity of 100%, a positive predictive value (precision) of 100%, and an F measure of 99.2%. The test results at the concept level had a sensitivity of 88.9% (95% CI 87.7% to 90.0%), a positive predictive value of 95% (95% CI 94.2% to 95.9%), and an F measure of 91.9% (95% CI 91.2% to 92.7%) - consequently, the authors came to the conclusion that such an automated information extraction system can be used to accurately extract EF for quality measurement [26].

4 Methods

In our project, we are using IBM Content Analytics (ICA) Studio 3.0, which utilizes a UIMA pipeline as depicted in Fig. 1. This pipeline includes a set of fundamental annotators. Note that the first two can not be changed, while the other can be configured according to ones needs.

Language Identification Annotator. Identifies the language of the document. This fundamental information can be utilized to branch in specialized parsing rule sets.

Linguistic Analysis Annotator. Applies basic linguistic analysis, such as POS, to each document.

Dictionary Lookup Annotator. Matches words from dictionaries with words in the text. Note that stemming, as well as the definition of synonyms is supported.

Named Entity Recognition Annotator. This annotator can only be activated or deactivated and not configured as of now. It extracts person names, locations and company names

Pattern Matcher Annotator. Identifies those pattern in the text that are specified via rules, e.g. in the ICA Studio

Classification Module Annotator. Performs automatic classification. It uses natural language processing as well as semantic analysis algorithms to determine the true intent of words and phrases. It combines contextual statistical analysis with a rule-based, decision-making approach

Custom Annotator. Custom annotators are essentially java programs that obey a given UIMA interface. This program has access to all annotations made by the previous annotators.

The interesting idea behind ICA Studio, the development suit behind ICA, is to efficiently include the human in the loop. It does that by offering the developer a quick and easy way to model his knowledge into an UIMA conform format.

Through a mixture of dictionaries, regular expressions and parsing rules, annotation schemes can quickly be realized and instantly tested. With the possibility to plugin custom annotators (Java, or C++ programs) into the UIMA pipeline, custom methods can easily be included and are able to access the realized annotation schemes. This opens a quick and easy way to fast prototyping and modular development, both while offering quality control.

In the following subsections we will describe and discuss some common computational methods that are in one way or another utilized by ICA.

4.1 Morphology

Most natural languages have some system to generate words and word forms from smaller units in a systematic way [19,27]. The seemingly infinity of words

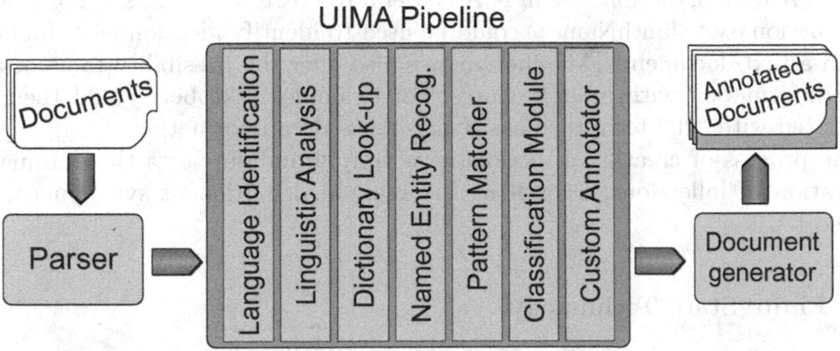

Fig. 1. The document processor architecture in IBM Content Analytics

in a language is produced by a finite collection of smaller units called *morphemes.* Simply put, morphology deals with the structure of words. These morphemes are either semantic concepts like *door*, *house*, or *green*, which are also called roots, or abstract features like *past* or *plural* [19]. Their realization as part of a word are then called *morph*, such as *door* or *doors.*

The information expressed with morphology varies widely between languages. In Indo-European languages for example, distinct features are merged into a single bound form [19]. These languages are typically called *inflectional languages.* Inflections do not change the POS category, but the grammatical function. Inflections and derivations convey information such as tense, aspect, gender or case.

When defining custom dictionaries, ICA Studio supports the manual definition of any custom morph or synonym, as well as providing the automatic generation of inflections. The canonical form is then defined as the *lemma.* For many supported languages, such as German and English, standard dictionaries are already build-in.

Custom dictionaries are most useful to define specific classes. For example a dictionary called "Defects", that includes various words (and their morphs) indicating a defect, such as "break", "defect", or "destroy", could be used in combination with other annotations to detect faulty products out of customer reports or forum entries.

4.2 Lexicography

The term "*computational lexicography*" can have different meanings. Hanks [19] listed two common interpretations:

1. Restructuring and exploiting human dictionaries for computational purposes
2. Using computational techniques to compile new dictionaries

In this paper we refer to the exploiting of human dictionaries for computational purposes.

The creation of dictionaries in ICA is useful to create word classes. For example a dictionary "MonthNames" could be used to identify mentioning of months within a text documents. The dictionaries also offer the possibility to associate other information with other features. For example "Oktober" could then be associated with "10" to in turn easily normalize date information.

The process of creating a dictionary is simply and supports the automatic generation of inflections, as well as the manual definition of synonyms if so desired.

4.3 Finite-State Technology

Many of the basic steps in NLP, such as tokenization and morphological analysis, can be carried out efficiently by the means of finite-state transducers [19]. These transducers are generally compiled from *regular expressions*, which is a formal language for representing sets and relations [19].

ICA studio utilizes regular expressions to realize character rules. This is useful to specify character patterns of interest such as dates or phone numbers. Those character sequences following these patterns can then be annotated and that way be used accordingly for relation extraction.

4.4 Text Segmentation

Text segmentation is an important step in any NLP process. Electronic text in its raw form is essentially just a sequence of characters. Consequently it has to be broken down into linguistic units. Such units include *words*, *punctation*, *numbers*, *alphanumerics*, etc. [19]. This process if also referred to as *tokenization*. Most NLP techniques also require the text to be segmented into sentences and maybe paragraphs as well [19].

ICA has this functionality included. Beside tokenization, which can be influenced by the means of special dictionaries or character rules, ICA splits the text into sentences and paragraphs. It migh be interesting to note, that neither character rules, nor dictionaries have to influence tokenization, it is a matter of choice.

4.5 Part-of-Speech Tagging

Most tasks NLP require the assignment of classes to linguistic entities (tokens) [27]. Part-of-Speech (POS), for instance, is an essential linguistic concept in NLP, and POS tagger are used to assign syntactic categories (e.g. noun, verb, adjective, adverb, etc.) to each word [27,28].

Automatic part-of-speech taggers have to handle several difficulties, including the ambiguity of word forms in their part-of-speech [29], as well as classification problems due to the ambiguity of periods ('.'), which can be either interpreted as part of a token (e.g. abbreviation), punctuation (full stop), or both [27].

ICA provides a POS tagger as part of the Linguistic Analysis annotator included in the UIMA Pipeline, and can overcome aforementioned classification difficulties by integration of the user. Default tagging can in later stages be influenced and improved by defining own types, such as real numbers or dates, by means of character rules for the disambiguation of punctations; custom dictionaries allow the user to assign the part-of-speech to special words or word classes - if needed - which can be later exploited in the process of designing parsing rules.

4.6 Information Extraction

The build-in tools of ICA follow the idea of information extraction (IE). Grishman (2003) [19, 27] defines IE as the process of automatically identifying and classifying instances of entities, relations and events in a text, based on some semantic criterion.

Typical task are *name-*, *entity-*, *relation-* and *event extraction* [27]. The first two are integrated into the *named entity annotator*, while the others can be defined within the studio.

The modeling of relations can be done by defining parsing rules, that can operate on different levels, such as phrase or entity, as well as different scopes, such as sentence, paragraph or document. These parsing rules can also automatically be derived out of sample text passages and manually changed and optimized as needed, speeding up the process.

An interesting functionality included in the ICA studio are the so called *normalizers*. They can be used to convert different formats of the same concept into one standardized format. For example: "12.10.1987" and "1987-10-12" describe the same concept - a date. The normalizers can also be used to overcome different points of reference, or units. For example: "100 pounds" could automatically be tagged with the normalized feature "45.359 kg".

5 Materials

In our experiment, we are interested to investigate the potential to support a medical doctor (dermatologist) in his research by applying NLP techniques to selected *medical sample records*. These records contain a number of structured information, such as *patient name* and *date of birth*, but most of it is unstructured information in the form of written text.

The big challenge when confronted with text written by medical doctors, is the large portion of abbreviations, that in turn are not standardized either.

The idea is to apply IE techniques, in order to associate patients with extracted information such as *diagnosis*, *therapies*, *medicaments*, changes in *tumor size*, and so fourth. These information are then used to create a chronological sequence to enable the medical professional to see trends and correlations - or in other words: apply *knowledge discovery*.

6 Lessons Learned

Realizing an annotation scheme is an iterative task. Even though we as humans know what information is important to us, for example the type and length of a therapy, formulating a rule to identify such information is not a trivial undertaking.

The simple and easy-to-use interface enables the developers to perform fast prototyping. The on-the-fly testing gives you fast feedback of the "quality" of your rules. On the other hand, a lot of basic - and needed - functionality, such as POS tagging and named entity identification, is already build in.

Currently we are investigating different ways to realize our annotation schemes. The big challenge in the biomedical domain is the high variance of medical "dialects" that exists between different doctors and even stronger between different hospital. The most promising approach so far is based on layers of annotation schemes. The lower the layer, the more reliable are the information. For example numbers (such as *"3"*, *"2.5"* or *"three"*) and explicit dates (such as *"12.10.2003"*) are on the lowest level, while length measurements (such as *"3 cm"* or *"1.3 mm"*) are build on numbers and are a few level above them. We also try to separate more or less standardized information from information that might be influence by "dialects". For example when detecting examinations in combination with location: "CT Thorso: ..." is a much more reliable link than "... Beside the previously mentioned CT we also looked at the thorso ...". We hope that this will render our overall annotation scheme more robust, flexible and maintainable.

7 Conclusion and Future Research

Given its complex nature, general solutions to text understanding are not available yet, however, the business need is here now. This problem can meanwhile only be successfully addressed with specialized approaches, such as rule-based or statistical annotation schemes, to the domain or customer needs. This in return shift the need to enable developers to quickly develop and test these annotation schemes. By speeding up the development process and providing rapid feedback, the developer can focus more time into building and improving the annotation schemes themselves, since there are many possible solutions for a problem, but most of them are imprecise. The quality of the annotations strongly depend on the skills of the developer formulating them.

In the future we would like to take advantage of the build in functionality of custom annotators and research possible topological or statistical methods while having an annotated input.

Acknowledgments. This work was supported by IBM Austria. We are grateful for the support of Niko Marek, Stefan Jaschke and Michael Grosinger.

References

1. Holzinger, A., Geierhofer, R., Modritscher, F., Tatzl, R.: Semantic information in medical information systems: Utilization of text mining techniques to analyze medical diagnoses. Journal of Universal Computer Science 14(22), 3781–3795 (2008)
2. Kreuzthaler, M., Bloice, M., Faulstich, L., Simonic, K., Holzinger, A.: A comparison of different retrieval strategies working on medical free texts. Journal of Universal Computer Science 17(7), 1109–1133 (2011)

3. Gregory, J., Mattison, J.E., Linde, C.: Naming notes - transitions from free-text to structured entry. Methods of Information in Medicine 34(1-2), 57–67 (1995)
4. Holzinger, A., Kainz, A., Gell, G., Brunold, M., Maurer, H.: Interactive computer assisted formulation of retrieval requests for a medical information system using an intelligent tutoring system. In: World Conference on Educational Multimedia, Hypermedia and Telecommunications, pp. 431–436. AACE, Charlottesville (2000)
5. Lovis, C., Baud, R.H., Planche, P.: Power of expression in the electronic patient record: structured data or narrative text? International Journal of Medical Informatics 58, 101–110 (2000)
6. Blandford, A., Attfield, S.: Interacting with information. Synthesis Lectures on Human-Centered Informatics 3(1), 1–99 (2010)
7. Holzinger, A.: On knowledge discovery and interactive intelligent visualization of biomedical data - Challenges in Human Computer Interaction & Biomedical Informatics (2012)
8. Beale, R.: Supporting serendipity: Using ambient intelligence to augment user exploration for data mining and web browsing. International Journal of Human-Computer Studies 65(5), 421–433 (2007)
9. Funk, P., Xiong, N.: Case-based reasoning and knowledge discovery in medical applications with time series. Computational Intelligence 22(3-4), 238–253 (2006)
10. Holzinger, A., Scherer, R., Seeber, M., Wagner, J., Müller-Putz, G.: Computational Sensemaking on Examples of Knowledge Discovery from Neuroscience Data: Towards Enhancing Stroke Rehabilitation. In: Böhm, C., Khuri, S., Lhotská, L., Renda, M.E. (eds.) ITBAM 2012. LNCS, vol. 7451, pp. 166–168. Springer, Heidelberg (2012)
11. Waldrop, M.M.: Natural-language understanding. Science 224(4647), 372–374 (1984)
12. Weizenbaum, J.: Eliza - a computer program for study of natural language communication between man and machine. Communications of the ACM 9(1), 36–45 (1966)
13. Turing, A.M.: Computing machinery and intelligence. Mind 59(236), 433–460 (1950)
14. Yndurain, E., Bernhardt, D., Campo, C.: Augmenting mobile search engines to leverage context awareness. IEEE Internet Computing 16(2), 17–25 (2012)
15. Erhardt, R.A.A., Schneider, R., Blaschke, C.: Status of text-mining techniques applied to biomedical text. Drug Discovery Today 11(7-8), 315–325 (2006)
16. Lee, W.B., Wang, Y., Wang, W.M., Cheung, C.F.: An unstructured information management system (uims) for emergency management. Expert Systems with Applications 39(17), 12743–12758 (2012)
17. Zins, C.: Conceptual approaches for defining data, information, and knowledge: Research articles. J. Am. Soc. Inf. Sci. Technol. 58(4), 479–493 (2007)
18. Boisot, M., Canals, A.: Data, information and knowledge: have we got it right? IN3 Working Paper Series (4) (2004)
19. Mitkov, R.: The Oxford Handbook of Computational Linguistics (Oxford Handbooks in Linguistics S.). Oxford University Press (2003)
20. Ferrucci, D., Lally, A.: Building an example application with the unstructured information management architecture. IBM Systems Journal 43(3), 455–475 (2004)
21. Nasukawa, T., Nagano, T.: Text analysis and knowledge mining system. IBM Systems Journal 40(4), 967–984 (2001)
22. Gotz, T., Suhre, O.: Design and implementation of the uima common analysis system. IBM Systems Journal 43(3), 476–489 (2004)

23. Mack, R., Mukherjea, S., Soffer, A., Uramoto, N., Brown, E., Coden, A., Cooper, J., Inokuchi, A., Iyer, B., Mass, Y., Matsuzawa, H., Subramaniam, L.V.: Text analytics for life science using the unstructured information management architecture. IBM Systems Journal 43(3), 490–515 (2004)
24. Holzinger, A., Simonic, K., Yildirim, P.: Disease-disease relationships for rheumatic diseases: Web-based biomedical textmining and knowledge discovery to assist medical decision making (2012)
25. Holzinger, A., Yildirim, P., Geier, M., Simonic, K.-M.: Quality-based knowledge discovery from medical text on the Web Example of computational methods in Web intelligence. In: Pasi, G., Bordogna, G., Jain, L.C. (eds.) Qual. Issues in the Management of Web Information. ISRL, vol. 50, pp. 145–158. Springer, Heidelberg (2013)
26. Garvin, J.H., DuVall, S.L., South, B.R., Bray, B.E., Bolton, D., Heavirland, J., Pickard, S., Heidenreich, P., Shen, S.Y., Weir, C., Samore, M., Goldstein, M.K.: Automated extraction of ejection fraction for quality measurement using regular expressions in unstructured information management architecture (uima) for heart failure. Journal of the American Medical Informatics Association 19(5), 859–866 (2012)
27. Clark, A., Fox, C., Lappin, S. (eds.): The Handbook of Computational Linguistics and Natural Language Processing. Blackwell Handbooks in Linguistics. John Wiley & Sons (2010)
28. Manning, C.D., Schütze, H.: Foundations of statistical natural language processing. MIT Press, Cambridge (1999)
29. Schmid, H.: Probabilistic part-of-speech tagging using decision trees (1994)

Designing Computer-Based Clinical Guidelines Decision Support by a Clinician

Ljiljana Trtica-Majnarić[1] and Aleksandar Včev[2]

[1] Department of Family Medicine, Department of Internal Medicine
[2] Department of Internal Medicine
School of Medicine, University J.J. Strossmayer Osijek, 31 000 Osijek, Croatia
ljiljana.majnaric@hi.t-com.hr

Abstract. Computer systems have long been promoted for their potential to improve the quality of health care, including their use in supporting clinical decisions. In this work, the need for developing the computer surveillance system, to support CV risk assessment procedure, according to the last update of the SCORE system of the European Society of Cardiology, is presented and documented. The key step, in transorming guidelines into the computer media, is designing the logical pathway diagram, to take structure and human reasoning into rules and recommendations provided by the guidelines. At this step, the role of the end user (clinician) is essential, to adjust human cognition with the computer-based information processing. The second benefit arises from the demand of the computer media for data standardisation and systematic documentation and screening of the whole population, as well, all together leading to the translation of a problem-solving approach, in a medical care domain, into a programed-practice approach. Beneficial is that programs allow follow-up, comparison, evaluation and quality improvement.

Keywords: cardiovascular diseases, risk assessment, guidelines, primary health care, computer, decision support, human reasoning, programed-practice approach.

1 Introduction

There is a growing interest of the health care systems across countries in Europe to improve quality of care and patients` outcomes [1]. The focus is on prevention and optimal management of cardiovascular diseases (CVD) - a leading cause of morbidity and mortality in both, developed and developing countries [2]. As a way to adopt preventive strategies, to combat the deleterious effects of widespread CVD, is thought to be through consistent implementation of clinical guidelines in family medicine and primary health care (PHC) [3]. However, despite much efforts done to date, their implementation in a PHC setting is still low [4]. Key barriers, identified by physicians, include insufficient familiarity and time, inability to adjust guidelines with patients preferences, overestimation of risk, inconsistency among numerous algorithms available and low outcomes expectations [4,5]. Some other issues of a

A. Holzinger and G. Pasi (Eds.): HCI-KDD 2013, LNCS 7947, pp. 25–34, 2013.

practical concern include unpreparedness of health care systems to adopt preventive programs at a large scale and lack of cost-effectiveness analyses regarding guidelines implementation strategies [3,6]. An assumption, presented here, is that computer-based clinical guidelines decision support systems could improve guidelines implementation and the usage justification, as much between different national health care systems as within practices inside a particular country.

2 Guidelines on CVD Prevention between Intentions and Practical Usefulness

During the last years, a number of guidelines on CVD prevention, dealing with dyslipidaemias, arterial hypertension, diabetes, or even prevention in general, have been issued in Europe [7]. Concerning the management of dyslipidaemias, guidelines are based on assessing the total (overall) 10-year risk for a fatal, or non-fatal CV event [8]. Risk graduation is important for physicians when making a decision on whether to recommend lipid-lowering drugs (mostly statins) to patients, or to provide them only with advises for healthy lifestyles. Such logic has been arising from the results of randomised studies showing much greater efficiency of statins, in preventing the onset of the first CV event, in high-risk, compared to intermediate- and low-risk patients [9]. Even only in the field of CV risk assessment, there is a flood of scores and estimates; according to the recent systematic review, more then 1900 publications have been identified [10]. Moreover, there is no uniformity between guidelines, in regard to risk factors and risk stratification strategies and in setting target values for lipids and other risk factors [4]. Is it reasonable, then, to expect their wide implementation in every day practice?

In order to overcome much of this inconsistency, two major joint updates, for the management of dyslipidaemias and CVD prevention in general, have recently been issued in Europe [7]. They are based on comprehensively presented knowledge, graded to assess the quality of evidence, allowing more accurate and credible recommendations. Being aware that guidelines should be as much as possible clear and easy-to-use, the expert bodies endeavour on concise messages, in the form of pocket, or PDA versions, or one page message. Even then, it is questionable whether guidelines can reach every working place and how much they are really accessible for routine use. Information in the guidelines are rather "flattened", weakly semantically structured and provided as a set of general rules. Because of lack of the topology, a physician (end user) has to take substantial efforts when trying to read through guidelines by wisdom. He/She is faced with a hard task of transforming tables and recommendations into his/her natural mental working processes, in order to produce reasonable patterns [11]. This is a complex task [12]. That`s why a computer mediation is needed to stay in between.

3 SCORE (Systemic COronary Risk Evaluation) Algorithm of the European Society of Cardiology (ESC) and Its Last Update

Due to its simplificity, SCORE algorithm has been widely used in many European countries [8]. Until recently, it was based on several traditional risk factors, including age, sex, total cholesterol level, systolic hypertension and smoking status. Similar to other available score systems, it suffers from some major shortcomings [8,13]. The first one is poor sensitivity for younger and middle-aged patients, who might have already acquired a cluster of risk factors. They are usually people having some other important risk factors, out of those included in the SCORE, such as diabetes, obesity, or chronic renal isufficiency. The opposite of this problem is the question of how to manage appropriately people of advanced age, who are often classified as high risk, despite only moderate risk factor make-up. This is due to the fact that age, by itself, is a very strong risk factor. In contrast, evidence indicates that there is only a limited effect of statin therapy in elderly population, accompanied with the increased risk of harmful side-effects. One more limitation of the available CV risk scores is that they do not account for the effect of treatment.

With the aim of improving the accuracy of risk prediction, the SCORE system has recently been modified, by adding HDL-cholesterol into the former model [8]. The reason is increased knowledge, supported by a high level of evidence, on a causal relationship between low HDL-cholesterol and increased cardiovascular risk [14]. Data are consistent with the view that HDL-cholesterol can improve CV risk estimation, especially in patients with cardiometabolic syndrome and already achieved LDL-cholesterol goal. In addition, several clinical and socio-demographic conditions and markers have been recommended for a reclassification adjustment of individuals at intermediate risk [8]. Based on these modifications, 4 different risk levels can now be recognised (very high, high, intermediate and low), instead of standard 3 (high, intermediate and low), which is associated with more LDL-cholesterol goal targeting refinement.

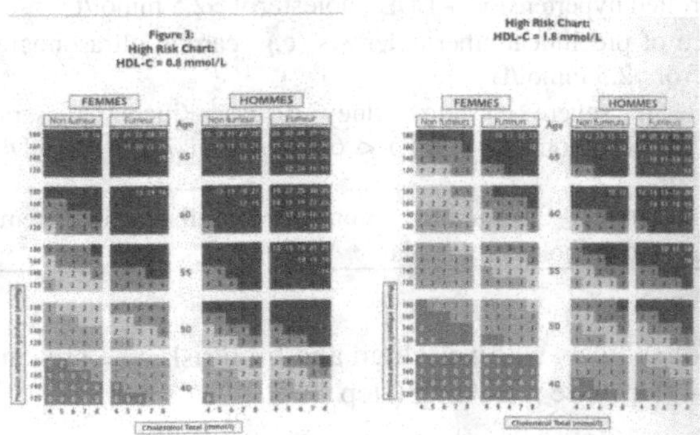

Fig. 1. The SCORE chart. A risk function with different HDL-cholesterol levels, in countries with high CV risk (European Society of Cardiology Guidelines for the management of dyslipidaemias, 2011).

4 The Logical Pathway Diagram - A Clinician`s View

The new SCORE guidelines, even in its pocket version, is the booklet encompassing a considerable amount of tables, charts and rules [8]. In this paper, there is an attempt to put structure and logic into the SCORE guidelines, preparing them for computer processing [15]. The ultimate goal is to help end users (medical doctors, practitioners) in their daily routines, by making guidelines to be mental-energy-saving and ever ready-to-use [7]. This is the first step on the pathway of transforming a weakly-structured diagnostic and prediction tool to the user-centered automated approach [15]. By speaking in terms of the computer language, the SCORE system can be considered as a knowledge base for a computer-based problem solving [16]. In this case, however, the knowledge base has already been elaborated by the expert team. The next step, in the computer programing process, is a logical pathway analysis [17]. At this time point of the logical processing, the clinician`s (user`s) expertise is crucial, to provide the meaningful results, by bringing the designer`s model into a harmony with the user`s mental model [18,19]. The resulting diagram (the human mental model) can be then used for engineering, in order to put the inferred mental procedure into practice [19].

Screening on high-risk patients for CVD, considered as the computer-based problem solving, should, in fact, answer the question: to whom to prescribe statin therapy?

Subjects with very high and high risk have already been provided by the SCORE, in the form of well defined clinical conditions, and do not need the CV risk calculation (Table 1.) [8].

Table 1. Criteria for very high and high CV risk and statin therapy, according to the SCORE (2011)

- Diabetes type 2 (type 1 if microalbuminuria) + LDL-cholesterol >1.8 mmol/L
- Uncontrolled hypertension + LDL-cholesterol >2.5 mmol/L
- Evidence of preclinical atherosclerosis (e.g. carotid ultrasonografy signs) + LDL-cholesterol >2.5 mmol/L
- Moderate to severe chronic kidney disease (increased serum creatinine, or glomerular filtration rate (GFR) < 60 ml/min/1.73 m^2) + LDL-cholesterol >1.8 mmol/L
- Total cholesterol ≥8.0 mmol/L - consider familial dyslipidaemias and causes of secondary hypercholesterolaemia

Identification of subjects with very high and high risk who are candidates for statin therapy includes a step-wise reasoning (step 1).

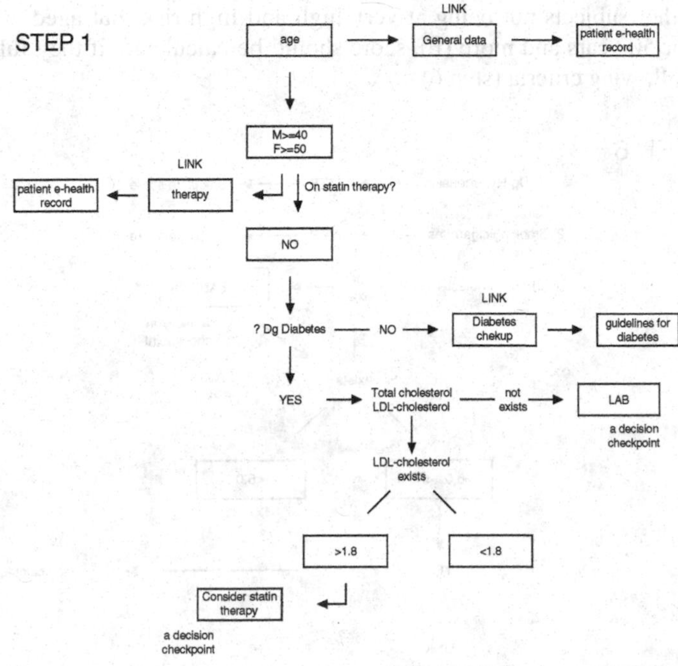

Rest subjects, selected according to the "age" criterium, but who did not fulfill other requested criteria, enter the next step of the logical pathway analysis (step 2).

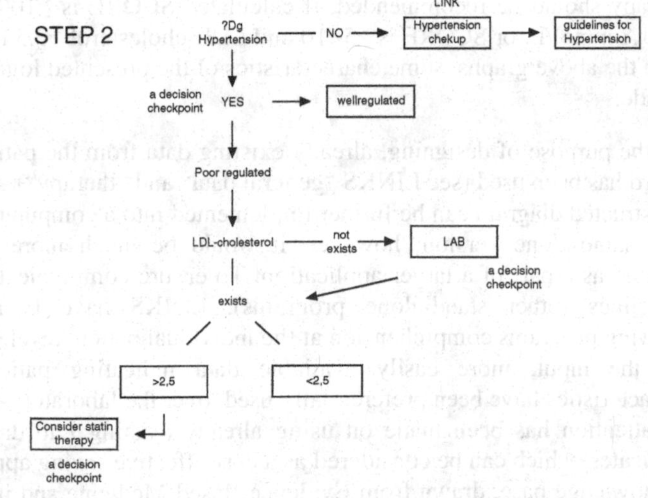

Subjects not selected in step 2 of the pathway analysis, enter step 3, and the rest enter step 4, and then step 5, until all criteria for very high and high CV risk have been exhausted (Table 1.).

For all other subjects not being at very high and high risk, but aged 40 years and more (M), or 50 years and more (F), score should be calculated, if they fulfill at least one of the following criteria (step 6)

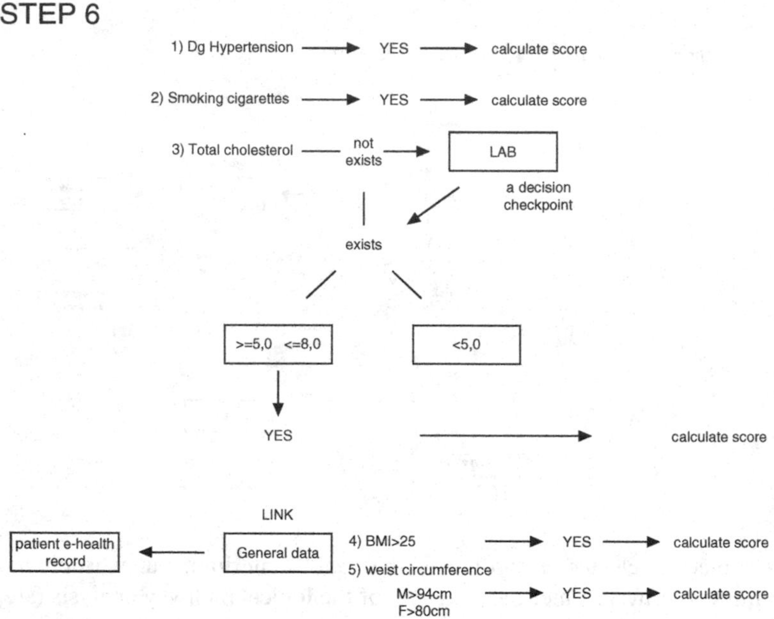

Statin therapy should be recommended, if calculated SCORE is ≥10% and LDL-cholesterol >1.8 mmol/L, or SCORE is ≥5<10 and LDL-cholesterol ≥2.5 mmol/L [8].

As seen on the above graphs, some characteristics of the presented logical pathway analysis include:

- For the purpose of designing, already existing data from the patient e-health record has been used (see LINKS "general data" and "therapy").
- Constructed diagram can be further implemented into a computer application as a stand-alone version; however, it would be much more efficient, if inserted as a part of a larger application. To ensure communication to other guidelines (other stand-alone programs), LINKS have been provided, allowing programs comprehension at the individual patient level.
- For the input, more easily available data indicating patients clinical characteristics have been preferentially used, over the laboratory-based data.
- An attention has been made on using already existing Lab data, to avoid duplicates, which can be considered as a cost-effective saving approach.
- A knowledge base, drawn from Evidence-Based-Medicine and integrated by the expert group, can ensure validity and credibility of the computer application at the large scale. To ensuring evaluation of changes in risk over time, including also the impact of treatment, trends and patterns in risk factors can be used [20].

- Finally, by using widely available age criteria (M≥40, F≥50) to start the input, it is ensured that all patients from the patient list can easily be evaluated and reclassified.

5 The Final Checkpoint

Before making a decision on whether or not to prescribe statins, all selected subjects, potential candidates for statin therapy, should be checked on liver functions and heart failure, in order to avoid adverse reactions on statins (step 7) [8]. This checkpoint can also be linked to the drug interaction checklist and the patient list of already prescribed drugs. Finally, a caution should be paid in prescribing statins for all subjects old 70 years and more (step 7).

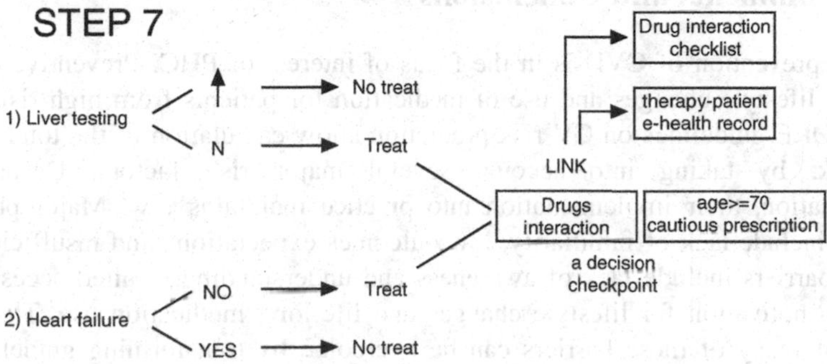

6 Computer-Based Guidelines on CV Risk Assessment in the Context of Human-Computer-Interaction

The expected benefit of the computer-based guidelines on CV risk assessment for the clinician`s every day work is not to enhance his/her strict adherence to the guidelines, the result of which might be an overwhelming promotion of drug use, but to have guidelines close to the clinician`s hand until there is a need for their use [7]. By getting the possibility of watching over the whole population that he/she is caring for and by being easily guided by the computer-based protocol, a clinician is likely to gain the capability of justifying his/her preventive work.

It is expected that the computer-based CV risk assessment tool, a schemata for which is presented in this work, will be able to meet the end user requirements for the simplificity and straightforwardness in application use [21]. By only a few "clicks" on the computer, a clinician will going through the large set of rules. This will not be a drop on the top of the overfull glass of the clinician`s daily work-load, but on the contrary, it will likely to save the time, tending to improve his/her professional self-confidence and self-esteem [22].

From the patient perspective, visual presentation of the protocol may add to the transparency of a doctor-patient dialog, as both are equally informed on the procedure, parameters required and the results which they have got [19,20]. This can provide them with the realistic framework for the partnership and sharing the decisions. In addition, it has already been proved that pure awareness on risk factors existence, without any other medical procedure used, can be a sufficient trigger for patients to change their lifestyles [4].

Not less important is the requirement of the computer system for systematic data collection and standardisation [19,20]. This, in turn, is the prerequisite to transforming the medical problem-solving into the programed practice approach, allowing programs comparison, combination, evaluation over time and quality improvement. In this context, this work can also provide a small contribution.

7 Comments and Conclusions

Primary prevention of CVD is in the focus of interest of PHC. Preventive activities involves lifestyle changes and use of medication for patients from high risk groups. The SCORE guidelines on CV risk prediction allow calculation of the total (overall) CV risk, by taking into account several major risk factors. Despite wide dissemination, their implementation into practice maintains low. Major physicians barriers include lack of familiarity, low outcomes expectations and insufficient time. Patient barriers include lack of awareness and understanding, limited access to care and low motivation for lifestyle changes and life-long medication use. The authors state that many of these barriers can be overcome by transforming guidelines into computer media. From the physician (end user) perspective, the emphasise is put on accessibility, simplificity and clearity of application and justification of a decision-making. From the patient perspective, perception of risk, access to the procedure, understanding and visibility, provide conditions for sharing decisions and better compliance with treatment.

Guidelines tend to bridge the gap between the scientific knowledge and its application. By speaking in computer language, guidelines may serve as a knowledge base for computation. The key step is designing the logical pathway diagram, in order to take structure and logic into numerous tables, rules and recommendations provided by the guidelines. At this step, the role of the end user (clinician) is essential, to allow adjustment of the human reasoning with the computer-based information processing.

The second major benefit of guidelines computerisation is the demand for data standardisation and systematic collection and the involvement of the whole target population by screening. This is a way to translate a problem-solving approach, in a medical care domain, into a programed-practice approach, allowing programs evaluation, comparison and follow-up. In terms of that, computerisation of guidelines is a prototype. On this background, research would be possible, based on using data from everyday practice.

References

1. Grol, R., Wensing, M., Jacob, A., Baker, R.: Quality assurance in general practice, the state of art in Europe 1993. Nederlands Huisartsen Genootschap, Utrecht (1993)
2. WHO. The global burden of disease 2004 update (2004), `http://www.who.int/healthinfo/global_burden_disease/2004_report_update`
3. Grimshow, J., Eccles, M., Thomas, R., MacLennon, G., Ramsay, C., Fraser, C., Vale, L.: Towards Evidence-Based quality improvement. Evidence (and its limitations) of the effectiveness of guideline dissemination and implementation strategies 1966-1998. J. Gen. Intern. Med. 21, S14–S20 (2006), doi:10.1111/j.1525-1497.2006.00357.x
4. Erhardt, L.R.: Managing cardiovascular risk: reality vs. perception. Eur. Heart J. 7(suppl. L), L11–L15 (2005), doi:10.1093/eurhearty/sui080
5. Hobbs, F.D.R., Jukema, J.W., Da Silva, P.M., McCormack, T., Catapano, A.L.: Barriers to cardiovascular disease risk scoring and primary prevention in Europe. QJM 103, 727–739 (2010), doi:10.1093/qjmed/hcq122
6. Petursson, H., Getz, L., Sigurdsson, J.A., Hetlevik, J.: Current European Guidelines for management of arterial hypertension: are they adequate for use in primary care? Modelling study based on the Norwegian HUNT 2 population. BMC Fam. Pract. 10(70), 1–9 (2009), doi:10.1186/1471-2296-10-70
7. Editorial. Guidelines on CVD prevention: confusing or complementary? Atherosclerosis 226, 299-300 (2012)
8. Committee for practice guidelines to improve the quality of clinical practice and patient care in Europe. ESC/EAS guidelines for the management of dyslipidaemias. Reprinted from Eur. Heart J. 32(14) (2011)
9. Brugts, J.J., Yetgin, T., Hoeks, S.E., Gotto, A.M., Shepherd, J., Westendorp, R.G., et al.: The benefits of statins in people without established cardiovascular disease but with cardiovascular risk factors: meta-analysis of randomised controlled trials. BMJ 338, b2376 (2009)
10. Ferket, B.S., Colkesen, E.B., Visser, J.J., Spronk, S., Kraaijenhagen, R.A., Steyerberg, E.W., Hunink, M.G.: Systematic review of guidelines on cardiovascular risk assessment: which recommendations should clinicians follow for a cardiovascular health check? Arch. Intern. Med. 170, 27–40 (2010)
11. Boring, R.L.: Human-Computer Interaction as cognitive science. In: Proceedings of the 46th Annual meeting of the Human Factors and Ergonomics Society (2001)
12. Hewett, T.T.: Human-Computer Interaction and cognitive psyhology in visualization education. In: Proceedings of the 46th Annual Meeting of the Human Factors and Ergonomics Society (2001)
13. Koenig, W.: Recent developments in cardiovascular risk assessment: relevance to the nephrologist. Nephrol. Dial. Transplant. 26, 3080–3083 (2011), doi:10.1093/ndt/gfr283
14. The European Atherosclerosis Society Consensus Panel. Triglyceride-rich lipoproteins and high-density lipoprotein cholesterol in patients at high risk of cardiovascular disease: evidence and guidance for management. Eur. Heart J. 32, 1345–1361 (2011)
15. Holzinger, A.: Weakly structured data in health-Informatics: the challenge for Human-Computer-Interaction. In: Baghaei, N., Baxter, G., Dow, L., Kimani, S. (eds.) Proceedings of INTERACT 2011 Workshop: Promoting and Supporting Healthy Living by Design, pp. 5–7. IFIP, Lisbon (2011)

16. Majumder, D.D., Bhattacharya, M.: Cybernetic approach to medical technology: application to cancer screening and other diagnostics. Kybernetes 29, 871–895 (2000)
17. Trtica-Majnarić, L.: Knowledge discovery and expert knowledge for creating a chart-model of a biological network - introductory to research in chronic diseases and comorbidity. In: Holzinger, A., Simonic, K.-M. (eds.) USAB 2011. LNCS, vol. 7058, pp. 337–348. Springer, Heidelberg (2011)
18. Holzinger, A., Scherer, R., Seeber, M., Wagner, J., Müller-Putz, G.: Computational sensemaking on examples of knowledge discovery from neuroscience data: towards enhancing stroke rehabilitation. In: Böhm, C., Khuri, S., Lhotská, L., Renda, M.E. (eds.) ITBAM 2012. LNCS, vol. 7451, pp. 166–168. Springer, Heidelberg (2012)
19. Holzinger, A.: On knowledge discovery and interactive intelligent visualization of biomedical data. In: Helfert, M., Fancalanci, C., Filipe, J. (eds.) Proceedings of the International Conference on DATA Technologies and Applications, pp. 5–16. DATA, Rome (2012)
20. Simonic, K.M., Holzinger, A., Bloice, M., Hermann, J.: Optimizing long-term treatment of rheumatoid arthritis with systemic documentation. In: Proceedings of the 5th International Conference on Pervasive Computing Technologies for Healthcare (PervasiveHealth), pp. 550–554. ISBN, Dublin (2011)
21. Holzinger, A.: Usability Engineering methods for software developers. Communications of the ACM 48(1), 71–74 (2005)
22. White, R.H., Mungall, D.: Outpatient management of warfarin therapy: comparison of computer-predicted dosage adjustment to skilled professional care. Ther. Drug Monit. 13, 46–50 (1991)

Opinion Mining on the Web 2.0 – Characteristics of User Generated Content and Their Impacts

Gerald Petz[1], Michał Karpowicz[1], Harald Fürschuß[1], Andreas Auinger[1], Václav Stříteský[2], and Andreas Holzinger[3]

[1] University of Applied Sciences Upper Austria, Campus Steyr, Austria
{gerald.petz,michal.karpowicz,harald.fuerschuss, andreas.auinger}@fh-steyr.at
[2] University of Economics, Prague
stritesv@vse.cz
[3] Medical University Graz, Medical Informatics, Statistics and Documentation, Austria
andreas.holzinger@medunigraz.at

Abstract. The field of opinion mining provides a multitude of methods and techniques to be utilized to find, extract and analyze subjective information, such as the one found on social media channels. Because of the differences between these channels as well as their unique characteristics, not all approaches are suitable for each source; there is no "one-size-fits-all" approach. This paper aims at identifying and determining these differences and characteristics by performing an empirical analysis as a basis for a discussion which opinion mining approach seems to be applicable to which social media channel.

Keywords: opinion mining, user generated content, sentiment analysis, text mining, content extraction, language detection, Internet slang, text mining.

1 Introduction and Motivation for Research

Opinion mining (some authors use "sentiment analysis" synonymously), deals with analyzing people's opinions, sentiments, attitudes and emotions towards different brands, companies, products and even individuals [1], [2]. The rise of the Web 2.0 and its user generated content led to many changes of the Internet and its usage, as well as a change in the communication processes. The user created content on the Web 2.0 can contain a variety of important market research information and opinions, through which economic opportunities as well as risks can be recognized at an early stage. Some of the challenges for qualitative market research on the Web 2.0 are on the one hand the variety of information and on the other hand the huge amount of rapidly growing and changing data.

Besides the typical challenges known from natural language processing and text processing, many challenges for opinion mining in social media sources make the detection and processing of opinions a complicated task:

A. Holzinger and G. Pasi (Eds.): HCI-KDD 2013, LNCS 7947, pp. 35–46, 2013.

- Noisy texts: User generated contents in social media tend to be less grammatically correct, they are informally written and have spelling mistakes. These texts often make use of emoticons and abbreviations or unorthodox capitalisation [3], [4].
- Language variations: Texts in user generated content typically contain irony and sarcasm; texts lack contextual information but have implicit knowledge about a specific topic [5].
- Relevance and boilerplate: Relevant content on webpages is usually surrounded by irrelevant elements like advertisements, navigational components or previews of other articles; discussions and comment threads can divert to non-relevant topics [5–7].
- Target identification: Search-based approaches to opinion mining often face the problem that the topic of the retrieved document does not necessarily match the mentioned object [5].

In the field of opinion mining, where language-specific tools, algorithms and models are frequently utilized, these challenges have quite an important impact on the properness of results, since the application of improper methods leads to incorrect or worse sentiment analysis results.

1.1 Objective and Methodology

The *objective* of this paper is to investigate the differences between social media channels and to discuss the impacts of their characteristics to opinion mining approaches. To attain this objective, we set up a methodology as follows:

(i) In the first step, we identify the most popular approaches for opinion mining in the scientific field and their underlying principles of detecting and analyzing text.

(ii) As a second step we identify and deduce criteria from literature to exhibit differences between the different kinds of social media sources regarding possible impacts on the quality of opinion mining.

(iii) Subsequently, we carry out an empirical analysis based on the deduced criteria in order to determine the differences between several social media channels. The social media channels taken into consideration in the third step are: social network services (*Facebook*), microblogs (*Twitter*), comments on weblogs and product reviews (*Amazon* and other product review sites).

(iv) In the last step, the social media source types need to be correlated with applicable opinion mining approaches based on their respective characteristics.

The next section gives a short overview about related work and approaches of opinion mining; section 3 describes the empirical analysis and discusses impacts of the characteristics of user generated content to opinion mining.

2 Background, Related Work

Opinion mining deals with different methods and algorithms from computational linguistics and natural language processing in order to find, extract and analyze people's opinions about certain topics.

2.1 Opinion Definition

Liu defines an opinion as a quintuple (e_i, a_{ij}, s_{ijkl}, h_k, t_l), where e_i is the name of an entity, a_{ij} is an aspect of e_i, s_{ijkl} is the sentiment on aspect a_{ij} of entity e_i, h_k is the opinion holder and t_l is the time, when the opinion is expressed. An entity is the target object of an opinion; it is a product, service, topic, person, or event. The aspects represent parts or attributes of an entity (part-of-relation). The sentiment is positive, negative or neutral or can be expressed with intensity levels. The indices *i*, *j*, *k*, *l* indicate that the items in the definition must correspond to one another [1].

2.2 Main Research Directions and Technical Approaches

Several main research directions can be identified [2], [8]: (1) *Sentiment classification*: The main focus of this research direction is the classification of content according to its sentiment about opinion targets; (2) *Feature-based opinion mining* (or aspect-based opinion mining) is about analysis of sentiment regarding certain properties of objects (e.g. [9], [10]) (3) *Comparison-based opinion mining* deals with texts in which comparisons of similar objects are made (e.g. [11]).

Opinion mining has been investigated mainly at three different levels: document level, sentence level and entity/aspect-level. Most classification methods are based on the identification of opinion words or phrases. The underlying algorithms can be categorized as follows: (1) Supervised learning (e.g. [12], [13]), (2) Unsupervised learning (e.g. [14]), (3) Partially supervised learning (e.g. [15]), (4) Other approaches / algorithms like latent variable models (hidden Markov model HMM [16]), conditional random fields CRF [17]), latent semantic association [18], pointwise mutual information (PMI) [19].

Due to the amount of different techniques, several researchers experimented with different algorithms and drew comparisions between them: [20–22].

2.3 Opinion Mining and Web 2.0

A couple of research papers focus explicitly on Web 2.0: A considerably amount of research work covers *weblogs*, e.g. [23–26], but most of them investigate the correlation between blog posts and "real life"-situations. Only a few papers evaluate techniques for opinion mining in the context of weblogs; there is no main direction of used techniques. Liu et al. [27] compare different linguistic features for blog sentiment classification, [28] experimented with lexical and sentiment features and different learning algorithms for identifying opinionated blogs. Surprisingly, little research work can be found about opinion mining in the area of *discussion forums* (e.g. [29], [30]). However, *microblogs* – in particular Twitter – seem to be quite attractive to researchers and a variety of papers focussing on microblogs have been published, e.g. [31–35]. The researchers mainly use supervised learning or semi-supervised learning as the dominant approach to mine opinions on microblogs. Despite the popularity of *social network services* like Facebook, relatively little research work about opinion mining in social networks can be found (e.g. [36], [37]). There are numerous research papers that deal with *product reviews*, and there is not one specific approach that seems to perform best. Many authors use text classification algorithms like SVM or Naïve Bayes and combine different techniques to increase the quality of opinion mining results. A promising technique could be LDA (e.g. [38], [39]). [40] proposed an LDA-based model that jointly identifies aspects and sentiments. This model (also e.g. the approach of [41], [42]) assumes that all of the words in a sentence cover one single topic.

3 Research Work and Results

We conducted an empirical analysis in order to find differences between social media channels. The following section describes the empirical analysis as well as the impacts of user generated content on opinion mining.

3.1 Empirical Analysis

Methodology of Survey. When starting the empirical analysis, it lends itself to asking the question of how an appropriate sample should to be drawn in order to conduct a representative survey. Basically, a random sample is reasonable, but it is actually a challenge to draw a random sample. Therefore, we have decided to draw a sample of self-selected sources and to make a kind of quota sampling. In order to avoid confounders, systematic errors and bias we define the following constraints: we focus on one specific brand / company (in our case: Samsung) and on a specific time period (in our case: between June, 15th 2011 and Jan, 28th 2013) for all sources in social media. Within this time period we conduct a comprehensive survey; if there are too many entries to perform a comprehensive survey, we draw a random sample of the entries. As we do not want to analyze the official postings of the company, we exclude these postings from the analysis. The data sets were labeled manually by four different human labelers. Before the labeling started, we discussed and defined rules for labeling in order to make the labeling consistent among the labelers [11]. The statistical calculations were carried out using SPSS.

The following sources have been surveyed in four different languages: social network service (Facebook; 410 postings), microblog (Twitter; 287 tweets), blog (387 blog posts), discussion forum (417 posts from 4 different forums) and product reviews (433 reviews from Amazon, and two product review pages). The collection of the data was performed manually for the discussion forums and automated using the API (Twitter, Facebook) and a Web-crawler for the other sources (Amazon).

Evaluation Criteria. In order to compare different social media channels, we need to determine indicators. These indicators – shown in table 1 – are derived from two sources: (i) criteria based on simple frequencies from content analysis, and (ii) criteria derived from the definition of opinions (see section 2.1):

Table 1. Evaluation criteria

Criteria	Description	Scale type [43]
Language	Describes the language used, e.g. English, German, etc.	Nominal
Number of words	How many words does a posting (e.g. blog posting, Facebook-post, product review, comment, etc.) contain?	Metric
Number of sentences	How many sentences does a posting contain?	Metric
Number of Internet slang abbreviations	How many typical Internet slang abbreviations (e.g. LOL, IMO, IMHO …) does the posting contain?	Metric
Number of emoticons	How many emoticons (e.g. ;-) :-) :-o …) does the posting contain?	Metric
Number of incorrect sentences	How many sentences contain grammatical and orthographical mistakes or typos per posting?	Metric
Subjectivity	Does the posting contain an opinion? Is the posting subjective or objective?	Nominal
Opinion holder	Is the opinion holder the author of the posting?	Nominal
Opinion expression	Is the opinion implicitly or explicitly formulated?	Nominal
Topic-related	Does the posting refer to the headline / overall topic?	Nominal
Aspect	Does the opinion refer to one or more aspects of the entity?	Nominal

Results of Survey. All in all we analyzed 1934 postings; in the following section we give a short overview on some key findings:

- *Length of postings*: As expected, the length of the postings differs between the social media channels. The average amount of words per posting is highest in product reviews (approx. 119 words), lowest in microblogs (approx. 14 words). Interestingly, the average amount of words per Facebook posting is only 19 words.
- *Emoticons and Internet slang*: Emoticons are widely used across all analyzed social media channels, with approximately every third (Facebook: 27.8%, Twitter: 24.4%, blogs: 27.6%) to fifth (discussion forums: 20.1%, product reviews: 15.5%)

posting containing them. Internet slang is not prominently featured in the analyzed channels, whereby no significant difference between them was detected. While Tweets contain the highest amount of typical abbreviations (20.2% of posting), they only occur in about 12.8% of all discussion forum posts, product reviews and blog comments. Surprisingly, only 8.3% of the analyzed Facebook comments feature Internet slang.

- *Grammatical and orthographical correctness*: Postings across all social media channels contain many grammatical as well as orthographical errors. The error ratio (number of incorrect sentences divided by number of sentences) is highest in Twitter (48.8%), Facebook (42.7%) and discussion forums (42.3%), and lowest in product reviews (37.2%) and blogs (35.4%). The detailed correlations between the variables were tested with Post-Hoc-tests / Bonferroni: product review / Twitter (p=0.002), Twitter / blog (p=0.0).
- *Subjectivity*: Across all analyzed channels 67.8% of the postings were classified as being subjective, as opposed to 18.1% objective ones. The remaining 14.1% of the postings contain both subjective and objective information. While the highest subjectivity can be detected on Twitter (82.9% of all analyzed Tweets), discussion forums not only features the fewest subjective posts (50.2%) but also the majority of objective ones (35.5%). Many of the postings in discussion forums do not contain an opinion, but questions, solution suggestions and hints how to solve a specific issue. An interesting discovery is the lack of exclusively objective product reviews – nearly two thirds (71.7%) of the analyzed reviews are solely subjective, while one quarter (25.4%) is based on both subjective and objective information. 2.9% of the reviews are rated as being objective. The detailed correlations between the variables were tested with Post-Hoc-tests / Bonferroni: Facebook / discussion forum (p=0.001), Twitter / product review (p=0.0), Twitter / blog (p=0.033), Twitter / discussion forum (p=0.0).

Table 2. Subjectivity in postings

Social media channel	Subjective	Objective	Subjective & objective
Microblog (Twitter)	82,9%	12,8%	4,3%
Product Review	71,7%	2,9%	25,4%
Blog	69,3%	19,6%	11,1%
Social Network (Facebook)	67,3%	26,1%	6,6%
Discussion forum	50,2%	35,5%	14,3%

- *Aspects and details*: As expected, the social media channels that tend to feature longer postings contain more details on certain aspects of entities. The detailed figures are exhibited in Table 4. While product review postings go into detail (39.6%) and contain aspects as well as opinions on entity-level (27.0%), Twitter and Facebook-postings mainly contain postings on entity-level (56.6%, 65.4%).

Table 3. Opinions about entites and aspects

Social media channel	Contains one or more aspects	Does not contain aspects	Contains opinion about entity and aspect
Discussion forum	60,6%	33,1%	6,3%
Blog	55,3%	39,1%	5,6%
Microblog (Twitter)	43,4%	56,6%	0%
Product Review	39,6%	33,4%	27,0%
Social Network (Facebook)	33,0%	65,4%	1,6%

- *Opinion holder*: The survey exhibited that in most cases the opinion holder is equal to the author of the posting; in Facebook, Twitter, product reviews and blogs between 95% and 97.6% of the postings reveal the author as the opinion holder. Only the postings in the discussion forums have a lower percentage (90.7%). 6.2% of the entries in discussion forums have several opinion holders, and 3.1% depict the opinion of another person.
- *Topic relatedness*: At the beginning of our survey we were curious about the users' "discipline" regarding the topic relatedness of their postings. Surprisingly, the postings in all the social media channels are highly related to the overall discussion topic. As shown in the following table, the highest relatedness can be found in discussion forums, which may be related to the presence of moderators and forum rules.

Table 4. Topic relatedness

Social media channel	Topic related	Not topic related	Topic and non-topic related content
Discussion forum	95.6%	3.4%	1.0%
Microblog (Twitter)	95.3%	4.7%	0%
Product review	93.1%	1.2%	5.8%
Blog	92.6%	6.3%	1.1%
Social Network (Facebook)	82.3%	16.6%	1.1%

Discussion of Survey. The criteria we used for the survey are often criticized in research papers for their ambiguity, e.g. subjective vs. objective. The team that conducted the survey exchanged their experiences and carried out multiple evaluations on the same sample set. There remains the question of how to conduct a survey that is both representative and accomplishable with manageable efforts. In our survey we used one brand from the electronic consumer market, but the results may vary depending on other market segments or genres.

3.2 Impact on Opinion Mining

Based on the empirical analysis the following impacts can be derived for the opinion mining process:

Impacts on Opinion Mining Process. Many research papers in the field of opinion mining assume grammatically correct texts [4], but as shown in the empirical analysis, user generated texts contain many mistakes, emoticons and Internet slang words. Therefore it is reasonable and necessary to preprocess texts from Web 2.0-sources. In some cases the text languages changed on the same channel, e.g. some Facebook postings on the German Facebook site are written in English, Turkish and other languages. In these cases the application of language detection methods is reasonable. In general, because of the grammatical mistakes, grammar-based approaches (e.g. [44], [45]) are not appropriate.

The above figures showed, that user generated texts contain Internet slang as well as emoticons. These text parts could be considered as input for feature generation to improve sentiment classification. Furthermore, people often use different names for the same object, e.g. "Samsung Galaxy S3" is also being called "Galaxy S3" or "SGS3", which makes the extraction of entities or aspects more difficult.

Characteristics and Impacts of Social Media Channels. The following table gives a short overview about the impacts of each investigated social media channel:

Table 5. Social media channels and their impacts

Social media channel	Impact
Discussion forum	The empirical analysis revealed, that discussions in forums are often organized in discussion threads, users respond to other user's questions and comments, and forum postings often contain coreferences – all these factors make opinion mining more difficult and a variety of approaches have to be adopted to discussion forums. More research work is required to evaluate, which methods perform best.
Microblog (Twitter)	The characteristics of Twitter can be summarized as follows: many grammatical errors, short sentences, heavy usage of hashtags and other abbreviations. That already led researchers to taking Twitter characteristics into consideration, e.g. Davidov et al. [46] use Twitter characteristics and language conventions as features, Zhang et al. [47] combine lexicon-based and learning-based methods for Twitter sentiment analysis. The usage of part-of-speech features does not seem to be useful in the microblogging domain (e.g. [48]).

Table 5. (*Continued.*)

Product review	Several researchers proposed models to identify aspects and sentiments; a few of them assume that all of the words in a sentence cover one single topic. This assumption may be reasonable for product reviews, but this assumption has to be questioned for Facebook, because there are often missing punctuations and it is - even for humans – not easy to detect the boundaries of sentences and to find out the meaning of expressions.
Blog	Many research papers that focus on blogs do not unfold how comments to the blog posts are taken into consideration. The comments to blog posts vary in terms of length, coreferences, etc., and thus can be very short answers when the user replies with a short answer or quite long texts when users discuss a topic controversially for instance. From our point of view, depending on the type of the blog (corporate blog vs. j-blog) both the blog posting and the blog comments can be interesting sources for opinion mining.
Social Network (Facebook)	Because users can interact with each other, respond to questions and the amount of grammatical mistakes, there are similar challenges like with discussion forums. More research work is required.

4 Conclusion and Further Research

This paper discusses the differences of social media channels including microblogs (Twitter), social network services (Facebook), weblogs, discussion forums and product review sites. A survey has been conducted to exhibit the differences of these social media channels, and implications for opinion mining have been derived. The survey covers only the contents related to one specific brand, because the authors wanted to emphasize the viewpoint of a company; of course, the results could be different in other genres (e.g. political discussions), which would require more empirical analysis. The work shows that the dominant approach to mine opinions on microblogs is supervised or semisupervised learning; while for product reviews a wide range of techniques is applied.

Further research work should be conducted: (i) Measure and compare the factual implications of the characteristics of social media on the performance of the different opinion mining approaches, and (ii) conduct more research work on alternative (statistical / mathematical) approaches.

Acknowledgements. This work emerged from the research projects *OPMIN 2.0* and *SENOMWEB*. The project *SENOMWEB* is funded by the European Regional Development fund (*EFRE, Regio 13*). *OPMIN 2.0* is funded under the program *COIN – Cooperation & Innovation. COIN* is a joint initiative launched by the Austrian Federal Ministry for Transport, Innovation and Technology (*BMVIT*) and the Austrian Federal Ministry of Economy, Family and Youth (*BMWFJ*).

References

1. Liu, B.: Sentiment analysis and opinion mining. Morgan & Claypool, San Rafael (2012)
2. Pang, B., Lee, L.: Opinion Mining and Sentiment Analysis. Foundations and Trends in Information Retrieval 2(1-2), 1–135 (2008)
3. Abbasi, A., Chen, H., Salem, A.: Sentiment analysis in multiple languages: Feature selection for opinion classification in Web forums. ACM Transactions on Information Systems (TOIS) 26(3), 12–34 (2008)
4. Dey, L., Haque, S.M.: Opinion mining from noisy text data. International Journal on Document Analysis and Recognition (IJDAR) 12(3), 205–226 (2009)
5. Maynard, D., Bontcheva, K., Rout, D.: Challenges in developing opinion mining tools for social media. In: Proceedings of @NLP can u tag #user_generated_content?! Workshop at LREC 2012 (2012)
6. Petz, G., Karpowicz, M., Fürschuß, H., Auinger, A., Winkler, S.M., Schaller, S., Holzinger, A.: On Text Preprocessing for Opinion Mining Outside of Laboratory Environments. In: Huang, R., Ghorbani, A.A., Pasi, G., Yamaguchi, T., Yen, N.Y., Jin, B. (eds.) AMT 2012. LNCS, vol. 7669, pp. 618–629. Springer, Heidelberg (2012)
7. Yi, L., Liu, B.: Web page cleaning for web mining through feature weighting. In: Proceedings of the 18th International Joint Conference on Artificial Intelligence, pp. 43–48. Morgan Kaufmann Publishers Inc., San Francisco (2003)
8. Kaiser, C.: Opinion Mining im Web 2.0 – Konzept und Fallbeispiel. HMD - Praxis der Wirtschaftsinformatik 46(268), 90–99 (2009)
9. Hu, M., Liu, B.: Mining Opinion Features in Customer Reviews. In: Proceedings of AAAI, pp. 755–760 (2004)
10. Liu, B., Hu, M., Cheng, J.: Opinion observer: analyzing and comparing opinions on the Web. In: Proceedings of the 14th International Conference on World Wide Web, New York, NY, USA, pp. 342–351 (2005)
11. Jindal, N., Liu, B.: Identifying Comparative Sentences in Text Documents. In: Dumas, S. (ed.) Proceedings of the 29th Annual International ACM SIGIR Conference on Research and Development in Information Retrieval, pp. 244–251. Association for Computing Machinery, New York (2006)
12. Zhang, T.: Fundamental Statistical Techniques. In: Indurkhya, N., Damerau, F.J. (eds.) Handbook of Natural Language Processing, 2nd edn., pp. 189–204. Chapman & Hall/CRC, Boca Raton (2010)
13. Pang, B., Lee, L., Vaithyanathan, S.: Thumbs up? Sentiment Classification Using Machine Learning Techniques. In: Proceedings of the ACL-2002 Conference on Empirical Methods in Natural Language Processing, pp. 79–86 (2002)
14. Liu, B.: Web data mining. Exploring hyperlinks, contents, and usage data, Corr. 2. print. Data-centric systems and applications. Springer, Berlin (2008)
15. Dasgupta, S., Ng, V.: Mine the Easy, Classify the Hard: A Semi-Supervised Approach to Automatic Sentiment Classification. In: Proceedings of the Joint Conference of the 47th Annual Meeting of the ACL and the 4th International Joint Conference on Natural Language Processing of the AFNLP, vol. 2, pp. 701–709 (2009)
16. Wong, T.-L., Bing, L., Lam, W.: Normalizing Web Product Attributes and Discovering Domain Ontology with Minimal Effort. In: Proceedings of the Fourth ACM International Conference on Web Search and Data Mining, pp. 805–814 (2011)
17. Choi, Y., Cardie, C.: Hierarchical Sequential Learning for Extracting Opinions and their Attributes. In: Proceedings of the ACL 2010 Conference Short Papers, pp. 269–274 (2010)

18. Guo, H., Zhu, H., Guo, Z., et al.: Domain Customization for Aspect-oriented Opinion Analysis with Multi-level Latent Sentiment Clues. In: Proceedings of the 20th ACM International Conference on Information and Knowledge Management, pp. 2493–2496 (2011)
19. Holzinger, A., Simonic, K.-M., Yildirim, P.: Disease-disease relationships for rheumatic diseases. Web-based biomedical textmining and knowledge discovery to assist medical decision making. In: IEEE 36th International Conference on Computer Software and Applications, pp. 573–580 (2012)
20. Cui, H., Mittal, V., Datar, M.: Comparative Experiments on Sentiment Classification for Online Product Reviews. In: Proceedings of AAAI-2006, pp. 1265–1270 (2006)
21. Chaovalit, P., Zhou, L.: Movie Review Mining: A Comparison Between Supervised and Unsupervised Classification Approaches. In: Proceedings of the 38th Annual Hawaii International Conference on System Sciences, pp. 112–121 (2005)
22. Moghaddam, S., Ester, M.: On the Design of LDA Models for Aspect-based Opinion Mining. In: Proceedings of the 21st ACM International Conference on Information and Knowledge Management, pp. 803–812 (2012)
23. Mishne, G., Glance, N.S.: Predicting Movie Sales from Blogger Sentiment. In: Proceedings of the 21st National Conference on Artificial Intelligence, pp. 11–14. AAAI Press, Boston (2006)
24. Sik Kim, Y., Lee, K., Ryu, J.-H.: Algorithm for Extrapolating Blogger's Interests through Library Classification Systems. In: Proceedings of the IEEE International Conference on Web Services, Beijing, China, September 23-26, pp. 481–488. IEEE (2008)
25. Liu, Y., Huang, X., An, A., et al.: ARSA: A Sentiment-Aware Model for Predicting Sales Performance Using Blogs. In: Proceedings of the 30th Annual International ACM SIGIR Conference on Research and Development in Information Retrieval, pp. 607–614. ACM Press, New York (2007)
26. Sadikov, E., Parameswaran, A., Venetis, P.: Blogs as Predictors of Movie Success. In: Proceedings of the Third International Conference on Weblogs and Social Media, pp. 304–307 (2009)
27. Liu, F., Wang, D., Li, B., et al.: Improving Blog Polarity Classification via Topic Analysis and Adaptive Methods. In: Proceedings of Human Language Technologies: The 2010 Annual Conference of the North American Chapter of the ACL, pp. 309–312 (2010)
28. Liu, F., Li, B., Liu, Y.: Finding Opinionated Blogs Using Statistical Classifiers and Lexical Features. In: Proceedings of the Third International ICWSM Conference, pp. 254–257 (2009)
29. Chmiel, A., Sobkowicz, P., Sienkiewicz, J., et al.: Negative emotions boost user activity at BBC forum. Physica A 390(16), 2936–2944 (2011)
30. Softic, S., Hausenblas, M.: Towards opinion mining through tracing discussions on the web. In: Social Data on the Web Workshop at the 7th International Semantic Web Conference (2008)
31. Go, A., Bhayani, R., Huang, L.: Twitter Sentiment Classification using Distant Supervision. CS224N Project Report, Stanford (2009)
32. Pak, A., Paroubek, P.: Twitter as a Corpus for Sentiment Analysis and Opinion Mining. In: Proceegins of the Seventh Conference on International Language Resources and Evaluation (LREC), Valletta, Malta, pp. 1320–1326 (2010)
33. Barbosa, L., Feng, J.: Robust Sentiment Detection on Twitter from Biased and Noisy Data. In: Proceedings of the 23rd International Conference on Computational Linguistics. Posters, pp. 36–44 (2010)

34. Bollen, J., Mao, H., Zeng, X.: Twitter mood predicts the stockmarket. Journal of Computational Science 2(1), 1–8 (2011)
35. Derczynski, L., Maynard, D., Aswani, N., et al.: Microblog-Genre Noise and Impact on Semantic Annotation Accuracy. In: 24th ACM Conference on Hypertext and Social Media (2013)
36. Thelwall, M., Wilkinson, D., Uppal, S.: Data Mining Emotion in Social Network Communication: Gender differences in MySpace. Journal of the American Society for Information Science and Technology 61(1), 190–199 (2010)
37. Bermingham, A., Conway, M., McInerney, L., et al.: Combining Social Network Analysis and Sentiment Analysis to Explore the Potential for Online Radicalisation. In: International Conference on Advances in Social Network Analysis and Mining, pp. 231–236 (2009)
38. Titov, I., McDonald, R.: A Joint Model of Text and Aspect Ratings for Sentiment Summarization. In: Proceedings of ACL-2008: HLT, pp. 308–316 (2008)
39. Titov, I., McDonald, R.: Modeling Online Reviews with Multi-grain Topic Models. In: Proceedings of the 17th International Conference on World Wide Web, pp. 111–120 (2008)
40. Zhao, W.X., Jiang, J., Yan, H., et al.: Jointly Modeling Aspects and Opinions with a MaxEnt-LDA Hybrid. In: Proceedings of the 2010 Conference on Empirical Methods in Natural Language Processing, pp. 56–65 (2010)
41. Brody, S., Elhadad, N.: An Unsupervised Aspect-Sentiment Model for Online Reviews. In: Proceedings of HLT 2010 Human Language Technologies: The 2010 Annual Conference of the North American Chapter of the Association for Computational Linguistics, pp. 804–812 (2010)
42. Jo, Y., Oh, A.: Aspect and Sentiment Unification Model for Online Review Analysis. In: Proceedings of the fourth ACM International Conference on Web Search and Data Mining, pp. 815–824 (2011)
43. Backhaus, K., Erichson, B., Plinke, W., et al.: Multivariate Analysemethoden, 12th edn. Eine anwendungsorientierte Einführung. Springer, Berlin (2008)
44. Moilanen, K., Pulman, S.: Sentiment Composition. In: Proceedings of the Recent Advances in Natural Language Processing International Conference, pp. 378–382 (2007)
45. Sayeed, A.B.: A Distributional and Syntactic Approach to Fine-Grained Opinion Mining. Dissertation, University of Maryland (2011)
46. Davidov, D., Tsur, O., Rappoport, A.: Enhanced sentiment learning using Twitter hashtags and smileys. In: Proceedings of the 23rd International Conference on Computational Linguistics. Posters, pp. 241–249 (2010)
47. Zhang, L., Ghosh, R., Dekhil, M., et al.: Combining Lexicon-based and Learning-based Methods for Twitter Sentiment Analysis. Technical Report HPL-2011-89 (2011)
48. Kouloumpis, E., Wilson, T., Moore, J.: Twitter Sentiment Analysis: The Good the Bad and the OMG? In: Proceedings of the Fifth International AAAI Conference on Weblogs and Social Media, pp. 538–541 (2011)

Evaluation of $SHAPD2$ Algorithm Efficiency Supported by a Semantic Compression Mechanism in Plagiarism Detection Tasks

Dariusz Adam Ceglarek

Poznan School of Banking, Poland
dariusz.ceglarek@wsb.poznan.pl

Abstract. This paper presents the issues concerning knowledge protection and, in particular, research in the area of natural language processing focusing on plagiarism detection, semantic networks and semantic compression. The results demonstrate that the semantic compression is a valuable addition to the existing methods used in plagiarism detection. The application of the semantic compression boosts the efficiency of the Sentence Hashing Algorithm for Plagiarism Detection 2 (SHAPD2) and the *w−shingling* algorithm. All experiments were performed on an available PAN–PC plagiarism corpus used to evaluate plagiarism detection methods, so the results can be compared with other research teams.

Keywords: plagiarism detection, longest common subsequence, semantic compression, sentence hashing, w-shingling, intellectual property protection.

1 Introduction

The main objective of this work is to present recent findings obtained in the course of research on more efficient algorithms used in matching longest common subsequences, as well as semantic compression. As signalled in the previous publications introducing Sentence Hashing Algorithm for Plagiarism Detection 2 ($SHAPD2$) [1], this algorithm is capable of providing better results than the most known alternatives operating on hash structure representing fragments (usually n-grams) of a text document. By better, author understand a certain set of features that the alternatives cannot deliver along with performance characteristics surpassing known competitors. More details will be provided further.

One of important domains of the longest sequence matching is plagiarism detection. It was observed that a solution that uses text hashing to detect similar documents can benefit greatly from the inclusion of a mechanism that makes it resilient to a number of techniques used by those inclined to commit an act of plagiarism.

The common plagiarism techniques include changing the word order in a plagiarised text, paraphrasing passages of a targeted work and interchanging original words with their synonyms, hyponyms and hypernyms. Not all of the above can be addressed at the moment, but the technique based of synonym

A. Holzinger and G. Pasi (Eds.): HCI-KDD 2013, LNCS 7947, pp. 47–58, 2013.

usage can be easily detected when one is equipped with sufficiently large semantic network that is a key requirement for the semantic compression.

Throughout the research activities author crafted a majority of the necessary tools and methods vital in order to establish how well an application of the semantic compression should level up the results of plagiarism detection efforts.

In order to provide a reasonable study three approaches were tested in detail:

- Sentence Hashing Algorithm for Plagiarism Detection ($SHAPD2$),
- $w - shingling$ using 4-grams
- $w - shingling$ using 6-grams

Each of them was used to gather results on detection of plagiarized documents. The experiment was twofold: firstly, test data was run across the unmodified implementations of the enlisted approaches, secondly the semantic compression was introduced in several steps gauging at each step the strength of the compression. The PAN-PC-10 plagiarism evaluation corpus [2] from Weimar University has been utilised in order to run a benchmark. PAN corpora have been used since 2009 in plagiarism uncovering contests, gathering researchers from this domain to evaluate their methods in comparison with others. The possibility to use the same data sets and measures allows to perceive the results as reliable.

The $SHAPD2$ algorithm has been already evaluated using Clough & Stephenson corpus [3] for plagiarism detection and the results were published in [4]. The results proved that the idea of combining sentence hashing and semantic compression improves method's performance in terms of both efficiency and results.

The first evaluation positions $SHAPD2$ as effective as $w - shingling$, or even more for specific cases, while overrunning $w - shingling$ when considering run time of comparisons of bigger document sets. In the course of experiments, Vector Space Model has been checked, too, and combined with semantic compression turned out to be ineffective when considering task precision - too many documents appeared similar when processed with semantic compression enabled. It may be argued that not all the alterations are possible to be discovered by the semantic compression. In the current form, used semantic networks do not store data on whole phrases that can be synonym or hyponym of any other given phrase. Such an extension should provide even better results, yet the amount of work needed to craft such a resource is at the moment beyond the grasp of the author.

The results obtained throughout the experiments show a great improvement of $SHAPD2$ augmented with semantic compression over $w - shingling$. What is more, time wise performance of $SHAPD2$ is far better than the $w - shingling$.

Article is structured as follows: related work section is given where the most important algorithms concerned with a longest common sequence finding are reviewed briefly. The discussion is accompanied by description of plagiarism detection methods and some of the most important initiatives addressing the issue of similarity of documents to one another. Following that, there is a brief presentation of $SHAPD2$ and the semantic compression. The next section is devoted to the experiments, test corpus used, implementations of benchmarked algorithms and the obtained results. All is summarised in the final section extended with plans of future research work.

2 Related Work

The core of the work presented in this article is dependent on the *SHAPD*2 algorithm that allows for robust and resilient computation of a longest common subsequence shared by one or many input documents. The task of matching a longest common subsequence is an important one in many subdomains of Computer Science. Its most naive implementation was deemed to have a time complexity of $O(m_1 * m_2)$ (where m_1 and m_2 are the numbers of terms in compared documents). The question whether it is possible to achieve significantly better results was stated first by Knuth in [5]. First affirmative answer was given in [6] with time complexity $O((m_1 * m_2)/log(m_2))$ for case when $m < n$ and they pertain to a limited sequence. One of the most important implementations of a search for the longest common subsequence is to be found in [7]. This work presents an application of the Smith-Waterman algorithm for matching a longest common subsequence in a textual data, which is the fastest algorithm that does not operate with text frames and their hashes. Other works such as [8] or [9] prove that better efficiency is yielded rather by careful engineering strategies than a fundamental change in time complexity. All of the above cited works use algorithms whose time complexity is near quadratic which results in drastic drop of efficiency when dealing with documents of considerable length.

It was first observed in [10] that introduction of a special structure that was later known as shingling or chunks (a continuous sequence of tokens in a document) can substantially improve the efficiency determining the level of similarity of two documents by observing a number of common shinglings. Subsequent works such as [11,12] introduce further extensions to the original idea. A number of works represented by publications such as [13] provided plausible methods to further boost measuring of the similarity between entities.

The importance of plagiarism detection is recognized in many publications. It may be argued that, it is an essential task in times, where access to information is nearly unrestricted and culture for sharing without attribution is a recognized problem (see [14] and [15]).

With respect to plagiarism obfuscation further explanations are necessary. Plagiarists often paraphrase or summarize the text they plagiarize in order to obfuscate it, i.e., to hide their offense. In the PAN plagiarism corpus a synthesizer, that simulates the obfuscation of a section of text *sx* in order to generate a different text section *sq* to be inserted into *dq*, has been designed on the basis of the following basic operations [16]:

- Random Text Operations. Given *sx*, *sq* is created by shuffling, removing, inserting, or replacing words or short phrases at random.
- Semantic Word Variation. Given *sx*, *sq* is created by replacing each word by one of its synonyms, hyponyms, hypernyms, or even antonyms.
- POS-preserving Word Shuffling. *sq* is created by shuffling words while maintaining the original sequence of parts of speech in *sx*.

It is obvious that these operations do not guarantee the generation of humanreadable text. However, automatic text generation is still a largely unsolved problem which is why we have approached the task from the basic understanding of content similarity in information retrieval, namely the bag-of-words model.

In order to reliably compare the results with other methods submitted to PAN-PC competitions in plagiarism detection, the same measures needed to be employed in the evaluation. PAN–PC competitions use the following indicators to determine methods' performance: precision, recall, granularity and *plagdet* score.

- precision and recall are standard measures used in Information Retrieval and calculated here in a respective way. Precision is determined as a ratio of correct plagiarism detections to a total number of reported potential plagiarised text fragments. Recall is a ratio of correctly reported plagiarism cases to a total number of plagiaries, existing in the corpus. An aim is to achieve both measures as close to 1.00 as possible.

$$precision = \frac{r_s}{|R|} \quad (1)$$

$$recall = \frac{r_s}{|S|} \quad (2)$$

  where: r_s is a number of correct plagiarism detections, R is a set of reported suspicious plagiarism cases, S is a set of plagiarism cases
- as it's possible, that some methods can report one plagiarised text passage as multiple plagiarism detections, a measure of granularity is introduced. It can be quantified as an average number of reported plagiarisms per one plagiarised text passage. $SHAPD2$ is always reporting subsequent or overlapping detections as a single plagiarism detection, hence achieving granularity of 1.00.
- in order to balance precision, recall and granularity in one synthetic indicator, a plagdet score has been introduced by PAN–PC authors. Plagdet is calculated from the following formula:

$$plagdet = \frac{H}{log_2(granularity)} = \\ = \frac{2 * precision * recall}{(precision + recall) * log_2(granularity)} \quad (3)$$

  where H is a harmonic mean of precision and recall.

3 $SHAPD2$ Algorithm and Semantic Compression

3.1 Sentence Hashing Algorithm for Plagiarism Detection - $SHAPD2$

$SHAPD2$ algorithm focuses on whole sentence sequences, calculating hash-sums for them. It also utilizes a new mechanism to organize the hash-index as well as to search through the index. It uses additional data structures such as correspondence list CL to aid in the process. The detailed description of $SHAPD2$ can be found in [1].

The $SHAPD2$ algorithm works with two corpora of documents which comprise the algorithm's input: a corpus of source documents (originals) $D = \{d_1, d_2, ..., d_n\}$, and a corpus of suspicious documents to be verified regarding possible plagiaries, $P = \{p_1, p_2, ..., p_r\}$. SHAPD2 focuses on whole sentence sequences. A natural way of splitting a text document is to divide it into sentences and it can be assumed that documents containing the same sequences also contain the same sentences.

Initailly, all documents need to be split into text frames of comparable length – preferably sentences, or in the case of longer sentences – split into shorter phrases (long passages of text without a full-stop mark such as different types of enumerations, tables, listings, etc.). A coefficient α is a user-defined value which allows to set the expected number of frames whis a longer sentence is split into.

Then, each sentence in the whole document has to undergo a set of transformations in a text-refinement process which is a standard procedure in NLP/IR tasks. The process of text-refinement starts from extracting lexical units (tokenization), and further text refinement operations include elimination of words from the so-called information stop-list, identification of multiword concepts, and bringing concepts to the main form by lemmatization or stemming. This is an especially difficult task for highly flexible languages, such as Polish or French (with multiple noun declination forms and verb conjugation forms). The last step in this procedure is the concept disambiguation (i.e. choosing right meaning of polysemic concept). As an output of the text-refinement process the system produces vectors containing ordered concept descriptors coming from documents.

In the next step, a hash table T is created for all documents from corpus D, where for each key the following tuple of values is stored: $T[k_{i,j}] =< i, j >$, (document number, frame number). A correspondence list CL is declared, with elements of the following structure: n_d – document number, m_l – local maximum, and n_l – frame number for local sequence match. For documents from the corpus P are also created indexes containing hash values for all frames coming from sentences from suspicious documents. As a result, every document from original corpus, as well as all suspicious documents, is represented by index as a list of sentence hashes.

Another data structure is the maxima array TM for all r documents in corpus P, containing records structured as follows: m_g – global maximum, n_g – frame number with global sequence match.

The comparison between corpus D and P is performed sequentially in phase 2 for all documents from corpus P. In phase 2 for all documents d_i from corpus P (containing suspicious documents), the correspondence list CL and maxima array TM are cleared. For each frame, set of tuples is retrieved from index table T. If there are any entries existing, it is then checked whether they point to the same source document and to the previous frame. If the condition is true, the local correspondence maximum is increased by one. Otherwise, the local maximum is decreased. After all of the frames are checked, table TM storing the correspondence maxima is searched for records whose correspondence maxima are greater than a threshold set e (the number of matching frames to be reported as a potential plagiarism). Frame and document number is returned in these cases.

The important distinction between those given above and the $SHAPD2$ is the emphasis on a sentence as the basic structure for a comparison of documents and a starting point of a procedure determining a longest common subsequence. Thanks to such an assumption, $SHAPD2$ provides better results in terms of time needed to compute the effects. Moreover, its merits does not end at the stage of establishing that two or more documents overlap. It readily delivers data on which sequences overlap, the length of the overlapping and it does so even when the sequences are locally discontinued. The capability to perform these, makes it a method that can be naturally chosen in the plagiarism detection, because such situations are common during attempts to hide plagiarism. In addition, it implements the construction of hashes representing the sentence in an additive manner, thus word order is not an issue while comparing documents.

The $w-shingling$ algorithm runs significantly slower when the task is to give a length of an extendedcommon subsequence. Due to the fixed frame orientation when performing such operating $w-shingling$ behaves in a fashion similar to the Smith-Waterman algorithm resulting in a significant drop of efficiency. Quality of the similarity measures is discussed in the experiments' part, and execution times for both algorithms are presented in Table 1.

Table 1. Processing time [s] for comparing 3000 documents with a corpus of n documents. Source: own elaboration.

n	1000	2000	3000	4000	5000	6000
w-shingling	5.680	8.581	11.967	16.899	23.200	50.586
SHAPD2	4.608	5.820	7.125	7.527	8.437	8.742

3.2 Semantic Compression

The idea of the global semantic compression has been introduced by the author in 2010 [17] as a method of improving text document matching techniques both in terms of effectiveness and efficiency. Compression of text is achieved by employing a semantic network and data on term frequencies (in form of frequency dictionaries). The least frequent terms are treated as unnecessary and they are replaced with more general terms (their hypernyms stored in semantic network). As a result, a reduced number of terms can be used to represent a text document without significant information loss, which is important from a perspective of processing resources (especially when one would like to apply a Vector Space Model [18] or [19]). Another feature of the emphasized concept level allows for capturing of common meaning expressed with differently worded sentences.

The semantic compression combines data from two sources: term frequencies from frequency dictionaries, and concept hierarchy from a semantic network. Usually, one extensive semantic network is used for a given language (e.g. *WiSENet* [20] semantic network converted from WordNet for English [21], *SenecaNet* for Polish [22]) and thus it is able to include linguistic knowledge covering multiple domains.

Source Document (Fragment). As no appearance of change in the character of the country within twenty or thirty miles was visible, and we had only two days' provisions left (not having expected the stream to extend so far), and the camp at sixty miles distant, we were obliged to leave the farther examination of the river to some future explorers; but we regretted it the less as, from the nature of the gravel and sand brought down by the stream, there seemed great probability that it takes its rise in large salt marshes similar to those known to exist 100 miles east of the Irwin.

Subject Document (Fragment). As no appearance of change in the character of the gravel sand brought down by the stream, there seemed worthy probability that it takes its rise in large salt marshes similar to those known to exist 100 miles east of the Irwin.

Resumed our commute; passed two parties of natives; a few of them constituted us some distance, and having overcome their first surprise, commenced talking in own language explorers; but we regretted it the less that, from the nature of the country within twenty or thirty miles was visible, and we had actually two days'provisions left (not having expected the stream to extend merely so), and the camp of seven hours we arrived at the river.

Comparison between Original and Subject Document after Semantic Compression. appearance change character country twenty thirty mile visible food left expected stream run camp adjective mile duty-bound leave adjective investigation river future someone regret nature cover courage bring stream seem suitable probability take rise large salt land similar known exist mile east Irvin.

Table 2. Similarity levels for a sample document (`suspicious-document02234.txt`) compared to original article (`source-document02313.txt`). Source: own elaboration.

method	no semantic compression	compression level 2000	compression level 1000
w-shingling	0.271	0.567	0.585
SHAPD2	0.229	0.433	0.508

Table 2 gives an illustrative example of difference between original and subject document before and after semantic compression. One has to understand that semantic compression is a lossy one. Yet, the loss of information is minimal by selecting the least frequent words and replacing them by more general terms, so their meaning remain as similar to the original as possible. The compression ratio can be tuned easily, by setting a number of concepts to be used to describe text documents. Experiments, that were conducted to measure quality of the method in Natural Language Processing tasks showed, that the number of words can be reduced to about 4,000 or even 3,000 without significant deterioration of classification results.

4 Experiments

The algorithm's performance has been evaluated using the PAN–PC plagiarism corpus, which is designed for such applications. Available test data sets, source and suspicious documents, have been downloaded and made available for the implementation of the algorithm. A series of program executions have been run in the course of the experiment with different level of semantic compression set. This enabled to adjust the compression strength to gain optimum results.

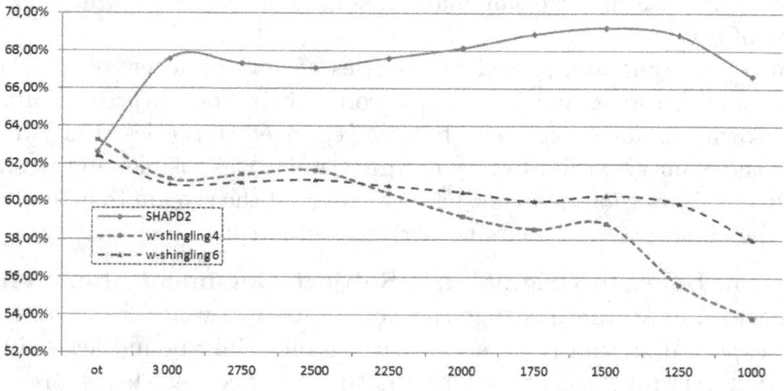

Fig. 1. Comparison of synthetic *plagdet* indicator: $SHAPD2$ versus w – *shingling* using 4-grams and 6-grams - for all cases of plagiarism in the PAN-PC corpus

In order to verify whether the semantic compression is a valuable addition to the already available technologies a set of test runs was performed on the mentioned corpus. The results are given in Tables 3 and 4. As one can easily see, the results prove that $SHAPD2$ cope better with the task than w – *shingling*.

Another step in the experiments was an introduction of the semantic compression of varying strength (understood as a number of concepts from the semantic network that were allowed to appear in final data). Data from this step are provided in Table 2.

Fig. 1 shows the difference in the value of the plagdet in the PAN-PC corpus of with the use of SHAPD2 and w-shingling algorithms. Is easy to see on Fig. 2 and 3 that the algorithm SHAPD2 produces better results of recall than w – *shingling*. The results of corpus matching using $SHAPD2$ method employing the semantic compression with a compression force threshold set to 1000 are especially worth noticing. Clough & Stevenson test corpus contains a subset of non-plagiarism documents (ie. participants were to read and rewrite the article using their knowledge, without any possibility to copy or modify the original), which stands for 40% of the corpus. The remaining documents were created by copying original articles and revising them to a certain extent: exact copy, light revision, heavy revision - 20% of documents each. The results presented in Table 3 show, that the method employing the semantic compression allows for achieving the results structure very similar to the original one.

One of the test cases, which reveals a major improvement in recognizing an evident plagiarism is demonstrated in Table 3. The semantic compression enabled to calculate similarity measures at around 0.6, while both methods ($w-shingling$ and $SHAPD2$) without semantic compression returned relatively low similarity measures (about 0.3).

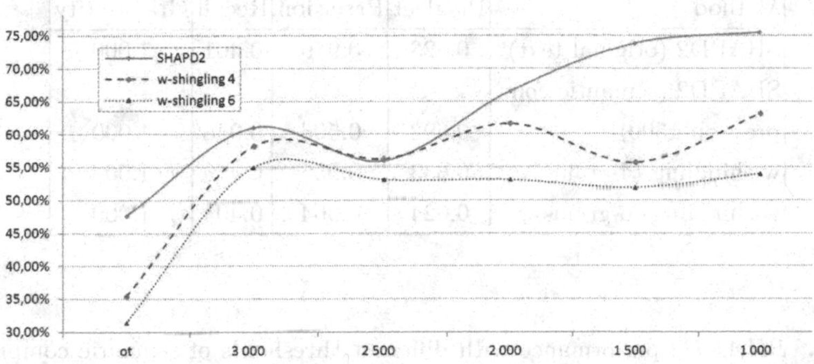

Fig. 2. Comparison: $SHAPD2$ versus $w-shingling$ using 4-grams and 6-grams - recall for simulated plagiarism cases in the PAN-PC corpus

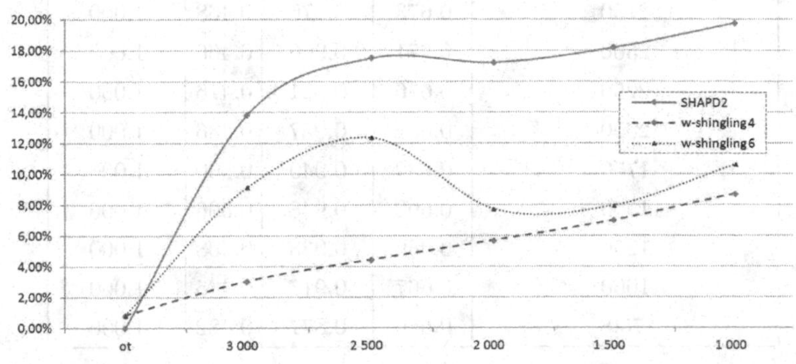

Fig. 3. Comparison: SHAPD2 vs w – *shingling* 4-grams and 6-grams - recall for all cases

Generalization, which is a key operation in semantic compression, entails minimal information loss, which cumulates together with reducing target's lexicon size. It does not appear to deteriorate results of the experiments employing $SHAPD2$ or $w - shingling$ method (see Table 3). It is especially visible in cases where the strong semantic compression was enabled (target lexicon size below 1500 words), causing some non-plagiarized articles on the same subject to be marked as highly similar and recognized as possible plagiaries (cf. Table 4). This suggest, that for certain use cases the compression threshold needs to be set carefully. Further research is necessary to identify the optimum compression threshold for individual information retrieval tasks.

Table 3. $SHAPD2$ and $w - shingling$ methods' performance on the PAN plagiarism corpus. The metrics used - precision, recall, granularity and overall mark: plagdet - correspond to International Competition on Plagiarism Detection scores [23].

Method	PlagDet	Precision	Recall	Granularity
SHAPD2 (original text)	0.623	0.979	0.401	1.000
SHAPD2 (semantic compression 1500)	0.692	0.506	0.945	1.000
w-shingling (4-grams)	0.633	0.955	0.419	1.000
w-shingling (6-grams)	0.624	0.964	0.404	1.000

Table 4. $SHAPD2$ performance with different thresholds of semantic compression. Source: own elaboration.

Semantic compression level	PlagDet	Precision	Recall	Granularity
none	0.626	0.979	0.401	1.000
3000	0.676	0.974	0.469	1.000
2750	0.673	0.970	0.468	1.000
2500	0.671	0.966	0.466	1.000
2250	0.676	0.961	0.476	1.000
2000	0.682	0.957	0.486	1.000
1750	0.689	0.949	0.500	1.000
1500	0.692	0.948	0.506	1.000
1250	0.689	0.933	0.508	1.000
1000	0.667	0.917	0.485	1.000
750	0.650	0.877	0.482	1.000

5 Summary

To summarize conducted research, the following should be emphasized:

- The $SHAPD2$ algorithm can be successfully employed in plagiarism detection systems, giving results of competitive quality when compared to $w - shingling$, while performing substantially better when taking into account time efficiency of the algorithms.
- Enabling semantic compression in $w - shingling$ and $SHAPD2$ methods improves the quality of plagiarism detection significantly. A combination of $SHAPD2$ and semantic compression returns results which structure is very close to experts' assessment.
- For certain information retrieval tasks, semantic compression strength needs to be adjusted carefully in order not to lose too much information, which may lead to deterioration of retrieval precision. Semantic compression threshold set to a level 1500–1250 words in the target lexicon seems to ba a safe value when used for plagiarism detection.
- The designed SHAPD2 algorithm - employing semantic compression - is strongly resilient to false-positive examples of plagiarism which may be an issue in cases when competitive algorithms are used.

In the near future, author plans to further develop various algorithms and reorganize of available assets so that the semantic compression can be applied in a automated manner to text passages without introduction of hypernyms disrupting user's experience. this might be achieved by introduction of information on concept relevance in current culture and prohibition on archaic concepts.

References

1. Ceglarek, D.: Single-pass Corpus to Corpus Comparison by Sentence Hashing. In: The 5th International Conference on Advanced Cognitive Technologies and Applications - COGNITIVE 2013. Xpert Publishing Services, Valencia (2013)
2. PAN-PC-10 Corpus (access March 11, 2013), http://pan.webis.de/
3. Clough, P., Stevenson, M.: A Corpus of Plagiarised Short Answers. University of Sheffield (2009), http://ir.shef.ac.uk/cloughie/resources/plagiarism_corpus.html
4. Ceglarek, D., Haniewicz, K., Rutkowski, W.: Robust Plagiary Detection Using Semantic Compression Augmented SHAPD. In: Nguyen, N.-T., Hoang, K., Jędrzejowicz, P. (eds.) ICCCI 2012, Part I. LNCS, vol. 7653, pp. 308–317. Springer, Heidelberg (2012)
5. Chvatal, V., Klarner, D.A., Knuth, D.E.: Selected combinatorial research problems. Technical report, Stanford, CA, USA (1972)
6. Masek, W.J., Paterson, M.S.: A faster algorithm computing string edit distances. Journal of Computer and System Sciences 20, 18–31 (1980)
7. Irving, R.W.: Plagiarism and collusion detection using the Smith-Waterman algorithm. Technical report, University of Glasgow (2004)
8. Grozea, C., Gehl, C., Popescu, M.: Encoplot: Pairwise sequence matching in linear time applied to plagiarism detection. In: Time, pp. 10–18 (2009)

9. Lukashenko, R., Graudina, V., Grundspenkis, J.: Computer-based plagiarism detection methods and tools: an overview. In: Proceedings of the 2007 International Conference on Computer Systems and Technologies, CompSysTech 2007, pp. 40:1–40:6. ACM, New York (2007)
10. Manber, U.: Finding similar files in a large file system. In: Proceedings of the USENIX Winter 1994 Technical Conference on USENIX, WTEC 1994 (1994)
11. Broder, A.Z.: Syntactic clustering of the web. Comput. Netw. ISDN Syst. 29(8-13), 1157–1166 (1997)
12. Andoni, A., Indyk, P.: Near-optimal hashing algorithms for approximate nearest neighbor in high dimensions. Commun. ACM 51(1), 117–122 (2008)
13. Charikar, M.S.: Similarity estimation techniques from rounding algorithms. In: Proceedings of the 34th Annual ACM Symposium - STOC 2002, pp. 380–388 (2002)
14. Ota, T., Masuyama, S.: Automatic plagiarism detection among term papers. In: Proceedings of the 3rd International Universal Communication Symposium, IUCS 2009, pp. 395–399. ACM, New York (2009)
15. Burrows, S., Tahaghoghi, S.M., Zobel, J.: Efficient plagiarism detection for large code repositories. Software: Practice and Experience 37(2), 151–175 (2007)
16. Potthast, M., Stein, B., Barrón-Cedeño, A., Rosso, P.: An Evaluation Framework for Plagiarism Detection. In: Proceedings of 23rd International Conference on Computational Linguistics - COLING, Beijing, pp. 997–1005 (2010)
17. Ceglarek, D., Haniewicz, K., Rutkowski, W.: Semantic compression for specialised information retrieval systems. In: Nguyen, N.T., Katarzyniak, R., Chen, S.-M. (eds.) Advances in Intelligent Information and Database Systems. SCI, vol. 283, pp. 111–121. Springer, Heidelberg (2010)
18. Manning, C.D., Raghavan, P., Schutze, H.: Introduction to Information Retrieval. Cambridge University Press (2008)
19. Erk, K., Pado, S.: A Structured Vector Space Model for Word Meaning in Context, pp. 897–906. ACL (2008)
20. Ceglarek, D., Haniewicz, K., Rutkowski, W.: Towards knowledge acquisition with WiseNet. In: Nguyen, N.T., Trawiński, B., Jung, J.J. (eds.) New Challenges for Intelligent Information and Database Systems. SCI, vol. 351, pp. 75–84. Springer, Heidelberg (2011)
21. Miller, G.A.: WordNet: a lexical database for English. Commun. ACM Communications of the ACM 38(11) (1995)
22. Ceglarek, D.: Architecture of the Semantically Enhanced Intellectual Property Protection System. In: Burduk, R., Jackowski, K., Kurzyński, M., Woźniak, M., Żołnierek, A. (eds.) CORES 2013. AISC, vol. 226, pp. 711–721. Springer International Publishing, Switzerland (2013)
23. Potthast, M., Gollub, T., Hagen, M., Kiesel, J., Michel, M., Oberlander, A., Tippmann, M., Barrón-Cedeño, A., Gupta, P., Rosso, P., Stein, B.: Overview of the 4th International Competition on Plagiarism Detection. In: Forner, P., Karlgren, J., Womser-Hacker, C. (eds.) CLEF 2012 Evaluation Labs and Workshop – Working Notes Papers (2012)

Using Hasse Diagrams for Competence-Oriented Learning Analytics

Michael D. Kickmeier-Rust and Dietrich Albert

Cognitive Science Section, Knowledge Management Institute
Graz University of Technology
Inffeldgasse 13, 8010 Graz, Austria
{michael.kickmeier-rust,dietrich.albert}@tugraz.at

Abstract. Learning analytics refers to the process of collecting, analyzing, and visualizing (large scale) data about learners for the purpose of understanding and pro-actively optimizing teaching strategies. A related concept is formative assessment – the idea of drawing information about a learner from a broad range of sources and on a competence-centered basis in order to go beyond mere grading to a constructive and tailored support of individual learners. In this paper we present an approach to competence-centered learning analytics on the basis of so-called Competence-based Knowledge Space Theory and a way to visualize learning paths, competency states, and to identify the most effective next learning steps using Hasse diagrams.

Keywords: Learning analytics, data visualization, Hasse diagram, Competence-based Knowledge Space Theory.

1 Introduction

Learning Analytics is a best practice and change on bringing together issues from human intelligence and computational intelligence, hence it fits perfectly in the HCI-KDD approach [1, 2]. In principle, the idea is to find theoretical frameworks, models, procedures, and smart tools to record, aggregate, analyze, and visualize large scale educational data. The principal goal is to make educational assessment and appraisal more goal-oriented, pro-active, and beneficial for students. In short, learning analytics is supposed to enable formative assessment of all kinds of information about a learner, on a large basis . Usually, the benefits are seen in the potential to reduce attrition through early risk identification, improve learning performance and achievement levels, enable a more effective use of teaching time, and improve learning design/instructional design [3]. Methods used for learning analytics and so-called "educational data mining" are extremely broad, for example, social network analyses, activity tracking, error tracking, keeping of e-Portfolios, semantic analyses, or log file analyses.

On the basis of this kind and amount of data, smart tools and systems are being developed to provide teachers with effective, intuitive, and easy to understand

A. Holzinger and G. Pasi (Eds.): HCI-KDD 2013, LNCS 7947, pp. 59–64, 2013.

aggregations of data and the related visualizations. There is a substantial amount of work going on this particular field; visualization techniques and dashboards are broadly available (cf. [4, 5, 6]), ranging from simple meter/gauge-based techniques (e.g., in form of traffic lights, smiley, or bar charts) to more sophisticated activity and network illustrations (e.g., radar charts or hyperbolic network trees).

A special challenge for visualizations, however, is to illustrate learning progress (including learning paths) and - beyond the retrospective view - to display the next meaningful learning steps/topics. In this paper we introduce the method of *Hasse diagrams* for structuring learning domains and for visualizing the progress of a learner through this domain.

2 Displaying Learning: Past, Present, and Future

A Hasse diagram is a strict mathematical representation of a so-called *semi-order*. The technique was invented in the 60s of the last century by *Helmut Hasse*; entities (the knots) are connected by relationships (indicated by edges), establishing a *directed graph*. The properties of a semi-order are (i) reflexivity, (ii) anti symmetry, and (iii) transitivity. In principle, the direction of a graph is given by arrows of the edges; per convention however, the representation is simplified by avoiding the arrow heads, whereby the direction reads from bottom to top. In addition, the arrows from one element to itself (reflexivity property) as well as all arrows indicating transitivity are not shown. The following image (Figure 1) illustrates such a diagram. Hasse diagrams enable a complete view to (often huge) structures. Insofar, they appear to be ideal for capturing the large competence spaces occurring in the context of assessment and recommendations of learning.

In an educational context, a Hasse diagram can display the non-linear path through a learning domain starting from an origin at the beginning of an educational episode (which may be a single school lesson but could also be the entire Semester). The beginning is shown as *{ }* (the empty set) at the bottom of the diagram. Now a learner might focus on three topics (K, P, or H); this, in essence, establishes three possible learning paths. After P, as an example, this learner might attend to topics K, A, or H next, which opens further three branches of the learning path until reaching the final state, within which all topics have been attended to (PKHTAZ).

In the context of formative learning analytics, a competence-oriented approach is necessary. Thus, a Hasse diagram can be used to display the competencies of a learner in the form of so-called *competence states*. A common theoretical approach to do so is *Competence-based Knowledge Space Theory* (CbKST). The approach originates from Jean-Paul Doignon and Jean-Claude Falmagne [7, 8] and is a well-elaborated set-theoretic framework for addressing the relations among problems (e.g., test items). It provides a basis for structuring a domain of knowledge and for representing the knowledge based on *prerequisite relations*. While the original *Knowledge Space Theory* focuses only on performance (the behavior; for example, solving a test item), CbKST introduces a separation of observable performance and latent, unobservable competencies, which determine the performance (cf. [9]). In addition, the approach is based on a probabilistic view of having or lacking certain competencies.

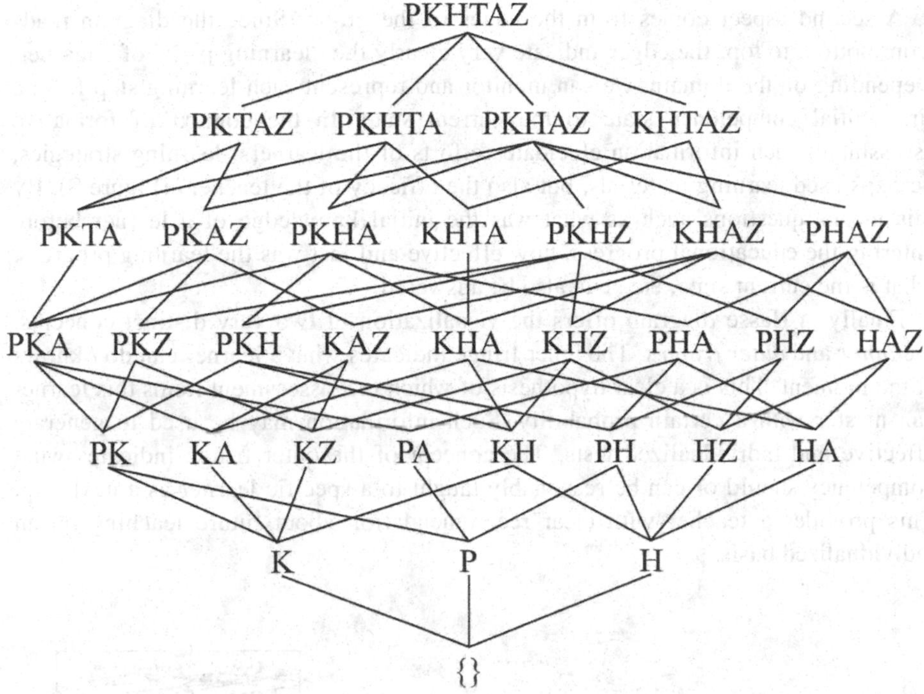

Fig. 1. An example for a directed graph, shown in form of a Hasse diagram

Very briefly, the idea is that usually one can find a natural structure, a natural course of learning in a given domain. For example, it is reasonable to start with the basics (e.g., the competency to add numbers) and increasingly advance in the learning domain (to subtraction, multiplication, division, etc.). As indicated above, this natural course is not necessary linear. On this basis, in a next step, we obtain a so-called competence space, the ordered set of all meaningful competence states a learner can be in. As an example, a learner might have none of the competencies, or might be able to add and subtract numbers; other states, in turn, are not included in the space, for example it is not reasonable to assume a learner has the competency to multiply numbers but not to add them. By the logic of CbKST, each learner is, with a certain likelihood, in one of the competence states. This allows displaying and coding of the state likelihoods for example by colors and thereby visualizing areas and set of states with high (or vice versa low) probabilities. An example is shown in Figure 3, where in the lower right part a color coded Hasse diagram is shown. The darker the colors, the higher the state probability. The simplest approach would be to highlight the competence state for a specific learner with the highest probability. The same coding principle can be used for multiple learners. This allows identification of various sub-groups in a class, outliers, the best learners, and so on (Figure 2).

A second aspect comes from the edges of the graph. Since the diagram reads from bottom to top, the edges indicate very clearly the "learning path" of a learner. Depending on the domain, we can monitor and represent each learning step from a first initial competence state to the current state. In the context of formative assessment, such information elucidates efforts of the learners, learning strategies, perhaps used learning materials, but also the efficacy of the teachers (Figure 2). By this means questions such as what was the initial knowledge of a learner before entering the educational progress, how effective and fast was the learning progress, what is the current state, etc., can also be answered.

Finally, a Hasse diagram offers the visualization of two very distinct concepts, the *inner* and *outer fringes*. The inner fringe indicates what a learner can do / knows at the moment. This is a clear hypothesis of which test/assessment items this learner can master with a certain probability. Such information may be used to generate effective and individualized tests. The concept of the outer fringe indicates what competency should or can be reasonably taught to a specific learner as a next step. This provides a teacher with clear recommendation about future teaching on an individualized basis.

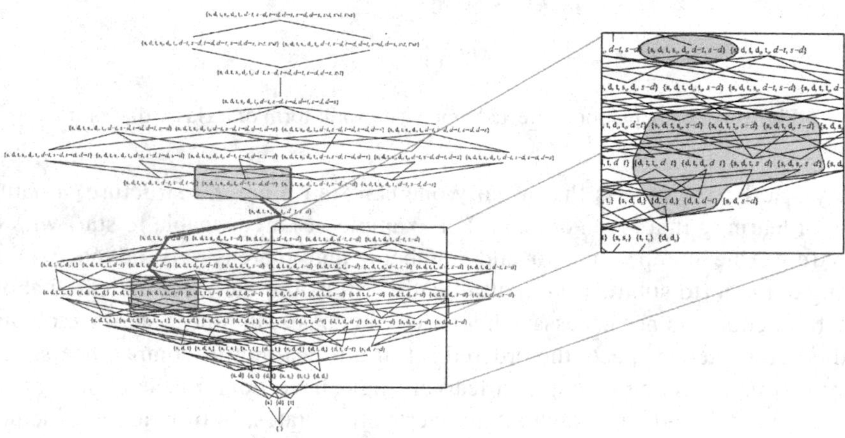

Fig. 2. Hasse diagram of a competence space; the left part illustrates an option to display a learning path from a starting state to the current competence state. The snapshot on the right illustrates a visualization of an individual (oval on the top) in comparison to the states 70 percent of the class are in.

The visualization in the form of Hasse diagrams was realized in the context of the European Next-Tell project (www.next-tell.eu) as part of the educational tool *ProNIFA*, which stands for probabilistic non-invasive formative assessment. The tool, in essence, establishes a handy user interface for services and functionalities related to learning analytics (in particular CBKST-based approaches). In principle the ProNIFA

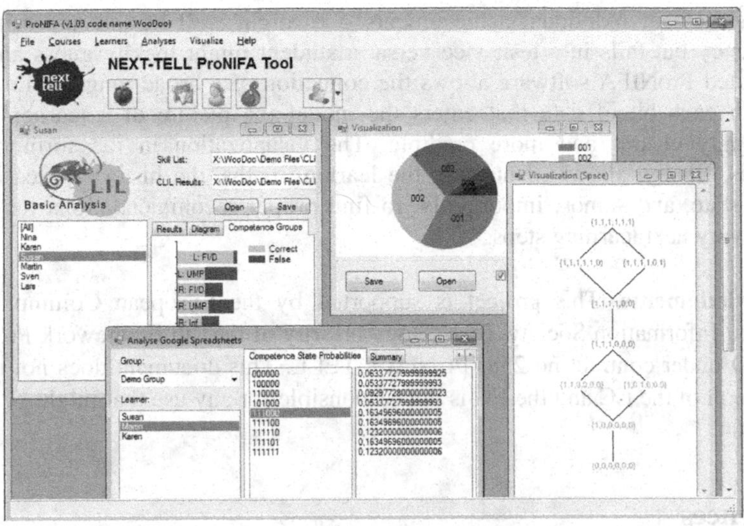

Fig. 3. Screen shots of the ProNIFA tool

tool is a front-end software for teachers and educators; the learning analytics services are running in the background on a server. ProNIFA provides several authoring, analysis, and visualization features. The tool is a Windows application that utilizes various interfaces and links to online-based contents. A distinct feature in the context of formative assessment is the multi-source approach. ProNIFA allows connection of the analysis features to a broad range of sources of evidence. This refers to direct interfaces (for example to *Google Docs*) and it refers to connecting, automatically or manually, to certain log files (cf. Figure 3).

3 Conclusion

There is no doubt that frameworks, techniques, and tools for learning analytics will increasingly be part of a teacher's work in the near future. The benefits are convincing – using the (partly massive) amount of available data from the students in a smart, automated, and effective way, supported by intelligent systems in order to have all the relevant information available just in time and at first sight. The ultimate goal is to formatively evaluate individual achievements and competencies and provide the learners with the best possible individual support and teaching.

The idea of formative assessment and educational data mining is not new but the hype over recent years resulted in scientific sound and robust approaches becoming available, and usable software products appeared.

The framework of CbKST offers a rigorously competence-based approach that accounts for the latent abilities of students. This is in line with the fact that educational policies in Europe are presently moving from a focus on knowledge to a focus on competency, which is reflected in revisions on curricula in the various countries. In addition, the probabilistic dimension allows teachers to have a more cautious view of individual achievements – it might well be that a learner has a competency but fails in a test; vice versa, a student might luckily guess an answer. The related ProNIFA software allows the collection of a broad range of information and with each bit of data that enters the model the picture of a learner becomes increasingly clearer and more credible. The visualization in the form of Hasse diagrams, finally, allows identifying the learning paths, the history of learning, the present state, and – most importantly, to find proper recommendations for the next and the very next learning steps.

Acknowledgments. This project is supported by the European Community (EC) under the Information Society Technology priority of the 7th Framework Programme for R&D under contract no 258114 NEXT-TELL. This document does not represent the opinion of the EC and the EC is not responsible for any use that might be made of its content.

References

1. Holzinger, A.: On Knowledge Discovery and interactive intelligent visualization of biomedical data - Challenges in Human–Computer Interaction & Biomedical Informatics. In: Helfert, M., Francalanci, C., Filipe, J. (eds.) Proceedings of the International Conference on Data Technologies and Application, DATA 2012, Rome, pp. 3–16. SciTec Press, Setubal (2012)
2. Holzinger, A., Yildirim, P., Geier, M., Simonic, K.-M.: Quality-based knowledge discovery from medical text on the Web Example of computational methods in Web intelligence. In: Pasi, G., Bordogna, G., Jain, L.C. (eds.) Qual. Issues in the Management of Web Information. ISRL, vol. 50, pp. 145–158. Springer, Heidelberg (2013)
3. Siemens, G., Gasevic, D., Haythornthwaite, C., Dawson, S., Buckingham Shum, S., Ferguson, R., Duval, E., Verbert, K., Baker, R.S.J.D.: Open Learning Analytics: an integrated & modularized platform: Proposal to design, implement and evaluate an open platform to integrate heterogeneous learning analytics techniques (2011), http://solaresearch.org/OpenLearningAnalytics.pdf
4. Ferguson, R., Buckingham Shum, S.: Social Learning Analytics: Five Approaches. In: Proceedings of the 2nd International Conference on Learning Analytics & Knowledge, Vancouver, British Columbia, Canada, April 29-May 02 (2012)
5. Duval, E.: Attention Please! Learning Analytics for Visualization and Re-commendation. In: Proceedings of the 1st International Conference on Learning Analytics & Knowledge, Banff, Alberta, Canada, February 27-March 1 (2011)
6. Dimitrova, V., McCalla, G., Bull, S.: Open Learner Models: Future Research Directions (Special Issue of IJAIED Part 2). International Journal of Artificial Intelligence in Education 17(3), 217–226 (2007)
7. Doignon, J., Falmagne, J.: Spaces for the assessment of knowledge. International Journal of Man-Machine Studies 23, 175–196 (1985)
8. Doignon, J., Falmagne, J.: Knowledge Spaces. Springer, Berlin (1999)
9. Albert, D., Lukas, J. (eds.): Knowledge Spaces: Theories, Empirical Research, and Applications. Lawrence Erlbaum Associates, Mahwah (1999)

Towards the Detection of Deception in Interactive Multimedia Environments

Hugo Plácido da Silva[1], Ana Priscila Alves[1], André Lourenço[1,2], Ana Fred[1], Inês Montalvão[3], and Leonel Alegre[3]

[1] IT - Instituto de Telecomunicações
Instituto Superior Técnico, Avenida Rovisco Pais, 1
1049-001 Lisboa, Portugal
{hugo.silva,anapriscila.alves,arlourenco,afred}@lx.it.pt
[2] Instituto Superior de Engenharia de Lisboa
Rua Conselheiro Emídio Navarro, 1
1959-007 Lisboa, Portugal
alourenco@deetc.isel.ipl.pt
[3] Ciência Viva - Agência Nacional para a Cultura Científica e Tecnológica
Alameda dos Oceanos Lote 2.10.01
1990-223 Lisboa, Portugal
imontalvao@cienciaviva.pt, lalegre@pavconhecimento.pt

Abstract. A classical application of biosignal analysis has been the psychophysiological detection of deception, also known as the polygraph test, which is currently a part of standard practices of law enforcement agencies and several other institutions worldwide. Although its validity is far from gathering consensus, the underlying psychophysiological principles are still an interesting add-on for more informal applications. In this paper we present an experimental off-the-person hardware setup, propose a set of feature extraction criteria and provide a comparison of two classification approaches, targeting the detection of deception in the context of a role-playing interactive multimedia environment. Our work is primarily targeted at recreational use in the context of a science exhibition, where the main goal is to present basic concepts related with knowledge discovery, biosignal analysis and psychophysiology in an educational way, using techniques that are simple enough to be understood by children of different ages. Nonetheless, this setting will also allow us to build a significant data corpus, annotated with ground-truth information, and collected with non-intrusive sensors, enabling more advanced research on the topic. Experimental results have shown interesting findings and provided useful guidelines for future work.

Keywords: Human-Computer Interaction, Deception, Educational Module, Biosignals, Pattern Recognition.

1 Introduction

Over the years, biosignals have seen an exponential growth in terms of application areas. Nowadays they are extensively used in a broad array of fields, ranging

A. Holzinger and G. Pasi (Eds.): HCI-KDD 2013, LNCS 7947, pp. 65–76, 2013.

from medicine and wellbeing, to entertainment and human-computer interaction [1–3]. Biomedical signal processing and knowledge discovery have been pivotal for this progress, since the extraction of meaningful information from the collected data, often in a multimodal approach, is fundamental [4–6]. An interesting property of some biosignal modalities is their relation with the psychophysiological state of the subject, mainly due to the activity of the sympathetic and parasympathetic branches of the Autonomic Nervous System (ANS) [7]; a classical, well known use of biosignals and this relation is the polygraph test [8, 9].

Invented in early 1920's, the so-called "lie detector" has remained mostly faithful to its original configuration, comprising the measurement of cardiovascular, electrodermal and respiratory parameters, and a Q&A protocol, which, in principle, enables an expert examiner to infer deception by analyzing differences between the patterns of the measured parameters when the subject responds to a set of questions known as relevant, and the patterns obtained for a series of control questions [9]. The latter are selected to be neutral in terms of psychophysiological stimuli, while the former are the potentially deceptive questions.

The validity of the polygraph test for its original purpose has always been surrounded by deep controversy, but the underlying psychophysiological principles in which the method is grounded, that is, the fact that some subjects may exhibit differentiated psychophysiological responses when engaged in deceptive processes, are still an interesting add-on for informal applications such as recreational and educational activities. However, the dependency on a human expert is major limiting factor in these contexts, although throughout the years several authors have proposed automated recognition methods [10, 11]. Furthermore, the instrument used to conduct polygraph tests consists of a physiological data recorder with a set of four sensors that are applied to the trunk and upper limbs of the subject [12], rendering it unpractical for such applications. These sensors measure heart rate, blood pressure, skin impedance, chest respiration, and abdominal respiration, and although for some parameters it is mandatory that the sensor is applied to the body of the subject (e.g. blood pressure), others can be measured in less intrusive ways (e.g. heart rate).

In this paper, we present a study and experimental setup, targeted at the detection of deception in a role-playing interactive multimedia environment that involves two character roles; one of the roles corresponds to the fictional character Pinocchio, which will be making use of deception when presented with a set of specific situations, while the other character will play the role of an Inspector, attempting to guess the situation in which the Pinocchio actually fakes his answer. Our goal is to build an educational interactive module that will be integrated in a science exhibition, to demonstrate the combination between psychophysiology, biosignals and knowledge discovery techniques to children, and create a competition where the Inspector tries to beat the machine (i.e. the automatic recognition algorithms).

We formulate the problem in a feature extraction and supervised learning framework; a feature space derived from the collected raw data is proposed, and we benchmark session-centric and answer-centric algorithms for pattern classification, exploring different combinations of the derived features to automatically detect the cases in which the Pinocchio is deceiving his answer to the Inspector. Our paper is organized as follows: in Section 2 we describe the experimental setup; in Section 3 a summary of the extracted features is presented; in Section 4 we describe the devised classification strategies; in Section 5 we present the evaluation methodology and preliminary results; and finally, in Section 6 we draw the main conclusions and present guidelines for future work.

2 Experimental Setup

In our approach, the Pinocchio is monitored using only biosignals collectable in an off-the-person approach, that is, that can be integrated in a surface or object with which the participant interacts with. We devised a setup that can monitor Blood Volume Pulse (BVP) and Electrodermal Activity (EDA) signals, which are both influenced by the regulatory activity of the ANS, while the Pinocchio is engaged in the interactive multimedia activity. Fig. 1 presents the proposed experimental setup, together with the workflow that we follow for the automatic biosignal analysis and classification process. The BVP and EDA signals are acquired at the Pinocchio's hand palm using a bioPLUX research biosignal acquisition system, in a configuration where the analog-to-digital conversion is performed at 1kHz and with 12 bits resolution per channel. These signals are then transmitted via Bluetooth wireless to a base station, where after pre-processing we perform feature extraction and classification.

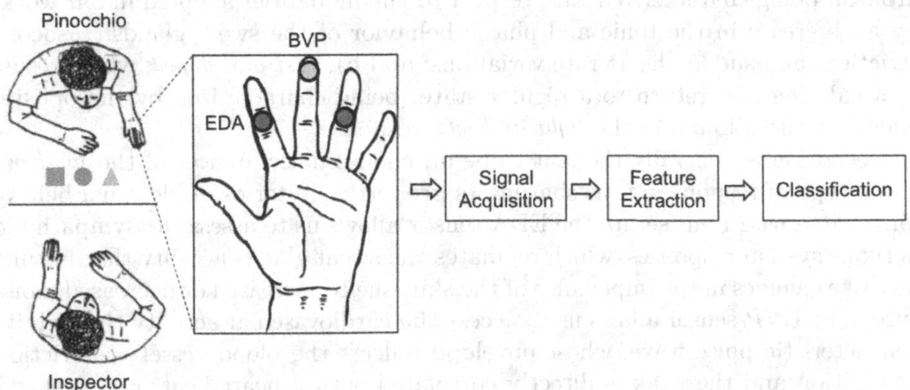

Fig. 1. Overall physical layout, sensor placement, and workflow. The EDA is placed on the second phalanx of the index and ring fingers, while the BVP is placed on the first phalanx of the middle finger.

The experimental protocol for the interactive multimedia activity was defined as follows: *Step 1)* the Pinocchio is presented with a set of objects placed before him in a random sequence; *Step 2)* the Pinocchio is asked to conceal an object of choice; *Step 3)* the Inspector is informed about the initial set of objects presented to Pinocchio; *Step 4)* the Inspector starts the interrogation by asking the neutral question *"What is your name?"*, to collect baseline data from the Pinocchio; *Step 5)* sitting face-to-face before the Pinocchio, the Inspector asks, in turn, whether a given object was concealed or not; *Step 6)* the Pinocchio is required to always reply with the sentence *"I have not concealed the object <name>"*; *Step 7)* in the end, the Inspector is required to guess for which object did the Pinocchio provide a deceiving answer.

For this paper, real-world data was collected within a total of 16 subjects that voluntarily participated in two variations of the experimental protocol, one that used 5 objects (9 subjects), and another that used 6 objects (7 subjects). To increase the intensity and potentiate arousal in the Pinocchio character, an intimidating movie-like interrogation room environment was created, involving an oscillating light bulb and a sound clip that was played every time the Pinocchio character provided an answer. Ground-truth annotations were performed using a manual record[1], and the question transition events were marked synchronously with the biosignal data using an analog switch connected to the biosignals acquisition system.

3 Feature Extraction

The bodily regulatory effects of the Autonomic Nervous System (ANS) to endosomatic or exosomatic stimuli are often typified in the reference literature as two different classes of responses [13, 14]: a) *fight-or-flight*, which result in positive arousal, being characterized with respect to the modalities adopted in our work, by an increase in the tonic and phasic behavior of the sweat glands, vasoconstriction, and sudden heart rate variations; and b) *rest-and-digest*, which result in a calming and return to a regular state, being characterized by the oposite effects to those found in the *fight-or-flight* response.

Deception is generally thought to be an endosomatic inducer of the *fight-or-flight* type of responses, thus sharing its properties in terms of biosignal behavioral patterns. In our setup, the EDA sensor allows us to assess the sympathetic nervous system responses, which regulates the sweat glands activity that in turn provoke changes in the impedance of the skin due to increased or decreased moisture. The BVP sensor allows us to access the cardiovascular activity through its characteristic pulse wave, whose envelope reflects the blood vessel constriction or dilation and the peak is directly correlated with a heart beat. Fig. 2 shows an example of the EDA, BVP, and Heart Rate (HR) time series for one of the experimental sessions, providing a glimpse of some of the changes that occur throughout the interactive multimedia activity.

[1] Given the experimental protocol, there will be only one object in which the Pinocchio will be deceiving the Inspector.

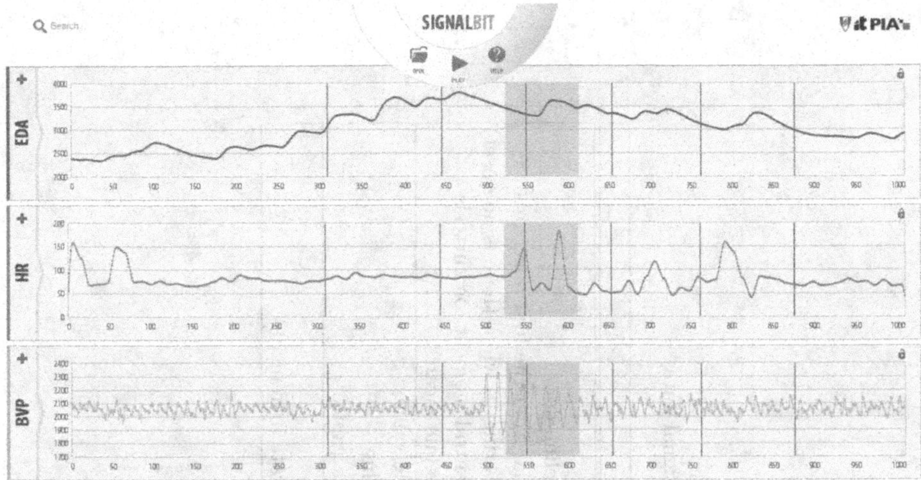

Fig. 2. Example of the EDA, HR, and BVP time series (from top to bottom), for one of the experimental sessions. Each pair of vertical markers delimits the timeframe within which the Pinocchio provides his answer after being queried by the Inspector, while the dark gray timeframe indicates the [-2; 5] seconds window around the time in which the Pinocchio started to state his answer and where we consider that he might have tried to deceive the Inspector.

In our preliminary approach to the problem, we followed a feature-based framework for the recognition of deceptive patterns in the biosignals. As such, after a pre-processing step where the raw EDA and BVP signals are filtered to remove power line noise and motion artifacts, we perform a feature extraction step, in which each time series is reduced to a set of features. Each of the time series can be denoted as a set of samples, $X[N] = \{x[n] : n \in N\}$, with $N = \{1, \cdots, t.f_s\}$, t being the total duration of the session (in seconds) and f_s the sampling rate of the biosignal acquisition system (in samples per second).

For the feature extraction process, we considered the time span of interest to correspond to the samples $X[Q] : Q = \{N[q] - 2.f_s, \cdots, N[q] + 5.f_s\}$, $N[q]$ denoting the sample index where the analog switch was triggered, signaling the beginning of Pinocchio's answer to question q. In general, the variations in the biosignals due to the ANS activity exhibit a high intra- and inter-subject variability. As such, we focused on deriving relative measurements from the signals; the time span Q enables the derivation of a set of relative measurements between a segment which we believe to correspond to a relaxed state, and a segment that may correspond to a deceptive state. We considered the samples $X[B] : B = \{N[q] - 2.f_s, \cdots, N[q]\}$ to be the relaxed or *baseline* segment (that is, the 2 seconds preceding the answer), and the samples $X[S] : S = \{N[q], \cdots, N[q] + 5.f_s\}$ to be the deceptive or *stimulatory* segment (that is, the 5 seconds immediately after the beginning of the answer).

Table 1. Features extracted from the multimodal biosignal data

Feature	Computation	Modality	Description
f_1	$max(X_{EDA}[Q]) - min(X_{EDA}[Q])$	EDA	Amplitude range
f_2	$\sum_{n \in Q}(x_{EDA}[n] - min(X_{EDA}[B]))$		Area above the baseline floor
f_3	$\sum_{n \in Q}(x_{EDA}[n] - x_{EDA}[Q[o_1]])$		Total area differential from the first onset $o_1 \in S$, detected while answering (0 when there is no onset)
f_4	$\beta_2(X_{EDA}[Q])$		Fourth statistical central moment, or kurtosis
f_5	$\sum_{n \in S} 1/f_s$, if $x_{EDA}[n] > \overline{X}_{EDA}[B])$		Time above the baseline mean
f_6	$max(X_{BVP}[Q])$	BVP	Maximum amplitude value
f_7	$\overline{X}_{BVP}[Q]$		Mean amplitude value
f_8	$min(X_{BVP}[Q])$		Minimum amplitude value
f_9	$\overline{X}_{HR}[R \subset Q]$	HR	Mean heart rate
f_{10}	$\overline{\Delta X}_{HR}[R \subset Q]$		Mean of the heart rate increments

Let's denote $X_{EDA}[N]$ and $X_{BVP}[N]$ as the pre-processed EDA and BVP time series respectively, and $X_{HR}[R] = \{\frac{60}{f_s.(N[r]-N[r-1])} : (r-1) >= 1\}$ as the tachogram, where $r \in R \subset N$ is the sample index where a heart beat was detected in $X_{BVP}[N]$. For a given answer to question q, we extract the features presented in Table 1, which were selected based both on the evidence found in the psychophysiology literature review [6, 7, 14–16], and also on the careful inspection of the biosignals collected using the previously described setup.

4 Classification of Deception

Given the educational dimension of our work, we are focusing on the use of conceptually simple algorithms, that can be easily explained to a child in laymen terms. For that purpose, we experimented with a simple heuristic method and also with a decision tree (DT) algorithm, in particular the CART (Classification and Regression Tree), which constructs a binary tree using in each node the feature that has the largest information gain [17–19].

4.1 CART Decision Tree

In this method we adopted a supervised learning approach, where the labeled training data was used to predict if a given answer was deceptive or not using the information contained in the extracted features. We build a matrix, where each line (or sample) corresponds to one answer, and each column corresponds to one of the features described in Section 3. The sample labels are defined as the ground-truth annotations collected during the experimental procedure on wether a given answer was deceptive or not. This method follows an answer-centric rationale, assuming that the feature vectors for each answer contain enough information discriminate between two classes (deceptive and non-deceptive).

We focused on binary trees, where the value of a feature is tested at each node, and a partition optimizing the separability of the different classes is created. The decision tree algorithm implicitly performs feature selection, and the intermediate nodes provide the conditions to separate the classes based on a common set of features shared by the samples of a given class (the leafs). In our case, the leaf nodes will indicate if the answer was deceptive or non-deceptive. For the context of our work we used the Python scikit-learn library [20].

4.2 Heuristic Method

Through a preliminary manual inspection of the data and feature space, we observed that a recurrent pattern among the set of answers provided by the Pinocchio (one per object), was that the deceiving answer would exhibit either the 1st or 2nd highest values (maximum) in one or more of the extracted features (Table 1). Furthermore, it is important to highlight again the high intra- and inter-subject variability of the signals, which we found to be reflected also in the relative measurements obtained from our feature extraction process.

For our heuristic method, we constructed a binary matrix where each line (or sample) corresponds to the whole session and where there are 20 columns, 2 per feature, associated respectively with the 1st and 2nd highest values of each feature. For a given feature, the column associated with the 1st highest value is marked as 1 if the answer where the maximum value is found corresponds to deceptive answer, otherwise it is marked as 0. For a given feature, the column associated with the 2nd highest value will be marked as 1 if the answer where the 2nd highest value was found corresponds to the deceptive answer, otherwise it is marked with 0. This method follows a session-centric rationale.

In this case, we performed feature selection as follows: a) we determine a score for each training sample, computed as the column-wise summation of the binary matrix, that is, we count the number of lines marked with 1 within each column; b) we determine the maximum scores over all training samples; c) we select the features in which the sum is greater than 50% the maximum score.

For the classification step we construct a new binary matrix where each line corresponds to an individual answer provided by the Pinocchio, and each of the columns represent the features selected. We mark with 1 the answers where the features occur as the 1st or 2nd highest value, and finally we perform a sum over each line of the matrix. The answer (or line) with the highest sum is classified as the deceptive answer. When the maximum value occurs in more than one answer, the deceptive answer is chosen randomly.

5 Results

Our algorithm was evaluated on the data collected according to the experimental setup and procedure described in Section 2. We benchmarked the algorithms described in Section 4 using a re-substitution method, where all the available records are used simultaneously as training and testing data, and the leave-one-out method, where each available record is isolated in turn as testing data and the remaining records are used as training data. The re-substitution method evaluates the specificity of each algorithm taking into account the available data, while the leave-one-out method evaluates their generalization ability.

Table 2. Accuracy of the evaluated algorithms when compared to the random guess probability of a human. The re-substitution (RS) method shows the specificity of each algorithm given the available data, while the Leave-One-Out (LOO) shows the generalization ability.

#Obj.	#Users	Accuracy				
		Human	Heuristic		Decision Tree	
			RS	LOO	RS	LOO
5	9	20.0%	61.2±5.6%	50.0±5.6%	100.0%	33.3±47.1%
6	7	16.7%	50.0±7.1%	28.6±20.2%	100.0%	14.3±35.0%

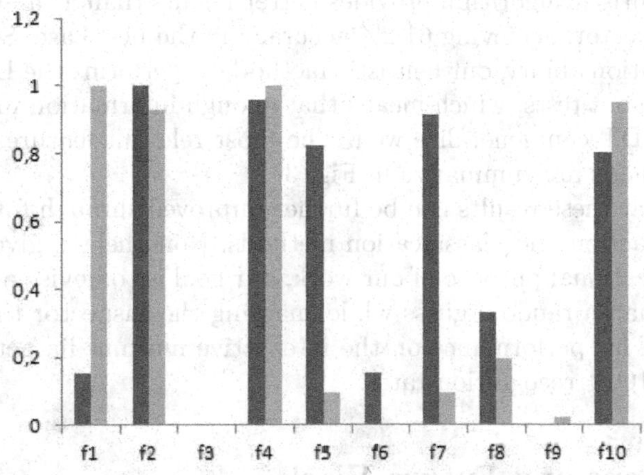

(a) Decision Tree

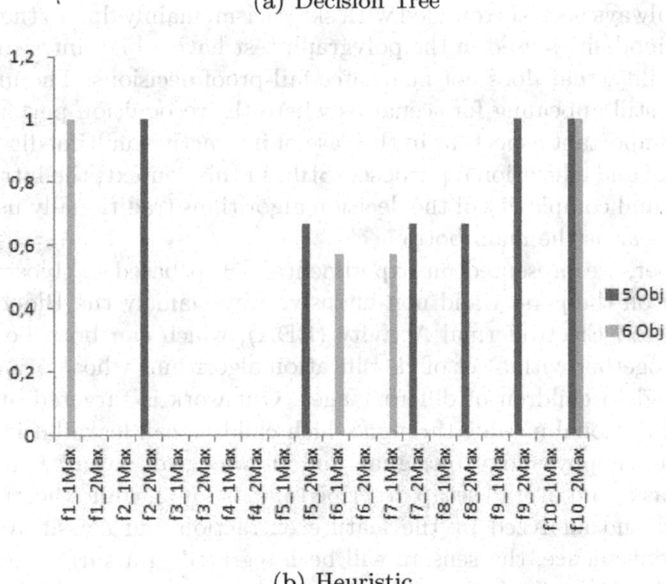

(b) Heuristic

Fig. 3. Histogram of the features selected when using the leave-one-out method for benchmarking the decision tree and the heuristic approaches

Table 2 summarizes the performance of each algorithm devised in our preliminary study when compared to a random guess, that is, the probability that the Inspector has of correctly identifying the answer in which the Pinocchio tried to deceive. As we can observe, with 100% accuracy in both the 5 and 6 objects version of the experimental protocol, the DT algorithm has better specificity, although the heuristic algorithm provides better results than a random guess issued by the Inspector, achieving 61.2% accuracy in the best case. Still, in terms of the generalization ability, our heuristic method outperforms the DT, as shown by the leave-one-out tests, which means that enough information must be gathered before the DT can generalize well. The most relevant features selected by each of the methods are summarize in Fig. 3.

We expect that these results can be further improved through future work on the feature extraction and classification methods. Nonetheless, given the recreational and educational purpose of our work, our goal is to devise a system that can be better than a random guess while enabling the Inspector to be competitive in terms of his performance on the interactive multimedia activity, rather than achieving 100% recognition rates.

6 Conclusions and Future Work

Detection of deception has been a recurrent research topic for decades, mainly associated with the recognition of lying behaviors in forensic scenarios. This application has always been surrounded with skepticism, mainly due to the fact that the biosignal modalities used in the polygraph test have a high intra- and inter-subject variability, that does not guarantee fail-proof decisions. The underlying principles are still appealing for scenarios where the recognition performance is not the most important aspect, as in the case of interactive multimedia activities for recreational and educational purposes. Still, in this context, the intrusiveness of the sensors and complexity of the decision algorithms traditionally used in the polygraph appear as the main bottlenecks.

In this paper, we presented an experimental setup based on biosignals collectable in an off-the-person and non-intrusive way, namely the Blood Volume Pulse (BVP) and Electrodermal Activity (EDA), which can both be acquired at the hand, together with a set of classification algorithms whose principles are easily explained to children of different ages. Our work is targeted at the creation of an educational module through which children can learn basic concepts related with psychophysiology, biosignals and knowledge discovery. Children will perform the tasks and overall setup described in Section 2, while their biosignals are monitored and analyzed by the feature extraction and classification algorithms. For convenience, the sensors will be integrated in a surface over which the child playing the role of Pinocchio only needs to rest one of the hands.

We presented preliminary results obtained from real-world data collected among 16 persons with which two different configurations of the role playing task were tested. Using a combination of feature extraction and classification algorithms, we were able to achieve results that are considerably better than a random guess performed by a human subject. Tests were performed using an heuristic session-centric algorithm, and an answer-centric algorithm based on decision trees. Experimental results have shown that while the decision tree approach is able to retain better that discriminative potential of the data, the heuristic method is able to generalize better for the available set of records. We consider the best results to be provided by the heuristic algorithm, since the leave-one-out evaluation revealed an average recognition accuracy of 50.0% for the case in which the experimental protocol consisted of 5 objects, and 28.6% for the case in which 6 objects were used.

Future work will be focused on increasing our real-world data corpus through additional data acquisition sessions in order to further validate our results, on further developing the feature extraction and classification techniques in order to reduce the number of sensors and improve the recognition rates, and on the introduction of a continuous learning method in which the automatic recognition algorithms are updated after a new participant interacts with the system.

Acknowledgments. This work was partially funded by Fundação para a Ciência e Tecnologia (FCT) under the grants PTDC/EEI-SII/2312/2012, SFRH/BD/65248/2009 and SFRH/PROTEC/49512/2009, and by Ciência Viva, the National Agency for Scientific and Technological Culture, under the project "Mentir de Verdade", whose support the authors gratefully acknowledge.

References

1. Helal, A., Mokhtari, M., Abdulrazak, B.: The Engineering Handbook of Smart Technology for Aging, Disability and Independence, 1st edn. Wiley-Interscience (September 2008)
2. Jun Kimura, M.D.: Electrodiagnosis in Diseases of Nerve and Muscle: Principles and Practice, 3rd edn. Oxford University Press, USA (2001)
3. Topol, E.: The Creative Destruction of Medicine: How the Digital Revolution Will Create Better Health Care. Basic Books (2012)
4. Chang, H.H., Moura, J.M.F.: Biomedical signal processing. In: Biomedical Engineering and Design Handbook. McGraw Hill (June 2009)
5. Holzinger, A., Stocker, C., Bruschi, M., Auinger, A., Silva, H., Gamboa, H., Fred, A.: On applying approximate entropy to ECG signals for knowledge discovery on the example of big sensor data. In: Huang, R., Ghorbani, A.A., Pasi, G., Yamaguchi, T., Yen, N.Y., Jin, B. (eds.) AMT 2012. LNCS, vol. 7669, pp. 646–657. Springer, Heidelberg (2012)
6. Silva, H., Fred, A., Eusebio, S., Torrado, M., Ouakinin, S.: Feature extraction for psychophysiological load assessment in unconstrained scenarios. In: Proceedings of the Annual International Conference of the IEEE Engineering in Medicine and Biology Society (EMBC), pp. 4784–4787 (2012)
7. Lewis, M., Haviland-Jones, J.M., Barrett, L.F. (eds.): Handbook of Emotions, 3rd edn. The Guilford Press (November 2010)
8. Marston, W.M.: Lie Detector Test. R.R. Smith (1938)
9. National Research Council: The Polygraph and Lie Detection. National Academies Press (2003)
10. Jiang, L., Qing, Z., Wenyuan, W.: A novel approach to analyze the result of polygraph. In: IEEE International Conference on Systems, Man, and Cybernetics, vol. 4, pp. 2884–2886 (2000)
11. Layeghi, S., Dastmalchi, M., Jacobs, E., Knapp, R.: Pattern recognition of the polygraph using fuzzy classification. In: Proceedings of the Third IEEE Conference on Fuzzy Systems, vol. 3, pp. 1825–1829 (1994)

12. Geddes, L.: Tühe truth shall set you free [development of the polygraph]. IEEE Engineering in Medicine and Biology Magazine 21(3), 97–100 (2002)
13. Cannon, W.B.: Bodily changes in pain, hunger, fear, and rage. D. Appleton and Co. (1915)
14. Kreibig, S.: Autonomic nervous system activity in emotion: A review. Biological Psychology 84, 394–421 (2010)
15. Picard, R.W.: Affective Computing, 1st edn. The MIT Press (July 2000)
16. van den Broek, E.L.: Affective Signal Processing (ASP): Unraveling the mystery of emotions. PhD thesis, University of Twente (2011)
17. Breiman, L., Friedman, J., Stone, C.J., Olshen, R.A.: Classification and Regression Trees, 1st edn. Chapman and Hall/CRC (January 1984)
18. Duda, R.O., Hart, P.E., Stork, D.G.: Pattern Classification, 2nd edn. John Wiley & Sons, New York (2001)
19. Theodoridis, S., Koutroumbas, K.: Patern recognition. Academic Press (1999)
20. Pedregosa, F., Varoquaux, G., Gramfort, A., Michel, V., Thirion, B., Grisel, O., Blondel, M., Prettenhofer, P., Weiss, R., Dubourg, V., Vanderplas, J., Passos, A., Cournapeau, D., Brucher, M., Perrot, M., Duchesnay, E.: Scikit-learn: Machine learning in Python. Journal of Machine Learning Research 12, 2825–2830 (2011)

Predictive Sentiment Analysis of Tweets: A Stock Market Application

Jasmina Smailović[1,2], Miha Grčar[1], Nada Lavrač[1], and Martin Žnidaršič[1]

[1] Jožef Stefan Institute, Jamova Cesta 39, 1000 Ljubljana, Slovenia
{jasmina.smailovic,miha.grcar,nada.lavrac, martin.znidarsic}@ijs.si

[2] Jožef Stefan International Postgraduate School, Jamova cesta 39, 1000 Ljubljana, Slovenia

Abstract. The application addressed in this paper studies whether Twitter feeds, expressing public opinion concerning companies and their products, are a suitable data source for forecasting the movements in stock closing prices. We use the term predictive sentiment analysis to denote the approach in which sentiment analysis is used to predict the changes in the phenomenon of interest. In this paper, positive sentiment probability is proposed as a new indicator to be used in predictive sentiment analysis in finance. By using the Granger causality test we show that sentiment polarity (positive and negative sentiment) can indicate stock price movements a few days in advance. Finally, we adapted the Support Vector Machine classification mechanism to categorize tweets into three sentiment categories (positive, negative and neutral), resulting in improved predictive power of the classifier in the stock market application.

Keywords: stock market, Twitter, predictive sentiment analysis, sentiment classification, positive sentiment probability, Granger causality.

1 Introduction

Trying to determine future revenues or stock prices has attracted a lot of attention in numerous research areas. Early research on this topic claimed that stock price movements do not follow any patterns or trends and past price movements cannot be used to predict future ones [1]. Later studies, however, show the opposite [2]. It has also been shown that emotions have an effect on rational thinking and social behavior [3] and that the stock market itself can be considered as a measure of social mood [4].

As more and more personal opinions are made available online, recent research indicates that analysis of online texts such as blogs, web pages and social networks can be useful for predicting different economic trends. The frequency of blog posts can be used to predict spikes in the actual consumer purchase quantity at online retailers [5]. Moreover, it was shown by Tong [6] that references to movies in newsgroups were correlated with their sales. Sentiment analysis of weblog data was used to predict movies' financial success [7]. Twitter[1] posts were also shown to be useful for predicting

[1] www.twitter.com

A. Holzinger and G. Pasi (Eds.): HCI-KDD 2013, LNCS 7947, pp. 77–88, 2013.

box-office revenues of movies before their release [8]. Thelwall et al. [9] analyzed events in Twitter and showed that popular events are associated with increases in average negative sentiment strength. Ruiz et al. [10] used time-constrained graphs to study the problem of correlating the Twitter micro-blogging activity with changes in stock prices and trading volumes. Bordino et al. [11] have shown that trading volumes of stocks traded in NASDAQ-100 are correlated with their query volumes (i.e., the number of users' requests submitted to search engines on the Internet). Gilbert and Karahalios [12] have found out that increases in expressions of anxiety, worry and fear in weblogs predict downward pressure on the S&P 500 index. Moreover, it was shown by Bollen et al. [13] that changes in a specific public mood dimension (i.e., calmness) can predict daily up and down changes in the closing values of the Dow Jones Industrial Average Index. In our preliminary work [14] we used the volume and sentiment polarity of Apple financial tweets to identify important events, as a step towards the prediction of future movements of Apple stock prices.

The paper follows a specific approach to analyzing stock price movements, contributing to the research area of sentiment analysis [15,6,16,17], which is aimed at detecting the authors' opinion about a given topic expressed in text. We use the term predictive sentiment analysis to denote the approach in which sentiment analysis is used to predict the changes in the phenomenon of interest. Our research goal is to investigate whether large-scale collections of daily posts from social networking and micro-blogging service Twitter are a suitable data source for predictive sentiment analysis. In our work we use the machine learning approach to learn a sentiment classifier for classification of financial Twitter posts (tweets) and causality analysis to show the correlation between sentiment in tweets and stock price movements. In addition, visual presentation of the sentiment time series for detection of important events is proposed. We analyzed financial tweets of eight companies (Apple, Amazon, Baidu, Cisco, Google, Microsoft, Netflix and Research In Motion Limited (RIM)) but due to space limitations, detailed analysis of only two companies (Google and Netflix) is presented in this paper.

The paper is structured as follows. Section 2 discusses Twitter specific text preprocessing options, and presents the developed Support Vector Machine (SVM) tweet sentiment classifier. The core of the paper is presented in Section 3 which presents the dataset collected for the purpose of this study, and the methodology developed for enabling financial market prediction from Twitter data. The developed approach proposes *positive sentiment probability* as an indicator for predictive sentiment analysis in finance. Moreover, by using the *Granger causality test* we show that sentiment polarity (positive and negative sentiment) can indicate stock price movements a few days in advance. Furthermore, since financial tweets do not necessarily express the sentiment, we have introduced sentiment classification using the *neutral zone*, which allows classification of a tweet into the neutral category, thus improving the predictive power of the sentiment classifier in certain situations. We conclude with a summary of results and plans for further work in Section 4.

2 Tweet Preprocessing and Classifier Training

In this work, we use a supervised machine learning approach to train a sentiment classifier, where classification refers to the process of categorizing a given observation (tweet) into one of the given categories or classes (positive or negative sentiment polarity of a tweet). The classifier is trained to classify new observations based on a set of class-labeled training instances (tweets), each described by a vector of features (terms, formed of one or several consecutive words) which have been pre-categorized manually or in some other presumably reliable way. This section describes the datasets, data preprocessing and the algorithm used in the development of the tweet sentiment classifier, trained from a set of adequately preprocessed tweets.

There is no large data collection available for sentiment analysis of Twitter data, nor a data collection of annotated financial tweets. For this reason, we have trained the tweet sentiment classifier on an available large collection of tweets annotated by positive and negative emoticons collected by Stanford University [18], approximating the actual positive and negative sentiment labels. This approach was introduced by Read [19]. The quality of the classifier was then evaluated on another set of actually manually labeled tweets.

To train the tweet sentiment classifier, we used a dataset of 1,600,000 (800,000 positive and 800,000 negative) tweets collected and prepared by Stanford University, where positive and negative emoticons serve as class labels. For example, if a tweet contains ":)", it is labeled as positive, and if it contains ":(", it is labeled as negative. Tweets containing both positive and negative emoticons were not taken into account. The list of positive emoticons used for labeling the training set includes :), :-), :), :D, and =), while the list of negative emoticons consists of :(, :-(, and : (. Inevitably this simplification results in partially correct or noisy labeling. The emoticons were stripped out of the training data for the classifier to learn from other features that describe the tweets. The tweets from this set do not focus on any particular domain.

The test data set collected and labeled by Stanford University contains tweets belonging to the different domains (companies, people, movies…). It consists of 498 manually labeled tweets, of which 182 were labeled as positive, 177 as negative and the others labeled as neutral. The tweets were manually labeled based on their sentiment, regardless of the presence of emoticons in the tweets.

As the Twitter community has created its own language to post messages, we explore the unique properties of this language to better define the feature space. The following tweet preprocessing options [18,20] were tested:

- **Usernames:** mentioning of other users by writing the "@" symbol and the username of the person addressed was replaced a unique token *USERNAME*.
- **Usage of Web Links:** web links were replaced with a unique token *URL*.
- **Letter Repetition:** repetitive letters with more than two occurrences in a word were replaced by a word with one occurrence of this letter, e.g., word *looooooove* was replaced by *love*.
- **Negations:** since we are not interested in particular negations, but in negation expressions in general, we replaced negation words (*not, isn't, aren't, wasn't, weren't,*

hasn't, haven't, hadn't, doesn't, don't, didn't) with a unique token *NEGATION*. This approach handles only explicit negation words and treats all negation words in the same way. Implicit negations and negative emotions presented in a tweet (e.g., A*void CompanyX*) are nevertheless handled to some extent by using unigrams and bigrams which assign negative sentiment to a word or a phrase in a tweet.

- **Exclamation and Question Marks:** exclamation marks were replaced by a token *EXCLAMATION* and question marks by a token *QUESTION*.

In addition to Twitter-specific text preprocessing, other standard preprocessing steps were performed [21] to define the feature space for tweet feature vector construction. These include text tokenization, removal of stopwords, stemming, N-gram construction (concatenating 1 to N stemmed words appearing consecutively) and using minimum word frequency for feature space reduction. In our experiments, we did not use a part of speech (POS) tagger, since it was indicated by Go et al. [18] and Pang et al. [22] that POS tags are not useful when using SVMs for sentiment analysis.

The resulting terms were used as features in the construction of TF-IDF feature vectors representing the documents (tweets). TF-IDF stands for term frequency-inverse document frequency feature weighting scheme [23] where weight reflects how important a word is to a document in a document collection.

There are three common approaches to sentiment classification [24]: (i) machine learning, (ii) lexicon-based methods and (iii) linguistic analysis. Instead of developing a Twitter-specific sentiment lexicon, we have decided to use a machine learning approach to learn a sentiment classifier from a set of class labeled examples. We used the linear Support Vector Machine (SVM) algorithm [25,26], which is standardly used in document classification. The SVM algorithm has several advantages, which are important for learning a sentiment classifier from a large Twitter data set: it is fairly robust to overfitting, it can handle large feature spaces [23,27] and it is memory efficient [28]. Given a set of labeled training examples, an SVM training algorithm builds a model which represents the examples as points in space separated with a hyperplane. The hyperplane is placed in such a way that examples of the separate categories are divided by a clear gap that is as wide as possible. New examples are then mapped into that same space and predicted to belong to a class based on which side of the hyperplane they are.

The experiments with different Twitter-specific preprocessing settings were performed to determine the best preprocessing options which were used in addition to the standard text preprocessing steps. The best classifier, according to the accuracy on the manually labeled test set, was obtained with the following setting: using N-grams of size 2, using words which appear at least two times in the corpus, replacing links with the *URL* token and by removing repeated letters in words. This tweet preprocessing setting resulted in feature construction of 1,254,163 features used for classifier training. Due to space limitations, the entire set of experimental results, including ten-fold cross-validation results, is not presented in the paper. Classifier testing showed that this preprocessing setting resulted in 81.06% accuracy on the test set, a result comparable to the one achieved by Go et al. [18].

3 Stock Market Analysis

This section investigates whether sentiment analysis on tweets provides predictive information about the values of stock closing prices. By applying the best classifier obtained with the process explained in Section 2, two sets of experiments are performed. In the first one, in which tweets are classified into two categories, positive or negative, the newly proposed sentiment indicators are calculated with the purpose of testing their correlation with the corresponding stock's closing price. We also present a data visualization approach used for detecting interesting events. In the second set of experiments the initial approach is advanced by taking into account the neutral zone, enabling us to identify neutral tweets (not expressing positive or negative sentiment) as those, which are "close enough" to the SVM's model hyperplane. This advancement improves the predictive power of the methodology in certain situations.

3.1 Data Used in the Stock Market Application

A large dataset was collected for these experiments. On the one hand, we collected 152,572 tweets discussing stock relevant information concerning eight companies in the period of nine months in 2011. On the other hand, we collected stock closing prices of these eight companies for the same time period. The data source for collecting financial Twitter posts is the Twitter API[2], i.e., the Twitter Search API, which returns tweets that match a specified query. By informal Twitter conventions, the dollar-sign notation is used for discussing stock symbols. For example, $GOOG tag indicates that the user discusses Google stocks. This convention simplified the retrieval of financial tweets. We analyzed English posts that discussed eight stocks (Apple, Amazon, Baidu, Cisco, Google, Microsoft, Netflix and RIM) in the period from March 11 to December 9, 2011. The stock closing prices of the selected companies for each day were obtained from the *Yahoo! Finance*[3] web site.

3.2 Sentiment and Stock Price Visualization

Using the best classifier obtained with the process explained in Section 2, we classified the tweets into one of two categories (positive or negative), counted the numbers of positive and negative tweets for each day of the time series, and plotted them together with their difference, the moving average of the difference (averaged over 5 days), and the daily stock closing price. The proposed visual presentation of the sentiment time series for Google can be seen in Fig. 1. Peaks show the days when people intensively tweeted about the stocks. Two outstanding positive peaks can be observed in August and October 2011. One is a consequence of Google buying Motorola Mobility and the other is due to recorded high increase in revenue and earnings for Google year-over-year.

[2] `https://dev.twitter.com/`
[3] `http://finance.yahoo.com/`

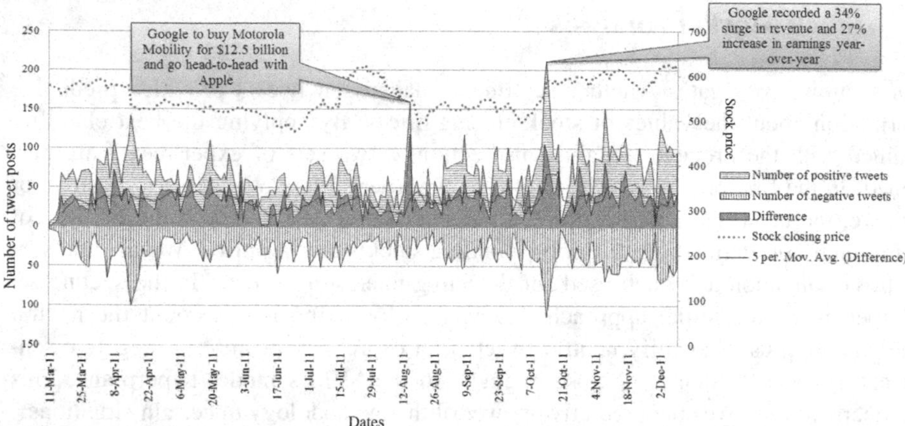

Fig. 1. Number of positive (green) and negative (red) tweet posts, their difference (blue), the moving average of the difference (averaged over 5 days), and the stock closing price per day for Google

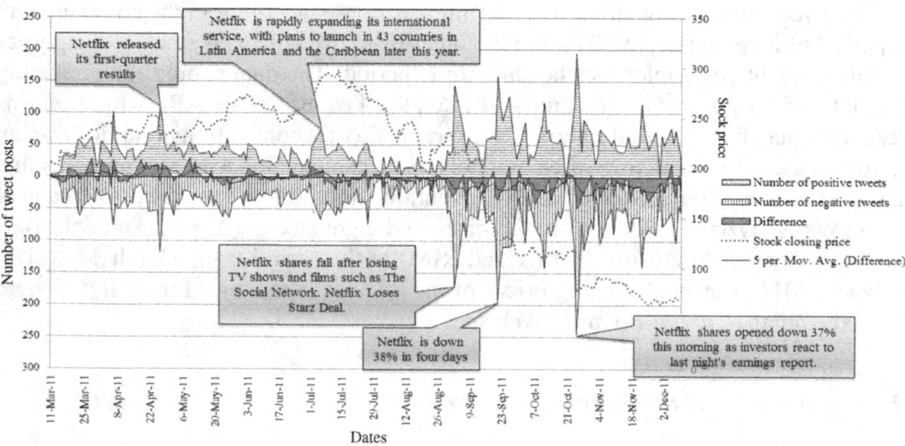

Fig. 2. Number of positive (green) and negative (red) tweet posts, their difference (blue), the moving average of the difference (averaged over 5 days), and the stock closing price per day for Netflix

This type of visualization can be used as a tool for easier and faster overview analysis of important events and general observation of trends. Relation of stock price and sentiment indicators time series provides insight into the reasons for changes in the stock price: whether they should be prescribed to internal market phenomena (e.g., large temporary buying/selling) or external phenomena (news, public events, social trends), which are expressed also in our sentiment indicators. It is interesting to observe how the tweet sentiment time series is often correlated with the stock closing price time series. For example, the sentiment for Netflix (Fig. 2) at the beginning of the year was mostly positive. As sentiment reversed its polarity in July, the stock closing price started to fall. For the whole second half of the year, sentiment remained mostly negative and the stock closing price continued to fall. On the other hand, a few days correlation between the sentiment and the stock closing price cannot be observed with the naked eye. For example, in Fig. 1 "Google buys Motorola" event has high amount of positive sentiments but stock price seems to drop. To calculate the real correlation we employ causality analysis, as explained in the next section.

3.3 Causality Analysis

We applied a statistical hypothesis test for stationary time series to determine whether tweet sentiment is related with stock closing price in the sense of containing predictive information about the values of the stock closing price or the other way around. To this end, we performed *Granger causality analysis* [29]. Time series *X* is said to Granger-cause *Y* if it can be shown that *X* values provide statistically significant information about future values of *Y*. Therefore, the lagged values of *X* will have a statistically significant correlation with *Y*. The output of the Granger causality test is the *p*-value. In statistical hypothesis testing, the *p*-value is a measure of how much evidence we have against the null hypothesis [30]; the null hypothesis is rejected when the *p*-value is less than the significance level, e.g., 5% ($p < 0.05$).

Positive Sentiment Probability. To enable in-depth analysis, we propose the positive sentiment probability sentiment indicator to be used in predictive sentiment analysis in finance. Positive sentiment probability is computed for every day of a time series by dividing the number of positive tweets by the number of all tweets on that day. This ratio is used to estimate the probability that the sentiment of a randomly selected tweet on a given day is positive.

Time Series Data Adaptation. To test whether one time series is useful in forecasting another, using the Granger causality test, we first calculated positive sentiment probability for each day and then calculated two ratios, which we have defined in collaboration with the Stuttgart Stock Exchange experts: (a) Daily change of the positive sentiment probability: positive sentiment probability today – positive sentiment probability yesterday, and (b) Daily return in stock closing price: (closing price today – closing price yesterday)/closing price yesterday.

Hypotheses Tested. We applied the Granger causality test in two directions, to test the following two null hypotheses: (a) "sentiment in tweets does not predict stock closing prices" (when rejected, meaning that the sentiment in tweets Granger-cause the values of stock closing prices), and (b) "stock closing prices do not predict sentiment in tweets" (when rejected, meaning that the values of stock closing prices Granger-cause the sentiment in tweets).

We performed tests on the entire 9 months' time period (from March 11 to December 9, 2011) as well as on individual three months periods (corresponding approximately to: March to May, June to August and September to November). In Granger causality testing we considered lagged values of time series for one, two and three days, respectively. The results indicate that in several settings sentiment of tweets can predict stock price movements. Results were especially strong for Netflix (Table 1), Baidu (all day lags for June-August and 2 and 3 day lags for March-May), Microsoft

(1 and 2 days lag for March-May and 1 day lag for the entire 9 months' time period), Amazon (2 and 3 days lag for September-November and 3 days lag for March-May) and RIM (all day lags for the entire 9 months' time period). For the given period, these companies had many variations in the closing price values (Baidu, Microsoft and Amazon) or a significant fall in the closing price (Netflix and RIM). On the other hand, the correlation is less clear for the other companies: Apple, Cisco and Google, which did not have many variations nor a significant fall in the closing price values for the given time period. This means that in the situations explained above, Twitter feeds are a suitable data source for predictive sentiment analysis and that daily changes in values of positive sentiment probability can predict a similar rise or fall of the closing price in advance.

Table 1. Statistical significance (*p*-values) of Granger causality correlation between daily changes of the positive sentiment probability and daily return of closing prices for Netflix

NETFLIX	Lag	Stocks = f(Tweets)	Tweets =f(Stocks)
9 months	1 day	0.1296	0.8784
March - May	1 day	0.9059	0.3149
June - August	1 day	0.1119	0.7833
September - November	1 day	0.4107	0.8040
9 months	2 days	**0.0067****	0.6814
March - May	2 days	0.4311	0.3666
June - August	2 days	0.2915	**0.0248****
September - November	2 days	**0.0007*****	0.9104
9 months	3 days	**0.0084****	0.6514
March - May	3 days	0.6842	0.3942
June - August	3 days	0.5981	**0.0734***
September - November	3 days	**0.0007*****	0.8464

*p < 0.1
**p < 0.05
***p < 0.001

Second Experimental Setup Results, Using the SVM Neutral Zone. In this section we address a three class problem of classifying tweets into the positive, negative and neutral category, given that not all tweets are either positive or negative. Since our training data does not contain any neutral tweets, we define a neutral tweet as a tweet that is "close enough" to the SVM model's hyperplane. Let us define the neutral zone to be the area along the SVM hyperplane, parameterized by t which defines its extent. Let d_{Pa} be the average distance of the positive training examples from the hyperplane and, similarly, let d_{Na} be the average distance of the negative training examples from the hyperplane. Then, the positive bound of the neutral zone is computed as

$$d_P(t) = t \cdot d_{Pa} \tag{2}$$

Similarly, the negative bound of the neutral zone is computed as

$$d_N(t) = t \cdot d_{Na} \tag{3}$$

If a tweet x is projected into this zone, i.e., $d_N(t)<d(x)<d_P(t)$, then it is assumed to bear no sentiment, i.e., that it is neutral. This definition of the neutral zone is simple and allows fast computation. Its drawback, however, is its lack of clear and general interpretation outside the context of a particular SVM classifier.

A series of experiments were conducted where the value for *t*, i.e., the size of the neutral zone, was varied and tested on manually labeled 182 positive and 177 negative tweets included in the test data set described in Section 2. Tweets were preprocessed using the best tweet preprocessing setting described in Section 2. With every new value of *t*, we calculated the accuracy on the test set and the number of opinionated (non–neutral) tweets, where the accuracy is calculated based on tweets that are classified as positive or negative by the SVM classifier described in Section 2. As a result of not taking neutral tweets into account when calculating the accuracy, the accuracy gets higher as we increase the *t* value (see Fig. 3). Hence, as we expand the neutral zone, the classifier is more confident in its decision about labeling opinionated tweets. As a negative side effect of increasing the neutral zone, the number of opinionated tweets is decreasing. We show this phenomena in Fig. 3, where also the number of opinionated tweets (classified as positive or negative) is plotted.

Accuracy is not a good indicator in the presented 3-class problem setting, therefore we experimentally evaluated the neutral zone according to its effect directly on stock price prediction. We repeated our experiments on classifying financial tweets, but now also taking into account the neutral zone. Since the Granger causality analysis showed that tweets could be used to predict movements of stock prices, we wanted to investigate whether the introduction of the neutral zone would further improve predictive capabilities of tweets. Therefore, every tweet which mentions a given company was classified into one of the three categories: positive, negative or neutral. Then, we applied the same processing of data as before (count the number of positive, negative and neutral tweets, calculate positive sentiment probablity, calculate daily changes of the positive sentiment probability and daily return of stocks` closing price) and Granger analysis test. We varied the *t* value from 0 to 1 (where t=0 corresponds to classification without the neutral zone) and calculated average *p*-value for the separate day lags (1, 2 and 3). Results for Google and Netflix are shown in Fig. 4 and Fig. 5.

From Fig. 3 it follows that it is reasonable to focus on the parts of the plots that correspond to narrow boundaries of the neutral zone, for which the number of opinionated tweets is still considerable (for example up to 0.6). Slightly below 0.6 at 10% increase in accuracy, the loss of opinionated tweets is only at 20%. As for Google, shown in Fig. 4, the introduction of the neutral zone proved to be beneficial, as we got more significant *p*-values and the average *p*-value initially dropped. The best correlation between sentiment in tweets and stock closing price is observed at t=0.2 where

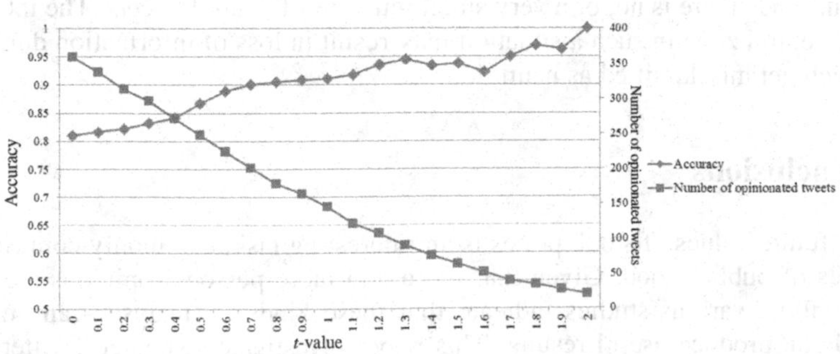

Fig. 3. The accuracy and the number of opinionated tweets while changing the value of *t*

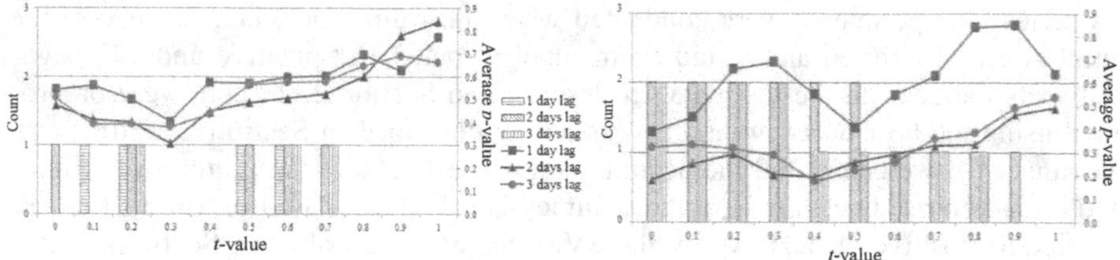

Fig. 4. Number of significant *p*-values (less than 0.1, column chart) and average *p*-values (line chart) for every day lag while changing the *t* value for Google

Fig. 5. Number of significant *p*-values (less than 0.1, column chart) and average *p*-values (line chart) for every day lag while changing the *t* value for Netflix

for every day lag we obtained a significant result and at *t*=0.3 where the average *p*-value is the lowest. Also for most of the other companies (Apple, Amazon, Baidu, Cisco and Microsoft), the neutral zone improved the predictive power of tweets, mostly for 2 and 3 days lags, given that the average *p*-value decreased when adding the neutral zone. Taking into account the number of significant *p*–values, the improvement was observed for Apple and Baidu, for Microsoft and Cisco there was a small and mixed improvement and for Amazon the number of *p*-values even dropped. In general, the best improvement was obtained with *t*=0.2.

For Netflix (see Fig. 5) and for RIM the neutral zone did not prove to be so useful. The neutral zone had the most positive effect on the results of Baidu and the least positive effects on the results for RIM. By analyzing these two extreme cases we observed that the closing price of Baidu had many variations and the RIM stock closing price has constant fall during a larger period of time. These observations of a relation between the impact of the neutral zone and the stock price variations hold also for most of the other companies.

In summary, it seems that when there is no apparent lasting trend of the closing price of a company, people write diverse tweets, with or without sentiment expressed in them. In such cases, it is desirable to use the neutral zone to detect neutral tweets in order to calculate the correlation only between the opinionated tweets and the stock closing price. On the other hand, it seems that once it is clear that the closing price of some company is constantly falling, people tend to write tweets in which they strongly express their opinion about this phenomenon. In this case, we might not need the neutral zone since there is no, or a very small number, of neutral tweets. The introduction of the neutral zone in such a situation may result in loss of information due to the tweets which get misclassified as neutral.

4 Conclusions

Predicting future values of stock prices is an interesting task, commonly connected to the analysis of public mood. Given that more and more personal opinions are made available online, various studies indicate that these kinds of analyses can be automated and can produce useful results. This paper investigates whether Twitter feeds are a suitable data source for predictive sentiment analysis. Financial tweets of eight companies (Apple, Amazon, Baidu, Cisco, Google, Microsoft, Netflix and RIM) were analyzed. The study indicates that changes in the values of positive sentiment probability can predict a similar movement in the stock closing price in situations where stock closing prices have many variations or a big fall. Furthermore, the introduced SVM neutral zone, which gave us the ability to classify tweets also into the neutral category, in certain situations proved to be useful for improving the correlation between the opinionated tweets and the stock closing price.

In future, we plan to experiment with different datasets for training and testing the classifier, preferably from a financial domain, in order for the classifier to be more finance adjusted since we are interested in this particular domain. Furthermore, we intend to expand the number of companies for further analysis to gain more insights in which situations our approach is most applicable. Finally, we plan to adjust our methodology to data streams with the goal to enable predicting future changes of stock prices in real-time.

Acknowledgements. The work presented in this paper has received funding from the European Community's Seventh Framework Programme (FP7/2007-2013) within the context of the Project FIRST, Large scale information extraction and integration infrastructure for supporting financial decision making, under grant agreement n. 257928 and by the Slovenian Research Agency through the research program Knowledge Technologies under grant P2-0103. The research was also supported by Ad Futura Programme of the Slovenian Human Resources and Scholarship Fund. We are grateful to Ulli Spankowski and Sebastian Schroff for their kind cooperation as financial experts in the stock analytics application presented in this paper.

References

1. Fama, E.: Random Walks in Stock Market Prices. Financial Analysts Journal 21(5), 55–59 (1965)
2. Kavussanos, M., Dockery, E.A.: Multivariate test for stock market efficiency: The case of ASE. Applied Financial Economics 11(5), 573–579 (2001)
3. Damasio, A.R.: Descartes error: emotion, reason, and the human brain. Harper Perennial (1995)
4. Nofsinger, J.R.: Social Mood and Financial Economics. Journal of Behavioral Finance 6(3), 144–160 (2005)
5. Gruhl, D., Guha, R., Kumar, R., Novak, J., Tomkins, A.: The predictive power of online chatter. In: Proceedings of the Eleventh ACM SIGKDD International Conference on Knowledge Discovery in Data Mining, pp. 78–87 (2005)
6. Tong, R.M.: An operational system for detecting and tracking opinions in on-line discussion. In: Working Notes of the ACM SIGIR 2001 Workshop on Operational Text Classification (OTC), pp. 1–6 (2001)
7. Mishne, G., Glance, N.: Predicting Movie Sales from Blogger Sentiment. In: AAAI Symposium on Computational Approaches to Analysing Weblogs AAAI-CAAW, pp. 155–158 (2006)
8. Asur, S., Huberman, B.A.: Predicting the Future with Social Media. In: Proceedings of the ACM International Conference on Web Intelligence, pp. 492–499 (2010)

9. Thelwall, M., Buckley, K., Paltoglou, G.: Sentiment in Twitter events. Journal of the American Society for Information Science and Technology 62(2), 406–418 (2011)
10. Ruiz, E.J., Hristidis, V., Castillo, C., Gionis, A., Jaimes, A.: Correlating financial time series with micro-blogging activity. In: Proceedings of the Fifth ACM International Conference on Web Search and Data Mining, pp. 513–522 (2012)
11. Bordino, I., Battiston, S., Caldarelli, G., Cristelli, M., Ukkonen, A., Weber, I.: Web search queries can predict stock market volumes. PLoS ONE 7(7), e40014 (2011)
12. Gilbert, E., Karahalios, K.: Widespread Worry and the Stock Market. In: Proceedings of the Fourth International AAAI Conference on Weblogs and Social Media, pp. 58–65 (2010)
13. Bollen, J., Mao, H., Zeng, X.: Twitter mood predicts the stock market. Journal of Computational Science 2(1), 1–8 (2011)
14. Smailović, J., Grčar, M., Žnidaršič, M., Lavrač, N.: Sentiment analysis on tweets in a financial domain. In: 4th Jožef Stefan International Postgraduate School Students Conference, pp. 169–175 (2012)
15. Das, S., Chen, M.: Yahoo! for Amazon: Extracting market sentiment from stock message boards. In: Proceedings of the 8th the Asia Pacific Finance Association Annual Conference, APFA (2001)
16. Turney, P.: Thumbs Up or Thumbs Down? Semantic Orientation Applied to Unsupervised Classification of Reviews. In: Proceedings of the Association for Computational Linguistics, pp. 417–424 (2002)
17. Liu, B.: Sentiment Analysis and Opinion Mining. Morgan and Claypool Publishers (2012)
18. Go, A., Bhayani, R., Huang, L.: Twitter Sentiment Classification using Distant Supervision. In: CS224N Project Report, Stanford (2009)
19. Read, J.: Using emoticons to reduce dependency in machine learning techniques for sentiment classification. In: Proceedings of the ACL Student Research Workshop, pp. 43–48 (2005)
20. Agarwal, A., Xie, B., Vovsha, I., Rambow, O., Passonneau, R.: Sentiment analysis of twitter data. In: Proceedings of the Workshop on Languages in Social Media, pp. 30–38 (2011)
21. Feldman, R., Sanger, J.: The Text Mining Handbook - Advanced Approaches in Analyzing Unstructured Data. Cambridge University Press (2007)
22. Pang, B., Lee, L., Vaithyanathan, S.: Thumbs up?: sentiment classification using machine learning techniques. In: Proceedings of the ACL-2002 Conference on Empirical Methods in Natural Language Processing, vol. 10, pp. 79–86 (2002)
23. Joachims, T.: Text Categorization with Support Vector Machines: Learning with Many Relevant Features. In: Nédellec, C., Rouveirol, C. (eds.) ECML 1998. LNCS, vol. 1398, pp. 137–142. Springer, Heidelberg (1998)
24. Pang, B., Lee, L.: Opinion mining and sentiment analysis. Foundations and Trends in Information Retrieval 2(1-2), 1–135 (2008)
25. Vapnik, V.: The Nature of Statistical Learning Theory. Springer, New York (1995)
26. Cortes, C., Vapnik, V.N.: Support-Vector Networks. Machine Learning 20, 273–297 (1995)
27. Sebastiani, F.: Machine learning in automated text categoriztion. ACM Computing Surveys (CSUR) 34(1), 1–47 (2002)
28. Chang, C.C., Lin, C.J.: LIBSVM: a library for support vector machines. ACM Transactions on Intelligent Systems and Technology (TIST) 2(3), 27 (2011)
29. Granger, C.W.J.: Investigating causal relations by econometric models and cross-spectral methods. Econometrica 37, 424–438 (1969)
30. Schervish, M.J.: P Values: What They Are and What They Are Not. The American Statistician 50(3), 203–206 (1996)

A UI Prototype for Emotion-Based Event Detection in the Live Web

George Valkanas and Dimitrios Gunopulos

Dept. of Informatics and Telecommunications, University of Athens, Athens, Greece
{gvalk,dg}@di.uoa.gr

Abstract. Microblogging platforms are at the core of what is known as the *Live Web*: the most dynamic, and fast changing portion of the web, where content is generated *constantly* by the *users*, in snippets of information. Therefore, the *Live Web* (or *Now Web*) is a good source of information for event detection, because it reflects what is happening in the physical world in a timely manner. Meanwhile, it introduces constraints and challenges: large volumes of unstructured, noisy data, which are also as diverse as the users and their interests. In this work we present a prototype User Interface (UI) of our *TwInsight* system, which deals with event detection of real-world phenomena from microblogs. Our system applies *i*) emotion extraction techniques on microblogs, and *ii*) location extraction techniques on user profiles. Combining these two, we convert highly unstructured content to thematically enriched, locational information, which we present to the user through a unified front-end. A separate area of the UI is used to show events to the user, as they are identified. Taking into account the characteristics of the setting, all of the components are updated along the temporal dimension. We discuss each part of our UI in detail, and present anecdotal evidence of its operation through two real-life event examples.

Keywords: Spatiotemporal analysis, Emotions, Live Web, Event detection.

1 Introduction

To turn data into information and, eventually, knowledge, one needs to rely on tools that facilitate these processes. Data and information visualization, and more recently visual analytics [1,2], are well-known techniques towards this direction. Their aim is to present information in a way that captures the underlying characteristics of the dataset, with a specific goal in mind. For example, graph visualization may be useful when searching for connectivity, distance properties or the existence of specific structure(s) [3]. Similarly, human mobility patterns [4] or trajectories of wildlife [5] can make much more sense, when presented on a map than as a raw sequence of 2 dimensional points.

Attempting to visualize microblogging data poses a lot more problems. Being at the core of what is coined as the *Live Web*, i.e., the most fast paced, ever changing, user generated portion of the contemporary Web, microblogs consist of highly noisy, unstructured information. The content is as diverse in terms of topics and language as the platform's users, who are numerous and increasing. For instance, Twitter – the most well-known microblogging service – now counts more than 200 million active users,

A. Holzinger and G. Pasi (Eds.): HCI-KDD 2013, LNCS 7947, pp. 89–100, 2013.

with an approximate 340 million "tweets" on a daily basis [1]. Moreover, their short length makes it hard to understand what is being discussed when seen out of context. Clearly, gaining useful insights from such voluminous data and presenting it in a user-friendly way is a challenging task at best.

Trying to visualize *all* microblogging information would almost certainly result in a lot of clutter, and would negate any advantages of data visualization. Therefore, a better alternative would be to visualize what is *important* or *significant* in the *aggregate*. Such information is usually the aftermath of an *event*. Event identification, and prompt notification of their occurrence, is an important task for crisis management, decision making and resource allocation, to name a few. However, we still lack the tools to present such information extracted from microblogging data in a meaningful way. We identify the main reasons to be: i) unstructured content, ii) high volume of data, iii) real-time nature of medium.

In this work, we present a User Interface (UI) prototype, which is the front end of *TwInsight*, an event detection system for the *Live Web*. Given the setting and our primary objective, *TwInsight* is essentially targeted at knowledge discovery from big data. *TwInsight* relies on Twitter data as its source of information and extracts emotional signals from the received microblogs. Therefore, our system is inherently related to sentiment analysis and text analytics methods. *TwInsight* also extracts information from a user's profile to map them to a location. Note that in both cases, the input data is in unstructured (textual) form. Combining these two, we are able to identify events by means of a high deviation of the emotional state of users, usually dispersed over a geographical region. Therefore, *TwInsight*'s UI amasses all that information and presents it in a unified way. The user is able to zoom-in / out of areas, to better understand how events impact each region separately. They can also alter the temporal granularity at which microblogs are processed, to gain insights on the lasting effects of each event. Towards this direction, we also visualize the emotional state of received microblogs on a world map. Therefore, the UI serves not only as a front-end to event detection, but also as a real-time, spatio-temporal hedonometer of the monitored users. Finally, identified events are presented in a dedicated area of the UI, together with a brief description. This way, a user will be notified about events of significance right away.

We integrate all of that information and present it through a custom Graphical User Interface (GUI). Given the temporal dynamics of our data source, all components are temporal, i.e., they change over time as needed. Our GUI is built in a way that reflects the functionality of each component in a contextualized manner. For example, the location extraction component places users on a Mercatorian map, whereas emotions are shown in separate windows as they change over time, individually for each monitored location. At the same time, the components interact with each other, so that the user can maximize the information they obtain.

Overall, our UI integrates three distinct facets of representing tweets:

1. Emotional analysis, grounded on influential notions from affective and cognitive theories from psychology. Sentiment analysis, text analytics, and classic data mining methods are utilized for this purpose.
2. Location extraction and geocoding of users, placing them on a map.

[1] `https://business.twitter.com/basics/what-is-twitter/`

3. Describing events, as a result of their temporal, spatial **and** emotional characteristics, thereby facilitating knowledge extraction from big data.

2 Related Work

Various systems and approaches have been proposed with the aim of identifying events from microblogging services. For example, [6] was one of the earliest works, dealing with the identification of earthquakes in Japan. Their approach was quite limited in the sense that they were solely interested in earthquakes, thereby looking for events of a very specific type. The authors also presented some pictures of a web-based earthquake alerting service, but they did not give any additional information. Their interface also looked rather simplistic, in that information was only posted to the website, without any further interaction.

The work presented in [7] discusses an event identification technique from microblogs, regardless of type. Moreover, in an extended version of their work[2], the authors presented a User Interface to their approach, but in that case limited themselves to a very specific event type: the SGE 2011 elections. A distinctive difference between their work and ours, is that we do not constrain ourselves to events of a particular type, when it comes to visualization. Most importantly, we are interested in providing as much information regarding the event as possible, to help users make informed decisions. We also map users to location, using custom location extraction techniques which is assumed to be known *a priori* in [7]. Overall, visualizing events and their related information is the basic purpose of our research.

TEDAS [8] is another research approach for event detection from Twitter data. However, events are extracted from whitelisted sources, which are practically news reporting agencies / channels. TwitterStand [9] does the exact same thing. Identifying events in such a way is trivial. Most importantly, though, it does not consider the users' reactions to these events, which we do. We are also highly interested in the location of the users and map them on a world map, which is not the case of [8].

Finally, the work in [10] has emotion extraction from tweets as its sole goal, and monitors users from specific locations in the United Kingdom. The authors then generate plots based on the extracted emotions and their variation over time, but do not try to correlate events with emotions in real time. Emotions have also been actively researched as a standalone discipline of psychology [11], as well as for their implications on usability and user experience issues [12]. Unlike these fields of research, emotions are for us a tool that we rely on, to identify events. Emotion extraction is only one of the three main components that we are interested in, which we monitor in real time. In particular, we extract emotions from received tweets, and attribute sudden changes to emotional states to external factors (events), which we try to identify automatically through subsequent analysis. We then present this information in a unified UI, allowing for multiple temporal and spatial scales to monitor events.

[2] www.hpl.hp.com/techreports/2011/HPL-2011-98.pdf

3 The TwInsight UI

We start with a brief overview of the processes that take place, before presenting the information to the user. Our only source of information is Twitter, and we receive tweets through the Gardenhose, which gives us a 10% access to all public tweets. Each received tweet undergoes the following two steps: i) emotion extraction to find out whether it conveys an emotion ii) location extraction, to associate the tweet with a target location. To identify events, we *aggregate* tweets over the temporal dimension, and process their collective information at the end of each aggregation interval. For example, every 10 minutes, we process the tweets that we received during the last 10' interval. This gives a trade-off between real-time reporting and computational resources.

In addition to the temporal dimension, events are identified based on their spatial coherence as well, which is why we need the loation extraction step. For event detection purposes, mapping users at the town level should be sufficient, although higher levels (e.g. county or state) could also be used, especially when aggregate information at the lower levels is not enough. Users are mapped to as precise a location as possible (e.g., town or suburb), subject to the textual information that they provide. For instance, a user indicating their location as "Athens, Greece", can be mapped at the town level. On the contrary, a user with "CA, USA", can only be mapped at a state level. However, we should find a way to present both of them to our UI. The reason is that the end-user may see some patterns at a coarser / more fine-grained level than what they are already monitoring, and they may want to switch between views.

Taking into account all of the above, we identify three major components that our User Interface should have, which we describe in detail in the following paragraphs:

i) Location Extraction / Geocoding of users
ii) Emotion extraction from tweets
iii) Event description / summarization

We consider the model of 6 basic emotions proposed by american psychologist Paul Ekman [11]. We also use a "Neutral" (or "None") emotion, to indicate the absence of an emotion, leading to a total of 7 target classes. Emotions have also been associated with specific colors, through color psychology and other emotional theories. Both the emotions and their mapping to colors shown in Table 1.

Table 1. Mapping of emotions to colors, and examples of corresponding tweets

Emotion	Color	Example
NEUTRAL	White	I am Dept. of Informatics & Telecommunications (Athens, Greece)
ANGER	Red	I hate it when I do something and everybody finds out! :@
DISGUST	Purple	RT Retweet this if you too are offended by #HoulaMassacre #Syria
FEAR	Yellow	I'm afraid this won't work out well
JOY	Green	Goaaaaaaaaaaaaaaal!!!!! Let's go @chelseafc!!! #cfc
SADNESS	Blue	I miss my baby :(
SURPRISE	Orange	@gvalk are you serious!?

3.1 Location Extraction

It has been generally observed that users in social settings are unwilling to provide their location. With the exception of 5% of all users, who provide highly accurate location information, through GPS, the rest give textual descriptions of their location or do not disclose it at all.

To address this problem, and be able to make use of the larger portion of the users, we have built an custom service to extract locations from textual information [13]. This process is also commonly known as *geocoding*. The reason we perform custom geocoding is that querying web services in real time, is not only cost ineffective, but also very time consuming, especially if we adhere to the *politeness* policy that web crawlers are expected to. This policy, basically, dictates that a web crawler should not request data from the same domain too aggressively, but should wait between consecutive requests. Typical values are between 10 seconds to 1 minute. Geocoding all of Twitter's currently active users (more than 200M users) with a 1 second interval would take more than 6 years to complete. For similar reasons, we opt for a custom map display.

Using online resources, such as the GeoNames database[3] and a dataset constructed by crawling Flickr places to derive an administrative hierarchy, we are not only able to geocode text locations, but we now have access to geodetic coordinates, i.e. (lat, lon), of these locations. It is worth noting that the hierarchy could also be constructed algorithmically (e.g. hierarchical clustering). Therefore, we can present this information to the user in an easily perceived 2D World Map. Geodetic coordinates can be transformed to cartesian ones (and vice-versa) using map projection equations, e.g. Mercatorian projections. The displayed world map is visualized using KML (Keyhole Markup Language) files, thereby conforming to OGC-compatible Open Standards.

The middle area of Fig. 4 is covered by the initial map that is displayed to the user[4]. The user is allowed to zoom in and out of areas, by selecting from a set of target countries, for which we currently perform emotion detection. Once a tweet has passed through emotional extraction and geocoding, it is displayed on the map in the following way: Given the location where the tweet was mapped to, we descend the hierarchy, in a random way, until we reach the lowest levels, i.e. a town or suburb in our case. If the tweet was mapped to a town / suburb in the first place, there is nothing more to do. Given the emotion of that tweet, we then color that region with the respective color of that emotion.

Currently, "coloring a town" means that we set the pixel corresponding to its location on the map to the color of the emotion. Although we only set a few pixels to that color, we expect that, in the aggregate, surges of emotions will become evident. Coloring pixels instead of broader areas has the advantage that we can update the UI easily and most importantly in real time. Newer emotions take precedence over older ones, and a town is always colored based on the most recent information. Finally, as time goes by, old town colorings are removed from the map, to accomodate for newer information, or simply returning it to the original color.

---

[3] `http://www.geonames.org/`

[4] The colors have been reversed, for printing efficiency.

3.2 Emotion Classification

As we have already described, upon receiving a tweet we cast it to one of the 7 emotions, shown in Table 1, according to the *conveyed* emotion, not the one that is being *invoked*. To understand the difference, consider the popular @chuck_facts Twitter account, that posts funny quotes about the famous actor. Such humorous tweets are used to *invoke* joy to the reader, but do not express any particular feeling of the poster.

In order to map tweets to emotions, we have built a decision tree classifier based on a gold standard of nearly 6700 tweets, tagged by human annotators. The basic feature set consists of the tweet's tokens, but we also consider features unique to the Twitter service, such as number of retweets, number of mentioned entities, whether external resources (URLs) are present, as well as emoticons, and so on. We augmented our initial dataset of emotional terms with additional resources, such as Affective Wordnet [14], and the moods dataset [15] from blogs, which associate specific terms with a given mood or emotion. Table 1 portrays an example tweet for each emotion, where we can easily identify terms, phrases, or even structural information to attribute the mapping.

In addition to displaying the emotions of tweets on the map, as discussed in the previous section, we also use a separate area of our UI for plotting the extracted emotions. The main reason is to provide a clearer view to the end-user, which will allow them to gain additional insights regarding some specific monitored locations. For instance, this approach makes it easier to observe how the emotional behavior of microbloggers varies over longer periods of time, therefore serving as a spatio-temporal hedonometer.

We use two distinct approaches to display this type of information to the user: The first one is to use cardiograms, whereas the second one is to use histograms. In the first case, all emotions of a specific location are shown in the same area. Fig. 1 shows an example of this visualization, in three distinct timestamps, for the United States. Similar cardiograms are displayed for other locations. This enables the end-user to understand the interplay – or lack thereof – of emotions experienced in that area.

For instance, neutral emotions (black line) make up most of the number of tweets that are received, with emotions of joy being second in line. An important observation is the difference between the trends of tweets conveying an emotion and the neutral ones. As a specific example, consider Fig. 1(b), where we can clearly see a distinctive surge in *neutral* tweets, right after the middle. However, this surge is not shared by tweets conveying an emotion, implying that we can avoid spurious bursts by using emotional theories. The figures also validate our intuition that we should rely on emotions to detect events, rather than use simpler aggregations: If the lines were identical (even if simply translated), there would be no merit in using emotional signals; monitoring the entire stream at once (i.e., tweet counting) should be sufficient. This is not the case, as the lines are different from one another.

The second approach, shown in Fig. 4 at the bottom of the screen uses histograms, and displays each emotion separately. Once again, the emotions are updated along the temporal dimension. This visualization allows for better understanding of how each emotion is varied in a specific area over time. It also gives a clearer view of the magnitude of each emotion at a specific point in time. Once more, emotions are colored appropriately, to give a semantic flavor to the visualization.

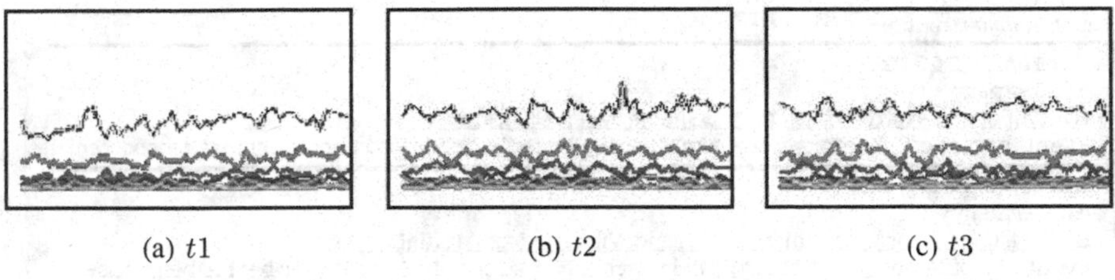

(a) $t1$ (b) $t2$ (c) $t3$

Fig. 1. Cardiograms of emotions, for three distinct timestamps in the United States

3.3 Event Detection

Event detection is our primary objective, and emotional cues are the means to achieve this goal. Following cognitive and affective theories [16], our event detection mechanism is based on the assumption that tweets which deviate significantly from the (aggregate) norm are the result of such events. We will not elaborate further on the mechanics behind event detection, as they go well beyond the scope of this paper. Suffice to say that we maintain an online sample of the received data, and that unlike other alternatives, we rely strictly on data streamed by Twitter and not on query-based solutions. Therefore, our system operates as though the stream was observed by the service itself.

Once an event is identified, we need to present it to the user, so that they are notified about it. Various event descriptions can be used ranging from simple ones (e.g., term frequency, TF-IDF) to more complex [17]. When it comes to presenting events to the user, we should provide as much information as possible. Consequently, a separate area of the UI is devoted to this purpose. As new events arrive, older ones are evicted from the list, for which we have already notified the end-user. The general information we currently show is:

- **Date:** The date and time (shown in UTC/GMT) when the event was identified. This is practically based on timestamps of incoming tweets.
- **At:** A description of the location, where the event was detected. These descriptions are based on what the user is currently monitoring. For instance, if they are monitoring at a country level, the “At” field would be “United Kingdom”, even if the event was identified in Manchester.
- **By:** A list of microbloggers who talked about the event. These are practically links to the original tweets, based on which the event was identified.
- **Event:** A list of terms describing the identified event. The description is used as a fast way for the end-user to know what is going on.

Fig. 2(a),(b) demonstrate a subset of the events shown to the user with respect to the Champions Leagufe final, an easily identifiable event in a dataset of tweets that we have. The figures are used to illustrate the fact that new events are added in the list.

Fig. 2(a) shows a distinctive (sub)event of the Champions League finals, which is the goal scored by Bayern’s football player Thomas Müller. The summary of the event clearly indicates that the event has been identified in Germany, which is only natural given that Bayern Munich is based in Germany, not to mention that the Allianz Arena stadium, where the final took place, is also in Germany.

Event Summarization

On: 19/05/12 20:22:27
At: Germany
By: @Sport_ist_Nord, @melj1213, @Shaun_Baker, @jankoning10
Event: bayern cfc championsleague mueller muller müller phooooto phooto thomas tooooooooor

On: 19/05/12 19:58:27
At: Germany
By: @silke73, @chrizzoandmaxim, @BrendaViloria_, @xx_MuffinMix_xD
Event: @HoffMotte @mezi2011 @sandyptschke76 @xColorMe_Purple aber bei bleiben chelsea cc

(a) Goal by Müller

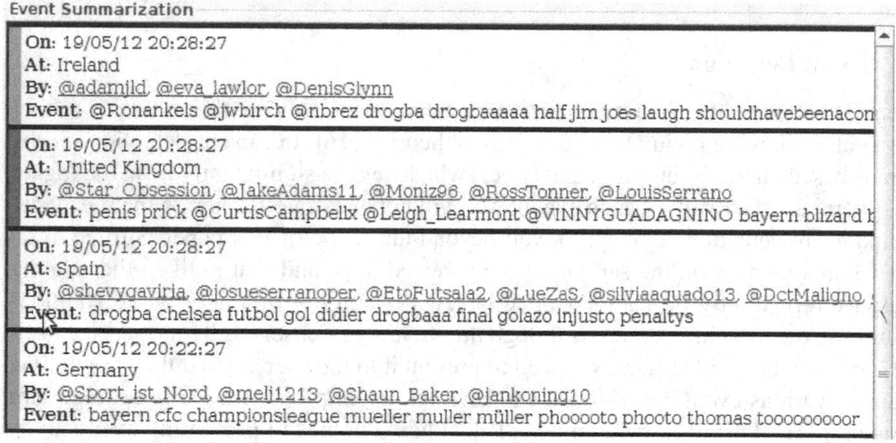

(b) Goal by Drogba

Fig. 2. Summary of two events shown to the end-user, regarding the Champions League Final

Similarly, Fig. 2(b) shows the (sub)event of the goal scored by Didier Drogba, Chelsea's football player. Notice that the previous event (Bayern's goal) is pushed down the list, to make room for the newly identified event(s) describing Chelsea's goal. The same event[5] is identified in *three* locations, because each one of them is being monitored separately by the user: Spain, United Kingdom and Ireland. Were we monitoring these locations collectively, the event would have been identified only once, but the "**At**" field would be different. Also notice that the user is promptly notified about both events[6], with respect to when the goals were scored.

Finally, an additional piece of information that we present to the user is the emotion associated with the event. This is again displayed as a color, to the far left of the event description. For instance, most of the events that we identified are associated with the color "Green", signifying "Joy". This is expected, as most of the tweets are cheerful about the goals scored, by the team they are supporting. The only exception is the goal scored by Drogba, that is associated with "Red", i.e., "Anger" – clearly not what one might expect. Most surprisingly, the event is identified in the United Kingdom, Chelsea's homeplace. However, if we look closely, we will see that the term "Bayern" is in the description of the

[5] We know its the same event due to the descriptions.

[6] `http://en.wikipedia.org/wiki/2012_UEFA_Champions_League_Final`

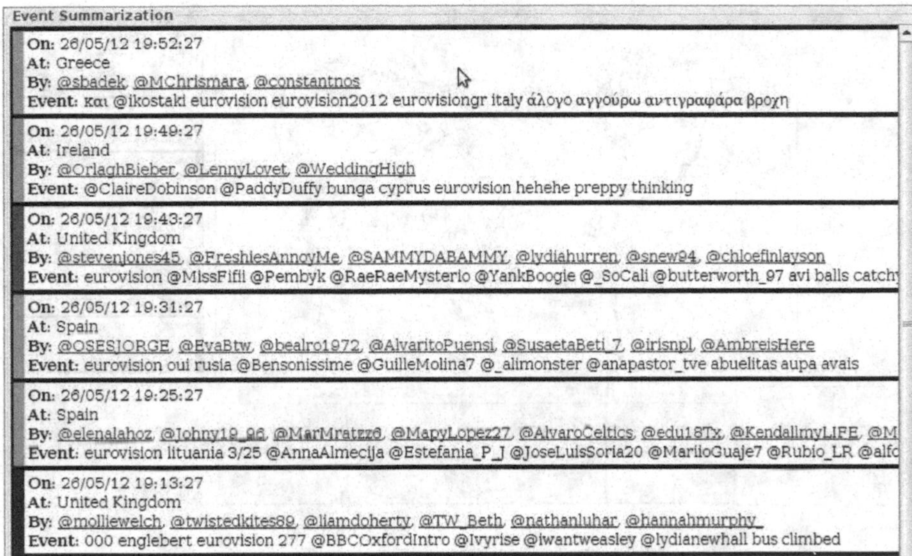

Fig. 3. List of identified events (and their summary), related to the eurovision contest

event and not "Chelsea" (or "goal", or "Drogba") , with some less than flattering words. It is useful to note, nevertheless, that these terms have come up during other runs of our approach, due to our sampling-based approach in event detection.

Fig. 3 shows a second list of events, with their respective summarization. The events all correspond to the Eurovision 2012 song contest final, which stirred up considerable discussions. Starting from the bottom and moving upwards the events list, we can easily verify[7] that the identified events describe the sequence in which the participating teams competed in the contest.

For instance, Engelbert opened the contest, which started at 19:00 GMT[8]. The singer was representing the United Kingdom, singing a ballad, thereby creating some moody feelings, as exemplified by the color "blue" (sadness) next to the event description. About 12 minutes later, we see an event containing the term "lituania", which was competing 4th in line. Hungary and Albania do not show up in this run, although Hungary received a very low ranking overall, and we did not see that many discussions concerning its participation. Lithuania was next, and with a maximum of 3 minutes per song, the user sees the event – as it is identified according to the discussions – right when it occurred. as shown in Fig. 3, we also identified discussions regarding Russia (which is written with a single "s" in Spanish), Cyprus and Italy.

3.4 Putting It All Together

In addition to the separate components that make up our system, the user is able to control the event detection process through a set of available options. These options are

[7] http://www.eurovision.tv/page/baku-2012/about/shows/final

[8] http://www.eurovision.tv/page/baku-2012/about/shows

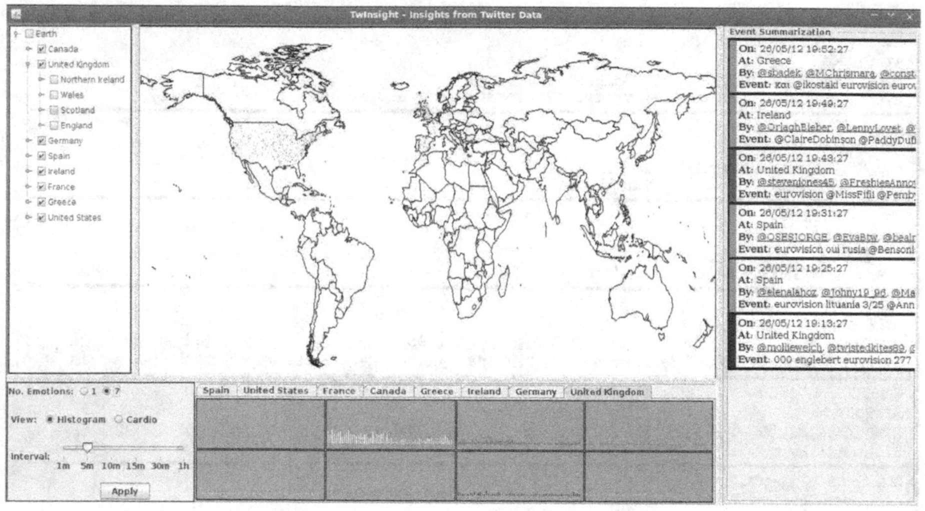

Fig. 4. The overall GUI that the user sees

available at all times, and can be altered while the system is running. Overall, our UI provides the following options that affect our system's functionality:

- Number of emotions against which tweets are classified.
- Monitored locations, including parents and children nodes in the hierarchy.
- Size of aggregation interval, i.e., number of minutes between two consecutive runs of our event detection mechanism.

Fig. 4 shows the complete version of our UI, as this is shown to the end-user. Observe the histogram type of monitoring the emotions at the lower section of the screen, that we have already discussed. Also note that emotions are shown on the world map, coloring specific pixels appropriately. The coloring result is more prominent in France, Ireland, Spain, the UK and the US. Regarding the United States, note that very few tweets are mapped in the state of Alaska. The reason is that, despite the arbitrary assignment to town locations, we descend the hierarchy from the initial geocoded location. Therefore, if a tweet was mapped to New York City, it will be mapped to a suburb of Manhattan or Brooklyn, but never to a city in Alaska. As the population in Alaska is far lower than other states, this is an indirect validation of our geocoding service.

The options available to the user can be seen at the left hand side of the UI. In the upper part of the options area, we display the areas that the user is allowed to select from. These are shown in a tree structure that reflects the hierarchy we currently rely on to identify events and display information. Note that event monitoring is currently performed at the country level and that it is also possible to select parent nodes separately from children node (e.g., United Kingdom is selected, but none of its children are). Therefore, only the countries appear in the bottom area of the UI, which shows the aggregate emotional state per region.

In the lower part of that area, we can clearly see options that affect the emotions that we monitor. Currently, the user is able to select between monitoring all 7 emotions, or just 1. The latter case is identical to monitoring the rate at which tweets arrive, regardless of any emotion that they may convey. We also allow various aggregation intervals: 1 minute, 5 minutes, 10 minutes, etc. By reducing the aggregation interval, users will be notified about events more timely, but more events will be generated. By contrast, increasing the aggregation interval will generate fewer events, but the user will be notified about them less promptly. They can also switch to monitoring a single "emotion", which is practically equivalent to monitoring the rate at which tuples are received. In the same part of the UI, we also clearly see the option to switch between histogram and cardiogram view of emotions.

An option not shown in this UI is that the user may select between real-time identification of events, by monitoring the Twitter stream, or replaying stored streams. The latter functionality can be used to improve all aspects of our sub-system, including our UI, as well as give access and insights to historical data. This is simply done by passing additional arguments when the system starts.

4 Conclusion

In this work, we presented a prototype UI of our *TwInsight* system, focusing on event detection from the *Live Web* using emotional signals. Our approach may also be seen as a (real-time) spatio-temporal hedonometer and as a tool to assess the importance of known events, and how these are perceived by the users. The playback functionality is a crucial step towards this direction.

Our front end operates under both a spatial and a temporal dimension, to help with these objectives, by converting unstructured, noisy and voluminous data from microblogs into meaningful information and visualizing it appropriately. We also showcased our system through two well known events found in a subset of real data from the Twitter stream.

We plan to enhance our system's functionality further, with advanced search capabilities such as keyword search and searching against a given time series. We will also investigate alternatives in displaying the events to the end user, taking into account their significance / impact in addition to their temporal dimension. Improved visualization techniques, such as coloring areas according to the proportionality of the sensed emotions, rather than pixel-based approaches will also be considered. Finally, we are working on a web-based version of our UI for easier public access.

Acknowledgements. The authors would like to thank the annotators of the data used for classification. This work has been co-financed by EU and Greek National funds through the Operational Program "Education and Lifelong Learning" of the National Strategic Reference Framework (NSRF) - Research Funding Programs: Heraclitus II fellowship and ARISTEIA - MMD" and the EU funded project INSIGHT.

References

1. William Wong, B.L., Xu, K., Holzinger, A.: Interactive visualization for information analysis in medical diagnosis. In: Holzinger, A., Simonic, K.-M. (eds.) USAB 2011. LNCS, vol. 7058, pp. 109–120. Springer, Heidelberg (2011)
2. Wong, P.C., Thomas, J.: Visual analytics. IEEE Computer Graphics and Applications 24(5), 20–21 (2004)
3. Broder, A., Kumar, R., Maghoul, F., Raghavan, P., Rajagopalan, S., Stata, R., Tomkins, A., Wiener, J.: Graph structure in the web. Comput. Netw. 33(1-6) (June 2000)
4. Stange, H., Liebig, T., Hecker, D., Andrienko, G.L., Andrienko, N.V.: Analytical workflow of monitoring human mobility in big event settings using bluetooth. In: ISA, pp. 51–58 (2011)
5. Andrienko, N., Andrienko, G., Gatalsky, P.: Towards exploratory visualization of spatio-temporal data. In: Third AGILE Conference on Geographical Information Science (2000)
6. Sakaki, T., Okazaki, M., Matsuo, Y.: Earthquake shakes twitter users: real-time event detection by social sensors. In: WWW, pp. 851–860 (2010)
7. Weng, J., Lee, B.-S.: Event detection in twitter. In: ICWSM (2011)
8. Li, R., Lei, K.H., Khadiwala, R., Chang, K.C.-C.: Tedas: A twitter-based event detection and analysis system. In: ICDE. IEEE (2012)
9. Sankaranarayanan, J., Samet, H., Teitler, B.E., Lieberman, M.D., Sperling, J.: Twitterstand: news in tweets. In: SIGSPATIAL-GIS, pp. 42–51 (2009)
10. Lansdall-Welfare, T., Lampos, V., Cristianini, N.: Effects of the recession on public mood in the UK. In: WWW Companion (2012)
11. Ekman, P., Friesen, W.V., Ellsworth, P.: Emotion in the human face: guide-lines for research and an integration of findings. Pergamon Press (1972)
12. Stickel, C., Ebner, M., Steinbach-Nordmann, S., Searle, G., Holzinger, A.: Emotion detection: Application of the valence arousal space for rapid biological usability testing to enhance universal access. In: Stephanidis, C. (ed.) Universal Access in HCI, Part I, HCII 2009. LNCS, vol. 5614, pp. 615–624. Springer, Heidelberg (2009)
13. Valkanas, G., Gunopulos, D.: Location extraction from social networks with commodity software and online data. In: ICDM Workshops, SSTDM (2012)
14. Bentivogli, B.M.L., Forner, P., Pianta, E.: Revising wordnet domains hierarchy: Semantics, coverage, and balancing. COLING, pp. 101–108 (2004)
15. Leshed, G., 'Jofish' Kaye, J.: Understanding how bloggers feel: recognizing affect in blog posts. In: CHI (2006)
16. Mikolajczak, M., Tran, V., Brotheridge, C., Gross, J.J.: Using an emotion regulation framework to predict the outcomes of emotional labour. Emerald, Bingley (2009)
17. Nichols, J., Mahmud, J., Drews, C.: Summarizing sporting events using twitter. In: IUI, pp. 189–198 (2012)

Challenges from Cross-Disciplinary Learning Relevant for KDD Methods in Intercultural HCI Design

Rüdiger Heimgärtner

Intercultural User Interface Consulting (IUIC)
Lindenstraße 9, 93152 Undorf, Germany
ruediger.heimgaertner@iuic.de

Abstract. In this paper, the challenges in cross-disciplinary learning relevant for using KDD methods in intercultural human-computer interaction (HCI) design are described and solutions are provided. For instance, reframing HCI through local and indigenous perspectives requires the analysis of the local and indigenous perspectives relevant for HCI design. This can be done by experts for intercultural HCI design with different cultural backgrounds and different focal points from relevant disciplines such as psychology, philosophy, linguistics, computer science, information science and cultural studies by using methods for intercultural HCI design. The most important goal is, therefore, to come to a common understanding regarding terminology, methodology and processes necessary in intercultural HCI design to be able to use them in the intercultural context. This also has implications for the use of KDD methods in the cultural context. Hence, the challenges evolved during the intercultural HCI design process are subject to analysis and some aspects useful in reaching the goal are suggested.

Keywords: learning, challenges, cross-disciplinary, culture, communication, HCI, intercultural, cultural studies, solutions, model, HCI design, cultural differences, culture, communication, understanding, empathy, intercultural communication, intercultural HCI design, empathy, KDD, knowledge, big data, discovery, interpretation, methods.

1 Challenges by Cross-Disciplinary Learning in Intercultural HCI Design

Much cultural background has to be considered when designing the functionality and the interaction for global devices [1]: intercultural HCI design comprises significantly more than merely the implementation of a catalogue of requirements for the user interface like considering different languages, colours or symbols [2]. Successful intercultural HCI design goes far beyond a regular design process by taking into account different mentalities, thought patterns and problem solving strategies that are anchored in culture, for example linear vs. non-linear differences [1], [3]. For example, usage patterns which do not occur in everyday life in the source country can arise in the target country due to different power structures [4], for example flat vs.

A. Holzinger and G. Pasi (Eds.): HCI-KDD 2013, LNCS 7947, pp. 101–111, 2013.

hierarchical ones. Moreover, the designer must know exactly what the user needs or wants (e.g. why, in which context, etc.) [5]. This knowledge can be determined most precisely by using inquiry approaches or methods based on communication [6]. Just using observation techniques or relying on expert opinions results in less reliable information although these results are very useful, too [7]. However, problems in intercultural communication, particularly those in requirement analysis, inhibit good usability for system design and the related user experience [8]. Moreover, as a result of the rapid development of computers in this age of information the possibilities for research have grown to such an extent that those numerically overwhelmed experts in the field cannot possibly understand all aspects in detail nor recognize all higher level interconnections. There are therefore more and more 'specialists' (lacking knowledge of interdependencies) on the one hand and more and more 'generalists' on the other, experts with putative knowledge of interdependencies (more specifically the dissemination of superficial interconnection knowledge without having done their own detailed studies to confirm or corroborate this knowledge). This also means that at this point the quality of human knowledge generally sinks.

In order to confront this problem we must take on this topic up in our research community and endeavor to discover and examine the gaps formed by missing interconnection knowledge. Especially the frictional loss in interdisciplinary context must be considered. It is here that divergent terminology, perspective and methodology easily lead to the loss of information and result in misunderstanding [9]. Trans-disciplinary communication serves as the basis for decisive action for several reasons. In intercultural communication differing meanings (i.e. disparate correlations), differing deduction patterns (i.e. logic traditions) and goals force the clarification of concepts and the assumption of a foreign perspective, which is the essence of trans-disciplinary communication (cf. [10]).

The reader or listener must also be prepared to modify all his assumptions, convictions, wishes and intentions (Web Of Belief, cf. [11]) or even revise them (Belief Revision, cf. [12]). In addition to that a certain competence in reception (previous knowledge and empathy) is necessary as well as accrued information (in order to make estimations) (cf. internal rules of communication and rules for communication and rules for relationships; cf. [13], [14]). In this case it is critical to remain technically precise, objective and deductively logical in order to avoid interpretation problems (due to differing vocabulary and use of speech, differing vocabulary associations, inclusion in expert discourse and conversational maxims) and interference stemming from disparate knowledge of the world, intentions and expectations (cf. [15], [16]). They can be avoided by the proper choice of methods for comprehensible examples and metaphors [17-19].

One effective speech or writing strategy is to clearly explain or even underscore all facts in the statements or expressions [9]. Some characteristics of experiential discourse are for instance personal letters, the portrayal of one's own adventures and experiences, subjective evaluations without recognizable reference to an objective system of values as well as a claim to subjective truth and objective reality (cf. [20]: objective vs. subjective social world (= inter-personal relationship)). Something happens at the level of incidents. The individual experience is processed as a mental concept at the same time on classification level. In accordance with the neo-positivists there is no experience without predication (cf. [21]); there are no protocol statements: empiricism is a sub-class of the experience for historical facts (i.e. a specific structured experience). According to Luther experience serves as a means to reduce complexity (proposition) which becomes certainty by way of familiarity yielding security (cf. securitas (security) vs. certitudo (confidence), cf. [22]). A passive experience is an event that one has experienced which has left a strong and enduring impression (cf. [9]). An active experience is connected with when the content of the action is oriented towards some goal. Academic rigor must always be especially conscious of its prerequisites. For different roles and world views make communication difficult and demanding. A human must therefore constantly plumb his possibilities and seek certainties in the world through his experiences with others. One generally looks for evidence (verification) rather than contradiction of one's own position (falsification). According to Popper the latter would be more sensible because it is logically irrevocable and avoids induction problems (cf. [23]). However, the effort of falsification can indeed prove to be much greater making verification the much more useful daily approach. It can nevertheless never yield absolute certainty (except if one were sure there were no further facts to validate). Resultantly many (putative) certainties (in the form of assumptions, convictions or axioms) are unconsciously or even consciously presupposed. Some examples would be Euclidian space (angle > 189°), laws of mechanics (v < 0.1) as well as types of logic. Finally, according to Gödel, there are statements, the validity of which cannot be determined [24]. Certainty demands that one can explain what one says. A negative example for this is the Lazarus story which has been discussed for 100 years. And texts are arbitrary (as paper is patient). Academic discourse in different disciplines tends to apply terminology as it fits their needs or so that they can say what they want to say most easily. Philosophers go a step further as definition conjurers and distinguish themselves at the meta-level of science theory (cf. [25]). The relationship of theory to experience remains important (cf. "concept police" vs. the possibility of conceptual distinction) which appears in activity related or decisive argumentation. Generally, better arguments lead to better decisions (on a purely formal level even completely independent of any moral content). If however 'qualia' (sensations like pain for example, cf. [26]) enter the equation, then the act of convincing by argumentation is taken ad absurdum (cf. [9]). When the argument cannot be followed, it loses validity and becomes ineffective. The validity of such arguments cannot be tested. At least in this case conviction without argumentation is possible [25]. It would be worth checking if there are not further cases in which conviction can be achieved without good arguments (i.e. threat and seduction, etc.). One can follow [25] in this sense when he postulates that belief and reason are part of activity on the same level. Here the concepts of trust and empathy come into play, which are more related to belief and intuition than to reason and argumentation taking strongly into account the principle of charity (cf. [27]). This is supported by the fact that most of conviction of other persons come from behaviour and outfit of the speaker rather than by verbal arguments (just 7%, cf. [28]). For this reason he demands the cultivation of a basic primal sense of trust springing from within rather than from without, with which to reach certainty through one's own experience. Often only those active experiences are considered legitimate justifications of decisions or substantiations of a claim which appear to be scientific. Especially in the inter-disciplinary context active experience and certainty do mean engaging with the internal certainties of the communicating disciplines (necessarily involving trust and empathic relationships).

2 Solving Problems in Intercultural HCI Design by Successful Intercultural (Cross-Disciplinary) Communication

Several levels of intercultural know-how contribute to successful intercultural HCI design. The communication level constitutes the basic level, followed by the levels of project management, software and usability engineering and HCI design itself on the way to successful intercultural HCI design (cf. Figure 1).

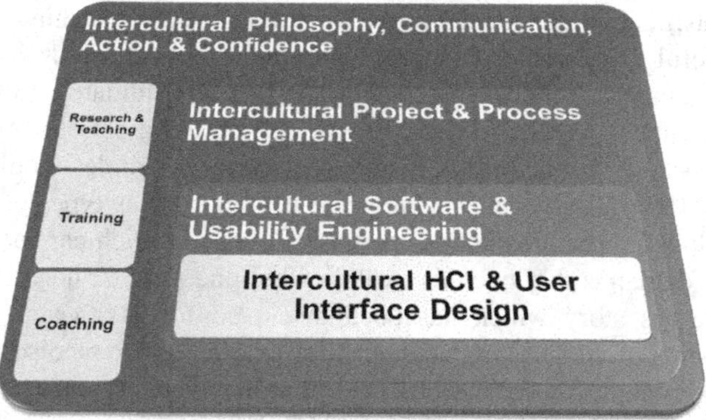

Fig. 1. Levels on the Way to Successful Intercultural HCI Design (Source: IUIC)

Hence, on all levels (strongly influenced by the philosophy of the respective cultures), intercultural communication skills at the basic level can contribute to the solution of problems raised on the upper levels by cultural differences.

2.1 Intercultural Communication Requires Mutual Understanding

For successful (intercultural) usability engineering, an adequate engineering process is necessary to ensure good usability (i.e. when the user understands the developer's device and is thus able to easily operate it satisfactorily), it is necessary that the developer understands the user [3], [29], because they have different points of view [7]. At least the following aspects of the user must be analysed in detail before the product can be developed:

- World view, Weltanschauung (metaphysical approach) of the end-user,
- General knowledge (procedural and factual knowledge) of the end-user,
- The context in which the product will be used by the end-user,
- The tasks the end-user intends to accomplish by using the product.

Only by considering these aspects, intercultural communication as an essential prerequisite for intercultural usability engineering, user interface design, and user experience will it be successful and can it lead to successful international product design.

2.2 Mutual Understanding Requires Common Conversation Code

Successful cross-disciplinary conversation and learning requires a mutual understanding of the participants in regard to terminology, methods and processes used. However, how to compare the perspectives of different cultural users? Even worse: what, if the test leaders and the analyzing experts are culturally different, too? This means that the intercultural HCI experts should have a common understanding of designing intercultural HCI by knowing the corresponding design processes and methods. However, a common understanding evolves over time and it is necessary to take into account many aspects while analyzing different examples of HCI such as current situation, ambience, mood, cognitive style, cognitive state, context, use case, cultural background, perspective, role, status, profession, experience, education, ability to be empathic, cultural dimensions, personal character, kind of language, use of language, idiolect, mother tongue, country of birth and primary residence. In addition to this short list, a detailed and partly verified model for intercultural HCI design has been worked out by the author [30], which is used as inspiration for thinking things through in this area and helps to elucidate the complex entanglement of the huge amount of data and information pieces forced by the number of aspects to consider in cultural context (like current situation, ambience, mood, cognitive style, cognitive state, context, use case, perspective, role, status, profession, experience, education, ability to be empathic, personal character, use of language). In addition, a compiled list of examples of indigenous HCI problems have been analyzed according to the criteria mentioned before and the results are stored in a matrix representing the cultural differences related to examples of HCI. From this matrix in conjunction with the model, rules can be derived to adjust the methods for intercultural HCI design such that the results and the design recommendations from applying the methods can be compared according to different cultures as well as applied for adequate use cases in the right situation and context. However, even if some evidence and rules have been obtained to narrow the challenges in analyzing perspectives from different disciplines relevant for HCI design using this matrix and model, the final analysis of the intercultural HCI design process and it's relating cultural differences is still outstanding.

2.3 Differences in the Alleged Common Conversation Code

One aim should be to research the basic cultural differences in this presupposed (and alleged) common conversation code while discussing and adapting well-known methods for their usage in intercultural HCI design. For instance, how to ensure understanding between test leaders as well as test leader and the users? A matrix should help to elucidate the complex entanglement of the huge amount of data and information pieces forced by the number of aspects to consider. A compiled list of examples of HCI problems can be analyzed according to the criteria mentioned before and the results should be stored in a matrix representing the disciplinary differences related to examples of HCI. The matrix can be enhanced using advanced statistical methods (e.g. structural equation models or neural networks etc.). For a similar approach to obtain models for intercultural HCI design, please refer to [30]. From this matrix, it should be possible to derive adjustment rules. These rules can be used to adjust the methods for intercultural HCI design such that the results and the design recommendations from applying the methods can be compared according to different cultures as well as applied for adequate use cases in the right situation and context.

3 Recommendations

Only assuming the perspective of a user by the HCI designer to grasp their needs depending on world view, general knowledge, context, and purpose of usage can lead to good user interfaces of high usability, thereby evoking excellent user experience design. Therefore, the aim is to research the basic cultural differences in this presupposed (and alleged) common conversation code while discussing and adapting well-known methods for their usage in intercultural HCI design.

3.1 Empathy as a Prerequisite for Mutual Understanding

Successful communication depends crucially on the capability for empathy of the people involved [15], [31], [32], [33]. Communication without empathy does not deliver the desired results [34]. This in turn assumes a certain level of confidence and trust (e.g. Principle of Charity, cf. [15]) and includes the knowledge of how to read between the lines of the counterpart's communication depending on culturally different rules. This includes the usage of linguistic rules, for example, Austin's felicity conditions [31] or Gricean maxims [35]. Therefore any literal translation of a conversation is prone to misinterpretation since the extension of the concepts can be different in different cultures ("linguistic relativity", cf. [36]). As context must also be taken into consideration, it is important to consider these aspects in communicating and focus on the intellectual horizon of the communication partner as widely as possible. This can occur in particular through personal and on-site communication and is particularly difficult over the phone due to the absence of mimical and gestical signals. Even more problems arise in intercultural communication compared to intra-cultural communication due to differing world views and the context in which clarification occurs. For this reason, the empathic capability to put oneself in someone else's situation is particularly important. The application of empathy in the end contributes to a successful communication supporting a mutual linguistic code. In particular, intercultural user experience designers must be able to put themselves in the position of the user in order to know and understand his or her intentions and needs, to ideally experience them, and to implement them in the product. Furthermore, in order to build up not only understanding but also the ability to put oneself in someone else's position, it is initially necessary to be on the same wavelength to find a connection to the other person. This requires the alignment of communication coding (vocabulary and grammar) and to achieve a situation where the other person wants to communicate. Thus a relationship is built up in such a way that future communication remains possible. If this connection is given, it is important to preserve access to the other person's knowledge base ("Web Of Belief", cf. [11]) using a mutual topic of conversation in order to examine the knowledge base of the counterpart in regard to extent, type, and quality. Only then it is possible to find the right "hook" in further conversations and consequently "fetch up" the other person's web of belief at the most relevant point to quickly pick up the same wavelength again. The web of belief contains beliefs and desires derived from premises, assumptions and facts using logical rules and are recursively formed by experience from birth. Through training intercultural competency, approaching the web of belief of a member of other cultures is possible. Thereby, exchange of experiences is very effective, trust can be conferred from one person to another by introducing the persons and critical interaction situations [10] can be weakened. This works, if it is clear how the other person thinks (i.e. what world view he or she holds i.e. which premises and assumptions about the world he or she has). This is necessary in order to make choices which are relevant to the job at hand and correct for successful communication with a continually expanding set of extra information. This is particularly the case in intercultural contexts. The ability to assess and understand the person's thinking patterns enables an adequate reaction to the people involved. In the same way, the leading and guiding of conversation, e.g., as facilitator or investigator is successfully supported. However, empathy also presupposes the capability to separate oneself from other persons to get the chance to recognize the differences to them and then to put oneself in their position. Within the intercultural context, this requires being aware of one's own cultural standards before it is possible to compare and recognize differences to other cultures [10].

3.2 Applying Empathic Capabilities in Intercultural HCI Design and Intercultural Usability Engineering

Global User Interfaces, which would suit all culture domains, users and contexts, do not yet exist, at least for technical if not for more fundamental reasons. Computers do not yet possess empathy (cf. the so-called „hard problems of AI" [37]). At the moment computers lack environmental data (through sensors), the complex processing patterns and the respective knowledge of the world needed to develop empathy. Furthermore, the cultural differences involved in the interaction of the user with the system must be integrated in such knowledge of the world so that the system can adjust for it respectively. Finally, even if these challenges were met the so-called 'bootstrapping' problem of adaptive systems would remain. Because the system is not yet acquainted with the user on his first encounter, the system cannot adapt to him. It is a matter of time until the system gets acquainted with the user and can adapt itself to him. At least the following areas must be considered in HCI, i.e. Human-Computer-Interaction: task, context/situation and tools used [38]. In this case the cognitive processes of the user differ from the results of studies or discrete situations due to his cultural and environmental conditioning and personal experience. The concept of the task intended (as well as the task itself) is no longer congruent. That requires the system (computer / machine / tool) to adapt as perfectly as possible to many different aspects, which however has not yet been possible to implement because of the multitude of aspects and the resultant complexity. As long as the above mentioned problems have not been completely solved, human beings must accordingly attune HCI to the intended cultural domain, user group and context [39]. To do so, the HCI designer must be able to immerse himself in these cultural domains, user groups and contexts in order to extract the relevant requirements for the HCI design.

3.3 Implications for Knowledge and Data Discovery (KDD) in Human Computer Interaction (HCI) Design

Empathy and mutual understanding explain why knowledge should be considered intercultural and why making sense is highly dependent on culture, which leads to an approach to better understand KDD as required in [40]. For example, it must be discussed in detail how software developers and UI designers could collaborate on formulating different cultural backgrounds in order to share the same knowledge. This means that the common conversion code also embraces methods and processes as well as a sufficiently similar interpretation of the outcome. In addition, software development and usability engineering are perceived quite differently in various parts of the world. Hence, integrating usability in the software development lifecycle is still a widely unresolved task, which is even more of a challenge in international projects. Challenges arising in mixed projects teams comprise culture, language, education, ethics, sense of hierarchy, quality, time and attitude towards teams, just to name a few. Moreover, achieving a common understanding of a project development process needs to be coordinated and agreed upon in every usability project applying KDD methods (as exemplified in [41]) no matter where project members come from (cultures or disciplines, cf. [9]).

4 Discussion and Conclusion

The question of whether the problems in understanding and communicating are more individual empathy or rather based on cultural differences remains open. The question "who gets on well with whom" must still be explicitly explored in relevant studies. It is, however, clear that for communication to be successful, both communication partners in a conversation must be open to each other and in this way be able to use certain basic empathic skills. Hence, empathy is an essential prerequisite for successful intercultural communication, which also promotes successful knowledge and data interpretation. In addition, according to [33], there should be overlapping cultural universals that can and should be used to derive recommendations for intercultural HCI design. Furthermore, I postulate that these universals can be found in matured matrices mentioned in section 2.2 before. However, the final analysis of such challenges to cross-disciplinary learning in intercultural HCI design (such as combining KDD and HCI in the cultural context) is still outstanding.

In sum, this paper touches on some problems concerning knowledge discovery in the cultural contexts. Many open questions must be explored and answered in detail in the future in order to obtain clues to better understand "knowledge and data discovery" in the cultural context. Even if existing examples of product design can be taken as models for intercultural HCI design, knowledge and data interpretation rules (and hence product/HCI design rules) differ from culture (discipline) to culture (discipline) because of different expectations by the members of the respective cultures (disciplines). Therefore, not only the inductive approach (from data collection on to building theories) but also the other way around in the form of a deductive approach (deriving hypothesis and testing them empirically) should be used. In this sense, this paper is a good start. It presents some thoughts on an approach to a common academic understanding and to adequate methods for intercultural HCI design across disciplines and thereby learning from each other to improve intercultural HCI design (e.g., by clarifying what knowledge is necessary to share and how this can be done in order to interpret it adequately and distribute the results comprehensively for all involved researchers from different disciplines).

References

1. Röse, K., Zühlke, D.: Culture-Oriented Design: Developers' Knowledge Gaps in this Area. In: 8th IFAC/IFIPS/IFORS/IEA Symposium on Analysis, Design, and Evaluation of Human-Machine Systems, September 18-20, pp. 11–16. Pergamon (2001)
2. Heimgärtner, R., Holzinger, A.: Towards Cross-Cultural Adaptive Driver Navigation Systems. In: Holzinger, A., Weidmann, K.-H. (eds.) Empowering Software Quality: How Can Usability Engineering Reach These Goals? 1st Usability Symposium, HCI&UE Workgroup, Vienna, Austria, November 8, vol. 198, pp. 53–68. Austrian Computer Society (2005)
3. Honold, P.: Culture and Context: An Empirical Study for the Development of a Framework for the Elicitation of Cultural Influence in Product Usage. International Journal of Human-Computer Interaction 12, 327–345 (2000)
4. Hofstede, G.H., Hofstede, G.J., Minkov, M.: Cultures and organizations: software of the mind. McGraw-Hill, Maidenhead (2010)
5. Holzinger, A.: Usability engineering methods for software developers. Commun. ACM 48, 71–74 (2005)
6. Rubin, J.: Handbook of usability testing: How to plan, design, and conduct effective tests. Wiley, New York (1994)
7. Heimgärtner, R., Holzinger, A., Adams, R.: From Cultural to Individual Adaptive End-User Interfaces: Helping People with Special Needs. In: Miesenberger, K., Klaus, J., Zagler, W.L., Karshmer, A.I. (eds.) ICCHP 2008. LNCS, vol. 5105, pp. 82–89. Springer, Heidelberg (2008)
8. Clemmensen, T.: Usability problem identification in culturally diverse settings. Information Systems Journal 22, 151–175 (2012)

9. Janich, P.: Kultur und Methode: Philosophie in einer wissenschaftlich geprägten Welt. Suhrkamp, Frankfurt am Main (2006)
10. Thomas, A., Kinast, E.-U., Schroll-Machl, S.: Handbook of intercultural communication and cooperation. Basics and areas of application. Vandenhoeck & Ruprecht, Göttingen (2010)
11. Quine, W.V.O.: Word and object. MIT Press, Cambridge (2004)
12. Gärdenfors, P., Rott, H.: Belief revision, Handbook of logic in artificial intelligence and logic programming: Epistemic and temporal reasoning, vol. 4. Oxford University Press, Oxford (1995)
13. Heimgärtner, R., Tiede, L.-W., Windl, H.: Empathy as Key Factor for Successful Intercultural HCI Design. In: Marcus, A. (ed.) HCII 2011 and DUXU 2011, Part II. LNCS, vol. 6770, pp. 557–566. Springer, Heidelberg (2011)
14. Liu, C.-H., Wang, Y.-M., Yu, G.-L., Wang, Y.-J.: Related Theories and Exploration on Dynamic Model of Empathy. Advances in Psychological Science 5, 014 (2009)
15. Dennett, D.C.: The intentional stance. MIT Press, Cambridge (1998)
16. Davidson, D.: 10. Belief and the Basis of Meaning (2001)
17. de Castro Salgado, L.C., Leito, C.F., de Souza, C.S.: A Journey Through Cultures: Metaphors for Guiding the Design of Cross-Cultural Interactive Systems. Springer Publishing Company, Incorporated (2012)
18. Evers, V.: Cross-cultural understanding of metaphors in interface design. In: Ess, C., Sudweeks, F. (eds.) Attitudes toward Technology and Communication (Proceedings CATAC 1998), Australia, pp. 1–3 (1998)
19. Nardi, B.A., Zarmer, C.L.: Beyond models and metaphors: visual formalisms in user interface design. In: Proceedings of the Twenty-Fourth Annual Hawaii International Conference on System Sciences, January 8-11, vol. 2, pp. 478–493 (1991)
20. Habermas, J.: Theorie des kommunikativen Handelns. Suhrkamp, Frankfurt/Main (1995)
21. Carnap, R.: Der logische Aufbau der Welt. Felix Meiner (1928)
22. Luther, M.: Martin Luther's Preface to the Epistle to the Romans. New Creation Publications, Blackwood (1995)
23. Popper, K.R., Eccles, J.C.: Das Ich und sein Gehirn. Piper, München (2005)
24. Hofstadter, D.R.: Gödel, Escher, Bach. Hassocks Harvester Press (1979)
25. Wohlrapp, H.: Wege der Argumentationsforschung (1995)
26. Heimgärtner, R.: Michael Tye: "Phenomenal Consciousness: The Explanation Gap as a Cognitive Illusion" (1999). Kritische Diskussion der zentralen Position. Philosophische Fakultät IV, Universität Regensburg, Regensburg, vol. BA (2001)
27. Lewis, D.: Radical interpretation. Synthese 27, 331–344 (1974)
28. Argyle, M., Schmidt, C.: Körpersprache & Kommunikation: Das Handbuch zur nonverbalen Kommunikation. Junferman, Paderborn (2005)
29. Nielsen, J.: Usability engineering. Kaufmann, Amsterdam (2006)
30. Heimgärtner, R.: Cultural Differences in Human Computer Interaction: Results from Two Online Surveys. In: Open Innovation, UVK, pp. 145–158 (2007)
31. Austin, J.L., von Savigny, E.: Zur Theorie der Sprechakte = (How to do things with words). Reclam, Stuttgart (2010)
32. Searle, J.R., Kiefer, F.: Speech act theory and pragmatics. Reidel, Dordrecht (1980)
33. Schwartz, S.H.: Universals in the content and structure of values: theoretical advances and empirical tests in 20 countries. Advances in Experimental Social Psychology 25, 1–65 (1992)
34. Stueber, K.R.: Rediscovering empathy: agency, folk psychology, and the human sciences. MIT Press, Cambridge (2010)

35. Grice, P.: Studies in the way of words. Harvard Univ. Press, Cambridge (1993)
36. Whorf, B.L.: Sprache - Denken - Wirklichkeit Beiträge zur Metalinguistik und Sprachphilosophie. Rowohlt, Reinbek bei Hamburg (2008)
37. Winograd, T., Flores, F.: Understanding computers and cognition: A new foundation for design. Addison-Wesley, Boston (2004)
38. Herczeg, M.: Software-Ergonomie: Grundlagen der Mensch-Computer-Kommunikation. Oldenbourg, München (2005)
39. Van Kleek, M., Shrobe, H.E.: A practical activity capture framework for personal, lifetime user modeling. In: Conati, C., McCoy, K., Paliouras, G. (eds.) UM 2007. LNCS (LNAI), vol. 4511, pp. 298–302. Springer, Heidelberg (2007)
40. Holzinger, A.: On Knowledge Discovery and Interactive Intelligent Visualization of Biomedical Data - Challenges in Human-Computer Interaction & Biomedical Informatics. In: 9th International Joint Conference on e-Business and Telecommunications (ICETE 2012), pp. IS9-IS20 (2012)
41. Heimgärtner, R., Kindermann, H.: Revealing Cultural Influences in Human Computer Interaction by Analyzing Big Data in Interactions. In: Huang, R., Ghorbani, A.A., Pasi, G., Yamaguchi, T., Yen, N.Y., Jin, B. (eds.) AMT 2012. LNCS, vol. 7669, pp. 572–583. Springer, Heidelberg (2012)

Intent Recognition Using Neural Networks and Kalman Filters

Pradipta Biswas, Gokcen Aslan Aydemir, Pat Langdon, and Simon Godsill

Department of Engineering
University of Cambridge, UK
{pb400,ga283,pml24,sjg}@eng.cam.ac.uk

Abstract. Pointing tasks form a significant part of human-computer interaction in graphical user interfaces. Researchers tried to reduce overall pointing time by guessing the intended target a priori from pointer movement characteristics. The task presents challenges due to variability of pointer movements among users and also diversity of applications and target characteristics. Users with age-related or physical impairment makes the task more challenging due to there variable interaction patterns. This paper proposes a set of new models for predicting intended target considering users with and without motor impairment. It also sets up a set of evaluation metrics to compare those models and finally discusses the utilities of those models. Overall we achieved more than 63% accuracy of target prediction in a standard multiple distractor task while our model can recognize the correct target before the user spent 70% of total pointing time, indicating a 30% reduction of pointing time in 63% pointing tasks.

1 Introduction

Human computer interaction has become an every-day aspect of lives as technological devices make their ways out of the laboratory. New input devices are being introduced to the market not only in the form of computers and smart phones, but televisions, kiosks in hospitals, even advertisement screens in airports. The availability of these devices to ordinary people increased the diversity of the population using them, which include impaired and elderly users as well as able bodied, expert and non-expert users. Most of the input devices require a pointing task which can be difficult to some users, especially users with motor impairments which can become an overwhelming activity that they want to avoid.

Several studies so far have examined the characteristics of cursor movement [14]. Algorithms to reduce the movement time to target were suggested taking Fitts' Law into account to determine the difficulty of a task such as increasing target size [8, 12], employing larger cursor activation regions, moving targets closer to cursor location, dragging cursor to nearest target or changing CD ratio[15]. It is certain that all of the proposed algorithms that reduce pointing time will perform better in the existence of a target prediction algorithm so that only correct or most probable targets could be altered dynamically. With this in mind, researchers have been proposing prediction methods.

A. Holzinger and G. Pasi (Eds.): HCI-KDD 2013, LNCS 7947, pp. 112–123, 2013.

One of the first algorithms for target prediction was suggested by Murata, which calculates the angle deviation towards all possible targets and select the target with minimum deviation. The results show that pointing time can be reduced by 25% using this algorithm [13]. Asano et.al.[1] points out that having more than one target on a particular movement direction results in poor performances of the afore mentioned algorithm, especially when dealing with target located far away. They used previous research results about kinematics of pointing tasks and showed that peak velocity and target distance has a linear relationship. They predict the endpoint with linear regression using the relationship between peak velocity and total distance to endpoint [1]. Lank et.al also employ motion kinematics where they assume minimum jerk law for pointing motion and fit a quadratic function to partial trajectory to predict endpoint [10].

However, cursor movement vary in characteristics for motor impaired users since they experience tremor, muscular spasms and weakness [7]. The velocity profile includes several stops and jerky movements. This needs to be taken into account when applying target prediction. State space filtering techniques are promising [10] in estimating intended targets as well as smoothing cursor trajectories since it is possible to model the movement, fine-tune or adapt the parameters for different users.

In the following sections the accuracy and sensitivity of Murata's algorithm [13] and a basic Kalman Filter framework will be investigated and compared using point-click task data from able-bodied and impaired users. This work on target prediction has considered the following aspects

1. There are not many earlier works on end point or target prediction for users with motor impairment, except Godsill's work [5] on particle filter and Wobbrock's work [15] on Angle Mouse.
2. Comparing different sets of features for target prediction as earlier work already found different set of features are more predictive in describing movement characteristics of users with motor impairment
3. Developing a set of metrics to compare different target prediction algorithms.

We have compared two techniques to detect user's intended target in a point and click tasks. We tried to analyze the movement trajectory of users and used different movement features to detect intended target. One method uses a neural network to detect whether the user is moving a pointer to reach a target or home on to the target and based on that predicts intended target. The other method parameterizes a Kalman Filter based on users' movement characteristics and uses this model to calculate probability of different targets. We have also proposed a set of metrics to compare intent recognition or end point prediction algorithms and used this metrics to compare our algorithms. It has been found that both neural network and Kalman filter have advantages and disadvantages based on their applications, and on average they can predict the correct target in more than 80% pointing tasks with approximately 60% accuracy.

The paper is organized as follows. The next section presents a brief literature survey on intent recognition algorithms. Section 2 presents the theories behind our models followed by implementation and validation details in section 3. Section 4 highlights the contribution of this work followed by conclusion in section 5.

2 Theory

Feature Calculation

We have used the following features as input to our models on intent recognition. This section explains the features in detail.

- Distance to Target: This feature is calculated as the straight line Euclidean distance from the point of movement to target
- Velocity: We have recorded the pointer position using the getCurrentPosition() API. The velocity is measured as the ratio of the Euclidian distance between two consecutive readings to the difference in timestamp in msec.
- Acceleration: The ratio of two consecutive velocity readings to the time difference between them in msec.
- Bearing Angle: This is calculated as the angle between two vectors- first being the previous cursor position to current cursor position and the other vector is current cursor position to target position.

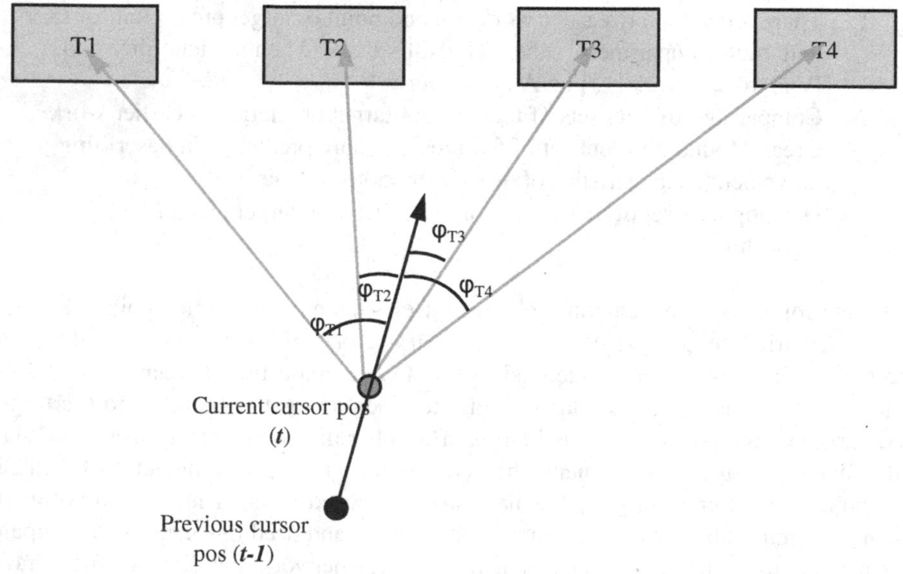

Fig. 1. Bearing Angle Calculation

Neural Network Based Model

Pointing tasks are traditionally modeled as a rapid aiming movement. In 1887, Woodsworth [16] first proposed the idea of existence of two phases of movements in a rapid aiming movement, main movement and homing phase, which was later formulated to predict pointing time by Paul Fitts. Fitts' law [4] study has been widely used in computer science to model pointing movement in a direct manipulation interface though its applicability for users with motor impairment is still debatable. However the existence of main movement and homing phase is generally accepted among all users' groups and was also supported in our previous studies.

Once a pointing movement is in homing phase, we can assume the user is pretty near to his intended target. The present algorithm tries to identify the homing phase and then predict the intended target. Previous work on analyzing cursor traces of users with a wide range of abilities concentrated on angle [15], velocity and acceleration profiles [6]. So we consider velocity, angle and acceleration of movement and with the help of a back-propagation neural network, try to identify the homing phase.

Neural Network is a mathematical model containing interconnected nodes (or neurons) inspired by biological neurons used as a classifier and pattern recognizer for complex data set. We have used a three layer Backpropagation network for this study. After the neural network predicts the homing phase, we predict the nearest target from current location towards the direction of movement as intended target. A simple version of the algorithm wIll be as follows

```
For every change in position of pointer in screen
      Calculate angle of movement
      Calculate velocity of movement
      Calculate acceleration of movement

      Run Neural Network with Angle, Velocity and Acceleration
      Check output
      If output predicts in homing phase
               Find direction of movement
               Find nearest target from current location towards
direction of movement
```

Kalman Filter Based Model

Kalman filters are used to estimate system state or unknown variables/parameters from noisy measurements in a two-step mechanism. At prediction step, current state estimates are calculated. At measurement state, these estimates are updated according to the measurements of observed variables [2]. Here in this paper, we explored the use of a Kalman Filter to estimate the cursor position and use the estimates to update a probability distribution for possible targets. We have analyzed cursor trajectories of users with and without motor impairment and found movements towards the target are often contaminated by noise either due to physical or situational impairment [3,6]. The noise or jitter in movement makes it difficult to estimate the right target. So we developed a model that can remove the jitter in movement through formalizing the noise in movement. We feed cursor positions to the model and it calculates new cursor position by removing the jitters in the movement. Based on these noise-free cursor locations, we try to estimate the intended target.

We have modeled the process of cursor movement as a "*nearly constant velocity (NCV)*" process, while the velocity is modeled as a Brownian motion which is characterized by small white noise acceleration [3,5,6]. The process model is explained in equation 1 below, where x is the horizontal and y is the vertical cursor positions respectively. Here Δt, $v_x[k]$, $v_y[k]$, $\upsilon_x[k]$, $\upsilon_y[k]$ represent time step, x velocity, y velocity, x and y velocity white noises respectively.

Process model:

$$x[k+1] = x[k] + \Delta t * v_x[k] + \upsilon_x[k]$$
$$y[k+1] = y[k] + \Delta t * v_y[k] + \upsilon_y[k]$$
$$v_x[k] = dB$$
$$v_y[k] = dB \tag{1}$$

Cursor position in horizontal and vertical directions is taken as measurements with white noise (Eqn.2). The noise characteristics can be personalized depending on the level of motion-impairment of users to account for jittery movements. Here, **z** is the measurement vector.

$$\mathbf{z}[k] = (x[k]\ y[k])' + \omega[k] \tag{2}$$

It is possible to assume that all targets are equally probable at the beginning of any pointing task and update the probabilities. For N targets, the probability of target *i* will be:

$$p_i = 1/N \tag{3}$$

Following Murata's method, which assumed that a cursor should travel the shortest path along the target axis, we considered angle of movement and distance to target for intent recognition. Our method constantly updates probabilities of each target to become the intended target according to the angle, distance and both angle and distance to targets at every measurement as given in Equation (4).

$$p_i[k+1] = p_i[k] * (1/distance\_to\_target)$$
$$p_i[k+1] = p_i[k] * (1/angle\_to\_target)$$
$$p_i[k+1] = p_i[k] * (1/distance\_to\_target) * (1/angle\_to\_target) \tag{4}$$

For the following sections KA will be used to indicate prediction using only angle to target, KD to indicate prediction using only distance to target and KAD to indicate prediction using them together. The full algorithm is as follows

```
Start with assigning 1/N probability for all targets in layout
Initialize Kalman Filter
For every change in position of pointer in screen

  Kalman filter measurement update
  Iterate Kalman filter to update position/velocity
  Update probabilities using distance and angle to target
  Normalize probability
  Find target with max probability
```

3 Evaluation Criteria

Once we have working algorithms that can predict intended target, we need to evaluate its performance so that we can compare and contrast its performance with prior research. So far there is not much consensus on the evaluation criteria of an intent recognition algorithm. We have defined the following three parameters to evaluate the quality of a target prediction algorithm.

Availability. In how many pointing tasks the algorithm makes a successful prediction.

For example, say a user has undertaken 10 pointing tasks. An algorithm that correctly predicts target in 7 of them is better than another one which is successful in predicting correct target in 5 pointing tasks.

Accuracy. Percentage of correct prediction among all predictions.

Any target recognition algorithm keeps on predicting target while the user is moving a pointer in a screen. It may happen that within a single pointing task, an algorithm initially predicts a wrong target but as the user gets closer to the intended target, the algorithm finds the correct target. So if an algorithm predicts an intended target 100 times among which 70 are correct, its accuracy will be 70%.

This metric complements the previous metric. For example, say we have an algorithm that predicts the nearest target from current pointer location as the intended target. This algorithm will have 100% availability as it will fire in all pointing task but pretty low accuracy while there are multiple targets available on the path of pointer.

Sensitivity. How quickly an algorithm can detect intended target.

For example, an algorithm that can find the intended target while the user crossed 90% of target distance is less sensitive than an algorithm that can predict correct target after the user crosses only 70% of target distance.

4 Implementation and Validation

We have implemented a bank of neural networks considering all possible combinations of three movement properties (bearing, velocity and acceleration of movement). We have used supervised learning to detect users' movement phases from movement features. Initially we trained the neural networks with the standard multiple distracter task. Later we used the same task to evaluate the networks following our evaluation criteria.

For the Kalman Filter based model, we have considered three different conditions of angle and distance to target. The result is used to find the best set of feature to detect the intended target for both kinds of models. The following section describes a user trial for training and validation phases followed by result and discussion.

User Trial

Participants

We have collected data from 23 users. The users used to operate computer everyday and volunteered for the study. The group of participants included users with a wide range of abilities in terms of visual and motor impairment. Age related impaired users were more than 60 years old and physically impaired users suffer from Cerebral Palsy or Spinabifida.

We trained the neural networks with 13 users among which five have age related or physical impairment (like cerebral palsy). Then we test the system with 10 participants among which five have age related visual and motor impairment. The gender was balanced to nearly 1:1 in both training and test cases.

Material

The study was conducted using a 20" screen with 1280 × 1024 resolution and a standard computer mouse.

Procedure

The task was like the ISO 9241 pointing task (commonly termed as Fitts' law task) with multiple distractors on screen (figure 2). Users need to click the button at the centre of the screen and then the target button appears with other distractors. We used five different target sizes (20, 40, 30, 50, 60 pixels) and source to target distances (100, 140, 180, 240, 300 pixels). The participants were instructed to click target for 5 minutes.

Fig. 2. Multiple Distractor task

Results

In the results section, we have used the following acronyms.

KBD	Kalman filter based on Bearing and Distance
KD	Kalman filter based on Distance
KB	Kalman filter based on Bearing
NVB	Neural Network using Velocity and Bearing
NVA	Neural Network using Velocity and Acceleration
NAB	Neural Network using Acceleration and Bearing
NVAB	Neural Network using Velocity, Acceleration and Bearing

Table 1 below shows the average percentage of availability and accuracy in different systems and figures 3 and 4 plot them in descending order.

Table 1. Availability and Accuracy

	Avaibality	**Accuracy**
KB	75.7	38.1
KBD	99.7	63
KD	97.7	54.8
NAB	63.6	62
NVA	64.2	63.2
NVAB	62	60.3
NVB	65.4	66.2

Figure 5 plots the sensitivity of the system. The x-axis shows the fraction of pointing time spent in a scale of 100. The y-axis represents the probability of correct target prediction in a scale of 0 to 100. The Kalman Filter based model starts with equal probability of all targets and gradually increase the probability of the correct target as the movement towards it continues. On the other hand, the Neural Network based model only fires (or turns available) when it detects a change in movement phases. In certain occasions it fails to detect this change of movement phases and so the availability is not always 100%.

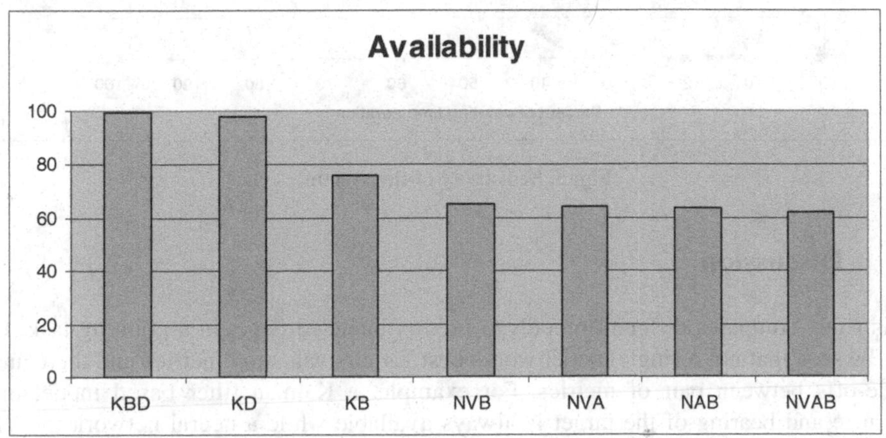

Fig. 3. Availability of different systems

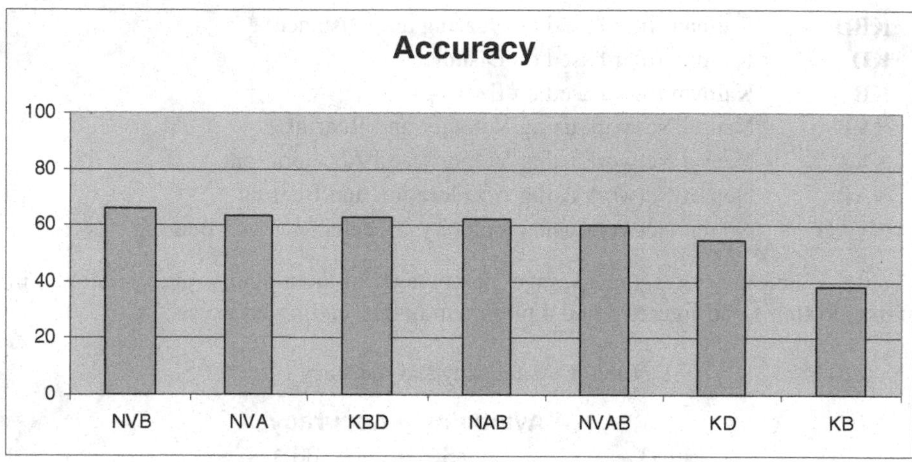

Fig. 4. Accuracy of different systems

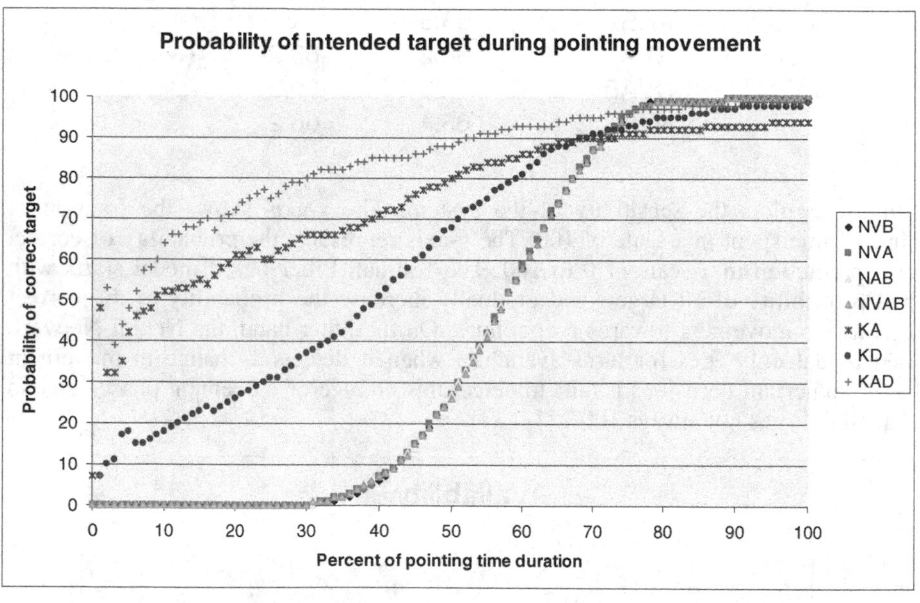

Fig. 5. Sensitivity of the system

5 Discussion

This study compares different models to predict intended target in a pointing task. It can be seen that not a single model works best for all evaluation metrics and there are trade-offs between pair of metrics. For example, a Kalman filter based model on distance and bearing of the target is always available while a neural network model based on velocity and bearing of movement is most accurate. The maximum accuracy we obtained is 63.20% as we have used a multiple distractor task with targets as small as 20 pixels in a circle even with a radius of 100 pixels. The accuracy is not great but not also worse than previous works. The sensitivity chart shows different prediction patterns between Kalman Filter and Neural Network based models. The probability of correct target increases gradually in case of Kalman Filter, while it takes a S-based pattern, similar to the activation function, in case of a Neural Network based model. Both KAD and NVB predict the correct target with more than 90% certainty within 70% of total pointing time.

6 Contributions

New Models. This paper proposes two new types of models based on Backpropagation Neural Network and Kalman Filters and compares them. In both cases we got more than 60% accuracy which is not worse than previous results [17]. For example, Ziebart's models [17] achieve more than 50% accuracy after crossing 70% of pointing time while our models have more than 70% accuracy at similar stage. Furthermore, lack of evaluation metrics makes it difficult to compare different methods in target prediction and most models do not publish results in as much detail as us. Additionally, most research on target prediction did not include users with age related or physical impairment.

Evaluation Metrics. This paper proposes a set of metrics to evaluate intent recognition algorithms and compare their performance for different applications. An application developer may choose an appropriate model or algorithm based on these evaluation metrics.
For example, if an interface has closely spaced small icons like a toolbar or menu, we should prefer an algorithm with high accuracy, perhaps compromising availability. However for large screen display or pointing through gesture, we should prefer an algorithm with high availability as otherwise users may not feel any effect of intent recognition. Choice of an algorithm also depends on the effect of intent recognition algorithms. For example, after predicting a target, we may simply highlight the target or proactively move the cursor on the button. In the first case, an algorithm with high availability is good as users can always get a feedback based on their movement. However in the case of proactive cursor movement based on intent recognition, we should choose an algorithm with high accuracy, otherwise a wrong prediction will frustrate users and in fact may increase pointing time.

Best Set of Features. We found that in case of Neural Network model, a model based on velocity and bearing of movement worked best, while for Kalman Filter model, bearing and distance to target was most indicative of intended target. Considering both model, we may conclude that bearing, velocity and remaining target distances are most indicative of intended target and future researchers can exploit this information to propose new models and algorithms.

7 Conclusions

This paper proposes a framework to evaluate intent recognition algorithms and then proposes two new types of algorithms based on Neural Network and Kalman filters. We compared the performances of the algorithms for people with a wide range of abilities and obtained more than 60% accuracy of prediction. This system will be particularly useful for Information Visualization and Assistive Interfaces if combined with screen magnification or zooming systems. For example, certain part of a graph or an image thumbnail among a set of other images can be zoomed in by detecting users' intent. In existing image search interfaces at Google Chrome or Internet Explorer, an image is only zoomed in after the user clicks on it, while using the present system, any thumbnail can be zoomed in without needing an explicit selection by user. This automatic zooming will be useful for image analysis (e.g. disease analysis from image of crops, analysis of CCTV image during a disaster etc). This technique will be more useful in assistive interfaces where the zooming level of the screen magnifier can be adjusted based on the intention of the user.

Acknowledgement. The work is funded by the EPSRC IUATC Grant.

References

1. Asano, T., Sharlin, E., Kitamura, Y., Takashima, K., Kishino, F.: Predictive Interaction Using the Delphian Desktop. In: Proceedings of the 186th Annual ACM Smposium on User Interface Software and Technology (UIST 2005), New York, pp. 133–141 (2005)
2. Bar-Shalom, Y., Li, X.R.: Estimation and Tracking; Principles, Techniques and Software. Artech House (1993)
3. Biswas, P., Langdon, P.: Developing multimodal adaptation algorithm for mobility impaired users by evaluating their hand strength. International Journal of Human-Computer Interaction 28(9) (2012) ISSN: 1044-7318
4. Fitts, P.M.: The Information Capacity of The Human Motor System in Controlling The Amplitude of Movement. Journal of Experimental Psychology 47, 381–391 (1954)
5. Godsill, S., Vermaak, J.: Models and Algorithms For Tracking Using Variable Dimension Particle Filters. In: International Conference on Acoustics, Speech and Signal Processing (2004)
6. Hwang, F., Keates, S., Langdon, P., Clarkson, P.J.: A submovement analysis of cursor trajectories. Behaviour and Information Technology 3(24), 205–217 (2005)
7. Keates, S., Hwang, F., Langdon, P., Clarkson, P.J., Robinson, P.: Cursor measures for motion-impaired computer users. In: Proceedings of the Fifth International ACM Conference on Assistive Technologies – ASSETS, New York, pp. 135–142 (2002)
8. Lane, D.M., Peres, S.C., Sándor, A., Napier, H.A.: A Process for Anticipating and Executing Icon Selection in Graphical User Interfaces. International Journal of Human Computer Interaction 19(2), 243–254 (2005)
9. Langdon, P.M., Godsill, S., Clarkson, P.J.: Statistical Estimation of User's Interactions from Motion Impaired Cursor Use Data. In: 6th International Conference on Disability, Virtual Reality and Associated Technologies (ICDVRAT 2006), Esbjerg, Denmark (2006)

10. Lank, E., Cheng, Y.N., Ruiz, J.: Endpoint prediction using motion kinematics. In: Proceedings of the SIGCHI Conference on Human Factors in Computing Systems (CHI 2007), New York, NY, USA, pp. 637–646 (2007)
11. Li, X.R., Jilkov, V.P.: Survey of Maneuvering Target Tracking.Part I: Dynamic Models. IEEE Transactions on Aerospace and Electronic Systems 39(4) (2003)
12. McGuffin, M.J., Balakrishnan, R.: Fitts' law and expanding targets: Experimental studies and designs for user interfaces. ACM Transactions Computer-Human Interaction 4(12), 388–422 (2005)
13. Murata, A.: Improvement of Pointing Time by Predicting Targets in Pointing With a PC Mouse. International Journal of Human-Computer Interaction 10(1), 23–32
14. Van Oirschot, H.K., Houtsma, A.J.M.: Cursor Trajectory Analysis. In: Brewster, S., Murray-Smith, R. (eds.) Haptic HCI 2000. LNCS, vol. 2058, pp. 127–134. Springer, Heidelberg (2001)
15. Wobbrock, J.O., Fogarty, J., Liu, S., Kimuro, S., Harada, S.: The Angle Mouse: Target-Agnostlc Dynamic Gain Adjustment Based on Angular Deviation. In: Proceedings of the 27th International Conference on Human Factors in Computing Systems (CHI 2009), New York, pp. 1401–1410 (2009)
16. Woodworth, R.S.: The accuracy of voluntary movement. Psychological Review 3, 1–119 (1899)
17. Ziebart, B., Dey, A., Bagnell, J.A.: Probabilistic pointing target prediction via inverse optimal control. In: Proceedings of the 2012 ACM International Conference on Intelligent User Interfaces (IUI 2012), New York, pp. 1–10 (2012)

HCI Empowered Literature Mining for Cross-Domain Knowledge Discovery

Matjaž Juršič[1,2], Bojan Cestnik[1,3], Tanja Urbančič[4,1], and Nada Lavrač[1,4]

[1] Jožef Stefan Institute, Ljubljana, Slovenia
[2] International Postgraduate School Jožef Stefan, Ljubljana, Slovenia
[3] Temida d.o.o., Ljubljana, Slovenia
[4] University of Nova Gorica, Nova Gorica, Slovenia
{matjaz.jursic,bojan.cestnik,tanja.urbancic,nada.lavrac}@ijs.si

Abstract. This paper presents an exploration engine for text mining and cross-context link discovery, implemented as a web application with a user-friendly interface. The system supports experts in advanced document exploration by facilitating document retrieval, analysis and visualization. It enables document retrieval from public databases like PubMed, as well as by querying the web, followed by document cleaning and filtering through several filtering criteria. Document analysis includes document presentation in terms of statistical and similarity-based properties and topic ontology construction through document clustering, while the distinguishing feature of the presented system is its powerful cross-context and cross-domain document exploration facility through bridging term discovery aimed at finding potential cross-domain linking terms. Term ranking based on the developed ensemble heuristic enables the expert to focus on cross-context terms with greater potential for cross-context link discovery. Additionally, the system supports the expert in finding relevant documents and terms by providing customizable document visualization, a color-based domain separation scheme and highlighted top-ranked bisociative terms.

Keywords: literature mining, knowledge discovery, cross-context linking terms, creativity support tools, human-computer interaction.

1 Introduction

Understanding complex phenomena and solving difficult problems often requires knowledge from different domains to be combined and cross-domain associations to be taken into account. These kinds of context-crossing associations are called bisociations [1] and are often needed for creative, innovative discoveries. Typically, this is a challenging task due to a trend of over-specialization in research and development, resulting in islands of deep, but relatively isolated knowledge. Scientific literature all too often remains closed and cited only in professional sub-communities. In addition, the information that is related across different contexts is difficult to identify with the associative approach, like the standard association rule learning approach [2] known from data mining and machine learning literature. Therefore, the ability of literature

A. Holzinger and G. Pasi (Eds.): HCI-KDD 2013, LNCS 7947, pp. 124–135, 2013.

mining methods and software tools for supporting the experts in their knowledge discovery process, especially in searching for yet unexplored connections between different domains, is becoming increasingly important.

The task of cross-domain literature mining has already been addressed by Swanson [3], [4], proving that bibliographic databases such as MEDLINE could serve as a rich source of hidden relations between concepts. By studying two separate literatures, the literature on migraine headache and the articles on magnesium, he discovered "Eleven neglected connections", identifying eleven linking concepts [3]. Laboratory and clinical investigations started after the publication of the Swanson's convincing evidence and have confirmed that magnesium deficiency can cause migraine headaches. This well-known example has become a golden standard in the literature mining field and has been used as a benchmark in several studies, including [5-8].

Literature mining supported discovery was successfully applied to problems, such as associations between genes and diseases [9], diseases and chemicals [10], and others. Smalheiser and Swanson [11] developed a web platform designed to assist the user in literature based discovery, which is in terms of detecting interesting cross-domain terms similar to our system. Holzinger et al. [12] describe several quality-oriented web-based tools for the analysis of biomedical literature, which include the analysis of terms (biomedical entities such as disease, drugs, genes, proteins and organs) and provide concepts associated with the given term.

Cross-domain literature mining is closely related to bisociative knowledge discovery as defined by Dubltzky et al. [13]. Assuming two domains of interest, a crucial step in cross-domain knowledge discovery is the identification of interesting bridging terms (B-terms), appearing in both literatures, which carry the potential of revealing the links connecting the two domains.

In this paper we present an online system CrossBee which helps the experts when searching for hidden links that connect two seemingly unrelated domains. As such, it supports creative discovery of cross-domain hypotheses, and could be viewed as a creativity support tool (CST). While CrossBee has been previously described [14], [15], these papers have not focused on its visual interface empowering the users in the bridging term discovery process, but have focused on its methodology and the heuristics included in the ensemble-based term ranking according to terms' bisociation potential, indicating the potential to act as bridging terms among two selected domains.

Creativity support tools are closely related to the field of human-computer interaction (HCI), as stated by Resnick et al. [16] when summarizing the aims of designing CSTs: "Our goal is to develop improved software and user interfaces that empower users to be not only more productive, but more innovative." Schneiderman [17], [18] provides a structured set of design principles for CSTs, which we follow in our implementation and use them for evaluation:

- *Support exploration.* To be successful at discovery and innovation, users should have access to improved search services providing rich mechanism for organizing search results by ranking, clustering, and partitioning with ample tools for annotation, tagging, and marking.
- *Enable collaboration.* While the actual discovery moments in innovation can be very personal, the processes that lead to them are often highly collaborative.

- *Provide rich history-keeping.* The benefits of rich history-keeping are that users have a record of which alternatives they have tried, they can compare the many alternatives, and they can go back to earlier alternatives to make modifications.
- *Design with low thresholds, high ceilings, and wide walls.* CST should have steep learning curve for novices (low threshold), yet provide sophisticated functionality that experts need (high ceilings), and also deliver a wide range of supplementary services to choose from (wide walls).

The main novelty of the presented system is ensemble-based ranking of terms according to their bisociative potential of contributing to novel cross-domain discoveries. This facility, together with numerous other content analysis and visualization options, distinguishes it as a powerful, user-friendly text analysis tool for cross-domain knowledge discovery support.

In the next section we present the main system functionality and a brief overview of the methodology, implemented in a contemporary workflow execution environment ClowdFlows. Section 3 presents a typical usage scenario and continues with some other system functionalities important for efficient human computer interaction in cross-context link discovery. In Section 4 we describe visual document clustering as implemented in our system. In Section 5 we summarize the most important features of the presented system and suggest some further work directions.

2 Main System Functionality and Methodology Overview

In cross-domain knowledge discovery, estimating which of the terms have a high potential for interesting discoveries is a challenging research question. It is especially important for cross-context scientific discovery such as understanding complex medi

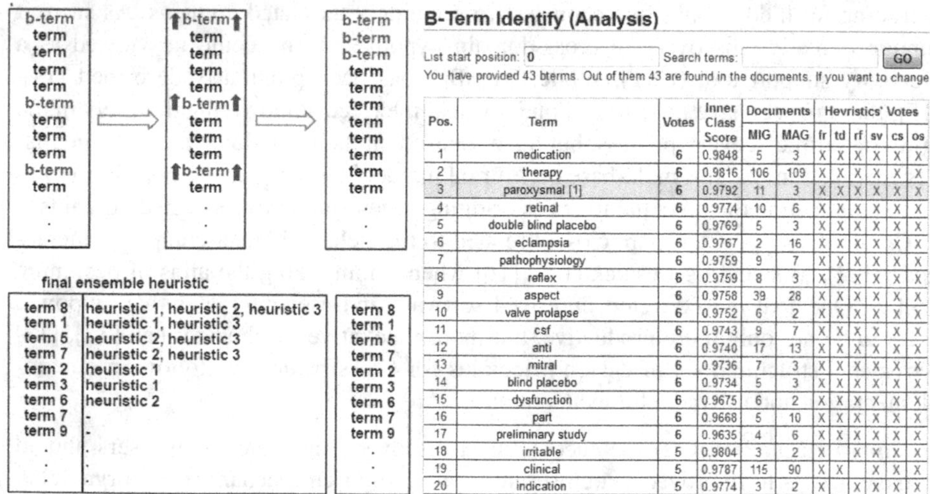

Fig. 1. Term ranking approach (illustrated at the left) and the actual CrossBee ensemble heuristic ranking page (at the right) indicating by a cross (X) which elementary heuristics have identified the term as potential B-term

cal phenomena or finding new drugs for yet not fully understood illnesses. Given this motivation, the main functionality of CrossBee is bridging term (B-term) discovery, implemented through ensemble-based term ranking, where an ensemble heuristic composed of six elementary heuristics was constructed for term evaluation. The ensemble-based ranking methodology, presented in more detail by Juršič et al. [14], [15], is illustrated in Fig. 1, showing the methodology of term ranking on the left and the ensemble ranked term list on the right side of the figure. The presented ranked list is the actual output produced by our system using the gold standard dataset in literature mining—the migraine-magnesium dataset [3].

We use workflow diagrams to present the cross-domain knowledge discovery methodology implemented in CrossBee. While the presented workflow diagrams are here used only as means to describe a conceptual pipeline of natural language processing (NLP) modules, the pipeline actually represent an executable workflow, implemented in the online cloud based workflow composition environment ClowdFlows [19].

The top-most level overview of the methodology, shown in Fig. 2, consists of the following steps: document acquisition, document preprocessing, outlier document detection, heuristics specification, candidate B-term extraction, heuristic terms scores calculation, and visualization and exploration. An additional ingredient shown in Fig. 2—methodology evaluation—is not directly part of the methodology, however it is an important step of the methodology development. A procedural explanation of the workflow from Fig. 2 is presented below.

1. *Document acquisition* is the first step of the methodology. Its goal is to acquire documents of the two domains, label them with domain labels and pack both domains together into the annotated document corpus (ADC) format.
2. The *document preprocessing* step is responsible for applying standard text preprocessing to the document corpus. The main parts are tokenization, stopwords labeling and token stemming or lemmatization.
3. The *outlier document detection* step is used for detecting outlier documents that are needed by subsequent heuristic specification. The output is a list (or multiple lists in case when many outlier detection methods are used) of outlier documents.
4. The *heuristic specification* step serves as a highly detailed specification of the heuristics to be used for B-term ranking. The user specifies one or more heuristics, which are later applied to evaluate the B-term candidates. Furthermore, each individual heuristic can be hierarchically composed of other heuristics; therefore an arbitrary complex list of heuristics can be composed in this step.

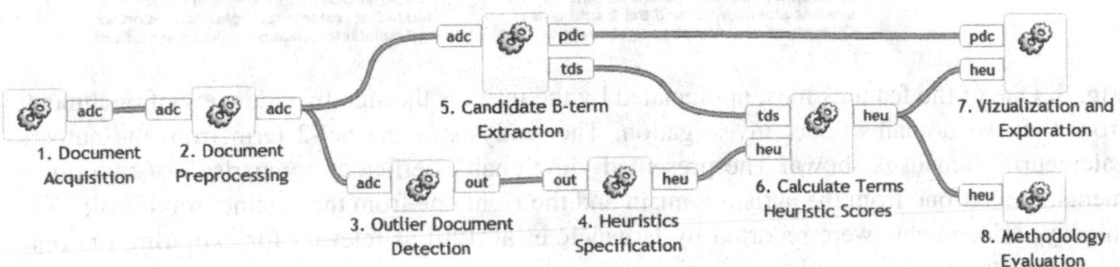

Fig. 2. Methodological steps of the cross-domain literature mining process

5. The *candidate B-term extraction* step takes care of extracting the terms which are later scored by the specified heuristics. There are various parameters which control which kind of terms are selected from the documents (e.g., the maximal number of tokens to be joined together as a term, minimal term corpus frequency, and similar). The first output is a list of all candidate B-terms (term data set TDS) along with their vector representations. The second output is a parsed document corpus (PDC) which includes information about the input documents from ADC as well as the exact data how each document was parsed. This data is needed by the CrossBee web application when displaying the documents since it needs to be able to exactly locate specific words inside a document, e.g., to color or emphasize them.
6. *Heuristic calculation* is methodologically the most important step. It takes the list of extracted B-term candidates and the list of specified heuristics and calculates a heuristic score for each candidate term for each heuristic. The output is structurally still a list of heuristics, however now each of them contains a bisociation score for each candidate B-term.
7. *Visualization and exploration* is the final step of the methodology. It has three main functionalities. It can either take the heuristically scored terms, rank the terms, and output the terms in the form of a table, or it can take the heuristically scored terms along with the parsed document corpus and send both to the CrossBee

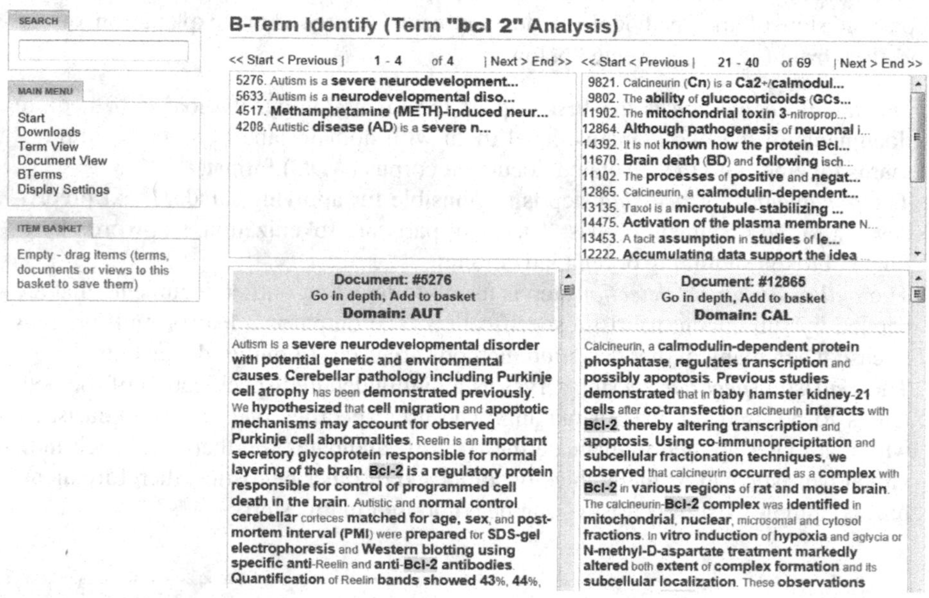

Fig. 3. One of the features most appreciated by the users is the side-by-side view of documents from the two domains under investigation. The analysis of the bcl-2 term from the autism-calcineurin domain is shown. The presented view enables efficient comparison of two documents, the left one from the autism domain and the right one from the calcineurin domain. The displayed documents were reported by Urbančič et al. [20] as relevant for exploring the relationship between autism and calcineurin.

web application for advanced visualization and exploration. Besides improved bridging concept identification and ranking, CrossBee also provides various content presentations which further speed up the process of bisociation exploration. These presentations include side-by-side document inspection (see Fig. 3), emphasizing of interesting text fragments, and uncovering similar documents. Finally, document clustering can be used for domain exploration (see TopicCircle visualization in Fig. 4).

8. An additional *methodology evaluation* step was introduced during the development of the methodology. Its purpose is to calculate and visualize various metrics that were used to assess the quality of the methodology. Requirement to use these facilities is to have the actual B-terms as golden standard B-terms available for the domains under investigation. The methodology was actually evaluated on two problems: the standard migraine-magnesium problem well-known in literature mining, and a more recent autism-calcineurin literature mining problem.

The CrossBee system has already been successfully applied to complex domains and resulted in finding interesting cross-domain links, when replicating the results of cross-domain migraine-magnesium literature mining by Swanson [3] and replicating the results in the area of autism by Urbančič et al. [20] and Petrič et al. [21].

We are mostly interested in the CrossBee heuristics quality from the end user's perspective. Such evaluation should enable the user to estimate how many B-terms can be found among the first 5, 20, 100, 500 and 2000 terms on the ranked list of terms produced by a heuristic [14]. The ensemble heuristic, performing ensemble voting of six elementary heuristic[1], resulted in very favorable results in the training domain (migraine-magnesium domain pair), where one B-term among the first 5 terms, one B-term—no additional B-terms—among the first 20 terms, 6 B-terms—5 additional—among the first 100 terms, 22 B-terms—16 additional—among first 500 terms and all the 43 B-terms—21 additional—among the first 2000 terms. Thus, e.g., if the expert limits himself to inspect only the first 100 terms, he will find 6 B-terms in the ensemble ranked term list. These results confirm that the ensemble is among the best performing heuristics also from the user's perspective. Even though a strict comparison depends also on the threshold of how many terms an expert is willing to inspect, the ensemble is always among the best.

In the autism-calcineurin domain pair, the ensemble finds one B-term among 20 ranked terms, 2 among 100 and 3 among 500 ranked terms. At a first sight, this may seem a bad performance, but, note that there are 78,805 candidate terms which the heuristics have to rank. The evidence of the quality of the ensemble can be understood if we compare it to a simple baseline heuristic, which represents the performance achievable using random sorting of terms which appear in both domains. The baseline heuristic discovers in average only approximately 0.33 B-terms before position 2000 in the ranked list while the ensemble discovers 5; not to mention the shorter term lists where the ensemble is relatively even better compared to the baseline heuristic.

[1] The voting mechanism and the exact description of the heuristics are out of the scope of this paper; more information on the baseline, elementary and ensemble heuristics is provided by Juršič et al. [14].

The above methodology evaluation provides evidence that the users empowered with the CrossBee functionality of term ranking and visualization are able to perform the crucial actions in cross-domain discovery faster than with conventional text mining tools.

3 Typical Use Case Scenario

This section presents a typical usage scenario, illustrated with an example from the autism domain, where the aim was to find new links with calcineurin, shown in Fig. 3.

The user starts a new session by selecting two sets of documents of interest and by regulating the parameters of the system. The required input is either a PubMed query or a file with documents from the two domains, where each line contains a document with exactly three tab-separated entries: (a) document identifier, (b) domain acronym, and (c) the document text. The user can also specify the exact preprocessing options, the elementary heuristics to be used in the ensemble, outlier documents identified by external outlier detection software, the already known bisociative terms (B-terms), and others. Next, the system starts actual text preprocessing, computing the elementary heuristics, the ensemble bisociation scores and term ranking. When presented with a ranked list of B-term candidates, the user browses through the list and chooses the term(s) he believes to be promising B-terms, i.e. terms for finding meaningful connections between the two domains. At this point, the user can inspect the actual appearances of the selected term in both domains, using the efficient side-by-side document inspection.

Other functionalities of our system support the expert in advanced document exploration supporting document retrieval, analysis and visualization. The system enables document retrieval from public databases like PubMed, as well as by querying the web, followed by document cleaning and filtering through several filtering criteria. Document analysis includes document presentation in terms of statistical and similarity-based properties, topic ontology construction through document clustering, and document visualization along with user interface customization which additionally supports the expert in finding relevant documents and terms of a color-based domains separation scheme and high-lighted top-ranked bisociative terms.

A rich set of functionalities and content presentations turn our system into a user-friendly tool which enables the user not only to spot but also to efficiently investigate cross-domain links pointed out by our ensemble-based ranking methodology. Document focused exploration empowers the user to filter and order the documents by various criteria. Detailed document view provides a more detailed presentation of a single document including various term statistics. Methodology performance analysis supports the evaluation of the methodology by providing various data which can be used to measure the quality of the results, e.g., data for plotting the ROC curves. High-ranked term emphasis marks the terms according to their bisociation score calculated by the ensemble heuristic. When using this feature all high-ranked terms are emphasized throughout the whole application thus making them easier to spot (see different font sizes in Fig. 3). B-term emphasis marks the terms defined as B-terms by the user (yellow terms in Fig. 3). Domain separation is a simple but effective option which colors all the documents from the same domain with the same color, making an obvious distinction between the documents from the two domains (different colors in Fig. 3). User interface customization enables the user to decrease or increase the intensity of the following features: high-ranked term emphasis, B-term emphasis and domain separation.

Note that the modular design of the system enabling new functionalities, in addition to the above described CrossBee functionalities; add to the fulfillment of the wide wall criterion, discussed when describing the TopicCircle document exploration facility.

4 Visual Document Clustering

Our system has the facility of clustering documents according to their similarity. Similarity between documents can be determined by calculating the cosine of the angle between two documents represented as *Bag of Words* (*BoW*) vectors, where the Bag of Words approach [22] is used for representing a collection of words from text documents disregarding grammar and word order. The BoW approach is used together with the standard *Tf·Idf* (term frequency inverse document frequency) weighting method. BoW representation of text documents is employed for extracting words with similar meaning. In the BoW vector space representation, each word from the document vocabulary stands for one dimension of the multidimensional space of text documents. Corpus of text documents is then visualized in form of Tf·Idf vectors, where each document is encoded as a feature vector with word frequencies as elements.[2]

The *cosine similarity*[3] measure, commonly used in information retrieval and text mining to determine the semantic closeness of two documents represented in the BoW vector space model, is used to cluster the documents. Cosine similarity values fall within the [0, 1] interval. Value 0 represents extreme dissimilarity, where two documents (a given document and the centroid vector of its cluster) share no common words, while 1 represents the similarity between two exactly identical documents in the BoW representation. For clustering, the standard k-means clustering algorithm is used.

The result of interactive top-down document clustering of the migraine-magnesium documents are presented on the left hand side of Fig. 4. At the first level, all the documents are split into one of the two domains: migraine and magnesium (top screenshot on right hand side of Fig. 4). At level 2, guided by the user, each of the two domains is further split into k sub-clusters, according to the user-selected k parameter. Each of the clusters is described by its most meaningful keywords (written inside each

---

[2] Elements of vectors are weighted with the Tf·Idf weights as follows: The i-th element of the vector containing frequency of the i-th word is multiplied with $Idf_i = \log(N/df_i)$, where N represents the total number of documents and df_i is document frequency of the i-th word (i.e., the number of documents from the whole corpus in which the i-th word appears).

[3] The cosine similarity is the dot product of BoW vectors, normalized by the length of the vectors: $\mathrm{CosSim}(vec_x, vec_y) = \mathrm{DotProduct}(vec_x, vec_y)/|vec_x| \cdot |vec_y|$. In the typical case, when the vectors are already normalized, the cosine similarity is identical to the dot product.

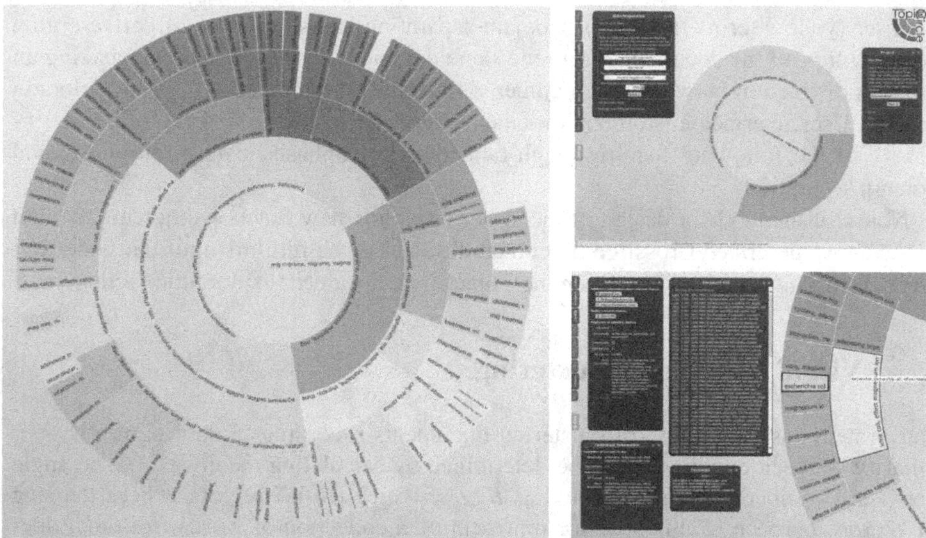

Fig. 4. The basic hierarchical cluster visualization is shown on the left along with two additional examples of screenshots of the application's data clustering functionalities on the right. The top-right screenshot presents data preprocessing and first splitting of documents to two sets (migraine and magnesium); the bottom-right screenshot presents zoomed-in view with a cluster selection, retrieved cluster documents and other contextual information.

cluster and displayed in detail when user moves the mouse over it). The bottom part of the right hand side of Fig. 4—the detailed information about a single document—shows one among many data representations, which support rich exploration, low threshold and wide walls as described by Schneiderman [17], [18].

The advantages of our new document cluster visualization, e.g., if compared to a well-established semi-automated cluster construction and visualization tool OntoGen [23], are that (a) the tool is not needed to download, (b) providing much more user friendly environment with especially low threshold for novice users to start exploring their data, and (c) providing wide walls with many different perspectives to the data—e.g. size of the cluster may be based on the number of sub clusters, included documents, or some other calculated property like similarity of the cluster to some query. Similarly is true for the color which may be used in a number of ways to help the user getting a better overview of the data. Fig. 5 presents an approach of using these properties (in this example we indeed use color) to better visualize cross-domain links which may be present in the data. When the user concentrates on a document in one domain he gets a suggestion of the similar clusters in both domains since all the similar clusters are emphasized with darker color. However, this is only one among many usages of the presented visualization for displaying additional rich cross-context aware information.

In terms of cross-context knowledge discovery, the top-down clustering approach enables the user to discover similar document sets within each domain, thus identifying potentially interesting domain subsets for further cross-domain link discovery

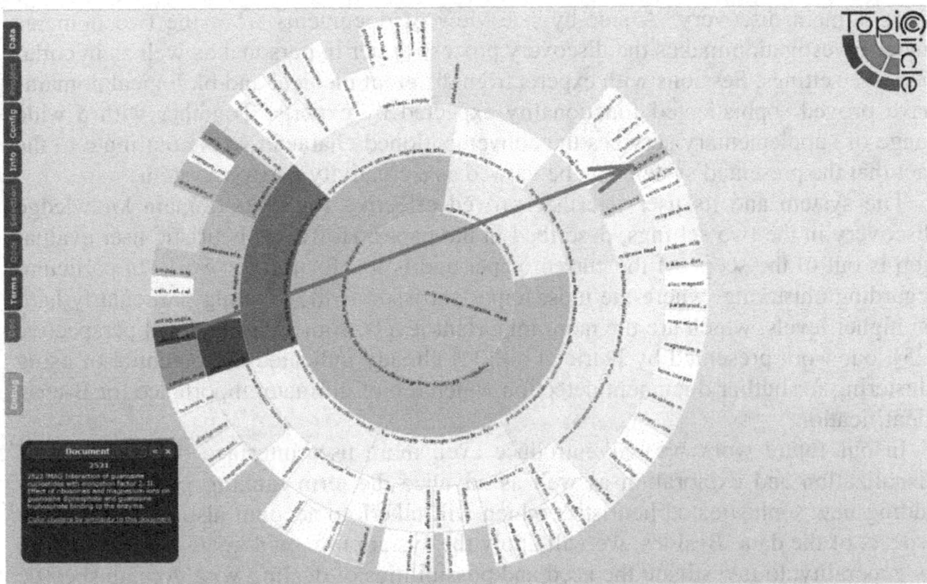

Fig. 5. Cluster colors can be used to show various information—in this case the cluster's similarity to a single selected document. The arrow shows similar clusters in two different domains, which can potentially indicate to a novel bisociative link between the two domains.

using our system. Note that in the example presented in Fig. 4, clustering has been performed for each domain separately, therefore not fully demonstrating the potential for cross-domain knowledge discovery. In our past work, however, we have shown that when using clustering on a document set joining documents from both domains, the differences between the clusters identified using similarity measures using 2-means clustering and the document clusters identified through the initial document labeling by class labels of the two domains can fruitfully serve to identify outlier documents which include an increased number of B-terms and thus a high potential for B-term identification [24].

5 Conclusion

The paper presents a system for cross-context literature mining which supports experts in advanced document exploration by facilitating document retrieval, analysis and visualization. The system has been designed as a creativity support tool, helping experts in uncovering not yet discovered relations between different domains from large textual databases. As this is a very time-consuming process in which estimating linking potential of particular terms as well as efficient selection and presentation of pairs of documents to be inspected is very important, user interface has been designed very carefully. It supports experts by features such as visual document clustering, color-based domain separation scheme, highlighted top-ranked bisociative terms and other functionalities, resulting in improved search capabilities needed for cross-domain discovery. A side-by-side view of documents from the two domains under investigation makes the discovery process easier in personal as well as in collaborative settings. Sessions with experts from different medical and biological domains have proved sophisticated functionality expected by experts. Together with a wide range of supplementary services the abovementioned characteristics contribute to the fact that the presented system can be viewed as a creativity support system.

The system and its user interface proved effective for cross-domain knowledge discovery in the two settings, described in the paper. However, heuristic user evaluation is out of the scope of the current paper and is left for further work. In particular regarding clustering, where the most important issue is the labeling particularly hard on higher levels, which are the more important levels from a navigational perspective [25], our work presented by Petrič et al. [24] already indicated the potential of using clustering for outlier document detection which are of ultimate importance for B-term identification.

In our future work we will introduce even more user interface options for data visualization and exploration as well as advance the term ranking methodology by adding new sophisticated heuristics which will take into account also more semantic aspects of the data. Besides, we will apply the system to new domain pairs to exhibit its generality, to investigate the need and possibilities of dealing with domain specific background knowledge, and last but not least to assist researchers in different disciplines on their way towards new scientific discoveries.

Acknowledgements. This work was supported by the Slovenian Research Agency and the FP7 European Commission projects MUSE (grant no. 296703) and FIRST (grant no. 257928).

References

1. Koestler, A.: The act of creation. MacMillan Company, New York (1964)
2. Agrawal, R., Mannila, H., Srikant, R., Toivonen, H., Verkamo, A.I.: Fast discovery of association rules. Advances in Knowledge Discovery and Data Mining, 307–328 (1996)
3. Swanson, D.R.: Migraine and magnesium: Eleven neglected connections. Perspectives in Biology and Medicine 31, 526–557 (1988)
4. Swanson, D.R.: Medical literature as a potential source of new knowledge. Bull. Med. Libr. Assoc. 78/1, 29–37 (1990)
5. Lindsay, R.K., Gordon, M.D.: Literature-based discovery by lexical statistics. Journal of the American Society for Information Science and Technology 50/7, 574–587 (1999)
6. Weeber, M., Vos, R., Klein, H., de Jong-van den Berg, L.T.W.: Using concepts in literature-based discovery: Simulating Swanson's Raynaud–fish oil and migraine–magnesium discoveries. J. Am. Soc. Inf. Sci. Tech. 52/7, 548–557 (2001)
7. Srinivasan, P.: Text Mining: Generating Hypotheses from MEDLINE. Journal of the American Society for Information Science and Technology 55/5, 396–413 (2004)
8. Urbančič, T., Petrič, I., Cestnik, B.: RaJoLink: A Method for Finding Seeds of Future Discoveries in Nowadays Literature. In: Rauch, J., Raś, Z.W., Berka, P., Elomaa, T. (eds.) ISMIS 2009. LNCS, vol. 5722, pp. 129–138. Springer, Heidelberg (2009)
9. Hristovski, D., Peterlin, B., Mitchell, J.A., Humphrey, S.M.: Using literature-based discovery to identify disease candidate genes. Int. J. Med. Inform. 74/2–4, 289–298 (2005)

10. Yetisgen-Yildiz, M., Pratt, W.: Using statistical and knowledge-based approaches for literature-based discovery. J. Biomed. Inform. 39/6, 600–611 (2006)
11. Smalheiser, N.R., Swanson, D.R.: Using ARROWSMITH: a computer-assisted approach to formulating and assessing scientific hypotheses. Computer Methods and Programs in Biomedicine 57/3, 149–153 (1998)
12. Holzinger, A., Yildirim, P., Geier, M., Simonic, K.-M.: Quality-based knowledge discovery from medical text on the Web Example of computational methods in Web intelligence. In: Pasi, G., Bordogna, G., Jain, L.C. (eds.) Qual. Issues in the Management of Web Information. ISRL, vol. 50, pp. 145–158. Springer, Heidelberg (2013)
13. Dubitzky, W., Kötter, T., Schmidt, O., Berthold, M.R.: Towards creative information exploration based on Koestler's concept of bisociation. In: Berthold, M.R. (ed.) Bisociative Knowledge Discovery. LNCS, vol. 7250, pp. 11–32. Springer, Heidelberg (2012)
14. Juršič, M., Cestnik, B., Urbančič, T., Lavrač, N.: Bisociative Literature Mining by Ensemble Heuristics. In: Berthold, M.R. (ed.) Bisociative Knowledge Discovery. LNCS, vol. 7250, pp. 338–358. Springer, Heidelberg (2012)
15. Juršič, M., Cestnik, B., Urbančič, T., Lavrač, N.: Cross-domain literature mining: Finding bridging concepts with CrossBee. In: Proceedings of the 3rd International Conference on Computational Creativity (2012)
16. Resnick, M., Myers, B., Nakakoji, K., Shneiderman, B., Pausch, R., Selker, T., Eisenberg, M.: Design Principles for Tools to Support Creative Thinking. In: Proceedings of the NSF Workshop on Creativity Support Tools, pp. 25–36 (2005)
17. Shneiderman, B.: Creativity support tools: accelerating discovery and innovation. Communications of the ACM 50/12, 20–32 (2007)
18. Shneiderman, B.: Creativity Support Tools: A Grand Challenge for HCI Researchers. In: Engineering the User Interface, pp. 1–9. Springer, London (2009)
19. Kranjc, J., Podpečan, V., Lavrač, N.: ClowdFlows: A cloud cased scientific workflow platform. In: Flach, P.A., De Bie, T., Cristianini, N. (eds.) ECML PKDD 2012, Part II. LNCS, vol. 7524, pp. 816–819. Springer, Heidelberg (2012)
20. Urbančič, T., Petrič, I., Cestnik, B., Macedoni-Lukšič, M.: Literature Mining: Towards Better Understanding of Autism. In: Bellazzi, R., Abu-Hanna, A., Hunter, J. (eds.) AIME 2007. LNCS (LNAI), vol. 4594, pp. 217–226. Springer, Heidelberg (2007)
21. Petrič, I., Urbančič, T., Cestnik, B., Macedoni-Lukšič, M.: Literature mining method RaJoLink for uncovering relations between biomedical concepts. Journal of Biomedical Informatics 42/2, 219–227 (2009)
22. Salton, G., Buckley, C.: Term weighting approaches in automatic text retrieval. Inf. Process Manag. 24/5, 513–523 (1988)
23. Fortuna, B., Grobelnik, M., Mladenić, D.: Semi-automatic Data-driven Ontology Construction System. In: Proceedings of the 9th International Multiconference Information Society, pp. 212–220 (2006)
24. Petrič, I., Cestnik, B., Lavrač, N., Urbančič, T.: Outlier Detection in Cross-Context Link Discovery for Creative Literature Mining. The Computer Journal 55/1, 47–61 (2012)
25. Muhr, M., Kern, R., Granitzer, M.: Analysis of structural relationships for hierarchical cluster labelling. In: Proceeding of the 33rd International ACM SIGIR Conference on Research and Development in Information Retrieval, pp. 178–185 (2010)

An Interactive Course Analyzer for Improving Learning Styles Support Level

Moushir M. El-Bishouty[1,2], Kevin Saito[1], Ting-Wen Chang[1], Kinshuk[1], and Sabine Graf[1]

[1] Athabasca University, Canada
[2] City for Scientific Research and Technological Applications, Egypt
{moushir.elbishouty,tingwenchang,kinshuk,sabineg}@athabascau.ca, kmsaito@ucalgary.ca

Abstract. Learning management systems (LMSs) contain tons of existing courses but very little attention is paid to how well these courses actually support learners. In online learning, teachers build courses according to their teaching methods that may not fit with students with different learning styles. The harmony between the learning styles that a course supports and the actual learning styles of students can help to magnify the efficiency of the learning process. This paper presents a mechanism for analyzing existing course contents in learning management systems and an interactive tool for allowing teachers to be aware of their course support level for different learning styles of students based on the Felder and Silverman's learning style model. This tool visualizes the suitability of a course for students' learning styles and helps teachers to improve the support level of their courses for diverse learning styles.

Keywords: Interactive course analyzer, visualization, learning styles, learning analytics, learning management systems.

1 Introduction

The widespread introduction of learning management systems (LMSs) – such as Blackboard and Moodle has meant that educational institutions deal with increasingly large sets of data. Each day, their systems amass ever increasing amounts of interaction data, personal data, systems information and academic information [1][2]. A combination of the availability of big datasets, the emergence of online learning on a large scale, and political concerns about educational standards has prompted the development of learning analytics research field. According to the call for papers of the 1st International Conference on Learning Analytics and Knowledge (LAK 2011), learning analytics is defined as the measurement, collection, analysis and reporting of data about learners and their contexts, for purposes of understanding and optimizing learning and the environments in which it occurs. Learning analytics is one of the fastest growing areas of technology enhanced learning (TEL) research [3].

A. Holzinger and G. Pasi (Eds.): HCI-KDD 2013, LNCS 7947, pp. 136–147, 2013.

The rise of big data in education mirrors the increase in take up of online learning. Currently, LMSs contain tons of existing courses but very little attention is paid to how well these courses actually support learners. In online learning, teachers build courses according to their teaching methods. Teaching methods vary. Some instructors lecture, others demonstrate or discuss; some focus on principles and others on applications; some emphasize memory and others understanding [4]. On the other hand, learners have different backgrounds, motivation and preferences in their own learning processes and web-based systems that ignore these differences have difficulty in meeting learners' needs effectively [5]. Learners who are given the freedom to explore areas based on their personal interests, and who are accompanied in their striving for solutions by a supportive, understanding facilitator not only achieve higher academic results but also experience an increase in personal values, such as flexibility, self-confidence and social skills [6]. Therefore, when designing instructional material, it is important to accommodate elements that reflect individual differences in learning. One of these elements is learning styles. Understanding a student's particular learning style and how to best meet the needs of that learning style is essential to perform better learning. Clay and Orwig [7] defined learning style as a unique collection of individual skills and preferences that affects how a person perceives, gathers and processes information. Learning styles affect how a person learns, including also the aspects of how a person acts in a learning group, participates in learning activities, relates to others, and solves problems. Basically, a person's learning style is the method that best allows the person to gather and to understand knowledge in a specific manner. Once a learner's particular learning style is detected, it is possible to identify ways to help in improving the learning process [8].

There are many models about learning styles in literature such as Kolb [9], Dunn & Dunn [10], Honey & Mumford [11], and Myers-Briggs [12]. This research paper utilizes the Felder and Silverman's Learning Style Model (FSLSM) [4] because of its applicability to e-learning and compatibility to the principles of interactive learning systems design [13]. In this model, Felder and Silverman proposed four dimensions of learning styles (active/reflective, sensing/intuitive, visual/verbal, and sequential/global) and teaching styles (active/passive, concrete/abstract, visual/verbal, and sequential/global), where each teaching style corresponds to (matches with) a learning style.

Many researches have been conducted to detect the learners' learning styles and provide recommendations and adaptations for online courses based on learning styles. Paredes & Rodríguez [14] presented a framework that collects explicit information about the students by means of the Index of Learning Styles (ILS) questionnaire developed by Felder and Soloman [15], adapts the course structure and sequencing to the student's profile and uses the implicit information about students' behavior gathered by the system during the course in order to dynamically modify the course structure and sequencing previously selected. Graf & Kinshuk [16] introduced a concept for LMSs with adaptivity based on learning styles. They used the open source LMS Moodle as a prototype and developed an add-on that enables Moodle to automatically provide adaptive courses that fit the learning styles of students. Mejía et al. [17] proposed an approach of an adaptation process that allows adjusting different types of resources to the user's preferences by means of the identification of the user's learning style in a LMS. Experimental results of evaluations of such adaptive systems indicated that providing adaptive courses based on students' learning styles plays an effective role in enhancing the learning outcomes by reducing learning time and increasing learners' satisfaction [16][18][19].

While most of the previous works focus on identifying students' learning styles and adapting courses based on the identified learning styles, our research is different in that we present a mechanism to analyze existing course contents in learning management systems and an interactive tool for teachers, which visualizes the suitability of a course for students' learning styles and helps teachers to improve their courses' support level for diverse learning styles. In the next section, the mechanism for analyzing course contents is presented; the interactive tool is illustrated in section 3, followed by the conclusions and the future plans of the research in section 4.

2 Course Analyzing Mechanism

This mechanism aims at analyzing an existing course structure and contents in LMS to measure its support level for diverse learning styles that allows teachers to be aware of how well their courses fit with their students' learning styles. The mechanism currently considers eleven types of learning objects (LOs), as listed below; however, from technical point of view, new types of LOs can easily be included in this mechanism, if required.

- *Commentaries*: provide learners with a brief overview of the section.
- *Content Objects*: are used to present the learning material of the course.
- *Reflection Quizzes*: include one or more open-ended questions about the content of a section. The questions aim at encouraging learners to reflect about the learned material.
- *Self-Assessment Tests*: include several close-ended questions about the content of a section. These questions allow students to check their acquired knowledge and how well they already know the content of the section through receiving immediate feedback about their answers.
- *Discussion Forum Activities*: provide learners with the possibility to ask questions and discuss topics with their peers and instructor. While a course typically includes only one or few discussion forums, the course can include several discussion forum activities as LOs that encourage learners to use the discussion forum.
- *Additional Reading Materials*: provide learners with additional sources for reading about the content of the section, including, for example, more detailed explanations.
- *Animations*: demonstrate the concepts of the section in an animated multimedia format.
- *Exercises*: provide learners with an area where they can practice the learned knowledge.

- *Examples:* illustrate the theoretical concepts in a more concrete way.
- *Real-Life Applications*: demonstrate how the learned material can be related to and applied in real-life situations.
- *Conclusions*: summarize the content learned in a section.

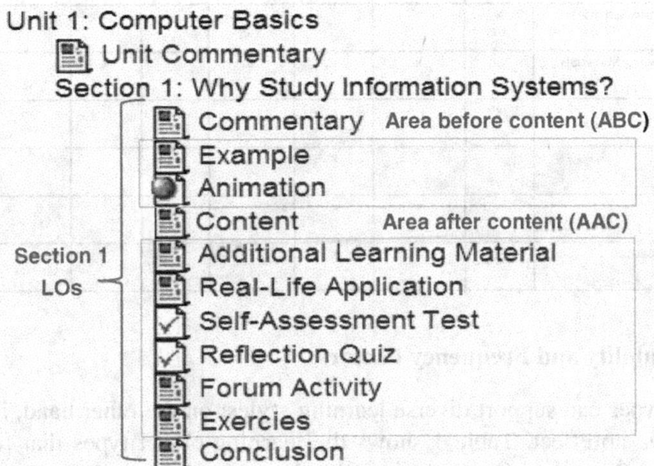

Fig. 1. Section Structure

Decisions on which types of LOs support particular learning styles and at which place/location in the course these types of LOs can support such learning styles are all based on literature from Felder and Silverman [4], who provide a clear description on how learners with particular learning styles prefer to learn.

It is assumed that a course consists of several units and a unit can (but does not have to) consist of several sections. One or more instances of the types of LOs described above can exist in each section. As shown in Fig. 1, a section may (but does not have to) start with a commentary. Subsequently, there is an area before content (ABC) that may include some LOs that aim at motivating the learners and making the section interesting for them. After this area, the content is presented. In the next area, namely area after content (AAC), different types of LOs may be presented. The conclusions of the section can be either right after the last content object or at the end of the section.

The presented mechanism recognizes how well a section of an existing course fits to each of the eight poles of FSLSM (i.e., active, reflective, sensing, intuitive, visual, verbal, sequential and global) by calculating the average of three factors: the availability, the frequency and the sequence of the learning objects in that section, as illustrated below. Consequently, the calculations are applied for each section and then summarized for each unit and for the whole course.

Table 1. The relation between the learning object types and the learning styles

Learning Object/Learning Style	Active	Reflective	Sensing	Intuitive	Visual	Verbal	Sequential	Global
Reflection Quizzes		x		x				
Self-Assessment Tests	x		x					
Discussion Forum Activities	x					x		
Additional Reading Materials		x		x		x		
Animations	x		x		x			
Exercises	x		x					
Examples		x	x					x
Real-Life Applications			x					x

2.1 Availability and Frequency Factors

Certain LO types can support diverse learning styles; on the other hand, it is possible that they have no effect. Table 1 shows the learning object types that fit with each learning style. Since learners with an active learning style prefer to learn by trying things out and discussing with others about the learned material, the availability of self-assessment tests, exercises, animations, and forum activities can support their learning. In contrast to active learners, reflective learners learn by thinking and reflecting about the material. Therefore, the existence of reflection quizzes, additional reading material, and examples fits their needs. Sensing learners prefer concrete material. They are more practical oriented and like to relate the learned material to the real world. Therefore, examples, exercises, animations, self-assessment tests and real-life applications can support their learning. Intuitive learners like abstract material such as concepts and theories, prefer open-ended questions, tend to be more creative, and like challenges. Intuitive learners can be supported by additional reading material and reflection quizzes. Visual learners can benefit from animations. Forum activities and additional reading material are both mostly text-based and therefore support better the learners with a verbal learning style. For sequential learners, providing guidance and a linear increase of complexity in learning is important to support their learning process. Therefore, no particular type of LO would support their learning process more than others. For global learners, it is important to get the big picture of the topic. Therefore, examples and real-life applications would support their learning style more effectively.

Based on the discussion above, the availability of types of LOs is considered as a factor to infer the learning styles that a section of the course fits well. The availability factor measures the existence of LO types that can support each learning style (ls) in a section with respect to all types of LOs that support the particular learning style. The availability factor is calculated using formula 1. On the other hand, the frequency factor measures the number of LOs in the section that support each learning style in respect to the frequency threshold. The frequency threshold represents the sufficient number of LOs in a section to fully support a particular learning style. This threshold is predefined and can be adjusted by the teacher if needed. If the number of LOs that support a particular learning style (ls) in a section is less than the value of the frequency threshold, then the frequency factor is obtained by formula 2, otherwise the frequency factor takes the value 1, which means a full frequency support level for that learning style.

The obtained values for both, the availability factor and the frequency factor, range from 0 to 1, where 1 indicates a strong suitability for the learning style and 0 means no support.

$$Ava_{ls} = \frac{(\#\ of\ existing\ LO\ types\ that\ support\ ls\ )}{(\#\ of\ LO\ types\ that\ support\ ls)} \quad (1)$$

$$Freq_{ls} = \frac{(\#\ of\ existing\ LOs\ that\ support\ ls\ )}{(frequency\ threshold\ )} \quad (2)$$

2.2 Sequence Factor

Not only the types but also the order and the position of the LOs affect the suitability of a course regarding different learning styles. The sequence factor measures the suitability of the sequence of LOs for different learning styles. An active learning style can be supported by the existence of self-assessment tests, exercises and animations at ABC, in order to spark students' interest in the content, or right after the content. The conclusion of the section can support an active learning style if it is located at its end. Since reflective learners prefer to read the content first before they can think and reflect about it through visiting other LOs, locating the conclusion right after the content, followed by additional learning material, examples and reflection quizzes can support a reflective learning style. Locating examples, animations and/or real-life applications at ABC fulfills sensing students' interest. Moreover, presenting them in addition to self-assessment tests and exercises right after the content and the conclusion at the end of the section can support a sensing learning style. Since intuitive learners like challenges, locating exercises in the beginning fits with their learning style. Additional reading material and reflection quizzes can support them if they are presented right after the content. Visual learners can benefit from animations if they are located before or right after the content. Forum activities and additional reading materials support verbal learners when they are presented right after the content. Due to the preference of linear increasing of complexity in learning, locating additional reading material, reflection quizzes, self-assessment tests, exercises, and animations right after the content can support sequential learners. For global learners, it is important to get the big picture of the topic. Therefore, presenting the conclusion right after the content followed by examples and real-life applications and locating the activities that require understanding of the material, including reflection quizzes, self-assessment tests, exercises and forum activities, towards the end of the section can support a global learning style.

The sequence factor is calculated for each LO according to its type, location (ABC or AAC) and position within ABC/AAC. It is determined according to how well this object type in that place fits with each of the eight learning styles of FSLSM. The sequence factor for each learning style is calculated using formula 3. In this formula, f_{ls} (LO) =1, if the LO is suitable for that learning style at that location, and f_{ls} (LO) =0 otherwise. n is the number of LOs in the section. The weight w represents how well the position of a learning object in AAC/ABC fits to the learning style; it is calculated by measuring how far the position of the LO is away from the content. Formula 3 represents the weighted mean of f_{ls}(LO). Its value ranges from 0 to 1, where 1 indicates a strong suitability for the learning style and 0 means no support.

$$Seq_{ls} = \frac{\sum_{i=1}^{n} f_{ls}(LO_i) \times w_i}{\sum_{i=1}^{n} w_i}\ ,\ 0 < w \leq 1 \tag{3}$$

3 Interactive Course Analyzer

Interactive course analyzer is a tool for visualizing the suitability of a course for students' learning styles. Furthermore, it allows the teacher to play around with the course structure (by adding, moving and/or removing LOs) showing the expected changes in the course support level for diverse learning styles. It aims at helping teachers to improve their course support level by making efficient modifications in the course structure to meet the need of different students' learning styles.

The interactive course analyzer tool is implemented as a client-server application. It is mainly developed using MySQL relational database management system and PHP scripting language. It is a stand-alone application that runs on the server side. It connects to a LMS database (Moodle as an example), retrieves the existing course structure and applies the mechanism introduced in section 2 to analyze the course contents with respect to learning styles [20].

The interface of the tool consists of two parts: the settings part and the visualization part. Fig. 2 illustrates a screenshot of the settings part of the user interface. The Analysis Settings area, shown at the left side of the screenshot, allows the teacher to switch between general and cohort visualization modes (that are explained in the next subsections). In the Course Structure area, the course structure is displayed in terms of units, sections, and a list of LOs in each section. The teacher is able to browse the course and select a particular unit/section by clicking on it. The Simulation Settings area, as displayed at the right side of Fig. 2, allows the teacher to simulate modifications in the course structure. By utilizing drag and drop functionality, the teacher can drag learning objects from the list of learning object types and place them in certain positions in the Course Structure area, drop learning objects from the Course Structure area to remove them, and/or move leaning objects from one place to another one in the Course Structure area. Once the teacher has completed the modifications on the course structure and wants to analyse how his/her modifications change the support level of the course, he/she can press on the Test button. Then the tool analyses the course structure and updates the visualization part respectively. Furthermore, in the Advanced Settings area, teachers can set the value of the frequency threshold and select the learning object types to be considered while analyzing the course support level.

The visualization part consists of four similar charts that show how well the course and the selected unit/section fit with students' learning styles. Two of the charts visualize the course support level before and after the modifications made by the teacher in the course structure. Similarly, the other two charts show the selected unit/section support level before and after the modifications.

In the following two subsections, the visualization of the General Mode and Cohort Mode is described in more detail.

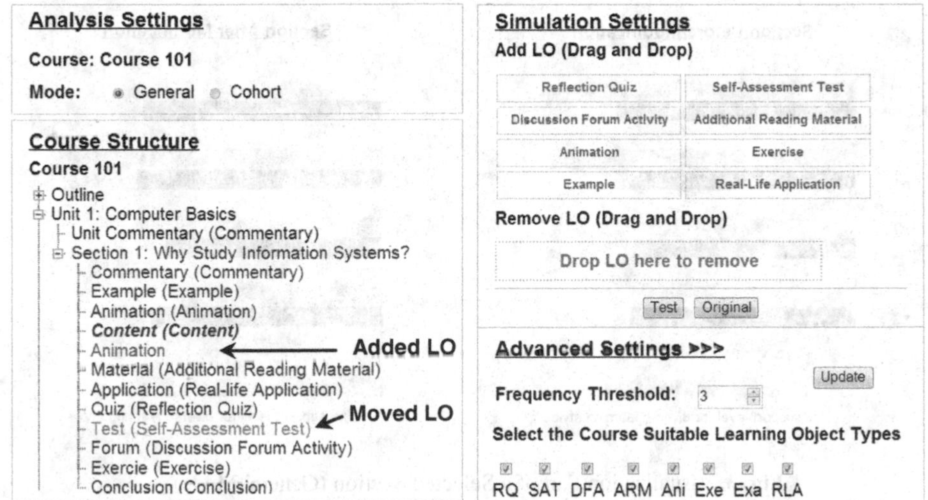

Fig. 2. Interactive Course Analyzer (Settings Part)

3.1 General Mode

This mode visualizes the support level of a course for diverse learning styles based on FSLSM. Fig. 3 illustrates a screenshot of the visualization part of the selected section in General Mode. Each chart consists of two parts. The upper part of a chart consists of a set of bars to show the strength of the harmony of the course/unit/section with each of the eight learning style poles (i.e., active, reflective, visual, verbal, sensing, intuitive, sequential and global), in terms of percentage (calculated by the average of the three factors illustrated in section 2). Each learning style dimension is represented by two horizontal bars, one for each pole, where the two poles show the two different preferences of the dimension. The longer the bar, the more the course/unit/section fits with the learning style. The lower part of a chart contains only one bar that shows the overall support level of the course/unit/section for diverse learning styles (calculated by the average of the support level of the eight poles). Once the teacher moves the cursor over any bar, a tooltip appears to display more details about the analysis factors illustrated in section 2.

For examples, as shown in Fig. 2, the teacher made two modifications in the selected section "Why Study Information Systems?" (displayed in brown): the "Animation" learning object was added (displayed in blue), and the "Self-Assessment Test" learning object was moved within the section (displayed in orange). After re-analyzing the course considering the two modifications and by comparing the two charts in Fig 3, it can be noticed that the section support levels for active, sensing and visual learning styles were improved (as shown in the right chart by the black arrows). Consequently, the overall section support level for diverse learning styles was improved as well (as shown at the bottom of the right chart).

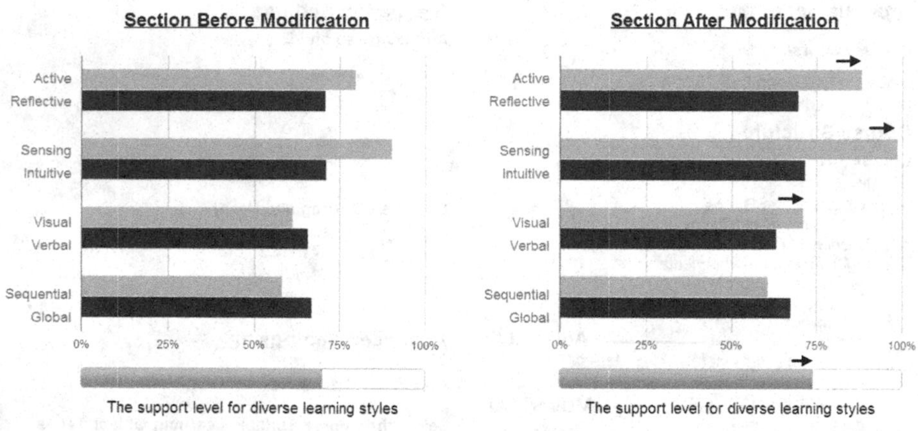

Fig. 3. Visualization Part of a Selected Section (General Mode)

3.2 Cohort Mode

Cohort Mode visualizes the support level of a course in respective to the learning styles of the cohort of students' enrolled in that course. Fig. 4 illustrates a screenshot of the visualization part of a selected section in Cohort Mode. The charts visualize the data about students' learning styles (which can be calculated, for example, through the ILS questionnaire [15] or by a tool like DeLeS [21]) in comparison with the course support level (calculated by the average of the three factors illustrated in section 2). As shows in Fig. 4, each learning style dimension in a chart contains two bars; the upper one shows the course/unit/section support level for each poles of that dimension (for examples, "active" on the right and "reflective" on the left); the lower bar shows the learning styles of the respective cohort of students, in terms of different levels varying from strong to balanced. In case that all students are fully supported, the bar will be displayed in green color, otherwise a gap will be shown in red. The intensity of the red color indicates the number of unsupported students.

For example, the chart at the left side (Fig. 4) shows that reflective, intuitive, verbal, global and sequential learners are well supported by the course. On the other hand, active, sensing and visual learners are not fully supported; there are gaps between the course support level for that learning styles and the learning styles of the respective cohort of students. Once the teacher moves the cursor over any bar, a tooltip appears to display more information about the level of support and the number of supported and unsupported students. Considering the example mentioned in the previous section and the teacher's modifications in that example, the chart at the right side (Fig. 4) shows that the gaps between the course support level and the learning styles of the respective cohort of students were eliminated, and the students were fully supported (as shown at the bottom of the right chart).

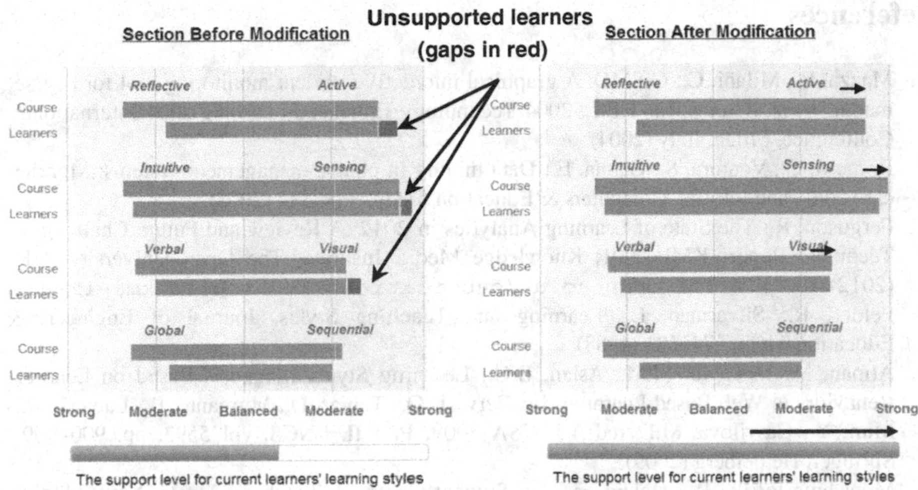

Fig. 4. Visualization Part of a Selected Section (Cohort Mode)

4 Conclusions and Future Work

This paper presents an interactive tool for analyzing existing course contents in learning management systems and providing teachers with information regarding how well their courses fit with students' learning styles based on the Felder and Silverman's learning styles model. A mechanism is proposed and utilized in the tool for identifying the course support level for diverse learning styles by calculating three factors: the sequence, the frequency and the availability of learning objects types in that course. The tool provides a teacher with an interactive graphical user interface, which can be used to analyze and visualize the course support level. Moreover, it allows the teachers to try out modifications in the course structure (e.g., adding learning objects) and see what impact such modifications have in terms of improving the support level of a course for students with different learning styles. It aims at helping the teacher to decide on what necessary modifications to make before implementing them "actually" in the LMS. The interface of the tool contains two modes: General Mode, which helps the teacher to improve the course support level for general learning styles and Cohort Mode that helps the teacher to improve the course support level for the current cohort of students enrolled in that course.

Plans for future research include providing teachers with automatic recommendations on how to best extend their courses to support more students with different learning styles and to fit the courses with the current cohort of learners.

Acknowledgements. The authors acknowledge the support of NSERC, iCORE, Xerox, and the research-related gift funding by Mr. A. Markin.

References

1. Mazza, R., Milani, C.: GISMO: A graphical interactive student monitoring tool for course management systems. In: T.E.L. 2004 Technology Enhanced Learning 2004 International Conference, Milan, Italy (2004)
2. Romero, C., Ventura, S., García, E.: Data mining in course management systems: Moodle case study and tutorial. Computers & Education 51(1), 368–384 (2008)
3. Ferguson, R.: The State of Learning Analytics in 2012: A Review and Future Challenges. Technical Report KMI-12-01, Knowledge Media Institute, The Open University, UK (2012), `http://kmi.open.ac.uk/publications/techreport/kmi-12-01`
4. Felder, R., Silverman, L.: Learning and Teaching Styles. Journal of Engineering Education 94(1), 674–681 (1988)
5. Atman, N., Inceoğlu, M.M., Aslan, B.G.: Learning Styles Diagnosis Based on Learner Behaviors in Web Based Learning. In: Gervasi, O., Taniar, D., Murgante, B., Laganà, A., Mun, Y., Gavrilova, M.L. (eds.) ICCSA 2009, Part II. LNCS, vol. 5593, pp. 900–909. Springer, Heidelberg (2009)
6. Motschnig-Pitrik, R., Holzinger, A.: Student-Centered Teaching Meets New Media: Concept and Case Study. IEEE Journal of Educational Technology & Society 5(4), 160–172 (2002)
7. Clay, J., Orwig, C.J.: Your learning style and language learning. Lingual links library. Summer Institute of Linguistic, Inc. (SIL), International version 3.5 (1999)
8. Onyejegbu, L.N., Asor, V.E.: An Efficient MODEL for Detecting LEARNING STYLE Preferences in a Personalized E-Learning Management System. Cyber Journals: Multidisciplinary Journals in Science and Technology, Journal of Selected Areas in Software Engineering, JSSE, May Edition (2011)
9. Kolb, A.Y., Kolb, D.A.: The Kolb Learning Style Inventory. Version 3.1, Technical Specification. Hay Group, Boston (2005)
10. Dunn, R., Dunn, K., Price, G.E.: Learning Style Inventory. Price Systems, Lawrence, KS (1996)
11. Honey, P., Mumford, A.: The Learning Styles Helper's Guide. Peter Honey Publications Ltd., Maidenhead (2006)
12. Myers, I.B., McCaulley, M.H.: Manual: A Guide to the Development and Use of the Myers-Briggs Type Indicator. Consulting Psychologists Press, Palo Alto (1998)
13. Kuljis, J., Liu, F.: A Comparison of Learning Style Theories on the Suitability for Elearning. In: Hamza, M.H. (ed.) The IASTED Conference on Web Technologies, Applications, and Services, pp. 191–197. ACTA Press (2005)
14. Paredes, P., Rodríguez, P.: A mixed approach to modelling learning styles in adaptive educational hypermedia. Advanced Technology for Learning 1(4), 210–215 (2004)

15. Felder, R.M., Soloman, B.A.: Index of Learning Styles Questionnaire. North Carolina State University (2012), http://www.engr.ncsu.edu/learningstyles/ilsweb.html
16. Graf, S., Kinshuk, K.: Providing Adaptive Courses in Learning Management Systems with Respect to Learning Styles. In: Richards, G. (ed.) The World Conference on E-Learning in Corporate, Government, Healthcare, and Higher Education (e-Learn), pp. 2576–2583. AACE Press, Chesapeake (2007)
17. Mejía, C., Baldiris, S., Gómez, S., Fabregat, R.: Adaptation process to deliver content based on user learning styles. In: International Conference of Education, Research and Innovation (ICERI 2008), pp. 5091–5100. IATED, Madrid (2008)
18. Popescu, E.: Adaptation provisioning with respect to learning styles in a web-based educational system: An experimental study. Journal of Computer Assisted Learning 26, 243–257 (2010)
19. Tseng, J.C.R., Chu, H.C., Hwang, G.J., Tsai, C.C.: Development of an adaptive learning system with two sources of personalization information. Computers & Education 51, 776–786 (2008)
20. El-Bishouty, M.M., Chang, T.-W., Kinshuk, K., Graf, S.: A Framework for Analyzing Course Contents in Learning Management Systems with Respect to Learning Styles. In: Biswas, G., et al. (eds.) The 20th International Conference on Computers in Education(ICCE 2012), pp. 91–95. Asia-Pacific Society for Computers in Education, Singapore (2012)
21. Graf, S., Kinshuk, K., Liu, T.-C.: Supporting teachers in identifying students' learning styles in learning management systems: An automatic student modelling approach. Educational Technology & Society 12(4), 3–14 (2009)

A Framework for Automatic Identification and Visualization of Mobile Device Functionalities and Usage

Renan H.P. Lima[1], Moushir M. El-Bishouty[2,3], and Sabine Graf[2]

[1] Federal University of Sao Carlos, Brazil
[2] Athabasca University, Canada
[3] City for Scientific Research & Technological Applications, Egypt
renan.lima90@gmail.com,
{moushir.elbishouty,sabineg}@athabascau.ca

Abstract. While mobile learning gets more and more popular, little is known about how learners use their devices for learning successfully and how to consider context information, such as what device functionalities/features are available and frequently used by learners, to provide them with adaptive interfaces and personalized support. This paper presents a framework that automatically identifies the functionalities/features of a device (e.g., Wi-Fi connection, camera, GPS, etc.), monitors their usage and provides users with visualizations about the availability and usage of such functionalities/features. While the framework is designed for any type of device such as mobile phones, tablets and desktop-computers, this paper presents an application for Android phones. The proposed framework (and the application) can contribute towards enhancing learning outcomes in many ways. It builds the basis for providing personalized learning experiences considering the learners' context. Furthermore, the gathered data can help in analyzing strategies for successful learning with mobile devices.

Keywords: Context modeling, visualization, mobile learning, device functionalities and their usage, personalization, ubiquitous learning analytics.

1 Introduction

The recent advances in mobile technologies have allowed the widespread use of mobile devices around the world for many different purposes. From educational point of view, learning can now take place anytime and anywhere using mobile devices (e.g., smartphones, tablets) to facilitate learner interaction and access learning contents with fewer restrictions of time and location [1-3]. Mobile settings bring important advantages for ubiquitous learning by providing a more flexible and authentic experience for learners.

The way people interact with devices is vital for their success. Looking at human computer interaction (HCI), it is apparent that interaction techniques are limited by

A. Holzinger and G. Pasi (Eds.): HCI-KDD 2013, LNCS 7947, pp. 148–159, 2013.

the technology available [4]. However, due to the increasing diversity of users, technical systems and usage contexts, many aspects are relevant in understanding users' acceptance beyond the ease of using a system and the perceived usefulness [5]. HCI benefits from situational context modeling (such as location, surrounding environment and/or state of the device), as an implicit input to the system [4]; for example contextual information can be used to adapt the system input and output to the current situation (such as: font size, volume, brightness, privacy settings, etc.).

In ubiquitous environments, factors such as a learner's context influence the learning process in a mobile ubiquitous environment [2] and should be considered in order to provide appropriate support for learners. While mobile learning gets more and more popular, we do not know much about how learners use mobile devices for learning. Furthermore, context information in terms of what device functionalities/features are available and frequently used by learners is typically not considered when learners are presented with learning activities.

Data gathered from the usage of the learners' devices provide important information about the user and his/her context [6-7] and can, together with the availability of device functionalities/features, be used to build a context profile [3]. In this paper, we introduce a framework that aims at building a comprehensive context profile, containing information about the devices a learner uses, the available functionalities/features on a learner's device as well as how frequently the respective device functionalities/features are used by the learner. Such functionalities and features include, for example, internet connection types, existing sensors, camera, keyboard, touch screen and so on. The proposed framework enables a system to automatically identify, monitor and visualize the availability and usage of device functionalities/features in mobile devices and desktop computers.

While some ubiquitous and mobile systems use the devices simply as source of user's location data and do not consider different types of information that they can provide, other systems (e.g., Phone Usage [8], Elixir [9]) do not consider how the user interacts with his/her device for user profiling – i.e. which functionalities/features he/she prefers to use and when. Our framework differs from others since it considers the availability and the usage of device functionalities/features to build a comprehensive context profile.

By creating such a comprehensive context profile, the basis is built for providing personalized learning experiences to mobile learners. By knowing the advantages and limitations of a currently used device, learning activities can be tailored to available and frequently used device functionalities/features. For example, online videos are only suggested if appropriate internet connectivity is available. Furthermore, if the usage data show that the learner does not use the respective device for inputting large amounts of text, learning activities that require writing significant amounts of text are not recommended. On the other hand, if features such as a camera or GPS are detected, learning activities that require such features can be suggested. In addition, data about the usage of device functionalities/features provide valuable information about how learners actually use different devices for learning and provide insights into what types of learning activities can facilitate mobile learning. Moreover, such usage data can help in analyzing successful learning strategies in mobile learning.

The paper is organized as follows: In section 2, related works are described. Section 3 introduces the proposed framework and Section 4 presents an application for Android phones developed based on the introduced framework. Section 5 concludes the paper.

2 Related Work

In the educational domain, the tracking and analyzing of device usage and the use of applications is particularly relevant and led to many educational institutions performing studies for this purpose. For example, the Educause Center for Applied Research (ECAR) has surveyed undergraduate students annually since 2004 about technology in higher education [10]. In 2012, ECAR collaborated with 195 institutions to collect responses from more than 100,000 students about their technology experiences. One of the key findings is that students want to access academic progress information and course material via their mobile devices [11]. Moreover, Curtin University [12] has presented a web survey since 2007, which was available to all students at the university for a two-week period through the student portal. The recent surveys [13] sought information on student access to the Internet off-campus, current and planned ownership and use of mobile devices, and perceptions as to how the learning experience at Curtin University might be enhanced with mobile devices, network services and online tools. Also, students were asked to report what they used their phones for (e.g., web access, SMS, MMS, and so on). The results of these surveys indicated that the majority of students were likely to have broadband (often wireless) access to the Internet off-campus, and use mobile devices, such as newer laptops and phones. The results also showed the familiarity of students with iTunes and iPhone apps (due to a high ownership of iPhones), and the ability to access wireless, take photos, send text (and some have the ability to record video and audio, and hold video conference) on a mobile device. Furthermore, the results indicated the ability to use Web 2.0 applications to create accounts, connect with others, communicate in web spaces, and indicate 'liking' and rating. Furthermore, Ally and Palalas [14] conducted a research study to determine the current state of mobile learning in Canada and to establish the direction Canada should take in the field of m-learning. The study surveyed and interviewed small, medium, and large organizations from fifteen different sectors across Canada on their use of mobile learning and the future direction in that area.

From the abovementioned studies, it can be seen that mobile learning gets more and more popular. Student ownership and familiarity of mobile devices have increased over recent years, with most students reporting that they own a computer and/or a mobile phone [10]. The technologies themselves have also changed over the years, with increasing ownership of laptop computers and smart phones, corresponding to decreasing ownership of desktop computer and simple mobile phones[15][16].

With the increasing popularity of mobile learning, there is also a need for understanding how learners actually use their devices for learning. In addition to manually track device usage, systems were presented to automatically identify such information. Trinder [17], for instance, proposed a tool to automatically collect usage log data from personal digital assistants (PDAs) handheld devices used by students to access the university's virtual learning environment (VLE). The tool recorded in detail when PDA applications (including formative assessment applications) were being used and overcame a number of technical barriers in securing this information for later analysis. It was anticipated that such log information would provide reliable timing information regarding PDA use. Automatic logging was considered to be preferable in contrast to asking the participants to keep a diary of how and when they used the PDA. A manual diary method adds extra user overhead and unless the event is recorded contemporaneously it may be forgotten. Automatic logging also allowed the collection of additional information such as when the PDA was hot-synced to a desktop machine, or when the student used the "beam" facility to exchange items with another PDA. Patterns and modes of their PDA use - considering a range of factors including overall duration of use, use as a function of time of day or time of week, and the complexity of use (e.g. frequency of application switching within a usage session) - were obtained and correlated with exam results and access to the University's VLE. In addition to the logging data, selected cohorts of students were also subject to questionnaire and interview. Practical experiments were conducted to examine how students used the loaned PDAs and data visualization tools and techniques of data analysis were introduced to show the findings. The results indicated the benefits to students of general PDA use and specific use of the formative assessment quiz application. Another finding was that the bottlenecks to PDA use should be overcome to enhance its usage and student learning [18].

Moreover, several applications are developed in order to monitor the usage of smart mobile devices aiming at enchanting the device usage. Due to the advantages of open source, many of them are implemented for Android-powered mobile devices [19]. Android Status [20], for instance, features CPU memory usage, process list, mobile network information, Internet connections, network information, Wi-Fi Status, storage (i.e., SD card) usage, routing information, and others. Elixir [9] is another application with highly configurable widgets. It displays information about battery, CPU usage, memory usage, internal and external storage, display, Wi-Fi status, mobile network, location services (i.e., GPS), Bluetooth, sync, airplane mode, sensors, etc. CenceMe [21] is a different type of application, which not only tracks sensor-related information from users' mobile devices but also interprets this information in the context of social activities and allows sharing of this information through social networking applications. It automatically infers people's sensing presence (e.g., dancing at a party with friends) and then shares this presence through social network portals such as Facebook.

In contrast to the methods and applications mentioned above, the proposed framework in this paper illustrates a methodology to automatically detect the existence of types of device features and functionalities that the learner may use in his/her device for learning. It is not targeting a particular feature, or a certain device type or operation system. In addition, it monitors how the learner interacts with and uses each feature in order to understand and recognize the learner's preferences on how to use mobile devices for learning. Such context information opens up new possibilities for providing adaptive learning environment.

3 Framework

The proposed framework (Fig. 1) is a client-server generic framework designed to run on smartphones, tablets, laptops and desktop-computers. On the client side – running on the learners' devices – the application is divided into two parts. The first one is the User Part, where personal information about the user (i.e., learner) is gathered and usage data visualization is presented. This part has two components: the Personal Information Manager, allowing the learner to login and to provide personal information (e.g., login name, full name, email address and other characteristics). This information is grouped under the name of User Object. The second component is the Visualizer, which is responsible for providing a user interface to show the feature availability and usage information. The second part of the application is the Device Part, where device information about the availability and usage of features is collected. This is done by Feature Detectors and the Tracker. Each Feature Detector is associated to a feature (e.g., internet connection types, keyboard presence, available touch screen) and able to discover whether the respective feature is available in that device. Table 1 presents all features currently considered in the framework. The data collection from all Feature Detectors is used to create the Device Object, which includes information to describe the tracked device in terms of its features. Furthermore, the Feature Detectors provide information about the usage of the respective features to the Tracker. The Tracker gathers the learner's device usage data by receiving updates from the Feature Detectors and stores this information on the server. However, if there is no connection to the server available, the Tracker creates a local backup file with the gathered data and postpones the online storage until connection to the server is available. Moreover, there is a component which belongs to both parts of the client side, the Detection Manager. The Detection Manager allows the learner to select what features the system should track. In order to do so, the Detection Manager detects what features are available in the device by calling the Feature Detectors when the learner starts the application for the first time and displays these features for selection. The set of information about the features selected by the learner is grouped under the name of Particular Device Object and represents a subset of the information in the Device Object. Furthermore, the Detection Manager encapsulates the user id, the device id and the Particular Device Object and sends all this information to the Tracker.

The server side mainly manages the database and the identification of devices and users, and is divided into two parts. Part A communicates with the User Part of the client and consists of the following components: the User Profiler, which is responsible for receiving the user information as a User Object from the client and for storing it in the database. The second component is the User Profile Structure, which stores what personal information of the learner is considered in the framework; and the third component is the XML Creator which processes the availability and usage data and generates a XML file which is used by the Visualizer.

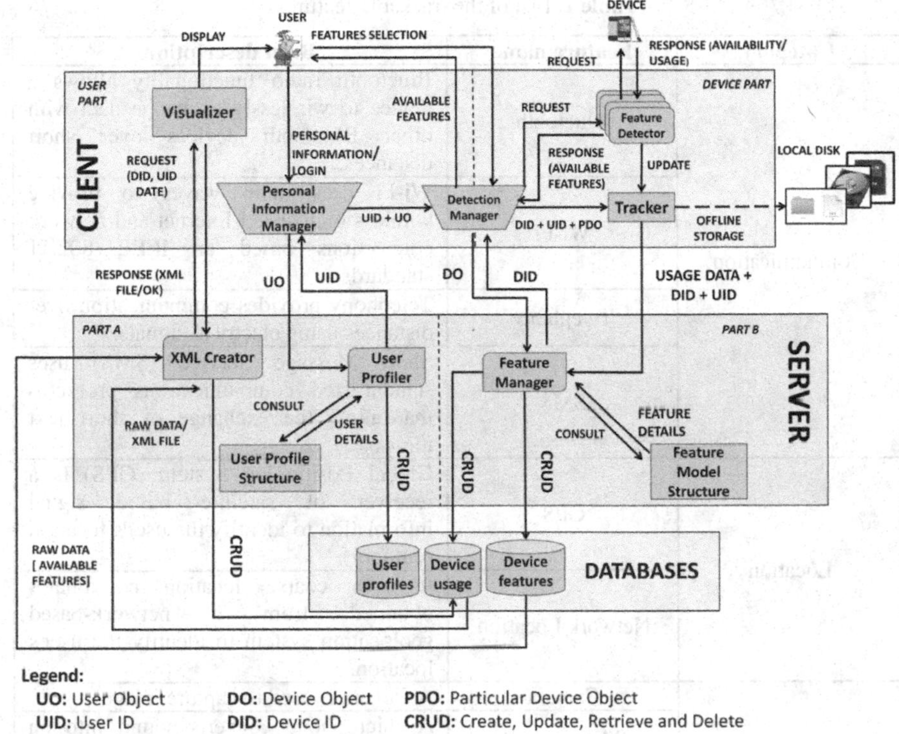

Fig. 1. The proposed framework architecture

Part B of the server side is responsible for the communication with the Device Part of the client. It is composed of the Feature Manager and the Feature Model Structure. The former is responsible for receiving the device information in form of a Device Object, including the available features on the device, and its usage data from the client, and for storing this information in the database. The Feature Model Structure describes the device features considered in the system. Moreover, there is a database, divided into three parts: First, it stores the device information, including the device id and which features are available to be used. In other words, it stores a set of Device Objects. Second, it stores user information (e.g., login name, full name, email address and other characteristics), in other words, a set of User Objects. Third, it combines the data relating a device to a learner and stores the usage data of the device by the learner.

Table 1. List of the trackable features

Category	Feature name	Brief description
Communication	Bluetooth	Bluetooth radio functionality allows a device to wirelessly exchange data with other Bluetooth devices over short distances.
	Wi-Fi	Wi-Fi uses radio waves to provide wireless high-speed Internet and network connections based on IEEE 802.11 standards
	Telephony	Telephony provides communication over distances using electrical signals.
	SMS	Short Message Service (SMS) uses standardized communications protocols that allow the exchange of short text message.
Location	GPS	Global Positioning System (GPS) is a receiver of satellite based signal information to identify the user's location (outdoor).
	Network Location	It uses coarse location coordinates obtained from a network-based geolocation system to identify the user's location.
Sensors	Camera	A camera is used to capture images.
	Microphone	A microphone converts sound into an electrical signal.
	Barometer	A barometer is a tool for measuring atmospheric pressure.
	Compass	A compass is a magnetometer, which provides directional readings.
	Gyroscope	A Gyroscope is a tool for measuring or maintaining orientation based on the principles of angular momentum.
	Light	A light sensor measures the light level.
	Proximity	A proximity sensor is used to measure how close an object is to the device.
	Accelerometer	An accelerometer is a tool that measures proper acceleration.
Input	Soft Keyboard	A soft keyboard is an on-screen virtual keyboard.
	Hard Keyboard	A hard keyboard is a real (hardware) device.
	Touchscreen	A touchscreen is an electronic visual display that allows the user to control the device through simple or multi-touch gestures.

4 Implementation for Android Phones

As mentioned above, the proposed framework is designed for tracking and analyzing the availability and usage of device features from different devices such as mobile phones, tablets and desktop-computers. In this section, we introduce an application for mobile phones, running on the Android operation system.

This application – the Usage Observer – aims at providing the possibility for learners to select features on their devices, monitor the usage of those features and view data about feature availability and usage in a user-friendly way. Moreover, learners can allow teachers to access these data. The application is composed of seven interfaces: login, main, feature selection, personal information/preferences, tracker management, feature usage visualization and administration. Some of the major interfaces can be seen in Fig. 2 and Fig. 3.

The login and registration process is managed by the Personal Information Manager and the Detection Manager. If the learner does not have an account yet, he/she needs to go through the registration process. In the first step, the learner can provide personal information and preferences (as shown in Fig. 2b), leading to the creation of a User Object by the Personal Information Manager. This interface (as well as the other user interfaces) is implemented as an Android component extension called Activity.

In the second step of the registration process, the learner registers his/her device by accessing the Detection Manager, where he/she can select which features should be tracked (Fig. 3a). This interface presents the device's available features, identified by the Feature Detectors, and allows the learner to select which features he/she wants to be tracked. As a result, a Particular Device Object is created containing information on those selected features. The Feature Detectors are implemented as one single service, called Detection Service. Services are an Android component extension, which run in the background and typically perform long-term tasks. If a learner is using different devices, the registration for the user account has to be done only once but whenever the learner is using a new device, this device has to be registered by performing the second step of the registration process. Once the learner and his/her device are registered, he/she can login by providing the correct username and password (Fig. 2a) and is presented with the main interface (Fig. 3b), where he/she can navigate through the application.

The main interface allows the learner to access the personal information entered in the registration process (through the Preferences symbol) and the feature selection interface (through the Features symbol) in order to change personal information and selected features. Furthermore, the main interface provides access to the tracking and visualization interfaces. The tracking interface presents raw data about the tracked features (accessible through the Tracking symbol) and the visualization interface shows user-friendly visualizations about the availability and usage of features on the learner's devices (accessible through the Visualize symbol). Administrators and teachers can additionally access the administrator interface (through the Admin symbol), which provides them with additional visualization interfaces to compare usage information from different learners and/or different devices.

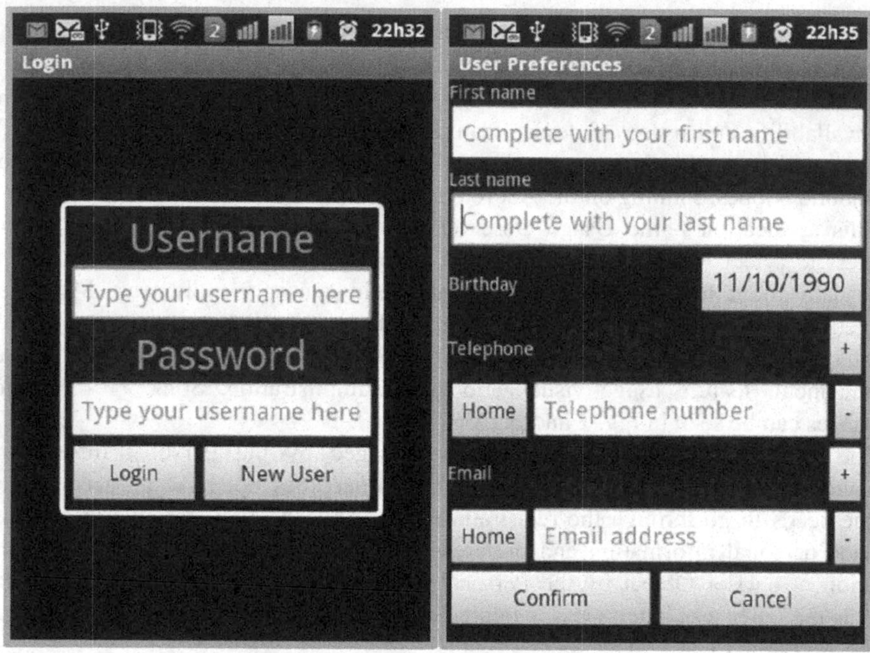

Fig. 2. Application interfaces – 2a: Login interface; 2b: Personal information interface

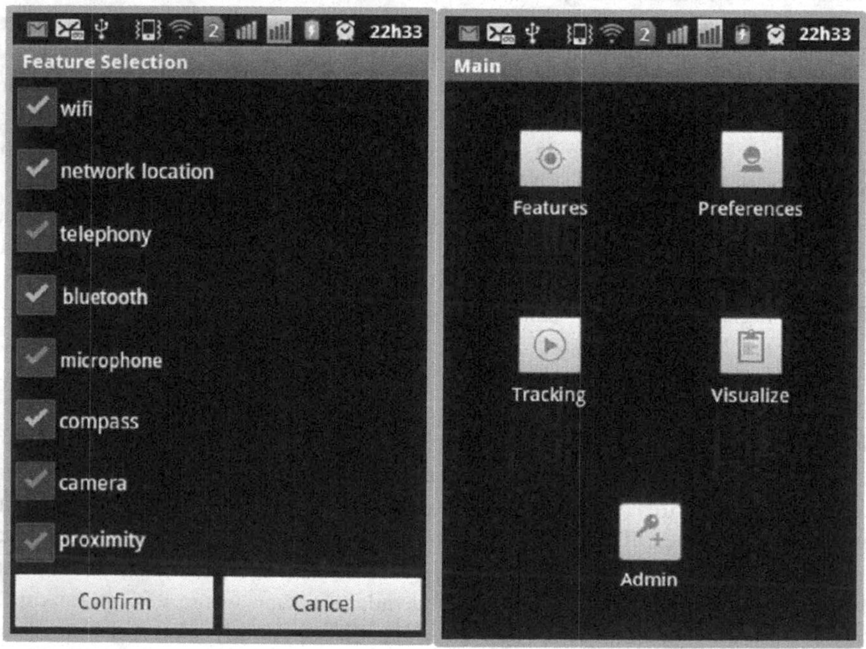

Fig. 3. Application interfaces – 3a: Feature selection interface; 3b Main interface (access point of the other interfaces)

While learners are using their devices, the system tracks the usage of the selected features. This is done by the Tracker Service, which represents part of the Feature Detectors and part of the Tracker component in the framework. The Tracking Service is responsible for collecting data about the usage of the previously selected features. One of the biggest development challenges was to implement the algorithm for the Tracking Service to track the usage of the selected features. The Android platform does not provide any direct way of collecting this type of data for certain features [22]. After evaluating several different approaches, we implemented this service by reading the system generated logs and gathering the timestamps related to the use of each selected feature. Hence, the Tracker component in the framework works as a logger which means the developed Tracker service collects log data provided by the system and sends it to the framework's server side. The server is then responsible for parsing the received log data and for obtaining the selected feature's usage data and statistics based on the timestamps contained in the log data. This task is performed by the Feature Manager component, which then stores the new information in the database.

5 Conclusions and Future Work

This paper introduces a framework that aims at building a comprehensive context profile through detecting available features of a device (e.g., keyboard, touch screen, internet connection, camera, GPS, and so on) and tracking the usage of these features by its users. Furthermore, it visualizes the gathered data in a user-friendly way. The proposed framework is designed for different devices such as smartphones, tablets and desktop-computers.

The gathered data about availability and usage of device features can be used in many ways to improve mobile learning and has high potential to help in enhancing learning outcomes of mobile learners. First, the gathered information is the basis for extending learning systems with advanced adaptive and intelligent capabilities that allow personalizing user interfaces and providing learners with adaptive course structures and recommendations based on availability and previous usage of device features. Second, the gathered data provides insights into what features are frequently available and used, which gives information about what kind of learning activities are most useful for facilitating mobile learning, as well as gives teachers feedback on how suitable their courses are for mobile learning. Third, the gathered data can be used to analyze successful learning with mobile and desktop-based devices and provide learners with personalized support and suggestions on how to improve their learning using different devices.

Based on the proposed framework, an application is implemented to run on Android devices. It is a client-server application which stores and analyzes the collected usage data on the server side. In order to keep the user's privacy, the application allows the learner to decide which features to track and when. In addition, the learner has the ability to delete his/her recent or entire usage history. In our future work, we plan to complete the development of the proposed framework, with focus on the server-side components and the visualization component. We also plan to implement applications for other devices such as tablets, desktop computers, other mobile phones, etc. Furthermore, we plan to perform an experiment to evaluate the system efficacy and usefulness for learners and teachers. Such evaluation will also help us in identify limitations of our system which can then be addressed in future work.

Acknowledgements. The authors acknowledge the support of NSERC, iCORE, Xerox, Mitacs, and the research-related gift funding by Mr. A. Markin.

References

1. Graf, S., MacCallum, K., Liu, T., Chang, M., Wen, D., Tan, Q., Dron, J., Lin, F., Chen, N., McGreal, R., Kinshuk, K.: An infrastructure for Developing Pervasive Learning Environments. In: IEEE International Workshop on Pervasive Learning, pp. 389–394. IEEE Press (2008)
2. Tortorella, R., Graf, S.: Personalized Mobile Learning Via An Adaptive Engine. In: IEEE International Conference on Advanced Learning Technologies, pp. 670–671. IEEE Press (2012)
3. Chen, G.D., Chang, C.K., Wang, C.Y.: Ubiquitous learning website: Scaffold learners by mobile devices with information-aware techniques. Computers & Education 50(1), 77–90 (2008)
4. Schmidt, A.: Implicit Human Computer Interaction Through Context. Personal Technologies 4(2&3), 191–199 (2000)
5. Ziefle, M., Himmel, S., Holzinger, A.: How usage context shapes evaluation and adoption criteria in different technologies. In: International Conference on Applied Human Factors and Ergonomics, San Francisco, pp. 2812–2821 (2012)
6. Ogata, H., Li, M., Hou, B., Uosaki, N., El-Bishouty, M.M., Yano, Y.: SCROLL: Supporting to Share and Reuse Ubiquitous Learning Log in the Context of Language Learning. In: World Conference on Mobile and Contextual Learning, pp. 40–47 (2010)
7. Roman, M., Campbell, R.H.: A User-Centric, Resource-Aware, Context-Sensitive, Multi-Device Application Framework for Ubiquitous Computing Environments. Technical Report (2002), `http://gaia.cs.uiuc.edu/papers/new080405/AppFramework1.doc` (accessed on April 15, 2013)
8. PhoneUsage, `https://play.google.com/store/apps/details?id=com.jupiterapps.phoneusage&hl=en` (accessed on April 15, 2013)
9. Elixir, `https://play.google.com/store/apps/details?id=bt.android.elixir` (accessed on April 15, 2013)
10. Smith, S.D., Salaway, G., Caruso, J.B.: The ECAR study of undergraduate students and information technology. EDUCAUSE Center for Applied Research (2009), `http://www.educause.edu/library/ERS0906` (accessed on April 15, 2013)
11. ECAR study of undergraduate students and information technology, `http://www.educause.edu/library/resources/ecar-study-undergraduate-students-and-information-technology-2012` (accessed on April 15, 2013)

12. Oliver, B., Nikoletatos, P.: Building engaging physical and virtual learning spaces: A case study of a collaborative approach. In: Same Places, Different Spaces, The Annual Australian Society for Computers in Learning in Tertiary Education Conference, pp. 720–728 (2009)
13. Oliver, B., Whelan, B.: Designing an e-portfolio for assurance of learning focusing on adoptability and learning analytics. Australasian Journal of Educational Technology 27(6), 1026–1041 (2011)
14. Ally, M.,Palalas, A.:State of Mobile Learning in Canada and Future Directions (2011), http://www.rogersbizresources.com/files/308/Mobile_Learning_in_Canada_Final_Report_EN.pdf (accessed on April 15, 2013)
15. Kennedy, G.E., Judd, T.S., Churchward, A., Gray, K., Krause, K.-L.: First year students' experiences with technology: Are they really digital natives? Australasian Journal of Educational Technology 24(1), 108–122 (2008)
16. Algonquin College,a new era of connectivity at Algonquin College: Collaborative approach to Mobile Learning Centre, a first in Canada, http://www.algonquincollege.com/PublicRelations/Media/2011/Releases/MobileLearningCentreNewsRelease.pdf (accessed on April 15, 2013)
17. Trinder, J.J.:Mobile learning evaluation: the development of tools and techniques for the evaluation of learning exploiting mobile devices through the analysis of automatically collected usage logs - an iterative approach, PhD thesis (2012), http://theses.gla.ac.uk/3303/(accessed on April 15, 2013)
18. Trinder, J.J., Magill, J.V., Roy, S.: Using automatic logging to collect information on mobile device usage for learning. In: Pachler, Kukulska-Hulme, Vavoula (eds.) Research Methods in Mobile and Informal Learning. Peter Lang Publishing Group (2009)
19. 35 Android Apps to Monitor Usage Stats and Tweak System Utilities, http://android.appstorm.net/roundups/utilities-roundups/35-android-apps-to-monitor-usage-stats-and-tweak-system-utilities/ (accessed on April 15, 2013)
20. Android Status, https://play.google.com/store/apps/developer?id=androidstatus (accessed on April 15, 2013)
21. Miluzzo, E., Lane, N., Fodor, K., Peterson, R., Lu, H., Musolesi, M., Eisenman, S., Zheng, X., Campbell, A.: Sensing Meets Mobile Social Networks: The Design, Implementation and Evaluation of the CenceMe Application. In: 6th ACM Conference on Embedded Network Sensor Systems, pp. 337–350 (2008)
22. Sensors Overview, http://developer.android.com/guide/topics/sensors/sensors_overview.html (accessed on April 15, 2013)

Crowdsourcing Fact Extraction from Scientific Literature

Christin Seifert[1], Michael Granitzer[1], Patrick Höfler[2], Belgin Mutlu[2], Vedran Sabol[2], Kai Schlegel[1], Sebastian Bayerl[1], Florian Stegmaier[1], Stefan Zwicklbauer[1], and Roman Kern[2]

[1] University of Passau
Innstrae 33, 94032 Passau, Germany
{firstname.lastname}@uni-passau.de
http://uni-passau.de
[2] Know-Center
Inffeldgasse 13/6, 8010 Graz, Austria
{phoefler,bmutlu,vsabol,rkern}@know-center.at
http://know-center.at

Abstract. Scientific publications constitute an extremely valuable body of knowledge and can be seen as the roots of our civilisation. However, with the exponential growth of written publications, comparing facts and findings between different research groups and communities becomes nearly impossible. In this paper, we present a conceptual approach and a first implementation for creating an open knowledge base of scientific knowledge mined from research publications. This requires to extract facts - mostly empirical observations - from unstructured texts (mainly PDF's). Due to the importance of extracting facts with high-accuracy and the impreciseness of automatic methods, human quality control is of utmost importance. In order to establish such quality control mechanisms, we rely on intelligent visual interfaces and on establishing a toolset for crowdsourcing fact extraction, text mining and data integration tasks.

Keywords: triplification, linked-open-data, web-based visual analytics, crowdsourcing, web 2.0.

1 Introduction

Scientific publications foster an extremely valuable body of knowledge and can be seen as the roots of our civilisation. However, with the exponential growth of written publications, comparing facts and findings between different research groups and communities becomes nearly impossible. For example, Armstrong et al. [1] conducted a meta-study on the improvements in ad-hoc retrieval concluding that *".. there is little evidence of improvement in ad-hoc retrieval technology over the past decade. Baselines are generally weak, often being below the median original TREC system"*. It could be expected, that more such results can be found in other tasks/disciplines. Efforts to overcome these issues are twofold.

A. Holzinger and G. Pasi (Eds.): HCI-KDD 2013, LNCS 7947, pp. 160–172, 2013.

First, various researchers developed benchmarking frameworks (e.g. TIRA [2]) for comparing different techniques and methodologies. Second, research data is published along with the regular papers. While both approaches are very important for the future of the research process, published papers usually do not make their research data explicit and reusable. In a recent paper, Holzinger et al. [3] showed that large-scale analysis of scientific papers can lead to new insights for domain experts. The authors showed that many disease have a strong statistical relationship to rheumatic diseases – however, full interpretation of automatically generated results must be performed by domain experts. It stands to reason that if domain experts are given adequate tools to master the enormous amount of scientific information, new, interesting, and helpful scientific knowledge could be obtained.

In this paper, we present a conceptual approach and first implementation for creating an open knowledge base of scientific knowledge mined from research papers. We aim to leverage research findings into an explicit, factual knowledge base which is re-usable for future research. This requires to extract facts from scientific papers, which are mostly published in Portable Document Format (PDF). Because the extracted facts need to be highly accurate and the automatic methods are imprecise, human quality control is of utmost importance. In order to establish quality control, we rely on visual interfaces and on establishing a toolset for crowdsourcing fact extraction, text mining and data integration tasks. While our conceptual approach aims to extract facts from different sources, we emphasise fact extraction and integration from tabular data in this paper (see [4] for more details on the CODE project in general). In particular, we:

1. Define a semantic format for expressing empirical research facts in the novel RDF Data Cube Vocabulary.
2. Discuss the process for extracting tabular data from research papers and integrating the single facts in those tables into the defined format. We show a first prototype along with a heuristic evaluation.
3. Discuss collaborative visual analytics applications on top of data cubes to foster extraction, integration and analysis of facts. We provide the visual analytics vocabulary, an ontology that supports semi-automatic mapping of data cubes to visualisations and sharing of the full state of visual analytics applications among collaborators.

Section 2 gives an overview of our approach, including a formal definition of the language for representing facts. Sections 3 to 5 then describe the core steps of our approach (table extraction, table normalisation and linking, web-based visual analytics). A summary and directions of future work are given in section 6.

2 Approach

Our approach comprises three steps: fact extraction from unstructured data sources, fact aggregation and integration, and fact analysis and sensemaking (see Fig. 1). The approach follows a classical KDD application, extraction of facts, integration with existing knowledge bases and subsequent analysis. However,

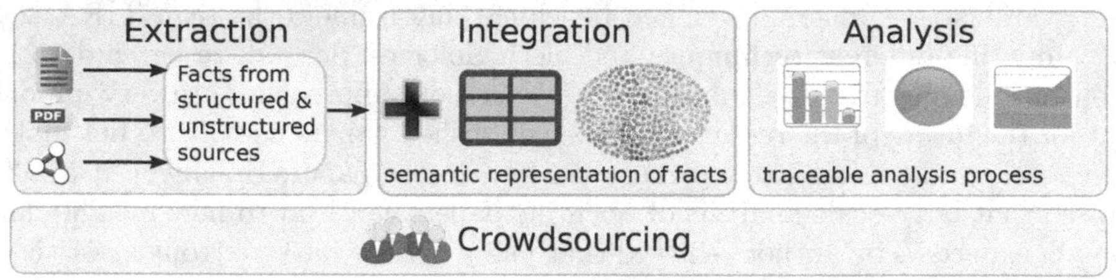

Fig. 1. The overall process for acquiring knowledge from scientific literature. All steps are semi-automatic involving users for quality control, data enrichment and discovery.

every single step is imperfect, adding errors to the final results. In order to overcome these issues, we emphasise crowdsourcing mechanisms through visual interfaces that engage users into quality control and enrichment processes. In order to conduct in-depth analysis of factual scientific knowledge we have to define a consistent semantic format for representing facts. We use the RDF Data Cube Vocabulary to achieve this. Given the RDF Data Cube Vocabulary, we have to address the following three questions in order to unleash the full potential for empirical facts:

1. How to extract data cubes from scientific papers? ⇒ section 3
2. How to integrate data cubes with existing linked data sources? ⇒ section 4
3. How to conduct visual analysis on extracted, integrated data cubes in a collaboratively? ⇒ section 5

As described above, we focus on empirical research data, which is mostly represented in tables of scientific publications. Thus, we have to identify tables in unstructured texts, normalise these tables and link the table structure and content to the Linked Data Cloud. All processing steps involve approaches, which do not always return perfect results when performed fully automatically. Thus, we require user involvement for quality control, data set enrichment and knowledge discovery. Hence, we developed visual interfaces in order to conduct the different steps semi-automatically. In the following we motivate the choice of representing facts as RDF data cubes (section 2.1).

2.1 Semantic Representation of Facts

In order to conduct in-depth analysis of factual scientific knowledge, we have to define the format for representing facts. In the last decade, Semantic Web languages have emerged as a general format of expressing explicit knowledge and associated facts. In their most basic form, RDF (Resource Description Format)[1], information is represented as Subject, Predicate and Object patterns. Together with languages of higher expressiveness, e.g., RDFS (RDF Schema)[2] and OWL (Web Ontology Language)[3], the Semantic Web standard enables reasoning

[1] http://www.w3.org/RDF/
[2] http://www.w3.org/TR/rdf-schema/
[3] http://www.w3.org/TR/owl2-overview/

capabilities and sophisticated querying mechanisms. While those languages provide powerful tools for expressing knowledge, it is not feasible to automatically create general models of scientific knowledge via text mining methods [5]. Consider for example the body of mathematical knowledge and axiomatic proofs. It seems to be unfeasible to extract and automatically reason over such a body of knowledge. Hence, we restrict our approach to *empirical facts*, i.e. statistically-based discoveries justified through experiments or observations. Examples are gene sequences using micro-array data, benchmarking of algorithms, and clinical trials.

Empirical facts can be represented using statistical languages. For example the SDMX standard (Statistical Data and Metadata eXchange) [6] defines the vocabulary and its relationships. In its most general form, an experiment results in a set of *observations* expressed by dependent and independent variables. The observation (dependent variable, response variable, measured variable) is the *measurement* (e.g. counts, real-valued number etc.) of the experimental outcome and has a *value* and a *unit*. The independent variables (explanatory variable, predictor variable) describes the observable circumstances of the measurement.

While the SDMX standard provides a sophisticated way for exchanging statistical data and metadata, it does not foster publishing and re-using this data in web-based environments. However, for integrating statistical data from various sources by a community, this is an essential prerequisite. As a remedy, the W3C published a draft for representing parts of the SDMX standard in RDF, which resulted in the so-called RDF Data Cube Vocabulary[4]. The vocabulary combines the statistical and database viewpoints extending the SDMX vocabulary with data warehousing concepts. In particular it defines the following elements, fully expressible in RDF:

Data Cubes represent a statistical data set containing empirical facts of interest. It can be formally defined as a triple $C = (\mathbf{D}, \mathbf{M}, \mathbf{O})$, of dimensions $\mathbf{D}$, measures $\mathbf{M}$, and observations $\mathbf{O}$.

Dimensions represent independent variables and serve to identify the circumstances of an observation (e.g. time, location, gender). Formally we define a Dimension as tuple $\mathbf{D} = (Values, Type, Rel, MD)$. $Values$ is the set of possible values in this dimension, $Type$ the type of dimension (i.g. ordinal, nominal, or real). MD is additional metadata for that dimension (e.g. human readable label, description) and Rel is a set of relationships of the form $(Predicate, Object)$ linking to other dimensions or semantic web concepts. For example, a dimension could be the type of "supervised machine learning algorithm" with the nominal values "Probabilistic Classifiers", "Linear Classifiers" and "Non-Linear Classifiers" and the relation that supervised machine learning is a sub-concept of "Machine Learning Algorithm".

Measures represent dependent variables and identify the observation made (e.g. blood pressure, accuracy of an algorithm). Formally, we define a Measure as a tuple $\mathbf{M} = (Unit, Attributes, MD)$ where $Unit$ determines the unit of the measure (e.g., kg, percent), $Attributes$ depicts features relevant for using the measure, like scaling factors and the status of the observation (e.g.

[4] `http://www.w3.org/TR/vocab-data-cube/`

estimated, provisional) and MD is a set of additional metadata (e.g., human readable labels, descriptions). For example, the accuracy of a classifier could be a measure provided as percentage of classifier decisions made.

Observations represent concrete measurements consisting of dimension values and measurements. Observation form the data points to be analysed. Formally it is defined as the tuple $O = (d_1 \dots d_n, m_1 \dots m_n)$ where $d_i \in D_i.Values$ is the value of the $i-th$ dimension. m_k is the $k-th$ measurement where the value of m_k is in the range of the specified type $M_k.Unit$. For example, the tuple $(SVM, DataSet1, 0.9, 0.8)$ could be an example tuple specifying that Support Vector Machines (SVM) achieve 0.9 precision and 0.8 recall on DataSet1.

Semantic Web languages provide powerful means for defining concepts and for identifying and re-using functional dependencies (e.g. taxonomic relationships) among dimensions. Moreover, we can enrich data cubes given some background knowledge. Consider for example the dimension city and the "number of inhabitants" as observation. Through freely available geospatial knowledge bases (e.g., geonames[5]), we can enrich the dimension "city" with additional dimensions like the state they are located in. Along such newly created dimensions we provide new means for correlating independent variables with the dependent ones resulting in powerful data sets for applying existing data mining techniques [7].

3 Extracting Tables from PDF

The Portable Document Format (PDF) provides not only textual information, but also layout information of characters and figures. While PDF's guarantee device-independent display of information, they do not provide structural information. Moreover, it is not guaranteed that word spaces are present. Hence, to extract text and structural elements from PDFs we had to analyse the layout of characters and merge them into usable blocks on different levels. By using a stack of clustering algorithms on layout and format features (e.g. font size, position), characters are merged to words, words to sentences and sentences to blocks. Then, we assign particular types to blocks using Conditional Random Fields. Example types are figure, caption, table, main text. The algorithms turned out to be robust on scientific literature achieving precision and recall between 0.75 and 0.9 (see [8] for more details and our online demo[6]). Still, the extraction remains imperfect, especially for complex tables, and the results can not be directly used for subsequent analysis without quality control.

4 Normalising and Linking Tables

As outlined in the previous section, fully automatic table extraction is imperfect and tables hardly occur in a format suitable for automatic linking. For instance, tables in papers provide dimensions in rows and/or columns and often have

[5] `http://geonames.org`

[6] `http://knowminer.at:8080/code-demo/`

Table 1. Example table and the abstract table model (not normalised). 3 dimensions, 2 measures and 4 observations. Values of dimension occur in columns and rows. Note, that there are no headings with the name of the dimensions.

		Data Set 1	
		Precision	Recall
SVM	Stopwords	0.7	0.8
Naive Bayes	No-Stopwords	0.9	0.8
		Data Set 2	
		Precision	Recall
SVM	Stopwords	0.75	0.4
Naive Bayes	No-Stopwords	0.3	0.4

		Dimension D_1 for Observations $O_1 \ldots O_2$	
		Measure M_1	Measure M_2
D_2 value of O_1	D_3 value of O_1	M_1 value of O_1	M_2 value of O_1
D_2 value of O_2	D_3 value of O_2	M_1 value of O_2	M_2 value of O_2
		Dimension D_1 for Observations $O_3 \ldots O_4$	
		Measure M_1	Measure M_2
D_2 value of O_3	D_3 value of O_3	M_1 value of O_3	M_2 value of O_3
D_2 value of O_4	D_3 value of O_4	M_1 value of O_4	M_2 value of O_4

Table 2. A normalised table and the corresponding abstract table model

Algorithm	Stopwords?	Precision	Recall
SVM	yes	0.7	0.8
Naive Bayes	no	0.9	0.8
MaxEnt	yes	0.95	0.7

D_1	D_2	M_1	M_2
D_1 of O_1	D_2 of O_1	M_1 of O_1	M_2 of O_1
D_1 of O_2	D_2 of O_2	M_1 of O_2	M_2 of O_2
D_1 of O_3	D_2 of O_3	M_1 of O_3	M_2 of O_3

merged cells, as shown in the example table 1. Hence, before creating an RDF Data Cube, the extracted tables have to normalised. Normalised means, that each column is either a dimension or a measure and that the first row contains the dimension and measure names while all subsequent rows contain the observation-value for the dimension/measure. A normalised table is comparable to a database table. Table 2 shows the normalised version of table 1.

After normalising a table, the last step is linking dimensions and measures to concepts in the Linked Data cloud and linking single values of dimensions to Linked Data concepts if possible. Due to the ambiguity of the textual labels this requires solving a disambiguation problem by using services like DBpedia SpotLight [9] or Sindice[7]. However, the context of the disambiguation problem is different: every value of a dimension is an instance of the concept for that dimensions. For example, “Berlin” as as value is an instance of the class “City”. This knowledge may be exploited for enhancing the disambiguation process itself. Given dimensions and dimension values which are represented using

[7] `http://sindice.com/`

Semantic Web concepts, we are able to add additional dimensions exploiting semantic predicates like taxonomic or mereotopological relationships or identify such relationships between existing columns. Hence, we get new, enriched data sets that allow to conduct analysis beyond the originally intended scope. Such analysis can be done automatically or by using visual interfaces. As outlined in the next section, we emphasise the latter one. Both steps, table normalisation and linking, are performed semi-automatically, including users' background knowledge for correcting and refining the automatic methods. The next section describes the prototypical implementation of this semi-automatic approach.

4.1 Prototypical Implementation

We have realised the table normalisation and linking through a web-based prototype. The prototype allows to upload a PDF, extracts the table structures from the PDF and allows to specify dimensions and measures along with their disambiguated concepts. The prototype is shown in Fig. 2 and is available online[8]. In the prototype, the semiautomatic fact extraction is performed in three steps.

Step 1: Table extraction: User selects a PDF or EXCEL file to upload. Plain text and tables are automatically extracted.

Step 2: Table refinement and annotation: User can remove wrongly as tables identified text blocks. User indicates cells containing measures and dimensions by selecting a set of adjacent cells. After that the user defines the attribute for that dimension/measure. A pop-up window shows suggestions for entities from the LOD cloud, the user can then accept or reject those links, thus annotating the cell semantically. Currently, the Sindice service is called for suggestions.

Step 3: Data Cube Export: The user defines the URI context for publishing the data, the data is automatically normalised (cmp. section 2.1), enriched with provenance information and stored. Normalisation can be done fully automatically in this step because the table is fully annotated after step 2.

Provenance plays an important role in each step. We use the PROV-O Ontology[9] for storing provenance information. Provenance information includes the origin of the data, who extracted and/or transformed the data.

4.2 Heuristic Evaluation of the Prototype

We performed a heuristic evaluation of the prototype for table extraction. The evaluation was performed by two experts which were not part of the development team. We used heuristic evaluation measures as proposed by Nielsen [10]. The evaluators were given two different scientific articles and performed the task of generating data cubes from tables from the PDF's using the prototype described in the previous section. The heuristic evaluation revealed the following issues:

[8] `http://zaire.dimis.fim.uni-passau.de:8080/code-server/demo/dataextraction`

[9] `http://www.w3.org/TR/prov-o/`

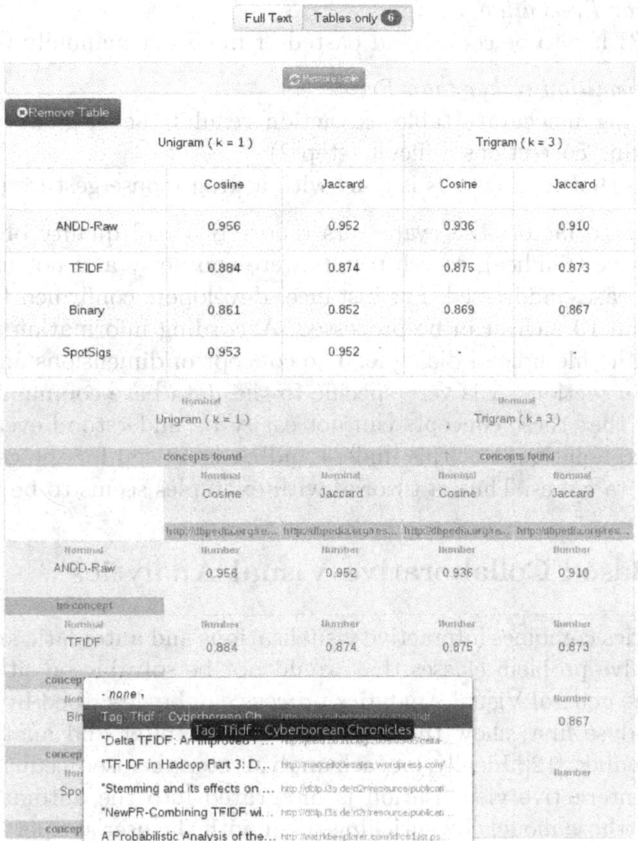

Fig. 2. Screenshot of the user interface, top - table after automatic extraction. bottom - manually annotated table and suggestions for corresponding LOD concepts.

Factor 1: Visibility of System Status

– problems with Firefox while file type test, unhelpful error message (step 1)
– uploading large files (20 pages), no progress after the file upload (step 1)
– after finishing the task there is no indicator on how to proceed (step 3)

Factor 2: Match Between System and the Real World

– when adding semantics to tables the notions of "dimensions" and "measures" are hard to understand (step 2)
– after annotating multiple dimensions, differences not visible anymore (step 2)
– not clear what the word "context" means when lifting data to RDF (step 3)

Factor 3: User Control and Freedom

– wizard does not allow users to move one step backward to perform corrections
– after selecting that a cell has no semantics, no undo possible (step 2)

Factor 5: Error Prevention
– context URI has to be copied and pasted or filled out manually (step 3)

Factor 6: Recognition rather than Recall
– when showing inaccurate table extraction results the original table is not shown making corrections difficult (step 2)
– not obvious that linking cells is done with a drag mouse gesture (step 2)

Apart from these factors the evaluators reported a bad quality of the disambiguation service (Sindice), search results were too noisy and not useful. Some issues can be easily addressed. For instance, developers confirmed that no files with more than 10 pages can be processed. According information needs to be presented on the file upload dialogue. The concept of dimensions and measures as presented in section 2.1 is very specific to the data base community. Evaluators expected that these concepts can not easily be understood even if there is an abstract explanation. But this understanding is crucial for the generation of meaningful data cubes. Thus, a tutorial with examples seems to be indicated.

5 Web-Based Collaborative Visual Analytics

Visual Analytics combines interactive visualisations and automatic analysis with the goal to solve problem classes that would not be solvable be either of them alone [11]. The general Visual Analytics process can be described by the famous mantra: "Analyse first, show the important, zoom, filter and analyse further, details on demand" [12]. Ideally, on each analysis step the user feedback collected through the interactive visualisation is integrated into the automatic models, thus adapting those models by enriching them with the user's expert knowledge.

Visual analytics applications have proven to be successful in many areas over the last few years [13], yet unresolved challenges remain, such as the traceability of the analytic process and the collaborative aspects. The traceability is especially important in enterprise context where decisions are made based on visual analytics processes and these processes have to be documented for quality assurance. As for collaborative aspects one crucial question remaining to be answered is how to combine the feedback of different experts, who might potentially disagree [14]. Furthermore, if we extend the notion of collaborative visual analytics to allow off-line collaboration, meaning that not all collaborators are working on the problem at the same time but sequentially, provenance information has to be included in the process. The CODE project aims at supporting collaborative, traceable, asynchronous visual analytics processes (asynchronous refers to the kind of collaboration), thus needing to resolve three core issues:

1. **Interactive Web-Based Visualisations:** Visualisations present the underlying data on a level suitable for sensemaking by humans and hide unimportant details from users. For web-based collaborative analytics HTML5 based multiple coordinated view (MCV) applications are necessary.

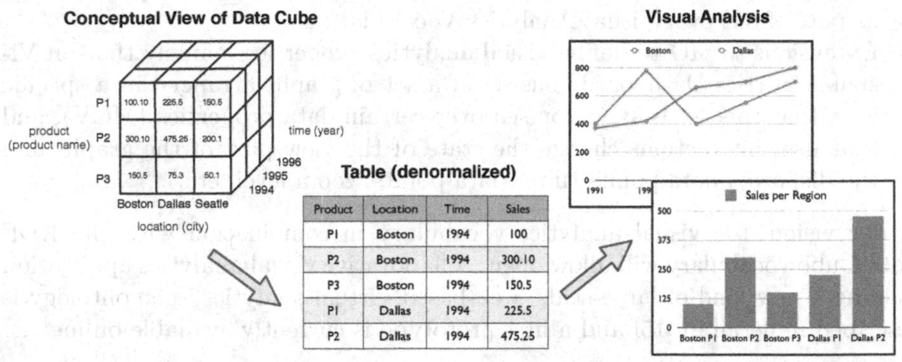

Fig. 3. Mapping from data cubes to visualisations supported by semantic vocabularies

2. **Mapping of Data Cubes to Visualisations:** Gluing the data to visualisations is a key step in any visual analytics application. Semantic description can be re-used for describing data cubes and their matching dimensions to support such mappings. For example, a data set with two nominal dimensions can not be visualised in a scatterplot. Fig. 3 depicts an example.
3. **Sharing the State of Visual Analytics Applications:** Insights generated through visual analytics are based on the visual state of an application. In order to foster collaboration (both real-time and asynchronous collaboration), users have to be able to share and persist the state of a visual application.

While a broad range of current HTML visualisation frameworks exist, we focus on creating a multiple coordinated views (MCV) framework using existing JavaScript visualisation libraries. We developed an OWL Ontology to support data mapping and to share the state of the visual analytics application. The ontology aims to ease interface development on Linked Data as well as to ensure discoverability of visual components and their use. It bridges between two standards, namely the RDF Data Cube Vocabulary for representing aggregated data sets and the Semantic Science Integrated Ontology (SIO)[10] for describing visual components. Also, we have defined a complete vocabulary to specify a visual analytics application as data sets, operations (e.g. filters) and a set of visualisations for displaying the data. The Visual Analytics Vocabulary[11] consists of the following parts:

1. Description of data points and data sources taken from the RDF Data Cube Vocabulary as well as a description of charts and their visual elements taken from the SIO.

[10] `http://semanticscience.org/ontology/sio.owl`

[11] Available at `http://code-research.eu/ontology/visual-analytics.owl`

2. A mapping between RDF Data Cubes and the Statistical Graph Ontology as part of our own Visual Analytics Vocabulary.
3. Extensions to SIO to define visual analytics properties, namely that an Visual Analytics Dashboard consists of a set of graphs arranged in a specific way, that this set is synchronised over certain data properties (MCV), and that user interactions change the state of the view port of the graphs and the displayed data points (filter data points, zoom level, etc.)

In our vision, the visual analytics vocabulary in combination with the RDF Data Cube vocabulary will allow open, collaborative visual analytics application yielding a new kind of large-scale, web-based virtual analytics. The ontology is described in detail in [15] and a first prototype is currently available online[12].

6 Conclusion and Future Work

We presented a conceptual approach for creating a open knowledge base of scientific knowledge mined from research publications. We detailed the necessary steps and concluded that all three steps (extraction, integration, and analysis) can not be done fully automatically. Our work is a first step in the direction of a large scale visual analytics application for analysis of results published in scientific papers. Domain experts can upload papers, extract findings and compare findings from different publications and making informed judgements.

The presented approach includes crowdsourcing concepts in the following two ways: First, extracted cubes are stored in a central repository and can be shared among users. These stored cubes represent not only facts from papers, but also users' background knowledge integrated through the semi-automatic annotation process. Cube sharing allows collaborative visual analysis, potentially leading to insights that otherwise not possible. Second, in the table linking step, the user is presented with suggestions from a disambiguation service. User's decisions of accepting, rejecting or creating new concepts are collected and are planned to be used to improve the link suggestions. This user feedback to the link suggestion service, which is technically a table disambiguation service, is currently not exploited, but a service integrating those suggestions is under development.

While the conceptual approach seems sound, our prototypes are not fully usable by domain experts, yet. The developed prototypes for each steps need to be effectively combined into a consistent web application to allow flawless and user-friendly semi-automatic fact extraction and analysis for scientific papers. Further, cubes extracted by other users can be combined with own extracted cubes allowing collaborative analysis.

We presented a heuristic evaluation of a prototype of web-based user interface for table extraction. We will improve the prototype addressing issues detected by the heuristic evaluation. For improving the quality of the automatic suggestions our own disambiguation service is developed within the CODE project aiming at providing high-quality suggestions for LOD entities and replacing Sindice in the final version.

[12] `http://code.know-center.tugraz.at/vis`

Acknowledgements. The presented work was developed within the CODE project funded by the EU Seventh Framework Programme, grant agreement number 296150. The Know-Center is funded within the Austrian COMET Program under the auspices of the Austrian Ministry of Transport, Innovation and Technology, the Austrian Ministry of Economics, Family and Youth and by the State of Styria. COMET is managed by the Austrian Research Promotion Agency FFG.

References

1. Armstrong, T.G., Moffat, A., Webber, W., Zobel, J.: Improvements that don't add up: ad-hoc retrieval results since 1998. In: Proceedings of the 18th ACM Conference on Information and Knowledge Management, CIKM 2009, pp. 601–610. ACM, New York (2009)
2. Gollub, T., Stein, B., Burrows, S.: Ousting Ivory Tower Research: Towards a Web Framework for Providing Experiments as a Service. In: Hersh, B., Callan, J., Maarek, Y., Sanderson, M. (eds.) 35th International ACM Conference on Research and Development in Information Retrieval (SIGIR 2012), pp. 1125–1126. ACM (2012)
3. Holzinger, A., Simonic, K.M., Yildirim, P.: Disease-disease relationships for rheumatic diseases: Web-based biomedical textmining an knowledge discovery to assist medical decision making. In: 2012 IEEE 36th Annual Computer Software and Applications Conference (COMPSAC), pp. 573–580 (2012)
4. Stegmaier, F., Seifert, C., Kern, R., Höfler, P., Bayerl, S., Granitzer, M., Kosch, H., Lindstaedt, S., Mutlu, B., Sabol, V., Schlegel, K., Zwicklbauer, S.: Unleashing semantics of research data. In: Proceedings of the 2nd Workshop on Big Data Benchmarking (2012)
5. Hazman, M., El-Beltagy, S.R., Rafea, A.: A survey of ontology learning approaches. International Journal of Computer Applications 22, 36–43 (2011); Published by Foundation of Computer Science
6. Gillman, D.W.: Common metadata constructs for statistical data. In: Proceedings of Statistics Canada Symposium 2005: Methodological Challenges for Future Information needs Catalogue no. 11-522-XIE (September 2005)
7. Paulheim, H., Fürnkranz, J.: Unsupervised generation of data mining features from linked open data. In: International Conference on Web Intelligence and Semantics, WIMS 2012 (2012)
8. Kern, R., Jack, K., Hristakeva, M.: TeamBeam – Meta-Data Extraction from Scientific Literature. D-Lib Magazine 18 (2012)
9. Mendes, P.N., Jakob, M., García-Silva, A., Bizer, C.: Dbpedia spotlight: shedding light on the web of documents. In: Proceedings of the 7th International Conference on Semantic Systems. I-Semantics 2011, pp. 1–8. ACM, New York (2011)
10. Nielsen, J.: Enhancing the explanatory power of usability heuristics. In: Proceedings of the SIGCHI Conference on Human Factors in Computing Systems, CHI 1994, pp. 152–158. ACM, New York (1994)
11. Thomas, J.J., Cook, K.A. (eds.): Illuminating the Path: The Research and Development Agenda for Visual Analytics. IEEE Computer Society (2005)

12. Keim, D.A., Mansmann, F., Thomas, J.: Visual analytics: how much visualization and how much analytics? SIGKDD Explor. Newsl. 11, 5–8 (2010)
13. Keim, D.A., Mansmann, F., Oelke, D., Ziegler, H.: Visual analytics: Combining automated discovery with interactive visualizations. In: Boulicaut, J.-F., Berthold, M.R., Horváth, T. (eds.) DS 2008. LNCS (LNAI), vol. 5255, pp. 2–14. Springer, Heidelberg (2008)
14. Heer, J., Agrawala, M.: Design considerations for collaborative visual analytics. In: IEEE Symposium on Visual Analytics Science and Technology, VAST 2007, pp. 171–178 (2007)
15. Mutlu, B., Hoefler, P., Granitzer, M., Sabol, V.: D 4.1 - Semantic Descriptions for Visual Analytics Components (2012)

Digital Archives: Semantic Search and Retrieval

Dimitris Spiliotopoulos[1], Efstratios Tzoannos[1], Cosmin Cabulea[2], and Dominik Frey[3]

[1] Innovation Lab, Athens Technology Centre, Greece
{d.spiliotopoulos,e.tzoannos}@atc.gr
[2] New Media, Innovation, Deutsche Welle, Bonn, Germany
cosmin.cabulea@dw.de
[3] Documentation and Archives, Suedwestrundfunk, Baden-Baden, Germany
dominik.frey@swr.de

Abstract. Social media, in the recent years, has become the main source of information regarding society's feedback on events that shape the everyday life. The social web is where journalists look to find how people respond to the news they read but is also the place where politicians and political analysts would look to find how societies feel about political decisions, politicians, events and policies that are announced. This work reports on the design and evaluation of a search and retrieval interface for socially enriched web archives. The considerations on the end user requirements regarding the social content are presented as well as the approach on the design and testing using a large collection of web documents.

Keywords: digital archives, social networks, user experience, big data.

1 Introduction

News, events and views are part of the everyday people's interaction. In the times of social media, people communicate their thoughts and sentiments over the events and views that are presented to them via social networks [1, 2]. Their views are opinionated and can be viewed and responded to by the rest of the community. Organizations that are involved in the process of collecting and processing the people's views, such as Broadcasters and Political Analyst Groups, try to harvest as much of the content as possible on an everyday basis. Instead of blindly searching for possible responses to news one by one, they use specialized tools to collect, analyze, group, aggregate, fuse and deliver the people's opinions that are relevant to their analyses [3]. Broadcasters use the social network information for two purposes. The first is for the classification of importance of events or entities (persons, locations, etc.) reported in the news. The second is to further refine their search on opinions for specific entities or events that have exhibited some kind of importance, for example opinions about a certain person that were very diverse or polarized.

The target of this work is an important problem of the recent years: big data and the approach that has to be adopted by the HCI researcher in order to create an interactive system for users to appreciate and reason the results of the big data

A. Holzinger and G. Pasi (Eds.): HCI-KDD 2013, LNCS 7947, pp. 173–182, 2013.

analysis. It presents the exploratory usability studies, in which user groups and usability requirements are identified and the follow up report on the archivist and end user findings.

This work reports on the design and evaluation of a user interface for search and retrieval of archived web documents. We are particularly interested in the user experience design and testing involved with the aim of building a final user interface prototype that will enable both archivists and end users to search into collections of archived content and retrieve, social web information such as opinions, sentiments, key peoples' names, and so on.

The next session provides the related work followed by a discussion on the user requirements for the Search and Retrieval Application interface (SARA) and the core functionalities. Then, the user experience methodology is described along with preliminary results and challenges faced. Next, the design considerations along with certain technical approaches are presented. Finally, the user interface prototype is presented along with the analyzed results on the archivist and end user activity.

2 Related Work

Web archiving is about striving to preserve a complete and descriptive snapshot of the available web data for the future. In the always-changing Web, a dynamic approach of content selection and appraisal is important to ensure that the web data that will be archived are of high quality, present a complete description of the selected area of context and that this description is persistently retained in the resulting archives.

For content and language analytics, social networks are a major part of the Semantic Web [4]. Social network information is the focus of major research because of the vast variety of content authors and the potential of the analyses that can be performed [5]. The crawling process collects the data according to parameters set by the archivist. The data quality at this point can be measured using specific crawling strategies [6]. The collected content is analyzed and annotated with semantic meta-information. The semantic data are archived as meta-data for the web data.

Identified *entities* are the most common result of linguistic analysis and the task of searching for entities in social networks involves the recognition of entities and relations [7]. The *news* domain is a large area of application for sentiment and includes many sources such as news web sites, blogs, RSS feeds and social media [8, 9] Entities are the most important element for creating training sets for sentiment analysis [10]. They are used to describe the ontologies needed for sentiment analysis for texts in both generic and specialized domains such as the arts [11]. Moreover, entities and relations can be modeled for ontological approaches beyond traditional polarity sentiment, for example for modeling emotion recognition by relating entities with human emotional states such as arousal and pleasantness [12].

Especially for the Social Web, from the moment that social networks such as Twitter provided an API for collecting information, sentiment analysis can be performed in a multitude of ways. Saif et al. have used features like semantic concept to predict sentiment on Twitter data sets [13] while other works present ways of analyzing sentiment using hashtags as feature annotation [14]. Applying sentiment analysis on collected texts requires that the text be processed to ensure that is valid and clean such as all data are for the intended language, the non-textual segments from the collection process are removed and so on. Petz et al. presented a process model for preprocessing such corpora [15].

The design and development of dedicated search and retrieval interfaces for accessing archived content is a dedicated task in the area of web archiving [16]. Hearst et al. presented the requirements for a successful search interface where they stressed the importance of successful faceting and browsing of the content using metadata [17].

Semantic search [18] can be performed over data that include semantic metadata and involves the use of complex semantic queries. Methods have been proposed to simplify the complexity of the semantic queries by translating keywords to formal queries [19]. However, from the end user point, the complexity of the semantic queries should be left out of the user interaction flow by providing the means to make semantic search simple of all users as a step towards maximizing usability [20]. Such simplicity can be achieved by providing ways to faceting through metadata after a simple initial search, allowing the user to dive deeper into the semantic context rather instead of requiring complex semantic queries to be entered at the initial search [21]. Including both browsing and refining qualities on the faceted metadata can further optimize faceting [22].

Usability is a major factor for measuring success for semantic web applications [23] while success is measured taking in to account the human factor, the user [24]. Understanding the user behavior behind the search workflow is paramount to designing a successful search interface [25]. Studying the intentions of the users and their expectations of the search and retrieval process can be the basis of a successful user centered design [26]. Works on usability testing suggest that a search interface should empower the users during the whole information retrieval process [27].

There are many usability evaluation methods that can be deployed for evaluating web interfaces. Evaluating usability entails both usability inspection and usability testing methods to be applied and a successful usable design should be a product of iterative usability evaluation that involves the users in all stages of the development lifecycle [28].

3 The Rationale behind the SARA User Interface

SARA is an integral part of the ARCOMEM project [29], the purpose of which is to leverage the wisdom of the people for web content selection and appraisal for digital preservation. The typical system process for the collection and analysis of the web and social web data is as follows:

(a) The crawling process collects web pages [30] and social web sources [31], based on initial seed lists and keywords that collectively describe a domain (e.g. EU economic crisis).

2. During the online phase the system analyzes the collected data and produce information regarding the sources, dates, reputation statistics, etc. The data are stored in an appropriate document store [32].
3. The semantic analysis, the offline phase, analyses the web documents and the social data and produces the semantic information. The semantic data are stored in a Resource Description Framework (RDF) structure in order to be easily searchable by semantic queries.
4. A high-level analysis is performed on the socially-derived data from which statistics like opinion mining, cultural analysis, entity analysis, are obtained.

Users that comment on news, events and entities are part of the data that are collected and analyzed in order to select the most important opinions, views, key players and roles that are, in effect, the targets of interest of the communities. Entities, such as people and locations, are identified and their importance as well as the user opinions on them is examined. Related entities are also discovered and further analyzed. Information from the social networks is linked to other web sources in order to provide a complete picture of the events that shaped the opinions of the readers.

The aim of the SARA web interface is to provide the means for the archivists and journalists to semantically search the vast amount of data (raw, semantic, social, analytics), retrieve and visualize the content so that all semantic links between the user search and the data are retained. Data include:

- Web resources (text, images, videos)
- Semantic information (sentiments, opinions)
- Entities (people, locations, events, etc.)
- Social network sources (statistics, user name, location, activity, etc.)

The above data had already been processed by the analysis modules of the ARCOMEM system so that more information about the semantic relations and the social web analytics has been stored. A non-definitive list of such data for a specific set of search parameters is:

- List of most relevant social media posts for one or more entities
- List of most diverse social media posts
- Topic detection
- The most influential users from social media
- Entity evolution information

The images and videos are content types that are not processed semantically but directly to provide indication of duplication of documents (i.e. news articles that contain the same video or picture may also be duplicates) or information on the evolution of entities over time. An example for the latter are pictures depicting the same entity at different points in time that provide explicit verification that the entity description has evolved, e.g. Cardinal Francis was formerly Cardinal J.M. Bergoglio.

4 User Experience Considerations

The user requirements collection was initially based on the expected functionalities by the two main user groups, Broadcasters and Parliament Archivist. The initial list of requirements that was based on the conceptual design of such web interface was very long, because of all the possible results expected from the analyses of the web data. The target users were overwhelmed by both the broad potential uses of the analyzed social information but also from the, at that time, unknown usefulness of each bit of that vast amount of data. That realization has made obvious the fact that the core advantage of the ARCOMEM approach, that is the content diversity and social web data potential, was also the main problem to solve regarding the actual design of the user interface. The provision, type and quality of the analyzed data would have to drive the user interface design and interaction process. In order to tackle this, it was decided that the best approach would have to be a combined focus group discussion on the user-system interaction and a heuristic approach during the requirements gathering. In our case, the classic low-fidelity prototyping would have had minimal, if any, success, since the user interaction would be driven by the actual content.

During the focus group discussions the archivists, which are experts in search and retrieval interfaces, were presented with possible approaches using examples of real web interfaces that are used for archived documents, like the Europeana portal [33]. A quick breakthrough came when it was realized that the social media content itself as well as the analyzed semantic information from the social media was the most controversial part of the user interaction. The users were very interested in using the semantic data for their search but were not fully aware of why that information was there and where it was derived from. It was also obvious that different levels of importance could be assigned to the types of semantic data.

The user perception on the importance of the types of semantic information for search and retrieval of web documents was an open research question as well as a pre-requisite for the user experience design of the SARA interface. A series of experiments were run on first time users using an early demo interface populated with semantic information [34]. The populated data were carefully selected and several delivery options in the user interface were explored (in text, separate lists, tag clouds, facets, etc.). The think aloud approach was used to get the subjective user feedback but also a simplistic analysis of the user path selections was done by logging the user clicks and time. The above approach has led to an initial set of functional and non-functional requirements that were used to create the high-quality interface prototype using a small subset of content data for the formative evaluation.

5 Design and Technical Challenges

The design of the interface prototype was based on the archivist and journalist end-user feedback. The expected type of information from the retrieval process by both main user groups was formalized. The obligatory entities and opinions were first on the list but so was marking and ranking of the most reputable sources. The journalists reported that events, both standalone (if unprocessed) or aggregated (if such process was available) were at the top of their priority. It was also asked that Twitter users, blog posters and other entities that report events should be analyzed for their reputation status, location and any other values that would assign them a "trustworthy source" for the reported event. This requirement bears similarities to an earlier study where journalists placed importance to the type of user that reports in the social media or the web, assigning the "eyewitness" versus "non-eyewitness" and the "journalist/blogger" versus "ordinary individual" values [35]. Apart from the above, the users asked for lists of relevant Twitter posts for each web resource. A web resource can be a web page, blog post, wiki page, and so on, but not only. Web resources are comprised of several web objects such as Twitter posts, blog comments, and other pieces of information that bear direct semantic relation to the bound entities and events.

The semantic information was chosen to be indexed using the named entities as the basic starting point. Each entity may have other entities associated with it, opinions, participate in an event, and so on. This approach provides the advantage that a complete list of attributes is always available for each entity. But it also requires several hops through the RDF storage via SPARQL queries in order to collect it. Moreover, SPARQL is slow and inefficient for such queries and does not support full text search. RDF storage search does not offer functionalities like faceting, hit highlighting, lemmatization, stemming, etc. In order to allow for fast indexing, it was decided that a full text search engine should be deployed.

In order to minimize response times, it was decided to fully populate the Solr index offline. For this purpose, it was needed to migrate most of the existing information from the RDF triple store to the Full Text Search Engine beforehand. The best practice for this was to convert the RDF triple store into a set of flattened documents that are compatible with the Solr schema. These flattened documents had to be populated one by one through an indexing process with values retrieved from the RDF triple store.

For this task, a custom indexing module in Java and Groovy was implemented. Parts of Groovy dynamic programming language were added to create a kind of domain specific language. This small domain specific language (a set of classes) offers to an external user the possibility to easily change indexing rules on the fly.

There are several approaches for the indexing procedure. Nevertheless, the most appropriate due to the large volume of the knowledge base was to apply an incremental formulation of Solr documents and index them one by one. This practice, though more time consuming, has minimal requirements for memory in both indexer and Solr Handlers while indexing. Moreover, it can be stopped anytime during indexing and resumed afterwards, without affecting the whole process.

Each Solr document is representing an RDF web resource instance. It starts by retrieving a set of web resource-ids Iterator from the RDF triple store. Then, the process iterates through the ids of the web resources, and, for each id, all the required fields referring to this web resource instance are retrieved by SPARQL queries. Once the Solr document is fully populated it is inserted into Solr. The type of fields of the Solr schema has been appropriately configured for the purposes of this case.

6 The SARA Prototype User Interface

The interface prototype was releases and a small scale domain on the Greek economic crisis was indexed. This domain was analyzed and tested manually in order to produce a large set of semantic data that are free of statistical errors. The obvious purpose for this was that the user feedback should be unaffected by potential error from the semantic analysis.

Fig. 1. Results of semantic search within a political domain

Figure 1 depicts a typical search results page. Social media derived data are used for search refinement and for the document opinion and trending visual preview. The entity terms that were extracted from the social media are used both as search parameters as well as facets.

The search results may be sorted by modality and social network source. Each search result contains the aggregated opinion of the web resource textual data (top right) and the social network source and opinion trending (bottom left). The user may perform semantic or full text search and retrieve all web resources that contain the list of search terms. The faceting is used to drill into the search parameters.

Figure 2 shows the web resource page that lists, apart from the raw content:

- Entities and events extracted from the document text
- Related events and latest social posts that are related to the search results (bottom left),
- Tag cloud that is dynamically comprised of entities and events retrieved from the search results (top left) and can be used for quick follow-up search,
- Twitter posts ranked by opinion polarity (far bottom),
- The most influential users relevant to the web resource content (bottom right),
- Timeline depicting the opinion over time for the associated entities of the resource.

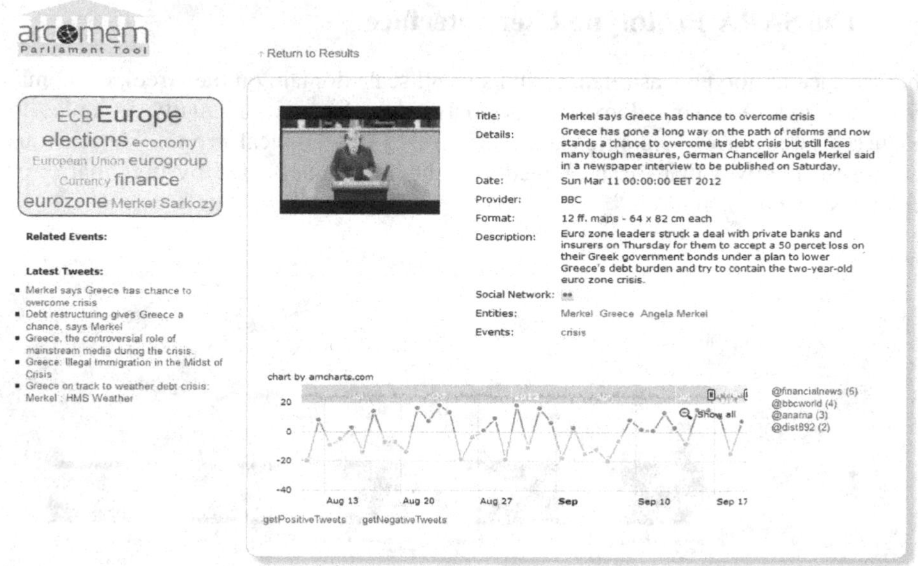

Fig. 2. A web resource page, enriched with social information

From the logs of the user interface usage, there are indications that two main user types are identified: the archivist and the end-user. This is a significant change from the initial assumption that user types such as a journalist and a political analyst end user are different types. The archivist proceeds in a well structured manner and goes through several items of the same hierarchical level in order to fulfill a sense of completeness for the archived data. The end-user, on the other hand, employs a more aggressive approach that drills down into the content, entities and opinions, in order to collect several examples that establish and validate a news story.

Archivist. Views the content and evaluate availability and completeness of the selected web documents for digital preservation. Expert user, highly trained to locate missing groups of information, web resources and semantic.

End-User. The generic user type. It includes researchers in news reporting, such as broadcasters and journalists, as well as researchers in other fields that are interested in the social web information. The latter may be policy makers, political/parliamentarian assistants, students of social sciences, law, and so on.

7 Conclusion and Further Work

This work reported on the user experience design considerations and findings for the design of a search and retrieval web interface for socially-aware web preservation. The next iteration of the design will involve the end users more actively. Transcribed scenarios will guide the users to perform certain actions in order to evaluate their interaction with the system. Free form search and retrieval will also be monitored in order to assess the efficiency of the current design and identify factors that may lead to usability performance optimizations by letting the users exploit the full potential of the social content. User training sessions are expected to provide new insight into how archivists and journalist end users extensively access the interface to its full potential. The expected results could be very revealing as to the nature of the archivist search behavior, eventually leading to a more refined user experience for both user types.

Acknowledgements. The work described here was partially supported by the EU ICT research project ARCOMEM: Archive Communities Memories, www.arcomem.eu, FP7-ICT-270239.

References

1. Schefbeck, G., Spiliotopoulos, D., Risse, T.: The Recent Challenge in Web Archiving: Archiving the Social Web. In: Int. Council on Archives Congress, Brisbane, Australia, August 20-24 (2012)
2. Anderson, R.E.: Social impacts of computing: Codes of professional ethics. Social Science Computing Review 10(2), 453–469 (1992)
3. Golberg, J., Wasser, M.: SocialBrowsing: Integrating social networks and web browsing. In: Proc. CHI 2007, San Jose, USA, April 28–May 3 (2007)
4. Musial, K., Kazienko, P.: Social Networks on the Internet. World Wide Web 16, 31–72 (2013)
5. Torre, L.: Adaptive systems in the era of the semantic and social web, a survey. User Model. User-Adapt. Interact. 19(5), 433–486 (2009)
6. Denev, D., Mazeika, A., Spaniol, M., Weikum, G.: The SHARC framework for data quality in Web archiving. VLDB 20(2), 183–207 (2011)
7. You, G., Park, J., Huang, S., Nie, Z., Wen, J.-R.: SocialSearch+: enriching social network with web evidences. World Wide Web (2013)
8. Godbole, N., Srinivasaiah, M., Skiena, S.: Large-scale sentiment analysis for news and blogs. In: Proceedings of the International Conference in Weblogs and Social Media (2007)
9. Ruiz-Martinez, J.M., Valencia-Garcia, R., Garcia-Sanchez, F.: Semantic-Based Sentiment analysis in financial news. In: Proc. 1st Int. Workshop on Finance and Economics on the Semantic Web (FEOSW 2012) in conjunction with 9th Extended Semantic Web Conference (ESWC 2012), Heraklion, Greece, May 27-28 (2012)
10. Kumar, A., Sebastian, T.M.: Sentiment Analysis: A Perspective on its Past, Present and Future. IJISA 4(10), 1–14 (2012)
11. Baldoni, M., Baroglio, C., Patti, V., Rena, P.: From tags to emotions: Ontology-driven sentiment analysis in the social semantic web. Intelligenza Artificiale 6(1), 41–54 (2012)
12. Zhang, X., Hu, B., Chen, J., Moore, P.: Ontology-based context modeling for emotion recognition in an intelligent web. World Wide Web (2013)
13. Saif, H., He, Y., Alani, H.: Semantic Sentiment analysis of Twitter. In: Cudré-Mauroux, P., et al. (eds.) ISWC 2012, Part I. LNCS, vol. 7649, pp. 508–524. Springer, Heidelberg (2012)
14. Mukherjee, S., Malu, A., Balamuralli, A.R., Bhattacharyya, P.: TwiSent: A Multistage System for Analyzing Sentiment in Twitter. In: Proc. 21st ACM Int. Conf. on Information and Knowledge Management (CIKM 2012), Maui, USA, October 29–November 02, pp. 2531–2534 (2012)
15. Petz, G., Karpowicz, M., Fürschuß, H., Auinger, A., Winkler, S.M., Schaller, S., Holzinger, A.: On text preprocessing for opinion mining outside of laboratory environments. In: Huang, R., Ghorbani, A.A., Pasi, G., Yamaguchi, T., Yen, N.Y., Jin, B. (eds.) AMT 2012. LNCS, vol. 7669, pp. 618–629. Springer, Heidelberg (2012)
16. Lesk, M.: Understanding Digital Libraries. Morgan Kaufmann (2004)

17. Hearst, M., English, J., Sinha, R., Swearingen, K., Yee, P.: Finding the Flow in Web Site Search. Communications of the ACM 45(9), 42–49 (2002)
18. Guha, R., McCool, R., Miller, E.: Semantic Search. In: Proc. 12th Int. Conf. on World Wide Web, pp. 700–709 (2003)
19. Wang, H., Zhang, K., Liu, Q., Tran, T., Yu, Y.: Q2Semantic: A lightweight keyword interface to semantic search. In: Bechhofer, S., Hauswirth, M., Hoffmann, J., Koubarakis, M. (eds.) ESWC 2008. LNCS, vol. 5021, pp. 584–598. Springer, Heidelberg (2008)
20. Lei, Y., Uren, V.S., Motta, E.: SemSearch: A Search Engine for the Semantic Web. In: Staab, S., Svátek, V. (eds.) EKAW 2006. LNCS (LNAI), vol. 4248, pp. 238–245. Springer, Heidelberg (2006)
21. Makela, E., Hyvonen, E., Sidoroff, T.: View-Based User Interfaces for Information Retrieval on the Semantic Web. In: Proc. ISWC-2005 Workshop on End User Semantic Web Interaction (November 2005)
22. Ziang, J., Marchionini, G.: Evaluation and Evolution of a Browse and Search Interface: Relation Browser++. In: Proc. Conf. on Digital Government Research, pp. 179–188 (2005)
23. Nedbal, D., Auinger, A., Hochmeier, A., Holzinger, A.: A Systematic Success Factor Analysis in the Context of Enterprise 2.0: Results of an Exploratory Analysis Comprising Digital Immigrants and Digital Natives. In: Huemer, C., Lops, P. (eds.) EC-Web 2012. LNBIP, vol. 123, pp. 163–175. Springer, Heidelberg (2012)
24. Calero Valdez, A., Schaar, A.K., Ziefle, M., Holzinger, A., Jeschke, S., Brecher, C.: Using mixed node publication network graphs for analyzing success in interdisciplinary teams. In: Huang, R., Ghorbani, A.A., Pasi, G., Yamaguchi, T., Yen, N.Y., Jin, B. (eds.) AMT 2012. LNCS, vol. 7669, pp. 606–617. Springer, Heidelberg (2012)
25. Teevan, J., Alvarado, C., Ackerman, M.S., Karger, D.: The perfect search engine is not enough: a study of orienteering behavior in directed search. In: Proc. SIGCHI Conf. on Human Factors in Computing Systems (CHI 2004), pp. 415–422 (2004)
26. Chen, Z., Lin, F., Liu, H., Liu, Y., Wenyin, L., Ma, W.: User Intention Modeling in Web Applications Using Data Mining. World Wide Web: Internet and Web Information Systems 5, 181–191 (2002)
27. Taksa, I., Spink, A.H., Goldberg, R.R.: A task-oriented approach to search engine usability studies. Journal of Software 3(1), 63–73 (2008)
28. Holzinger, A.: Usability engineering methods for software developers. Communications of the ACM 48, 71–74 (2005)
29. ARCOMEM: Archive Communities Memories. FP7-ICT-270239, `http://www.arcomem.eu`
30. Faheem, M.: Intelligent crawling of Web applications for Web archiving. In: Proc. PhD Symposium of WWW, Lyon, France (April 2012)
31. Gouriten, G., Senellart, P.: API Blender: A Uniform Interface to Social Platform APIs. In: Proc. Developer Track of WWW, Lyon, France (April 2012)
32. WARC File Format specifications, `http://archive-access.sourceforge.net/warc/`
33. Europeana Cultural Collections Archive Portal, `http://www.europeana.eu/portal/`
34. Spiliotopoulos, D., Tzoannos, E., Stavropoulou, P., Kouroupetroglou, G., Pino, A.: Designing user interfaces for social media driven digital preservation and information retrieval. In: Miesenberger, K., Karshmer, A., Penaz, P., Zagler, W. (eds.) ICCHP 2012, Part I. LNCS, vol. 7382, pp. 581–584. Springer, Heidelberg (2012)
35. Diakopoulos, N., De Choudhury, M., Naaman, M.: Finding and assessing social media information sources in the context of journalism. In: Proc. 2012 ACM Annual Conf. Human Factors in Computing Systems (CHI-2012) (2012)

Inconsistency Knowledge Discovery for Longitudinal Data Management: A Model-Based Approach

Roberto Boselli, Mirko Cesarini, Fabio Mercorio, and Mario Mezzanzanica

Department of Statistics and Quantitative Methods, CRISP Research Centre, University of Milan-Bicocca, Milan, Italy
{roberto.boselli,mirko.cesarini,fabio.mercorio, mario.mezzanzanica}@unimib.it

Abstract. In the last years, the growing diffusion of IT-based services has given a rise to the use of huge masses of data. However, using data for analytical and decision making purposes requires to perform several tasks, e.g. data cleansing, data filtering, data aggregation and synthesis, etc. Tools and methodologies empowering people are required to appropriately manage the (high) complexity of large datasets.

This paper proposes the multidimensional RDQA, an enhanced version of an existing model-based data verification technique, that can be used to identify, extract, and classify data inconsistencies on longitudinal data. Specifically, it discovers fine grained information about the data inconsistencies and it uses a multidimensional visualisation technique for showing them. The enhanced RDQA supports and empowers the users in the task of assessing and improving algorithms and solutions for data analysis, especially when large datasets are considered.

The proposed technique has been applied on a real-world dataset derived from the Italian labour market domain, which we made publicly available to the community.

Keywords: Data Cleansing, Data Quality, Model Checking, Model based approach, Data Visualisation.

1 Introduction and Related Work

ICT based services have opened the frontiers for managing data in an unforeseeable manner. From one side, methodologies, infrastructures, and tools for handling and analysing huge amount of data are now available, from the other site a lot of datasets are collected in several scenarios e.g. (citing two antithetic examples) from sensor networks to information systems. Considering the latter, a lot of data are collected about people's everyday life (from public services to social networks), and can be used to deeply describe social, economic, and business phenomena.

Focusing on information systems, it is well known that their data quality is frequently very low [1] and, due to the "garbage in, garbage out" principle, dirty

A. Holzinger and G. Pasi (Eds.): HCI-KDD 2013, LNCS 7947, pp. 183–194, 2013.

data strongly affect the information derived from them (e.g., see [2]). Accessing the real data or alternative and trusted data sources is rarely feasible, therefore data cleansing is often the only viable solution. Hence, in this context the data elaboration activities strongly focus on data cleansing.

Works focusing on data quality improvement can be classified into the following paradigms: *Rules based error detection and correction* allows users to specify rules and transformations to clean the data, a survey can be found in [3]. Specifying rules can be a very complex and time consuming task. Furthermore, bug fixing and rules maintenance require a non negligible effort. *Machine learning methods* exploit learning algorithms for error localisation and correction. An algorithm can be used to identify errors and inconsistencies after a training phase. It is well known that these methods can improve their performance in response to human feedbacks, however the model resulting from the learning phase cannot be easily accessed and interpreted by domain experts. *Record linkage* [4] aims to bring together corresponding records from two or more data sources or finding duplicates within the same one. The record linkage problem falls outside the scope of this paper, therefore it is not further investigated. A detailed survey of *data quality tools* is not reported due to space limitations, the interested reader can refer to [3]. Despite a lot of research activities have been carried out, the development of cleansing routines and algorithms is still a very complex and error-prone task. Furthermore, the complexity arises when longitudinal data are considered. Longitudinal data (or panel data) refer to a set of repeated observations of the same object or subject at multiple time points. These observations can be ordered with respect to time, generating a *longitudinal data sequence*.

Focusing on the data visualisation task, it is widely recognized that data visualisation and visual exploration play a key role in the knowledge discovery process, since they provide an effective understanding of the data and of the information managed [5]. To this aim, a number of data visualisation techniques are currently explored in several contexts, from healthcare to management information systems (e.g., [6, 7, 8, 9]).

The idea behind this paper draws on the work of [10] where Finite State Systems (FSSs) are used to formalise the consistent evolution of longitudinal data sequences. The authors exploit an explicit model checking technique to verify if the data sequence evolution is consistent or not (according to the model semantics). At the end of the process, the input dataset is classified into two data sequence partitions: the consistent and the inconsistent one.

To this regard, let us consider the dataset showed in Tab. 1(a): it describes the events recorded by a mobile telephone operator for lawful interception purposes. Lawful Interception is a security process where a service provider or a network operator collects individuals intercepted data or communications on behalf of law enforcement officials, see [11] for more details. The data showed in Tab. 1(a) refers to events describing mobile phones connecting to cells (of a cellular network), performing calls, exchanging messages, and data packets. The data in the table can be seen as a log of the activities that a law enforcement agency can request for investigation. Every record reports information about: the MS-ID (Mobile Station

ID, i.e. an ID identifying the mobile phone involved); the BTS-ID (the ID of the base transceiver station to which the Mobile Phone is connected); and the Event-Type. For the sake of simplicity, we mapped the several existing event types to the following restricted set: *cell-in*, *cell-out*, and *traffic*. The *cell-in* event happens when a mobile phone starts being served by a BTS (Base Transceiver Station), e.g. the mobile phone is switched on or it enters into the BTS coverage area. The *cell-out* event takes place when the mobile phone is no longer served by the BTS where it has previously performed a *cell-in* (this can be due to the mobile phone being switched-off, or to the exit from the BTS coverage area). The *traffic* event is recorded when a call is initiated, or a message is sent or received, or some data are exchanged by the phone. The Timestamp value reports the call start time or the message/data packet send time. A mobile phone event sequence should evolve according to the automaton described in Fig. 1(a), unfortunately the real data do not fully comply: several *cell-in* can be found in the same cell (with no *cell-out* in between), several traffic events on a BTS have no previous *cell-in*, etc. This is mainly due to signal drop issues affecting the radio connections. The elapsed intervals should be computed for analysis purposes i.e., the intervals when a mobile phone is served by (and thus being into) a BTS. Unfortunately the inconsistencies previously described affect the intervals computation, therefore, cleansing algorithms dealing with large datasets are required.

This paper is organised as follows: in Sec. 2 the Robust Data Quality Analysis (RDQA) methodology is introduced; in Sec. 3 an enhanced version of the RDQA is presented, namely the Multidimensional RDQA; its usage in the labour market domain has been presented in Sec. 4; in Sec. 5 the results have been commented and some concluding remarks are drawn.

2 The Robust Data Quality Analysis

This paper extends the Robust Data Quality Analysis (RDQA), a model-based data verification technique presented in [10]. The RDQA is aimed to evaluate the quality of longitudinal data before and after a cleansing intervention.

With the aim of summarising the main RDQA concepts, let us introduce the concept of Finite State Event Dataset, which builds a bridge between the database and the event-driven system domains. A database record is portrayed as an event, i.e. a record content or a subset thereof is interpreted as the description of an external world event modifying the system configuration, and a time-ordered set of records is interpreted as an event sequence. This concept is formalised as follows.

Definition 1. *Let* $\mathcal{R} = (R_1, \ldots, R_n)$ *be a schema relation of a database, let* $e = (r_1, \ldots, r_m)$ *be an* event *where* $r_1 \in R_1, \ldots, r_n \in R_n$*, then* e *is a* record *of the projection* $(R_1, \ldots, R_m)$ *over* $\mathcal{R}$ *with* $m \leq n$*. A* total order *relation* $\sim$ *on events can be defined such that* $e_1 \sim e_2 \sim \ldots \sim e_n$*. An* event sequence *is a* $\sim$*-ordered sequence of events* $\epsilon = e_1, \ldots, e_n$*.*

A Finite State Event Dataset *(FSED) is an event sequence* ϵ *derived from a longitudinal dataset. A* Finite State Event Database *(FSEDB) is a database DB whose content is* $DB = \bigcup_{i=1}^{k} S_i$ *where* $k \geq 1$.

The Mobile Phone Tracking Example. The following example should clarify the matter. Let us consider the dataset introduced in Tab. 1(a). The information collected about a single mobile phone is the FSED, while the information of several of them is the FSEDB. An *event* e_i is composed by the attributes *MS-ID*, *Event Type*, *Cell-ID*, and *Timestamp*, namely $e_i = (MS-ID_i, EType_i, Cell-ID_i, Timestamp_i)$. Moreover, the total-order operator $\sim$ could be the binary operator $\leq$ defined over the event's attribute $Timestamp$, hence $\forall e_i, e_j \in E, e_i \leq e_j$ iff $Timestamp_{e_i} \leq Timestamp_{e_j}$. Finally, a simply consistency property could be *"if a mobile phone connects to cell A, then it will disconnect from A before connecting to a different cell"*.

We can model this consistency property through an FSS. The consistency model is graphically represented in Fig. 1(a). In our settings, the system state is composed by two variables, namely (1) the variable *cell*, which describes to which cell the mobile phone is connected and (2) the variable $state \in \{con, dis\}$, whereas *con* denotes a phone connected to a cell, *dis* otherwise.

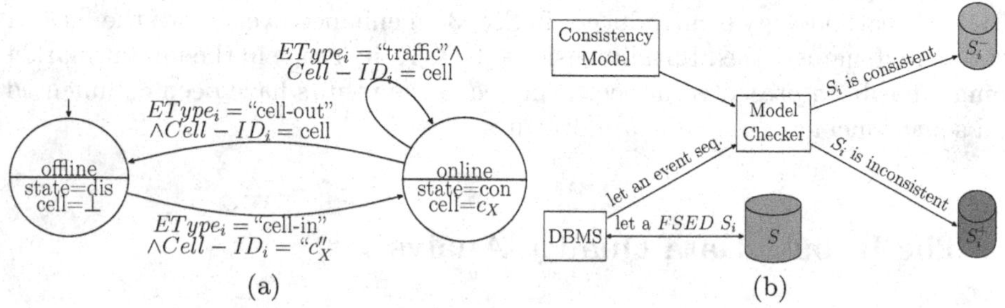

Fig. 1. (a) A Graphical representation of the consistency model for the Mobile Phone Tracking domain where the lower part of a node describes how the system state evolves when an event happens. (b) A Graphical representation describing a model checking based data consistency verification of a dataset.

The RDQA was conceived to evaluate the effectiveness of a cleansing routine (*clr* hereafter) on a specific dataset by addressing questions like *"what is the degree of cleanliness achieved through clr? Are we sure that clr does not introduce any error in the cleansed dataset? Which is the margin of improvement of clr (if any)?* The RDQA can be iteratively applied using several *clr* (improved) versions until a satisfactory data quality level is reached. The Fig. 2 describes a single RDQA iteration taking as input: a consistency model of the data, the source database S, and its cleansed instance C. The cleansing function is executed on each $S_i \in S$, generating the cleansed version $C_i \in C$ (an FSEDB is composed by several FSEDs, the latter are cleansed separately). A graphical representation of such approach is depicted in Fig. 1(b). The RDQA exploits a model-based verification function called *ccheck* to verify the consistency of each event sequence S_i and C_i. The output of a RDQA iteration is the Double Check Matrix (DCM), as shown in Tab. 1(b), which allows one to assign each (S_i, C_i) pair to a DCM cluster. Every line of Tab. 1(b) reports information about a DCM cluster. As an example, the Cluster 1 gives the number of FSEDs for which no error was found by both the *ccheck* and the *clr* (i.e., both *ccheck* and *clr* agreed that the original data were clean and no intervention was required). On the contrary, the Cluster 4 shows the number of FSEDs for which no error was found by *ccheck* on S_i even though a cleansing intervention took place, producing a result recognised as inconsistent by *ccheck* on C_i. The identification of such clusters helps discovering bugs in the *clr* or in the *ccheck* function.

Table 1. (a) An example of a Mobile Phone Tracking Dataset and (b) the Double Check Matrix are shown

(a)

Event-ID	Event Type	Cell-ID	Timestamp
01	cell-IN	3902	12/01/2011:08::35:00
02	traffic	3902	12/01/2011:11::00:05
03	traffic	3902	12/01/2011:13::10:15
04	traffic	3902	12/01/2011:18::45:55
05	cell-IN	40122	12/01/2011:22::00:00
...	...	...	...

(b)

Cluster	*What the ccheck function says about a sequence?*		
	Is S_i consistent?	$S_i \stackrel{?}{=} C_i$	Is C_i consistent?
1	Y	Y	Y
2	Y	Y	N
3	Y	N	Y
4	Y	N	N
5	N	Y	Y
6	N	Y	N
7	N	N	Y
8	N	N	N

Indeed, at each RDQA iteration the DCM acts like a *bug hunter* contributing to the improvement of the cleansing procedures and to better understand the domain rules. Clearly, this approach does not guarantee the correctness of the data cleansing process, nevertheless it helps making the process more robust with respect to data consistency.

3 The Multidimensional RDQA

The *ccheck* used in the RDQA works according to an on/off approach: it detects inconsistencies, but it doesn't provide any further information about the errors. Here it will be shown how the FSSs can be used to deeply investigate and classify the inconsistencies found. Since the data verification process is strongly affected by the actual data (i.e., the FSS verified by the model checker is instantiated according to the database data), some preliminary steps are required to identify "general" inconsistent patterns or properties: (1) to introduce an abstraction of the actual data, namely the symbolic data[1], and (2) to discover *all* the feasible

[1] The idea is not new and it is inspired by the *abstract interpretation* technique, see [12].

Table 2. (a) Error patterns for the Mobile Phone Tracking domain and (b) the values of its domain variables are provided

(a)

Error-Code	State	Inconsistent Event
1	$state$ = dis	$(cell - out, C_X)$
2	$state$ = dis	$(traffic, C_X)$
3	$state$ = con ∧ cell = "C_X"	$(cell - out, C_Y)$
4	$state$ = con ∧ cell = "C_X"	$(cell - in, C_X)$
5	$state$ = con ∧ cell = "C_X"	$(cell - in, C_Y)$
6	$state$ = con ∧ cell = "C_X"	$(traffic, C_Y)$

(b)

Variable Type	Variable	Domain Values
State Variables	state	con, dis
	cell	C_X, C_Y
Event data	cell	
	EType	cell-in, cell-out, traffic

inconsistency patterns affecting the *symbolic* data (3) by assigning a unique *error-code* to each of them. The error codes can then be used to classify the inconsistencies affecting the *real* data.

The Mobile Phone Tracking Example. Let us consider again the Mobile Phone Tracking example of Tab. 1(a) and let us focus on the inconsistent event sequences of two mobile phones: $Mob_1 = (cell - in, 03290), (cell - out, 03291)$ and $Mob_2 = (cell - in, 03120), (cell - out, 03288)$, whereas respectively the $EType_i$ and $Cell - ID_i$ attributes only are reported, and very short sequences are showed for the sake of simplicity. The inconsistencies found share a common characteristic: the *"cell-out"* has been made on a cell different from the one where the last *"cell-in"* took place. Then, we replace the actual *cell* domain data $D_{cell} = \{03120, 03288, 03290, 03291, \ldots\}$ (whose cardinality can be very high although finite) with a symbolic domain composed by a (small) set of symbols. In our example we can make an abstraction of the domain D_{cell} by using only two symbols, namely $D_{cell}^{symbolic} = \{C_X, C_Y\}$. The symbolic set cardinality has to be chosen according to the criteria described below. More formally, we define the following.

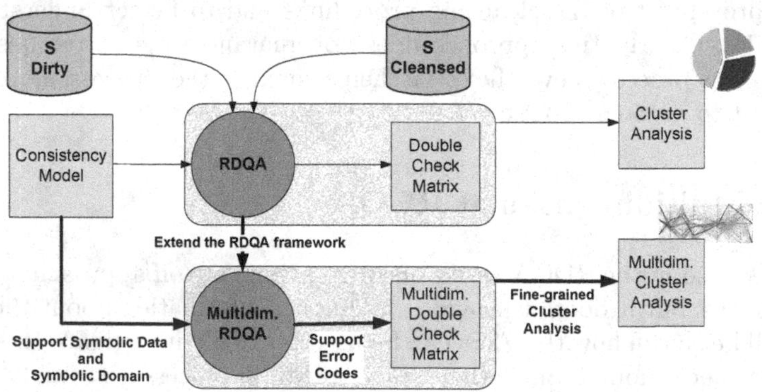

Fig. 2. An overview of the RDQA and the Multidimensional RDQA processes is given

Definition 2 (Symbolic Data and Symbolic Domain). *Let s be an FSS state and e be an event with respectively $s = x_1, \ldots, x_n$ state variables and $e = (r_1, \ldots, r_m)$ event attributes. Let D be a finite (although very large) attribute*

domain where $\{x_1, \ldots, x_{n'}\} \subseteq \{x_1, \ldots, x_n\}$ *and* $\{r_1, \ldots, r_{m'}\} \subseteq \{r_1, \ldots, r_m\}$ *are instances of D, i.e.,* $\{x_1, \ldots, x_{n'}\} \in D$ *and* $\{r_1, \ldots, r_{m'}\} \in D$.

An event e happening in the state s requires the evaluation of $x_1, \ldots, x_{n'}$ *and* $r_1, \ldots, r_{m'}$ *values, namely a configuration of* $n' + m'$ *different values of D. Then, we define the* Symbolic Domain *of D as a set of* different *symbols* $d_1, \ldots, d_{n'+m'}$, *called* Symbolic Data, *required to represent the values of D in the consistency model, i.e.* $D^{symbolic} = \{d_1, \ldots, d_{n'+m'}\}$.

Finally, some trivial conditions should be met before exploiting a Symbolic Domain rather than an Actual Domain: (p1) no total order relation is defined in the actual domain (or the total order relation is not considered for the scope of the analysis); (p2) No condition should compare a symbol to a non-symbolic value (e.g. $C_X = 03120$ in our example). Once the abstraction of the data has been identified, we can exploit the model-checking-algorithm to discover *all* the feasible inconsistency patterns affecting a *symbolic* event sequence. We will provide an example focusing on the Mobile Phone Tracking example, Tab. 2(b) shows the system variables, the events, and the domain values used. This task has been accomplished through the UPMurphi tool [13, 14] (i.e., a model-checking-based engine built upon the Murphi verifier). The results can be used as identification patterns for all the datasets that can be represented by the symbolic domain. The set of pairs <state values; event values> that lead to an inconsistency are shown in Tab. 2(a). Only a subset of the feasible pairs is reported, since the entries shown are reduced using some symmetry reduction techniques, e.g. $< state = con \wedge cell = C_X; cell - in, C_Y >$ and $< state = con \wedge cell = C_Y; cell - in, C_X >$ are symmetric then can be represented using only the first one according to the symmetry reduction technique [15]. Then, the model checking based identification allows one to univocally match a real data inconsistent sequence to a symbolic error pattern. The error patterns can be labelled and statistics about their occurrences on real data can be computed (e.g., by providing how the error pattern occurrences change from the dirty to the cleansed data).

The result of the process is an *enhanced* DCM, namely the Multidimensional DCM. Each DCM cluster is enriched with a square matrix having $n + 1$ rows and columns, where n is the number of error-codes detected, as shown in Eq. 1.

$$\forall l \in \{1, \ldots, 8\}\ Cluster^l_{n+1,n+1} = \begin{pmatrix} err_{0,0} & err_{0,1} & \cdots & err_{0,n} \\ err_{1,0} & err_{1,1} & \cdots & err_{1,n} \\ \vdots & \vdots & \ddots & \vdots \\ err_{n,0} & err_{n,1} & \cdots & err_{n,n} \end{pmatrix} \quad (1)$$

As an example, given a cluster of the DCM, an element of the matrix $err_{i,j}$ is a number $k \in \mathbb{N}$ if and only if k distinct event sequences have presented the error-code i in the original dataset and the error-code j in the cleansed one. For completeness we added the *dummy* error-code zero to represent consistent items i.e., we consider elements having $i = 0$ or $j = 0$ as consistent (e.g., $err_{0,0}$ is the number of sequences consistent before and after the cleansing intervention).

4 A Real-World Dataset: The Labour Market Domain

The scenario we are presenting focuses on the Italian labour market domain. According to the Italian law, every time an employer hires or dismisses an employee, or an employment contract is modified (e.g. from part-time to full-time, or from fixed-term to unlimited-term), a *Mandatory Communication* (i.e., an event) is sent to a job registry, managed at local level. The Italian public administration has developed an ICT infrastructure for recording *Mandatory Communications*, generating an administrative archive useful for studying the labour market dynamics (see, e.g., [16]). Each mandatory communication is stored into a record composed by the following attributes:

e_id: it represents an id identifying the communication;
w_id: it represents an id identifying the person involved;
e_date: it is the event occurrence date;
e_type: it describes the event type occurring to the worker career. Events types are the *start* or the *cessation* of a working contract, the *extension* of a fixed-term contract, or the *conversion* from a contract type to a different one;
c_flag: it states whether the event is related to a full-time or a part-time contract;
c_type: describes the contract type with respect to the Italian law. Here we consider *Limited*, i.e. fixed-term, and *unlimited*, i.e. unlimited-term, contracts.
empr_id: it uniquely identifies the employer involved.

Considering the terminology introduced in Def. 1, an FSED is the set of events for a given *w_id* (ordered by *e_date*), and the FSEDs union composes the FSEDB. The career event sequences can be considered as longitudinal data. Now we closely look to the careers consistency, where the consistency semantics is derived from the Italian labour law, from the domain knowledge, and from the common practice. Here are reported some constraints:

c1: an employee can have no more than one full-time contract at the same time;
c2: an employee cannot have more than K part-time contracts (signed by different employers), in our context we assume $K = 2$;
c3: an *unlimited term* contract cannot be extended;
c4: a contract extension can change neither the existing contract type (c_type) nor the part-time/full-time status (c_flag) e.g., a part-time and fixed-term contract cannot be turned into a full-time contract by an extension;
c5: a conversion requires either the c_type or the c_flag to be changed (or both).

For simplicity, we omit the descriptions of some trivial constraints e.g., an employee cannot have a *cessation* event when she/he is not hired, an event cannot be recorded twice, etc.

The UPMurphi tool allows us to build an FSS upon which the data verification tasks are performed. A worker's career model (i.e., the system state) is composed by three elements at a given time point: the list of companies for which the worker has an active contract ($C[]$), the list of modalities (part-time, full-time) for each contract ($M[]$) and the list of contract types ($T[]$). To give an example, $C[0] = 12$, $M[0] = PT$, $T[0] = unlimited$ models a worker having an active unlimited part-time contract with company 12. A graphical representation is showed in Fig. 3. Note that, to improve the readability, we omitted to represent *conversion* events as well as inconsistent states/transitions (e.g., a worker activating two full-time contracts), which are handled by the FSS generated by the UPMurphi model. We have done a map from the actual to the symbolic data as described in Sec 3 taking into account both the states and the events of Fig. 3. The attributes *c_type*, *e_type*, and *c_flag* are already bounded and we left them as is, while the *empr_id* attribute domain has been mapped on a symbolic set of 3 symbols $\{empr_x, empr_y, empr_z\}$ according to the process described in Sec. 3 and satisfying conditions p1 and p2.

5 Results and Concluding Remarks

We applied the Multidimensional-RDQA for evaluating the cleansing procedures used on people careers data. A subset of an Italian Region inhabitants and their careers data have been chosen. Such data consist of 1, 791, 282 mandatory communications describing the careers of 200, 000 people observed starting from the 1st January 2004 to the 31st December 2011. The output of this process is the DCM, as shown in Tab. 3, where each cluster has been enhanced with the error-code matrix. The results are shortly commented in the following list. The clusters labelled with (∗) represent the job careers dropped by the cleansing function (in spite of their consistency) cause they refer to workers living outside the investigated region and they are not in the scope of the analysis.

Table 3. The Double Check Matrix computed on the careers data of an Italian Region

Row Number	Cluster	*What ccheck function says about a sequence?*			*Careers Data*	
		Is S_i consistent?	$S_i \stackrel{?}{=} C_i$	Is C_i consistent?	#Careers	%
R1	1	Y	Y	Y	64,625	32.3
R2	2	Y	Y	N	0	0.00
R3	3	Y	N	Y	3,190	1.6
R4	4	Y	N	N	184	0.09
R5	*	Y	N	unknown	315	0.15
R6	5	N	Y	Y	0	0.00
R7	6	N	Y	N	1,054	0.52
R8	7	N	N	Y	116,216	58.1
R9	8	N	N	N	14,059	7.02
R10	*	N	N	unknown	357	0.17

Cluster 1: represents careers *already clean* that the *clr* did not modify. It provides an estimation of the *consistent careers* on the source archive.
Cluster 2: refers to careers considered consistent (by *ccheck*) before but not after the cleansing, although they have not been touched by *clr*. As expected this subset is empty.

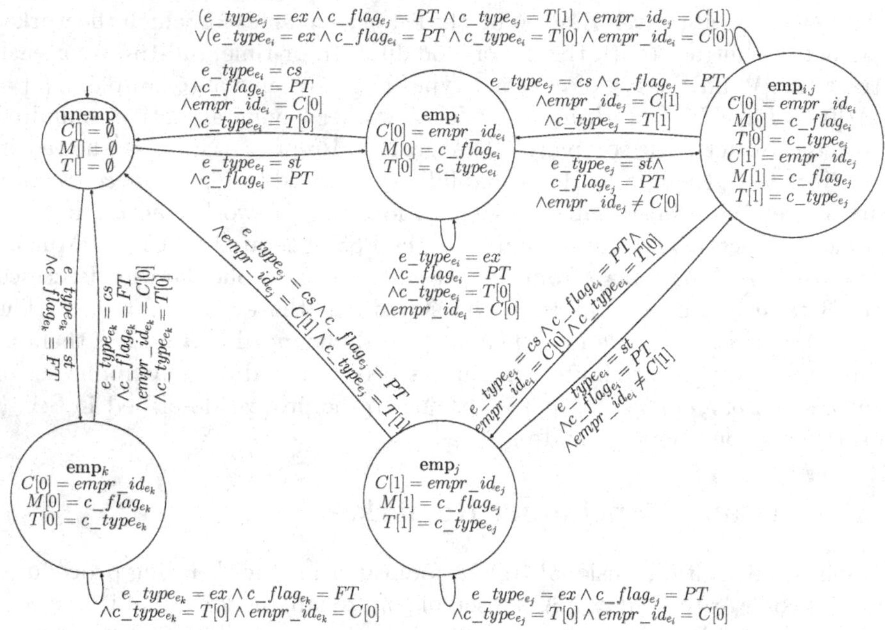

Fig. 3. An FSS graphical representation of a valid worker's career is shown, with st = $start$, cs = $cessation$, cn = $conversion$, and ex = $extension$

Cluster 3: describes consistent careers that have been *unexpectedly changed* by the clr. Note that, despite such kind of careers remain consistent after the cleansing intervention, the behaviour of clr needs to be investigated.
Cluster 4: represents careers originally consistent that clr has made inconsistent. These careers have strongly helped identifying and correcting bugs in the clr implementation.
Cluster 5: refers to careers recognised as not consistent before but consistent after cleansing (by *ccheck*), although they the clr did not modify them. This subset is empty, as expected.
Cluster 6: describes inconsistent careers that clr was able neither to detect nor to correct, and consequently they were left untouched.
Cluster 7: describes the number of (originally) inconsistent careers which *ccheck* recognises as *properly cleansed* by clr at the end.
Cluster 8: represents careers originally inconsistent which have been *not properly cleansed* since, despite an intervention of clr, the function *ccheck* identifies them still as inconsistent.

Finally, each DCM cluster has been further analysed by closely looking at its error-code matrix. To this aim, we exploited a well-suited multidimensional visualisation technique, namely the *parallel-coordinates* (abbrv: ||-coord see [17]).

Informally speaking, ||-coord allow one to represent an n-dimensional datum $(x_1, \ldots, x_n)$ as a polyline, by connecting each x_i point in n parallel y-axes. We used the ||-coord to plot the DCM and the error-code data by using four dimensions, namely $(l, i, err_{i,j}, j)$ which respectively represent the DCM row number, the error-code before the cleansing, the number or careers (i.e. the $err_{i,j}$ value in Eq. 1), and the error-code after the cleansing. Generally, ||-coord tools show their powerfulness when used interactively (i.e., by selecting ranges from the y-axes, by emphasising the lines traversing through specific ranges, etc).

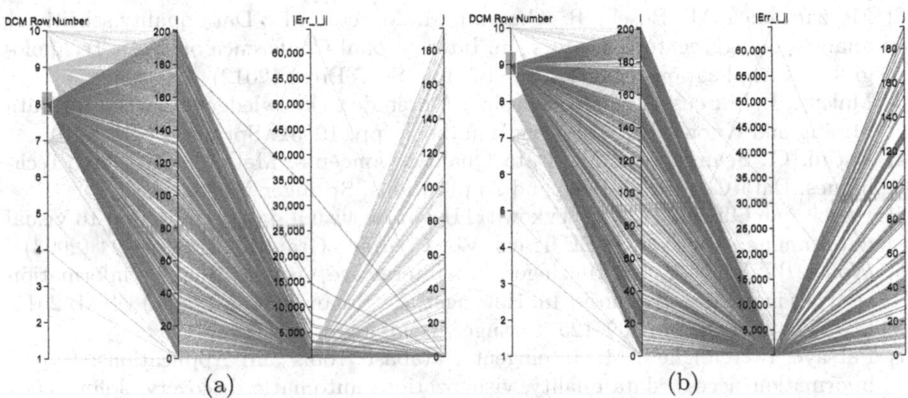

Fig. 4. Parallel Coordinates for (a) Row Number 8 and (2) Row Number 9 of the DCM in Tab. 3

For these reasons, the plot file has been made publicly available for downloading as detailed below. For the sake of completeness, we report two ||-coord graphs shown in Fig. 4. The Figure 4(a) shows the Cluster 7 of the DCM (i.e., the originally inconsistent careers correctly cleansed by the *clr* function). The figure explains how the error-codes are distributed on the original inconsistent data and their related frequencies. Differently, Fig. 4(b) highlights the Cluster 8 data (i.e. the careers improperly cleansed by the *clr*) which allows one to discover how an error-code in the source dataset evolves due to a wrong cleansing intervention. The error-code and the DCM data plotting played a key role in the analysis of the data evolution and in discovering cleansing issues (e.g., it was easy to identify the most relevant and more numerous cases and to prioritise them). Due to the space limitations, we can not further comment the results and how they have been used. An anonymous version of the analysed dataset, the Multidimensional RDQA outcomes and the ||-coord sheets have been made publicly available for download and demonstration[2], so that the results we present can be assessed, shared, and compared with other techniques.

Concluding Remarks. Our results show that the joint application of model-based verification approaches and visualisation techniques give the users the instruments to tackle the complexities of managing huge masses of data. Moreover, we presented a real-world scenario in the labour market domain where the verification and visualisation techniques have been successfully used. As a future

[2] `http://goo.gl/BTdES`. User: *hcikdd2013materials@gmail.com* Pass: *hci-kdd2013*.

work, other visualisation techniques will be investigated to empower the user managing large datasets.

References

[1] Fayyad, U.M., Piatetsky-Shapiro, G., Uthurusamy, R.: Summary from the kdd-03 panel. ACM SIGKDD Explorations Newsletter 5(2), 191–196 (2003)
[2] Mezzanzanica, M., Boselli, R., Cesarini, M., Mercorio, F.: Data quality sensitivity analysis on aggregate indicators. In: International Conference on Data Technologies and Applications (DATA), pp. 97–108. SciTePress (2012)
[3] Maletic, J., Marcus, A.: Data cleansing: A prelude to knowledge discovery. In: Data Mining and Knowledge Discovery Handbook, pp. 19–32. Springer, US (2010)
[4] Batini, C., Scannapieco, M.: Data Quality: Concepts, Methodologies and Techniques. Data-Centric Systems and Applications. Springer (2006)
[5] Ferreira de Oliveira, M.C., Levkowitz, H.: From visual data exploration to visual data mining: A survey. IEEE Trans. Vis. Comput. Graph. 9(3), 378–394 (2003)
[6] Wong, B.L.W., Xu, K., Holzinger, A.: Interactive visualization for information analysis in medical diagnosis. In: Holzinger, A., Simonic, K.-M. (eds.) USAB 2011. LNCS, vol. 7058, pp. 109–120. Springer, Heidelberg (2011)
[7] Parsaye, K., Chignell, M.: Intelligent Database Tools and Applications: Hyperinformation access, data quality, visualization, automatic discovery. John Wiley (1993)
[8] Clemente, P., Kaba, B., Rouzaud-Cornabas, J., Alexandre, M., Aujay, G.: Sptrack: Visual analysis of information flows within selinux policies and attack logs. In: Huang, R., Ghorbani, A.A., Pasi, G., Yamaguchi, T., Yen, N.Y., Jin, B. (eds.) AMT 2012. LNCS, vol. 7669, pp. 596–605. Springer, Heidelberg (2012)
[9] Simonic, K.M., Holzinger, A., Bloice, M., Hermann, J.: Optimizing long-term treatment of rheumatoid arthritis with systematic documentation. In: IEEE International Conference on Pervasive Computing Technologies for Healthcare, PervasiveHealth, pp. 550–554 (2011)
[10] Mezzanzanica, M., Boselli, R., Cesarini, M., Mercorio, F.: Data quality through model checking techniques. In: Gama, J., Bradley, E., Hollmén, J. (eds.) IDA 2011. LNCS, vol. 7014, pp. 270–281. Springer, Heidelberg (2011)
[11] European Telecommunications Standards Institute ES 201 671: Handover interface for the lawful interception of telecommunications traffic (2009)
[12] Clarke, E.M., Grumberg, O., Long, D.E.: Model checking and abstraction. ACM TOPLAS 16(5), 1512–1542 (1994)
[13] Della Penna, G., Intrigila, B., Magazzeni, D., Mercorio, F.: UPMurphi: a tool for universal planning on PDDL+ problems. In: ICAPS, pp. 106–113. AAAI Press (2009)
[14] Mercorio, F.: Model checking for universal planning in deterministic and nondeterministic domains. AI Communications 26(2), 257–259 (2013)
[15] Norris Ip, C., Dill, D.: Better verification through symmetry. Formal Methods in System Design 9(1), 41–75 (1996)
[16] Martini, M., Mezzanzanica, M.: The Federal Observatory of the Labour Market in Lombardy: Models and Methods for the Construction of a Statistical Information System for Data Analysis. In: Information Systems for Regional Labour Market Monitoring - State of the Art and Prospectives. Rainer Hampp Verlag (2009)
[17] Inselberg, A.: The plane with parallel coordinates. The Visual Computer 1(2), 69–91 (1985)

On Knowledge Discovery in Open Medical Data on the Example of the FDA Drug Adverse Event Reporting System for Alendronate (Fosamax)

Pinar Yildirim[1], Ilyas Ozgur Ekmekci[1], and Andreas Holzinger[2]

[1] Department of Computer Engineering, Faculty of Engineering & Architecture, Okan University, Istanbul, Turkey
{pinar.yildirim,oekmekci}@okan.edu.tr
[2] Institute for Medical Informatics, Statistics & Documentation
Medical University Graz, A-8036 Graz, Austria
andreas.holzinger@meduni-graz.at

Abstract. In this paper, we present a study to discover hidden patterns in the reports of the public release of the Food and Drug Administration (FDA)'s Adverse Event Reporting System (AERS) for alendronate (fosamax) drug. Alendronate (fosamax) is a widely used medication for the treatment of osteoporosis disease. Osteoporosis is recognised as an important public health problem because of the significant morbidity, mortality and costs of treatment. We consider the importance of alendronate (fosamax) for medical research and explore the relationship between patient demographics information, the adverse event outcomes and drug's adverse events. We analyze the FDA's AERS which cover the period from the third quarter of 2005 through the second quarter of 2012 and create a dataset for association analysis. Both Apriori and Predictive Apriori algorithms are used for implementation which generates rules and the results are interpreted and evaluated. According to the results, some interesting rules and associations are obtained from the dataset. We believe that our results can be useful for medical researchers and decision making at pharmaceutical companies.

Keywords: Open medical data, knowledge discovery, biomedical data mining, osteoporosis, drug adverse event, alendronate (fosamax), apriori algorithm, co-occurrence analysis.

1 Introduction

Open data is generally a big issue in every scientific community. Much has changed within the last years and Science has become part of our modern civilization and should be, and be seen to be, a public enterprise, not a private enterprise done behind closed laboratory doors [1]. Consequently, Open Data is considered an important issue in several scientific communities for some time now [2] and very recently it is in debate in the biomedical area [3], [4].

A. Holzinger and G. Pasi (Eds.): HCI-KDD 2013, LNCS 7947, pp. 195–206, 2013.

A big asset of open data in research is that it allows to build on the work of others more efficiently and helps to speed the progress of science [5]; because to build on previous discoveries, there must be trust in the validity of prior research – but most important, it facilitates trust between researchers and with the public, due to the fact that it allows secondary analyses that expand the usefulness of datasets and the resulting knowledge gained [6].

Postmarket surveillance for adverse events is an essential component of every national and regional health system for assuring drug safety [7]. The Food and Drug Administration (FDA) in USA is responsible not only for approving drugs for marketing but also for monitoring their safety after they reach the market. This function is carried out by the FDA's Office of Drug Safety, which maintains a spontaneous reporting database called the Adverse Event Reporting System (AERS). AERS receives adverse events information from two principal sources: mandatory reports from pharmaceutical companies on adverse events that had been spontaneously communicated to the firms, primarily by physicians and pharmacists: and adverse event reports that physicians, pharmacists, nurses, dentists and consumers submit directly to the FDA's MedWatch program. Since 1969, more than 2 million adverse event reports have been submitted to the FDA. In the United States, the estimated cost of morbidity and mortality related to adverse drug reactions (ADRs) is more than $75 million annually, and ADRs are among the top 10 leading causes of death [8].

The FDA can restrict distribution of a drug, and on rare occasions, it may request a drug's withdrawal from the market, or the manufacturer may voluntarily withdraw the drug [8]. In September 2004, Merck&Co, Inc, voluntarily withdrew rofecoxib (vioxx) from the global market because of an increased risk of cardiovascular events. Two months later, the FDA announced that manufacturers of isotretinoin will obtain registration of prescribers of isotretinoin, dispencing pharmacies, and patients who are prescribed the drug. The agency also announced the requirement of documentation of a negative pregnancy test result before isotretinoin is given to women capable of becoming pregnant. In April 2005, valdecoxib was withdrawn from the market because of serious dermatological conditions and an unfavorable risk vs benefit profile [9].

The aim of this study is to explore hidden relationship between alendronate (fosamax) and adverse events. Alendronate(fosamax) is the main medication for osteoporosis disease. Osteoporosis is widely recognised as an important public health problem because of the significant morbidity, mortality and costs associated with its complications, namely fractures of the hip, spine, forearm and other skeletal sites[10]. Across the whole of Europe, an estimated 3.1million fragility fractures occur each year in men and women age 50 years or over, including 620,000 cases of hip fracture, 490,000 clinical vertebral fractures and 574,000 forearm fractures. The incidence of fractures is highest amongst elderly white women, with one in every two women suffering an osteoporosis related fracture in their lifetime [11], [12], [13].

It is also known that there are some adverse events of drugs that are associated with a specific gender. Hence, we also considered that patient demographics such as age, gender may be associated with differential risk to alendronate(fosamax) and we present an approach for knowledge discovery on the adverse events of this drug. We analyzed the data from the FDA's AERS to discover associations between patient demographics, the adverse event outcomes (death, hospitalization, disability etc.) and adverse events of alendronate(fosamax). We aimed to capture some serious and useful information buried in event reports that are not easily recognized in clinical trials by researchers, clinicians and pharmaceutical companies.

2 Related Work

Novel data mining algorithms were developed and several studies were performed for knowledge discovery on drug adverse events associations. Tatoneti et al., mined the FDA's AERS for side-effect profiles involving glucose homeostatis and found a surprisingly strong signal for comedication with pravastatin and paroxetine[14].

Kadoyama et al., searched the FDA's AERS and carried out a study to confirm whether the database could suggest the hypersensitivity reactions caused by anticancer agents, paclitaxel, docetaxel, procarbazine, asparaginase, teniposide and etoposide. They used some data mining algorithms such as proportional reporting ratio (PRR), the reporting odds ratio(ROR) and the empirical Bayes geometric mean (EBGM) to identify drug-associated adverse events and consequently, they detected some associations[15].

Tamura et al., reviewed FDA AERS to assess the bleeding complications induced by the administration of antiplatelets and to attempt to determine the rank-order of the association. According to their results, both aspirin and clopidogrel were associated with haemorrhage, but the association was more noteworthy for clopidogrel; however, for gastrointestinal bleeding complications, the statistical metrics suggested a stronger association for aspirin than clopidogrel[16].

Gandhi et al., investigated the FDA's AERS to detect cardiovascular thromboembolic events associated with febuxosat. They implemented Bayesian statistics within the neural network architecture to identify potential risks of febuxosat. Their study indicated continuing combination cases of cardiovascular thrombotic events associated with the use of febuxosat in gout patients [17].

3 Methods

Data Sources

Input data for our study are taken from the public release of the FDA's AERS database, which covers the period from the third quarter of 2005 through the second of 2012. The data structure of AERS consists of 7 datasets: patient demographic and administrative information (DEMO), drug/biologic information (DRUG), adverse events (REAC), patient outcomes (OUTC), Report Sources (RPSR), drug therapy start and end dates (THER), and indications for use/diagnosis (INDI).The adverse events in REAC are coded using preferred terms (PTs) in the Medical Dictionary for Regulatory Activities (MedDRA) terminology. All ASCII data files are linked using ISR, a unique number for identifying an AER. Three of 7 files are linked using DRUG_SEQ, a unique number for identifying a drug for an ISR [18]. The outline of 7 datasets is shown in Fig.1 [19].

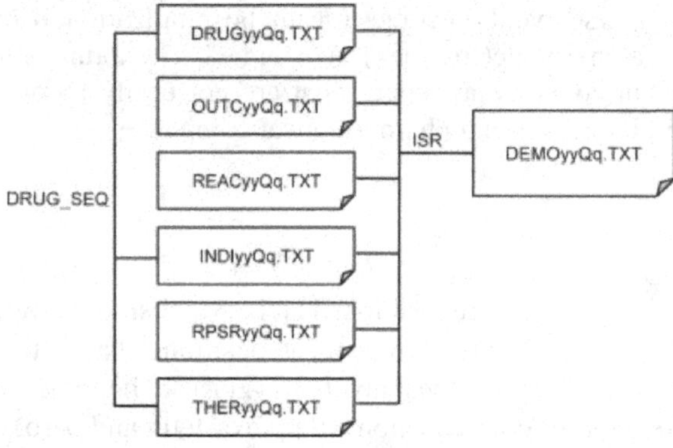

Fig. 1. Data structure of AERS

Association Rule Mining and Apriori Algorithm

Frequent sets play an essential role in studies on data mining to find interesting patterns from databases, such as association rules, correlations and sequences. Association rule mining explores hidden association or correlation relationships among a large set of data items. With massive amounts of data continuously being collected and stored, many industries are becoming interested in mining association rules from their database. The discovery of interesting association relationships among huge amounts of business transaction records can help in many business decision making processes[20].

Association rule mining methods try to find frequent items in databases. Given a set of transactions T(each transaction is a set of items), an association rule can be expressed in the form $X \Rightarrow Y$, where *X* and *Y* are mutually exclusive sets of items[21].

The rule's statistical significance is measured by *support*, and the rule's strength by *confidence*. The *support* of the rule is defined as the percentage of transactions in *T* that contain both *X* and *Y*, and may be regarded as , the probability of $X \cup Y$ (X union Y). The confidence of the association rule is the ratio of the support of the itemset $X \cup Y$ to the support of the itemset *X*, which roughly corresponds to the conditional probability *P(Y/X)*.

Association rule generation is usually split up into two steps:

- Find all combinations of items whose supports are greater than a user-specified minimum support(threshold). The combinations are called frequent itemsets.
- Use the items from frequent itemsets to generate the desired rules. More specifically, the confidence of each rule is computed, and if it is above the confidence threshold, the rule is satisfied [21],[22].

The Apriori algorithm is the most efficient algorithm for discovering association rules in large database. The pseudo code for Apriori algorithm is as follows:

```
Ck = Candidate itemset of size k
Lk = Frequent itemset of size k
L1 ={Frequent items};
        for (k = 1; Lk! = Ø; k++) do begin
                Ck+1 = candidates generated from Lk

                for each trancation t in database do
                        #increment the count of all candidates in Ck+1 that are
                        contained in t
                        Lk+1 = candidates in Ck+1 with minSupport
                end

        Return UkLk;
```

4 Experimental Results

The FDA's AERS datasets are taken from FDA's web site. The datasets in different time periods were combined into a single dataset and imported into Microsoft SQL Server 2012 database management system as database tables. Then, alendronate(fosamax) related records were selected to create a dataset for association analysis. In total, 9229 alendronate(fosamax) associated adverse event reports were collected from the whole database. Considering mostly seen adverse events, most frequent adverse events with alendronate were searched. Table 1 list the number of co-occurrences of adverse events with alendronate (fosamax) and Figure 2 shows the graphical representation of them. Based on the number of co-occurrences, fall has the highest frequency with alendronate, followed by osteonecrosis, pain, femur fracture, pneumonia, dyspnoea and anaemia in this order.

Table 2 also lists and Figure 3 shows the graphical representation of the number of co-occurrences of adverse event outcomes with alendronate(fosamax). Based on the number of co-occurrences, mostly seen outcome is OT(Other), followed by HO(Hospitalization), DS(Disability), LF(Life-Threatening), RI(Required Intervention to Prevent Permanent Impairment/Damage) and CA(Congenital Anomaly) in this order.

Normalisation is an important concern for studies based on data mining because some entities may have some variations such as synonyms and brand names. These names should be detected and normalized. In our study, alendroante has also some variations. For example, fosamax and adronat are the variations of alendronate. These variations were found by searching Drugbank database and mapped to one specific name. Drugbank is one of the biggest resource for drugs and currently contains >4100 drug entries, corresponding to >12000 different trade names and synonyms [23].

Table 1. The number of co-occurrences of adverse events with alendronate(fosamax)

Adverse event	N (the number of co-occurrences)
Fall	593
Osteonecrosis	556
Pain	553
Femur fracture	504
Pneumonia	498
Dyspnoea	473
Anaemia	463
Anxiety	457
Hypertension	448
Depression	446

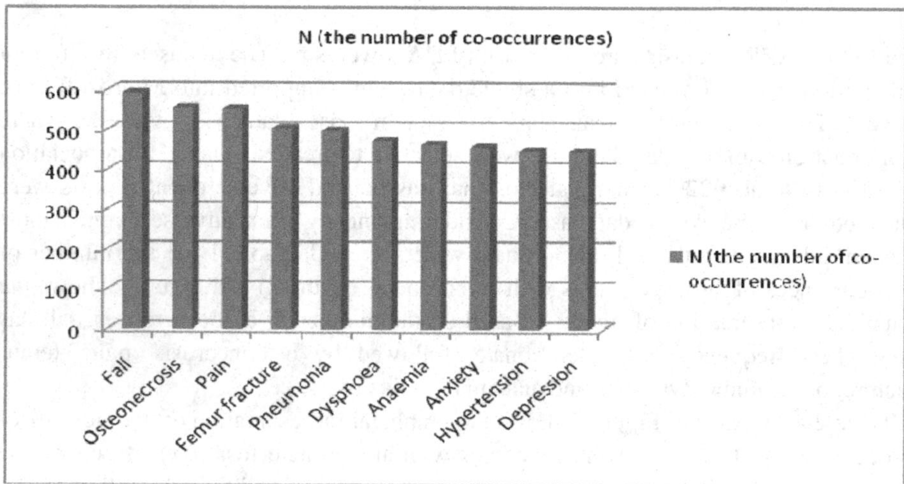

Fig. 2. The number of co-occurrences of adverse events with alendronate(fosamax)

Table 2. The number of co-occurrences of adverse event outcomes

Outcome	N(the number of co-occurrences)
OT(Other)	2821
HO(Hospitalization)	2745
DS(Disability)	1700
DE(Death)	1130
LF(Life-Threatening)	507
RI(Required Intervention to Prevent Permanent Impairment/Damage)	291
CA(Congenital Anomaly)	35

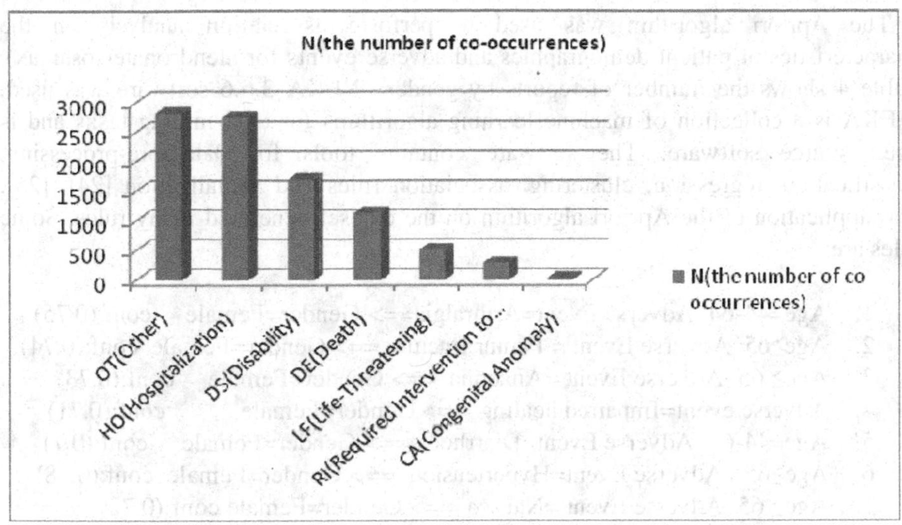

Fig. 3. The number of co-occurrences of adverse event outcomes

Table 3. Age Categories

Age	Category
0-6	Preschool child
7-12	Child
13-24	Young
25-43	Adult
44-64	Middle aged
65≤	Aged

Table 4. The number of reports by gender

Gender	The number of reports
Female	6149
Male	2821
NULL	249
Unknown	8
NS	2

After normalizing, the dataset was created and it contains patient demographics such as age, gender and adverse events. The attributes of the dataset were directly collected from database and age was categorized (Table 3). The dataset consists of three attributes: age, gender and the adverse events and 9229 instances.

The Apriori algorithm was used to perform association analysis on the characteristics of patient demographics and adverse events for alendronate(fosamax). Table 4 shows the number of reports by gender. WEKA 3.6.6 software was used. WEKA is a collection of machine learning algorithms for data mining tasks and is open source software. The software contains tools for data pre-processing, classification, regression, clustering, association rules and visualization [24], [25]. The application of the Apriori algorithm on the dataset generated many rules. Some rules are;

1. Age=44-64 Adverse Event=Arthralgia==> Gender=Female conf:(0.75)
2. Age≥65 Adverse Event = Femur fracture ==> Gender=Female conf:(0.74)
3. Age≥ 65 Adverse Event =Anaemia ==> Gender=Female conf:(0.73)
4. Adverse event=Impaired healing ==> Gender=Female conf:(0.71)
5. Age=44-64 Adverse Event=Diarrhoea ==> Gender=Female conf:(0.7)
6. Age≥65 Adverse Event=Hypertension ==> Gender=Female conf:(0.68)
7. Age≥ 65 Adverse Event =Nausea ==> Gender=Female conf:(0.72)
8. Age ≥65 Adverse Event=Depression ==> Gender=Female conf:(0.64)
9. Age=44-64 Adverse Event=Anxiety ==> Gender=Female conf:(0.63)
10. Age ≥65 Adverse Event=Osteonecrosis ==> Gender=Female conf:(0.63)

We also used the Predictive Apriori algorithm to compare the results of the Apriori algorithm. Similarly to the Apriori algorithm, the Predictive Apriori algorithm generates frequent item sets, but it uses a dynamically increasing minimum support threshold for the best rules concerning a support-based corrected confidence value. A rule is added is: the expected predictive accuracy of this rule is among the "n" best and it is not subsumed by a rule with at least the same expected predictive accuracy [26].The application of the Predictive Apriori algorithm on the dataset generated many rules. Some rules are;

1. Gender=Null Adverse Event=Depression 8 ==> Age ≥65 8 acc:(0.96017)
2. Age= 13-24 Adverse Event =Femur fracture 3 ==> Gender=Female 3 acc:(0.7783)
3. Age= 7-12 Adverse Event=Femur fracture 3 ==> Gender=Female 3 acc:(0.7783)
4. Age= 25-43 Adverse Event=Fall 30 ==> Gender=Female 26 acc:(0.71601)
5. Gender=Unkown Adverse Event=Pain 2 ==> Age= 44-64 2 acc:(0.69808)
6. Age= 13-24 Adverse Event=Depression 2 ==> Gender=Female 2 acc:(0.69808)
7. Age= 13-24 Adverse Event=Pneumonia 2 ==> Gender=Female 2 acc:(0.69808)
8. Age= 13-24 Adverse Event=Pain 2 ==> Gender=Male 2 acc:(0.69808)
9. Age ≥65 Adverse Event=Femur fracture 241 ==> Gender=Female 179 acc:(0.69727)
10. Age= 44-64 Adverse Event=Anaemia 229 ==> Gender=Female 167 acc:(0.68664)

The results of both Apriori and Predictive Apriori algoritms revealed some patterns in the dataset. According to the results of Apriori algorithm, some adverse events such as arthralgia, anaemia and diarrhoea have strong associations with middle aged patients (between 44 and 64). For example, rule 1 means that the possibility of arthralgia with middle aged and female patients is 75% (confidence). On the other hand, hypertension, nausea and depression are seen in patients over 65 years old.

Alendronate (fosamax) has some well known adverse events such as pain, femur fractures and nausea. The worst side effect is osteonecrosis of the jaw (ONJ), which is rare. Based on our results, osteonecrosis associated with patients over 65 years old. In addition, some events such as femur fracture, depression and anaemia are seen in the the rules which both Apriori and Predictive Apriori algorithms generated.

Medical researchers and clinicians can investigate our patterns and conduct clinical trials to discover new ideas for the assessment of drug safety of alendronate.

5 Discussion

The FDA's AERS database is considered an important resource, but some limitations were pointed out. First, the reports in the FDA's AERS contain errors, duplicate entries and missing data resulting in misclassifications. For example, patient age was not reported in many reports. Second, the structure of some datasets belonging to specific time periods are not compatible to the others. This means that reporting patterns and database structure changed over time. To overcome problems with data quality, we used some preprocessing methods. We omitted or corrected some records containing errors and missing data. For example, we omitted some records containing missing age or adverse events. Despite some limitations, the FDA's AERS database is a rich source to identify some important associations between drugs and adverse events [18].

Our study has both advantages and disadvantages. There are some web based tools to analyze the FDA's AERS database. The Drugcite processes the datasets between 2004 and 2012 and generates similar results with our study. We used this tool to find most ranked adverse events with alendronate(fosamax) and compared with our results. According to Drugcite, three most frequent adverse events are femur fracture, osteonecrosis and fall. On the other hand, our study highlighted that fall, osteonecrosis and pain are the top events. Because of some data quality problems with the datasets in different time periods above, we integrated the datasets which covers the third quarter of 2005 through the second of 2012. We think that this can result in some differences in both studies. Comparing our study and Drugcite tool, we found some statistical results for adverse events and outcomes with alendronate(fosamax) and implemented apriori algorithm to discover some association rules between patient demographics and adverse events. Drugcite is an useful tool which provides detail information about drugs and analyzes the FDA's AERS database, but it has no data mining functions. Therefore, we can conclude that our results are more informative and better than Drugcite's outputs.

Apart from the FDA's AERS database, medical records created in hospital information systems may be an important resource to determine drug adverse events and their outcomes. Therefore, a way to approach which includes both issues in the FDA's AERS and medical records can be used to reveal serious risks of a drug.

In this study, we found most ranked adverse events of alendronate(fosamax) and some relationships with patient demographics. Then, we searched some websites providing drug information such as http://www.nlm.nih.gov/medlineplus to compare our results. This comparison highlighted that some adverse events of alendronate(fosamax) are well known but some of them such as anxiety and depression have not been recognized. We also implemented both apriori and predictive apriori algorithms and discovered that anxiety has some associations with middle aged patients (between 44 and 64) and depression is seen over patients over 65 years old. This highlights that alendronate(fosamax) may cause psychological problems. We think that medical researchers should consider the results of our study and do clinical studies to determine the risks of the drug.

6 Conclusion

The aim of biomedical research is to discover new and useful knowledge to make contributions to better diagnosis, decision making and treatment [27],[28],[29]. The FDA's AERS is one of the big resources to reach this goal. In this study, we focused on knowledge discovery for alendronate(fosamax) drug and carried out a study based on patient demographics and adverse events relationships in the AERS reports. Alendronate(fosamax) is widely used for the treatment of osteoporosis and this disease is one of the major health problem in the world. Therefore, there is a big scientific interest in treatment and prevention of this disease.

We utilized database and computational techniques and then applied Apriori and Predictive Apriori algorithms to our dataset to obtain some rules. Our results show that some adverse events of alendronate(fosamax) have not known and patient demographics can have relationships with these events of the drug. Medical experts, researchers and pharmaceutical companies can explore and interprete these relationships. In conclusion, we believe that our study can make important contributions to postmarketing information in the assessment of drug safety of alendronate(fosamax).

Acknowledgements. We cordially thank Dr. Cinar Ceken for her help and medical expert advice. We also thank the SouthCHI'13 reviewers for their thorough review and helpful comments to further improve our paper.

References

[1] Boulton, G., Rawlins, M., Vallance, P., Walport, M.: Science as a public enterprise: the case for open data. The Lancet 377, 1633–1635 (2011)

[2] Rowen, L., Wong, G.K.S., Lane, R.P., Hood, L.: Intellectual property - Publication rights in the era of open data release policies. Science 289, 1881–1881 (2000)

[3] Thompson, M., Heneghan, C.: BMJ OPEN DATA CAMPAIGN We need to move the debate on open clinical trial data forward. British Medical Journal 345 (2012)

[4] Hersey, A., Senger, S., Overington, J.P.: Open data for drug discovery: learning from the biological community. Future Medicinal Chemistry 4, 1865–1867 (2012)

[5] Walport, M., Brest, P.: Sharing research data to improve public health. The Lancet 377, 537–539 (2011)

[6] El Emam, K., Arbuckle, L., Koru, G., Eze, B., Gaudette, L., Neri, E., Rose, S., Howard, J., Gluck, J.: De-identification methods for open health data: the case of the Heritage Health Prize claims dataset. J. Med. Internet Res. 14(1), 1–24 (2012)

[7] Hochberg, A.M., Reisinger, S.J., Pearson, R.K., O'Hara, D.J., Hall, K.: Using Data Mining to Predict Safety Actions From FDA Adverse Event Reporting System Data. Drug Information Journal 41, 633–643 (2007)

[8] Ahmad, S.R.: Adverse Drug Event Monitoring at the Food and Drug Administration. J. Gen. Intern. Med. 18(1), 57–60 (2003)

[9] Wysowski, D.K., Swartz, L.: Adverse Drug Event Surveillance and Drug Withdrawals in the United States, 1969-2002. Arch. Intern. Med. 165, 1363–1369 (2005)

[10] Hoblitzell, A., Mukhopadhyay, S., You, Q., et al.: Text mining for bone biology. In: HPDC 2010 (2010)

[11] Honeywell, M., Philips, S., Vo, K., et al.: Teriparatide for Osteoporosis: A Clinical Review. Drug Forecast 28(11), 713–716 (2003)

[12] Cummings, S.R., Melton, L.J.: Epidemiology and outcomes of osteoporotic fractures. Lancet 2002 359(9319), 1761–1767 (2002)

[13] Carmona, R.H.: Bone Health and Osteoporosis: A Report of the U.S. Surgeon General (2004)

[14] Tatonetti, N.P., Denny, J.C., Murphy, S.N., Fernald, G.H., Krishnan, G., Castro, V., Yue, P., Tsau, P.S., Kohane, I., Roden, D.M., Altman, R.B.: Detecting Drug Interactions From Adverse –Event Reports:Interaction Between Paroxetine and Pravastatin Increases Blood Glucose Levels. Clin. Pharmacol. Ther. 90(1), 133–142 (2011)

[15] Kadoyama, K., Kuwahara, A., Yamamori, M., Brown, J.B., Sakaeda, T., Okuno, Y.: Hypersensitivity Reactions to Anticancer Agents:Data Mining of the Public Version of the FDA Adverse Event Reporting System, AERS. Journal of Experimental & Clinical Cancer Research 30(93), 1–6 (2011)

[16] Tamura, T., Sakaeda, T., Kadoyama, K., Okuno, Y.: Aspirin and Clopidogrel-associated Bleeding Complications: Data Mining of the Public Version of the FDA Adverse Event Reporting System, AERS. International Journal of Medical Sciences 9(6), 441–446 (2012)

[17] Gandhi, P.K., Gentry, W.M., Bottorff, M.B.: Cardiovascular thromboembolic events associated with febuxostat:Investigation of cases from the FDA adverse event reporting system database. Seminars in Arthritis and Rheumatism 12, 1–5 (2012)

[18] Kadoyama, K., Miki, I., Tamura, T., Brown, J., Sakaeda, T., Okuno, Y.: Adverse Event Profiles of 5-Fluorouracil and Capecitabine: Data Mining of the Public Version of the FDA Adverse Event Reporting System, AERS, and Reproducibility of Clinical Observations. Int.J.Med.Sci. 9(1), 33–39 (2012)

[19] Kadoyama, K., Sakaeda, T., Tamon, A., Okuno, Y.: Adverse event Profile of Tigecycline:Data Mining of the Public Version of the U.S. Food and Drug Administraion Adverse Event Reporting System. Biol. Pharm. Bull. 35(6), 967–970 (2012)

[20] Han, J., Micheline, K.: Data mining: concepts and techniques. Morgan Kaufmann, San Francisco (2001)

[21] Zhu, A., Li, J., Leong, T.: Automated Knowledge Extraction for Decision Model Construction: A Data Mining Approach. In: AMIA 2003 Symposium Proceedings, pp. 758–762 (2003)
[22] Kim, J., Washio, T., Yamagishi, M., Yasumura, Y., Nakatani, S., Hashimura, K., Hanatani, A., Komamura, K., Miyatake, K., Kitamura, S., Tomoike, H., Kitakaze, M.: A Novel Data Mining Approach to the Identification of Effective Drugs or Combinations for Targeted Endpoints:Application to Chronic Heart Failure as a New Form of Evidence-based Medicine. Cardiovascular Drugs and Therapy 18, 483–489 (2004)
[23] Drugbank, `http://www.drugbank.ca` (last access: January 02, 2013)
[24] Hall, M., Frank, E., Holmes, G., Pfahringe, B., Reutemann, P., Witten, I.E.: The WEKA data mining software: an update. ACM SIGKDD Explorations Newsletter 11(1) (2009)
[25] WEKA: Weka 3: Data Mining Software in Java (last access: May 02, 2012), `http://www.cs.waikato.ac.nz/ml/weka`
[26] Scheffer, T.: Finding Association Rules That Trade Support Optimally against Confidence. In: Siebes, A., De Raedt, L. (eds.) PKDD 2001. LNCS (LNAI), vol. 2168, pp. 424–435. Springer, Heidelberg (2001)
[27] Holzinger, A., Simonic, K.M., Yildirim, P.: Disease-disease relationships for rheumatic diseases Web-based biomedical textmining and knowledge discovery to assist medical decision making. In: IEEE COMPSAC 36th International Conference on Computer Software and Application, pp. 573–580. IEEE, New York (2012)
[28] Holzinger, A., Yildirim, P., Geier, M., Simonic, K.-M.: Quality-based knowledge discovery from medical text on the Web Example of computational methods in Web intelligence. In: Pasi, G., Bordogna, G., Jain, L.C. (eds.) Qual. Issues in the Management of Web Information. ISRL, vol. 50, pp. 145–158. Springer, Heidelberg (2013)
[29] Holzinger, A., Scherer, R., Seeber, M., Wagner, J., Müller-Putz, G.: Computational Sensemaking on Examples of Knowledge Discovery from Neuroscience Data: Towards Enhancing Stroke Rehabilitation. In: Böhm, C., Khuri, S., Lhotská, L., Renda, M.E. (eds.) ITBAM 2012. LNCS, vol. 7451, pp. 166–168. Springer, Heidelberg (2012)

Random Forests for Feature Selection in Non-invasive Brain-Computer Interfacing

David Steyrl, Reinhold Scherer, and Gernot R. Müller-Putz

Graz University of Technology, Institute for Knowledge Discovery, BCI-Lab,
Inffeldgasse 13/IV, 8010 Graz, Austria
{david.steyrl,reinhold.scherer,gernot.mueller}@tugraz.at
http://bci.tugraz.at

Abstract. The aim of the present study was to evaluate the usefulness of the Random Forest (RF) machine learning technique for identifying most significant frequency components in electroencephalogram (EEG) recordings in order to operate a brain computer interface (BCI). EEG recorded from ten able-bodied individuals during sustained left hand, right hand and feet motor imagery was analyzed offline and BCI simulations were computed. The results show that RF, within seconds, identified oscillatory components that allowed generating robust and stable BCI control signals. Hence, RF is a useful tool for interactive machine learning and data mining in the context of BCI.

Keywords: Random Forest, non-invasive Brain-Computer Interfaces, electroencephalogram, feature selection, machine learning.

1 Introduction and Motivation

System calibration and user training are key features for successful operation of imagery-based brain-computer interfaces (BCIs). BCIs are devices that bypass the normal neuromuscular output pathways and translate a users brain signal directly into action. Imagery-based refers to the use of mental imagery for encoding messages [1]. Historically, BCIs were developed with the aim of restoring communication in completely paralyzed individuals and replacing lost motor function. However, an additional hands-free communication channel can be useful in many applications. Please see [1] [2] [3] for reviews on BCI.

In this paper we focus on electroencephalogram-based (EEG) BCIs that are operated by motor imagery (MI), i.e., the kinesthetic imagination of motor action. MI modulates sensorimotor rhythm (SMR) over corresponding somatotopic brain areas [4]. Hence, different types of MI typically induce distinct somatotopically arranged EEG patterns. Finding characteristic EEG signatures that allow 24/7 on-demand access to BCI-based interaction and optimizing system complexity are our major research goals [5] [6]. Achieving these goals, however, is challenging for several reasons. Most important are the non-stationarity and inherent variability of the EEG [7]. EEG signals are the nonlinear superposition of the electrical activity of large populations of neurons measured non-invasively

A. Holzinger and G. Pasi (Eds.): HCI-KDD 2013, LNCS 7947, pp. 207–216, 2013.

from the scalp. Noise sources, such as introduced by changing mental and emotional states of users (e.g. stress and related inability to perform MI) or the degree of wear of the recording equipment (e.g. drying of conductive gel resulting in changing electrode impedances), can mask induced EEG patterns and cause wrong detections.

Conventional calibration and BCI training protocols require (i) recording of EEG activity patterns from the user prior to BCI use, (ii) analyzing the data to train a pattern recognition algorithm and (iii) performing online feedback training to condition the brain to reliably generate patterns. Recurrent adaptation is used to update the recognition algorithm to the users most recent brain patterns [1], [2], [8]. This procedure leads to co-adaptation of the human brain and the machine. Finding a low amount of EEG features that generalize well and are neurophysiological meaningful for characterizing MI is of up-most importance during this stage. A number of different optimization and feature selection methods exist and are in use [9], however, most of them require problem-specific parametrization and/or are computationally demanding [10]. The situation is complicated by the limited knowledge of brain functioning and its underlying mechanisms relevant for defining the optimal fitness function. Neuronal deficiency such as caused by stroke or amyotrophic lateral sclerosis (ALS) impact on brain activity and connectivity, and result in modified EEG patterns. Best performance is typically achieved by interactive optimization, i.e., by combining human intelligence and machine learning. Human experts interpret the results computed from the machine and select and adapt parameters based on their practical experience. Several iterations may be necessary for selecting parameters that lead to "optimal" performance in the context of the target application. For meaningful real-world use it would be helpful to get results on the fly, i.e., within a training session (minutes time range) and not between sessions (after hours or days). Moreover, data collection in BCI is time consuming and usually only small data-sets are available to train pattern recognition algorithms (typically <100 trials per MI class). Consequently, appropriate techniques have to be employed to to avoid over-fitting the data.

We recently started exploring the usefulness of the Random Forest (RF) [11] machine learning technique for classifying MI tasks [12]. RF is a method that constructs a forest of classification trees where each tree is grown with a bootstrap sample of the training data, and the split criterion at each tree node is selected from a random subset of all features. The final classification of an individual is determined by majority voting over all trees in the forest. RFs can handle large amount of features even when comparable low numbers of training samples are available, generate accurate and robust classifiers with an internal unbiased estimate of generalizability and are computational effective [11]. Moreover, RFs have a build-in function to estimate the contribution of features to correct classification. In this paper, we evaluate RFs as feature selection method in the context of MI-based BCIs. More precisely, we apply RFs to identify oscillatory EEG components that best discriminate, by using a linear discriminant analysis classifier (LDA) [13], between two distinct MI tasks and compare the results with standard components that are often being used. LDA was selected because linear hyperplane classifiers are typically used in the BCI field [9] [7]. Moreover, according to our practical experience, LDA achieves stable results in able-bodied and disabled individuals during real-time operation [3].

2 Materials and Methods

2.1 Random Forests

A RF classifier is an ensemble classifier consisting of many decision trees as base classifier [11]. Decision trees are classifiers which separate the feature space into rectangular regions through multiple comparisons with feature specific thresholds (see Fig. 1). Decision trees are iterative and greedily built from the root to the leaves [14]. At each node the split threshold is calculated with the associated training examples to optimize a split criterion (in the case of RFs the Gini-Index is used as split criterion [15]). Normally decision trees are not as accurate as comparable classifiers [14].

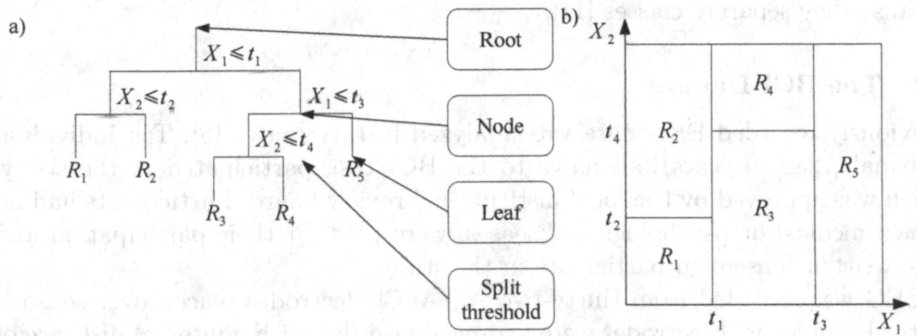

Fig. 1. (a) shows a possible binary decision tree and (b) shows the associated partitioning of the feature space

Nevertheless they are perfect for the use in a combined bagging and random subspace method - as RF is - because, with decision trees, these combined methods are simple to implement [11]. The accuracy of an ensemble classifier depends on the diversity and on the accuracy of the base classifiers, with an upper boundary of the generalization error of

$$PE^* \leq \frac{\overline{\rho}(1-s^2)}{s^2}, \tag{1}$$

where PE^* is the generalization error, $\overline{\rho}$ is the mean value of the correlation of the base classifiers and s is the accuracy of the base classifiers [11], [14]. To achieve a low generalization error two things are important, (i) high accuracy of the individual base classifiers, and (ii) low correlation between them. The correlation between the classifiers is the parameter that can be tuned. Therefore, RFs use two randomization processes during their construction to ensure the diversity of the decision trees. As the first step of bagging, a unique training set is compiled for each tree through random sub-sampling with replacement; this is also called the bootstrap step. Second, only a individual random subset of feature is used for the calculation of the split thresholds at each tree's nodes; this is also known as the random subspace method [16], [17]. These two methods combined cause the decision trees' diversity.

The class assignment of the RF classifier is chosen by a majority vote of all trees' decisions. This is the second part of bagging and is called the aggregating step. The majority decision smooths the single tree classifier models and prevents over-fitting [11].

Gini-Index Based Rating of Features. The random subspace method offers a simple but powerful rating method for the features' importance. The rating is based on the mean improvement in the GI per feature. All the differences in GI between two subsequent nodes are calculated and summed up for each feature separately. Features with high mean improvements over all trees are important, because they separate classes [11].

2.2 The BCI Dataset

Previously recorded EEG data was analyzed in this study [18]. Ten individual (6 female, 28±10 years), all naive to the BCI task participated in the study, which was approved by the local institutional review board. Participants had no known medical or psychological diseases, were paid for their participation and gave written consent to participate in the study.

EEG was recorded from thirty-two Ag/AgCl electrodes placed over sensorimotor brain areas. Electrodes were arranged grid-like with a interval distance of 2.5 cm, and centered according to the international 10-20 system positions C3, Cz and C4. Reference and ground electrodes were mounted on the left and right mastoid, respectively. Electrode impedance was less than 5 kOhm. The EEG was band pass filtered between 0.05 and 200 Hz (50 Hz Notch filter) and sampled at 1000 Hz (Synamps, Compumedics Germany GmbH, Singen, Germany).

A standard cue-guided paradigm was used to collect MI trials [19]. The timing is summarized in Fig. 2. The duration of each trial varied randomly between 10 and 11 s and started with a blank screen. At the beginning of a trial a fixation cross appeared in the middle of the screen. At second 2, a short warning tone was presented to focus the participant's attention. From seconds 3-4.25 an arrow (cue) was shown, indicating the mental task to be performed. An arrow pointing to the left, to the right, or downward indicated the imagination of a left hand, right hand, or foot movement, respectively. Participants were asked to continuously perform MI until the screen was cleared at second 8. The order of appearance of the arrows was randomized. Eight runs, each consisting of 30 trials (ten per class), were recorded on one day.

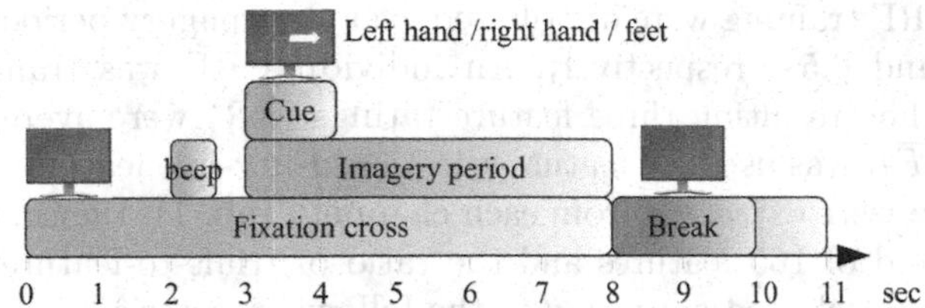

Fig. 2. Paradigm of the experiment. A standard cue paced paradigm with a motor imagery time of five seconds.

2.3 BCI Calibration and Feature Selection

One common and successful way to implement a BCI system is to use Fisher's linear disriminant analysis (LDA) [20] for the classification of logarithmic frequency components that are estimated from Laplacian re-referenced EEG channels [7], [18], [3]. Orthogonal source [21] or Laplacian re-referencing is a filtering technique used to enhance the signal to noise ratio that highlights the local activity of an electrode by subtracting broad activity present also in the four orthogonal nearest-neighboring electrodes. Typically Laplacian channels for electrode postitions C3, Cz and C4 are used for MI-based BCIs. Hence, in the remainder of this paper we focus on this three channels. Power spectral densities (PSD) were estimated from the three channels by applying Fast Fourier Transform (FFT) to 1 s EEG segments. One second segments represent a reasonable trade-off between PSD estimation errors and the time delay for providing feedback to the user on the ongoing brain activity.

The simplest way to encode information is to generate a binary control signal. Hence, the optimization task during the calibration phase in this study consists of identifying oscillatory EEG components in the Laplacian channels that maximize LDA classification between pairs of MI tasks during the 5 s imagery period (sustained activity). We apply RFs to select most discriminant features for LDA and compare the results with accuracies computed by using "standard" frequency bands 8-13 Hz and 16-24 Hz for each channel (two features per channel, six features in total). Matlab (Mathworks, Natick, MA, USA) and the RF package from [22] with recommended parameters (500 trees and $\sqrt{\#features}$ as number of features randomly chosen for calculation of the split threshold at each node) was used for the analysis.

To simulate BCI calibration and control, the dataset was split into train/test (run 1-5, 50 trials per class) and evaluation set (run 6-8, 30 trials per class). The former was used to identify most reactive frequency bands by RF and to train the LDA classifier and the latter was used to evaluate the performance of the BCI on unseen data by offline simulation of on-line feedback control.

Frequency Band Selection. To find frequency components that achieve good classification performance during the entire imagery period, three independent RFs were trained for each individual. The 1 s time windows used for feature extraction and RF training were distributed over the imagery period and started at 4.5 s, 5.5 s and 6.5 s, respectively. An individual RF was trained for each time segment. The resulting three feature ratings (FR) were averaged and the averaged value $\overline{FR}$ was used for feature selection. Fifty-five features in the range between 1-40 Hz were extracted from each channel (Tab. 1). Hence, one training example consisted of 165 features and the ratio of trials-to-features was about 1:3. Features were selected according to the following criteria:

- The single individual feature with the highest $\overline{FR}$ (condition BEST1).
- The six features with the highest $\overline{FR}$ (condition BEST6).
- All features with ratings that were higher that the mean plus one standard deviation of all ratings (condition ADAPTIVE).
- For each channel the single feature with the highest normalized rating $\overline{FR}_N$ (condition 1/CH). Normalized ratings $\overline{FR}_N$ were calculated by dividing individual ratings by the mean rating over all features. Only features with $\overline{FR}_N > 2$ were considered. When the > 2 criterion could not be met, then the feature with the highest rating $\overline{FR}$ was selected.
- For each channel the two features with the highest normalized ratings (condition 2/CH). Alternatively with the highest $\overline{FR}$.
- For each channel the three features with the highest normalized ratings (condition 3/CH). Alternatively with the highest $\overline{FR}$.
- For each channel components 8-13 Hz and 16-24 Hz (condition STANDARD).

Individual LDA classifiers were computed with the features identified for each of the above criteria. Note that only features extracted from the first time segment (4.5-5.5 s) were used to train the LDA. The analysis was repeated for each of the three class combinations.

BCI Simulation. The computed LDA classifier and identified frequency components were evaluated by computing a BCI simulation on the evaluation dataset. To get the time course of classification over the imagery period, the imagery period of each trial was subdivided into N=40 overlapping 1 s segments with time-lag of 100 ms, starting at second four. Frequency components were extracted from each segment and classified by LDA. The participant's accuracy time course was calculated by averaging the classification results per segment over the trials. The mean classification accuracy during imagery was computed by averaging the participant's accuracy time course.

Table 1. List of frequency components extracted from 1 s segments for each channel. Mu and beta frequency band components were computed by calculating the sum of corresponding FFT bins.

type	frequency ranges	amount of features
single bins	1-40 Hz in single Hz steps	40
mu range	6-8 Hz, 7-9 Hz, 8-10 Hz,... 12-14 Hz	7
beta range	14-19 Hz, 17-22 Hz, 20-25 Hz,... 35-40 Hz	8

3 Results

The duration of the frequency band selection, i.e.,the time required to train three RFs and apply the selection criteria, was about 3 s for each subject (Intel Core-i5 processor at 3 GHz).

The calculated BCI simulation accuracies, averaged over all subjects, are summarized in Tab. 2. Significant improvements (+6%, $p < 0.01$) of mean and median classification time course accuracies were found only for left hand vs. feet MI classification. Fig. 3 summarizes the selected frequency components for each channel and subject, and the mean classification accuracy, averaged over the 5-s imagery period, for conditions STANDARD and 1/CH, respectively. One can see that for channels C3 and C4 the majority of discriminative frequency components were found in the 8-13 Hz band range. For channel Cz most important features were predominantly found in the beta (13-30 Hz)and gamma (>30 Hz) band. Fig. 4 (a) and (b) illustrates the impact of the selected features on the classification time courses of participant 4 and 6, respectively. The horizontal line represents the 61% threshold level for chance classification [23]. Fig. 4 (a) shows an example in which classification was satisfactory, however, not stable during the imagery period before optimization. Fig. 4 (b) shows an example where standard bands did not lead to meaningful classification results. Found frequency components achieved satisfactory and stable performance.

Table 2. BCI simulation results. Mean/median classification accuracies, averaged over all the participants during the imagery period, for each condition and pair-wise classification task. Items marked with "*" are statistically significantly ($p < 0.01$) different.

Condition	Classification		
	left vs. right	left vs. feet	right vs. feet
STANDARD	59/58%	62/64%	63/60%
BEST1	57/53%	67/68%	64/67%
BEST6	60/56%	68/69%*	68/69%
ADAPTIVE	58/55%	66/67%	64/64%
1/CH	60/56%	68/70%*	66/68%
2/CH	60/56%	68/70%*	67/69%
3/CH	60/57%	69/70%*	66/68%

4 Discussion and Conclusion

System calibration and individual adaptation are crucial for enhancing MI-based BCI control. Optimization and adaptation, however, are usually computationally demanding and time consuming. Furthermore, best results are typically achieved by combining human intelligence and machine learning. The results of this study

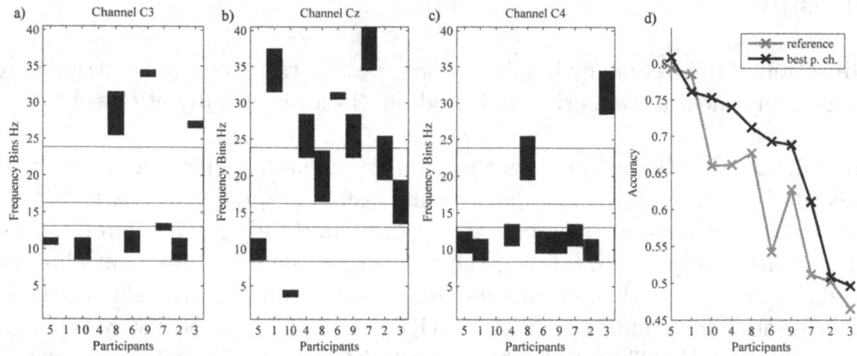

Fig. 3. Results for left hand versus feet MI for condition STANDARD and 1/CH. (a), (b) and (c) show the selected features for channels C3, Cz and C4, respectively. STANDARD bands are highlighted. (d) Mean accuracies, averaged over the 5-s imagery period, for each participant.

suggest that the use of RFs for feature selection and data mining leads to the identification of discriminative frequency components within a time period of seconds. This allows for meaningful interactive optimization. More sophisticated implementations of RF exist. We selected the basic form because of the low computational demands and the good performance that is generally achieved.

Classification accuracies computed with selected features are generally higher when compared to STANDARD (Tab. 2). However, only left hand vs. feet MI was found to be significantly higher ($p < 0.01$). These results suggest that the use of left hand vs. feet MI task in a 2-class BCI induces most distinct EEG patterns in naive users and should therefore preferably be used as mental strategy.

When looking at the selected frequency bands in Fig. 3, we find that broader bands were more often selected than 1 Hz bands. This is reasonable, given EEG oscillations are not "sine-waves" but result form the noisy superposition of large population of cortical neurons and are consequently not stable in time. For channels C3 and C4 mostly features in the standard 8-13 Hz band were selected. This further supports the usefulness of of the standard band 8-13 Hz definition. For channel Cz and the left hand vs. foot imagery task, beta band activity seems to contribute more to correct classification than mu band activity. A comparison of STANDARD and 1/CH band shows that twelve out of the 27 selected features were completely or partly outside the STANDARD feature range. Reducing the number of features from two per channel (STANDARD) to one per channel (1/CH) significantly improved the classification accuracy. This again confirms that feature selection (reduction) is essential for enhancing accuracy and the generalization capabilities of LDA classifiers.

Fig. 3 (d) shows that all participants except one (participant 1) benefited from optimization. The reason for the 2% drop in accuracy from 78% to 76% for this subject was that only one single band was selected for each channel. Adding information for the beta band leads to a performance increase.

Notably, the accuracies of five participants (number 10, 4, 8, 6, and 9; Fig.3(d)) increased by more than 6% and reached the 70% level, which is considered to be the lower boundary that allows meaningful 2-class BCI control. The maximum increase of approximately +15% was calculated for participant 6 (Fig.4(b)).

Concluding, feature optimization with RFs for classification of oscillatory EEG components with LDA is fast, leads to increased classification accuracies and hence to more stable and reliable control signals. These properties make RF a suitable tool for data mining and knowledge discovery in the context of BCI.

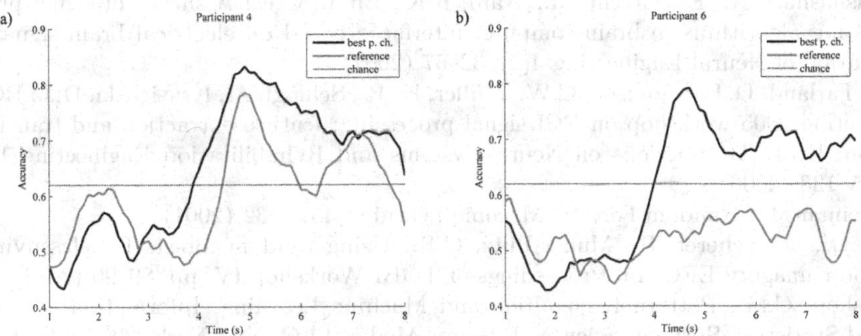

Fig. 4. Accuracy time courses for left hand versus feet MI for (a) participant 4 and (b) participant 6. The horizontal line represents the 61% level of chance classification. Visual cues were presented at second 3.

Acknowledgments. This work was supported by the FP7 Framework EU Research Project ABC (No. 287774). This paper only reflects the authors views and funding agencies are not liable for any use that may be made of the information contained herein.

References

1. Pfurtscheller, G., Neuper, C.: Motor imagery and direct brain-computer communication. IEEE Proceeding 89(5), 1123–1134 (2001)
2. Wolpaw, J.R., Birnbaumer, N., McFarland, D.J., Pfurtscheller, G., Vaughan, T.M.: Brain-Computer interfaces for communication and control. Clinical Neurophysiology 113, 767–791 (2002)
3. Scherer, R., Müller-Putz, G.R., Pfurtscheller, G.: Flexibility and practicality: Graz Brain-Computer Interface Approach. International Review on Neurobiology 86, 147–157 (2009)
4. Pfurtscheller, G., Neuper, C.: Motor imagery activates primary sensorimotor area in humans. Neuroscience Letters 239, 65–68 (1997)
5. Scherer, R., Faller, J., Balderas, D., Friedrich, E.V.C., Pröll, M., Allison, B., Müller-Putz, G.R.: Brain-Computer Interfacing: More than the sum of its parts. Soft Computing 17(2), 317–331 (2013)

6. Billinger, M., Brunner, C., Scherer, R., Holzinger, A., Müller-Putz, G.R.: Towards a framework based on single trial connectivity for enhancing knowledge discovery in BCI. In: Huang, R., Ghorbani, A.A., Pasi, G., Yamaguchi, T., Yen, N.Y., Jin, B. (eds.) AMT 2012. LNCS, vol. 7669, pp. 658–667. Springer, Heidelberg (2012)
7. Lotte, F., Congedo, M., Lécuyer, A., Lamarche, F., Arnaldi, B.: A review of classification algorithms for EEG-based brain-computer interfaces. Journal of Neural Engineering 4, R1–R13 (2007)
8. Faller, J., Vidaurre, C., Solis-Escalante, T., Neuper, C., Scherer, R.: Autocalibration and recurrent adaptation: Towards a plug and play online ERD-BCI. IEEE Transactions on Neural Systems Rehabilitation Engineering 20, 313–319 (2012)
9. Bashashati, A., Fatourechi, M., Ward, R.K., Birch, G.E.: A survey of signal processing algorithms in brain-computer interfaces based on electrical brain signals. Journal of Neural Engineering 4(2), 32–57 (2007)
10. McFarland, D.J., Anderson, C.W., Müller, K.-R., Schlögl, A., Krusienski, D.J.: BCI meeting 2005-workshop on BCI signal processing: feature extraction and translation. IEEE Transactions on Neural Systems and Rehabilitation Engineering 14, 135–138 (2006)
11. Breiman, L.: Random Forests. Machine Learning 45, 5–32 (2001)
12. Steyrl, D., Scherer, R., Müller-Putz, G.R.: Using random forests for classifying motor imagery EEG. In: Proceedings of TOBI Workshop IV, pp. 89–90 (2013)
13. Bishop, C.M.: Pattern Recognition and Machine Learning. Information Science and Statistics. Springer Science+Business Media, LLC, New York (2006); Jordan, M., Kleinberg. J., Schölkopf, B. (Series eds.)
14. Hastie, T., Tibshirani, R., Friedman, J.: Springer Series in Statistics: The Elements of Statistical Learning, 2nd edn. Springer (2009)
15. Breiman, L., Friedman, J.H., Olshen, R.A., Stone, C.J.: CART: Classification and Regression Trees. Wadsworth, Belmont (1983)
16. Breiman, L.: Bagging predictors. Machine Learning 24, 123–140 (1996)
17. Ho, T.K.: The random subspace method for constructing decision forests. IEEE Transactions on Pattern Analysis and Machine Intelligence 20, 832–844 (1998)
18. Müller-Putz, G.R., Scherer, R., Pfurtscheller, G., Neuper, C.: Temporal coding of brain patterns for direct limb control in humans. Frontiers in Neuroscience 4, 1–11 (2010)
19. Pfurtscheller, G., Müller-Putz, G.R., Schlögl, A., Graimann, B., Scherer, R., Leeb, R., Brunner, C., Keinrath, C., Lee, F., Townsend, G., Vidaurre, C., Neuper, C.: 15 years of BCI research at Graz University of Technology: current projects. IEEE Transactions on Neural Systems and Rehabilitation Engineering 14, 205–210 (2006)
20. Duda, R.O., Hart, P.E., Stork, D.G.: Pattern Classification. John Wiley and Sons, USA (2001)
21. Hjorth, B.: An on-line transformation of EEG scalp potentials into orthogonal source derivations. Electroencephalogr. Clin. Neurophysiol. 39, 526–530 (1975)
22. Jaiantilal, A.: Random Forest implementation for MATLAB (November 6, 2012), `http://code.google.com/p/randomforest-matlab/`
23. Müller-Putz, G.R., Scherer, R., Brunner, C., Leeb, R., Pfurtscheller, G.: Better than Random? A closer look on BCI results. International Journal of Bioelektromagnetism 10, 52–55 (2008)

End Users Programming Smart Homes – A Case Study on Scenario Programming

Gerhard Leitner[1], Anton J. Fercher[1], and Christian Lassen[2]

[1] Institute of Informatics-Systems
Alpen-Adria Universität Klagenfurt, Universitätsstrasse 65-67
9020 Klagenfurt, Austria
{gerhard.leitner,antonjosef.fercher}@aau.at
[2] Institute for Psychology
Alpen-Adria Universität Klagenfurt, Universitätsstrasse 65-67
9020 Klagenfurt, Austria
c.lassen@edu.uni-klu.ac.at

Abstract. Smart technology for the private home holds promising solutions, specifically in the context of population overaging. The widespread usage of smart home technology will have influences on computing paradigms, such as an increased need for end user programming which will be accompanied by new usability challenges. This paper describes the evaluation of smart home scenarios and their relation to end user programming. Based on related work a two-phase empirical evaluation is performed within which the concept of scenarios in the private home is evaluated. On the basis of this evaluation a prototype which enables the simulation of end user programming tasks was developed and evaluated in comparison to two commercial products. The results show that, compared to the commercial products, our approach has both, some advantages as well as drawbacks which will be taken into consideration in further development planned for the future.

Keywords: Smart Home, End user programming, Comparative Evaluation, Scenario Programming.

1 Introduction

The smart home experiences yet another hype in the context of population overageing. Expensive, high end systems are competing for the attention of potential customers with low cost off-the-shelf systems [1]. The advertising arguments to promote smart home systems are, independent of price segment, oftentimes emphasizing the good usability of the product's user interfaces. However, the promised features such as intuitiveness and ease of use even for non-experienced users have to be treated with caution when recalling the incarnation of unusability of the private home of the pre-smart home era, the VCR. This paper investigates usability problems in the context of smart home interaction by focusing on a specific feature of smart home systems - the possibility to program scenarios - which we suppose to be an extraordinary challenge, specifically for naïve

A. Holzinger and G. Pasi (Eds.): HCI-KDD 2013, LNCS 7947, pp. 217–236, 2013.

users. Scenarios enable users to switch devices in a combined way, examples frequently found in smart home advertising material are describe pre-programmed scenarios that can support daily routines (e.g. in the morning), help to control entertainment equipment in an integrated way or enable the simulation of the presence of a human being to discourage burglars. Each scenario consists of a combination of user triggered activities involving the control of devices (e.g. switches) and the provision of status information (e.g. from motion sensors). For example, a typical morning scenario starts in the bedroom where lights are switched on (or continuously dimmed to a predefined brightness level) and the blinds are opened. After a certain time delay the heating in the bathroom is activated or increased to a certain level and finally devices such as the coffee maker or toaster in the kitchen are starting their preprogrammed procedures to prepare breakfast just in time.

Although such functions are tempting, they require a lot of programming efforts to work in the intended way. Such programming requires a high level of computing skills, probably not found among average home owners or dwellers. To further disseminate smart home technology in private homes it is therefore necessary to enable persons with average or even lower than average computing skills to program, maintain and configure their smart home systems themselves. When taking a closer look on state-of-the-art smart home systems one can get the impression that they currently offer just what technology is capable of and not what people really need. Specifically the interfaces provides for programming and configuration exhibit several weaknesses. To investigate this problem a two-step study was designed with the goal to challenge the sensefulness of scenarios in the home in general, and to be able to specify the requirements for scenario programming of smart home tasks in particular. The study was carried out in the context of the project Casa Vecchia which aims at the development of technology for elderly in order to prolong independent living in their accustomed home.

The remainder of this paper is structured as follows. In the following section an overview on related work is given, afterwards the approach followed in the study is described in section 3. The outcome of the study is described and discussed in section 4, the paper finishes with a discussion and conclusion in section 5.

2 Related Work

The concept of a smart homes can be traced back for about three decades [2], but in contrast to industrial and public buildings, smart home technology didn't receive a broad acceptance in the private sector. One of the - of course multidimensional [3] - reasons is the anticipation of problems related to usability, which is, given the following examples, very comprehensible. Even a task as small as setting the clock on a VCR, which the former president of the USA, G.W. Bush, in the beginning of the 1990es formulated as a goal to achieve during his presidency with the words: *Actually, I do have a vision for the nation, and our goal is a simple one: By the time I leave office, I want every single American to be able to set the clock on his VCR.*[1] was still not reached in 2010, when Don Norman succinctly stated *"He failed..."*. [4]. Similar problems are exemplified in [5] or [6] describing people that are in the mercy of smart technology instead of controlling it. This is not only a phenomenon of low cost systems [7], even the legendary smart home of a certain Mr.Gates has to be maintained by qualified personnel and constitutes a Wizard of Oz system [8] rather than an environment that is really smart and enables end user control and maintenance in an adequate way.

These examples point to an important difference between industrial or governmental building and private homes. The former are based on systems which are developed, installed and maintained by professionals. End users touch these systems only to perform predefined standard tasks such as using presentation facilities in a seminar room. In the case of problems, which frequently can be observed on public events (e.g. blue screen on projector, no sound, wrong illumination), an administrator is available for help. The situation in the private home is different, although e.g. [1] observes some parallels, because in most private home there is also one person who takes over the main responsibility for technical issues. One reason for taking those things into one's own hands is the necessity to keep costs low. Consumer markets increasingly also follow this strategy to assign more and more tasks to the consumers with the purpose to increase their margins. A TV documentation named *the costumer as servant* strikingly illustrates that tasks such as buying and assembling furniture, self service checkouts in supermarkets, e-government etc. determine our lives. And because programmable microprocessors are meanwhile an integral part of many household appliances such as alarm clocks, TVs, VCRs, etc.[9] and personal customer services are continuously replaced by online self-services end user programming is already an essential part of our lives and will grow in importance. By the end of 2012 it was estimated by [10] that already 90 million end users do some sort of programming, 13 million of them are traditional programmers. Another difficulty for private home owners is that the access to qualified personnel can be considered as a specific challenge [11,12], because professionals are costly, rarely available and not willing or able to provide adequate information. Finally, there is a difference between work related and private settings in other aspects, such as privacy, observable on the fact that people want there homes explicitly "NOT" accessible from anywhere [10] or avoid systems that are too invasive [9]. Considering all these aspects it a difficult task to enhance the user experience of smart home systems.

One possibility is to enhance automated functions in the home by the usage of, for example, artificial intelligence methods that can use routines (or scenarios) as the basis for automation [13,15,14]. When taking a look back in the history of domestic technology, examples of such automated functions can be found. About twenty years ago a TV set had to be programmed manually to set time, channels and other basic settings. These things are nowadays done automatically over

[1] `http://www.electapres.com/political-quotes/`

satellite or internet connection. A basic level of self organization, for example components which perform their basic setup automatically, would be of similar benefit in smart home systems. By observing a frequent behavior of a person and the devices that are involved in this sequence a smart home system could also automate the support of routines (e.g. the morning activity described in section 1). Despite the advantages of automation, explicit interaction and manipulation will still be necessary in the future as lifestyles, daily routines [6,16] as well as the buildings themselves (structure, room layout, technical equipment, etc.)[17] are probably changing over time. Therefore end user manipulation has to be considered and supported, should be possible without the need of training [18] and designed in a way that even computing novices are able to program their systems in an easy and adequate way [18,19]. A big leap forward in that direction would be to invest efforts in the standardization of smart home technology. Compared to a state-of-the-art car, where tens of computers are smoothly inter-operating, the typical private home constitutes a wild mix of different devices and subsystems. Initiatives to enhance the interoperability are observable, for example respective publications [20], platforms such as OSGI [21] and initiatives striving for a standardization of interfaces, e.g. Universal Remote console [22] point in the right direction. However, interfaces for end user configuration, control and maintenance are, in general, still on a level not satisfactory at all. It seems that many vendors of smart home systems are trying to push proprietary solutions instead of investing their efforts to support standardization [23]. Changes in this philosophy as well as moving away from traditional computing paradigms are an important prerequisite for future developments [10].

Several barriers can negatively influence or even prevent end users from programming, cf. e.g.[24]. To reduce such barriers the approach we followed was to provide an interface which is as natural and intuitive as possible. This goal can be achieved by touch interaction [25], because this form of direct manipulation has parallels to the usage of devices and functions also available in conventional homes. Direct manipulation also has advantages in regard to usability [9], a user can immediately see the results of his or her actions and assess whether the achieved result is acceptable or not and reverse it [1]. The design of the study described in the next sections considered both, the theoretical aspects discussed above as well as the related literature addressing the concrete realization of prototypes, such as an approach based on a felt board and felt symbols used in [9], magnetic elements [28] for the low fidelity prototypes as well as [26,27,25] serving as a basis for the realization of the high-fidelity digital prototype. The specification of the scenarios was based on scientific literature [29,6] as well as smart home information and advertising material from different brands and manufacturers.

3 Method

The study is separated into two stages. The prestudy was designed to identify whether smart home scenarios are of interest for home dwellers at all or if these are just representations of, to quote Shneiderman [30], *the old computing (what computers or - in our context - smart home systems can do)*, in other words, a technical solution in search of a problem.

3.1 Prestudy

In the prestudy a paper and pencil based evaluation of daily activities was performed with the goal to identify routines and other aspects that would relate to scenarios. The target size of the participant sample was twenty, finally the data of eighteen people could be used for further processing. The participants are characterized by an average age of 30,78 years with a standard deviation of 13,25 years. Women and men were equally distributed in the sample. Fourteen participants lived in apartments, three in a family house and one in a students home. Regarding their marital status eight participants were singles, five lived with their partners, three with their family with children and two in a residential community. Occupations range from students over persons employed in different areas (technical or administrative) to already retired persons. The prestudy consisted of two phases, whereas the first phase was focused on activities in the home in general and the second aimed at the identification of devices in the household that are associated to the activities and routines identified in phase one. Phase one was performed as a structured interview without informing the participants about the objective of the study (i.e. smart home scenarios) in order to not to lead their thoughts in a certain direction. In the second phase the participants were asked to concretize typical routines based on provided material in the form of paper cards showing electrical devices typically available in a home. Moreover a time grid was provided to ease the description of activities in connection to the depicted devices. A picture of the material used is provided in Fig. 1. The whole evaluation lasted about 1,5 hours per participant and was performed individually. The evaluation took place on diverse locations, for example, the participants' homes or in offices at the university. The participants didn't get any refund for the participation.

3.2 Main Study

The results of the prestudy revealed that routines are frequent and that technology supported scenarios would be useful in regard to the needs the participants expressed. Therefore the next step was to conceptualize and implement a prototype to enable the programming of scenarios. The prototype had to meet several basic requirements. First, the participants of the prestudy expressed their joy of using the card sorting method, because it constitutes an intuitive and direct way to express needs and opinions. Based on this experience the prototype should provide a high level of directness similar to conventional card sorting. Second, the prototype should be based on state-of-the art technology which is already used in current and will probably be used also in future home settings. So the target platforms would be embedded PCs, smartphones or tablets operated with either Linux, Android, iOS or Windows. The resulting prototype was developed in Android and customized to a 10 inch Samsung Galaxy Tab. The central layout

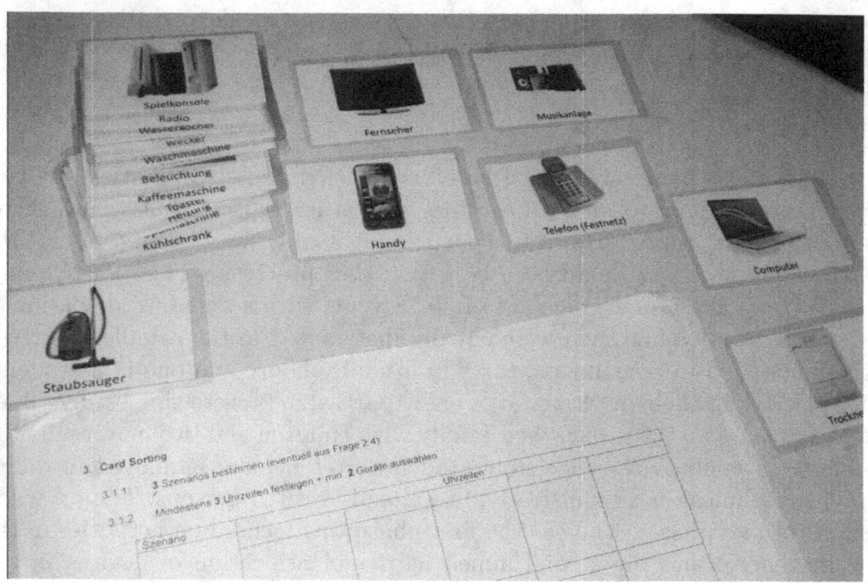

Fig. 1. Material provided for the performance of the pretest

is based on dragging and dropping icons of household devices which represent digital versions of the paper cards used in the prestudy. A screenshot of the prototype is shown in Fig. 2.

However, to be able to judge the quality of the developed prototype it was necessary to apply adequate measures. Besides the usage of a standard instrument to evaluate the subjective opinions (we used the german version of the user experience questionnaire (UEQ)) [31] objective measures were collected on the basis of a comparative evaluation. Our prototype was compared to two smart home systems available on the end consumer market. We assumed (and the assumption is supported by the literature, cf. e.g. [11,7,8]) that most probably owners of relatively cheap off-the-shelf smart home systems rather than owners of high priced ones would be in the need of programming their smart home systems themselves. Therefore the two systems investigated are taken from the low price segment. The first system (in the following named OSS) is sold online over electronics retailing platforms, the other one is offered by german and austrian energy providers (in the following named EPS). Both systems address the private home owner and advertising material provided by the distributors emphasizes the good usability of the products. The target number of participants for the study was again 20, finally the data of 17 subjects could be used for further analysis. Because of a possible bias we decided not to invite people who participated in the prestudy. In contrast to the prestudy which was not place-bound the main study had to be performed in a lab room at the university for the following reason. To be able to perform the study as efficiently as possible three comparable touch operated devices were pre-configured, each of them

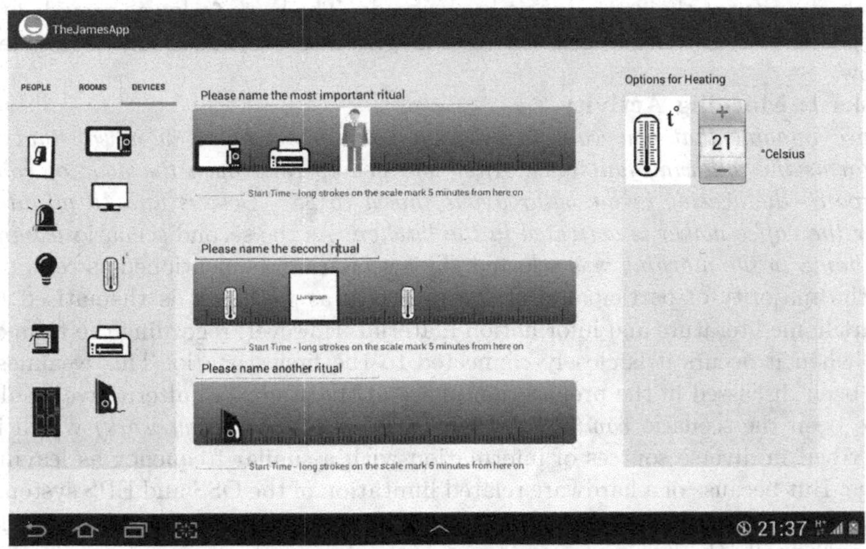

Fig. 2. The prototype for the main study. On the left hand side a tabbed container is provided which contains the source elements needed to program a scenario. Most important elements are the devices, but we also provided elements representing two other categories, rooms and people, because these could also be relevant for the configuration of a scenario. With Drag&Drop the elements can be positioned in one of the three containers in the center which include a time grid to configure three different scenarios.

running one of the systems to be evaluated. The devices were an Apple iPad running the OSS system, a Samsung Galaxy Tab running our prototype and an Asus embedded PC with touchscreen running the EPS system. During the evaluation this setting facilitated an easy switch between the systems. As in the prestudy one participant at a time performed the evaluation. The average age in the sample was 23,58 years with a standard deviation of 2,45 years. Six men and eleven women participated in the study, all of them were students, whereas nine study computer sciences, six psychology and two education sciences. After welcoming and introducing the participants to the study objective (in contrast to the prestudy the participants were fully informed) they had to perform two tasks with the goal to program scenarios with each of the three systems in an alternating sequence. The participants received 10 Euro as compensation.

The scenarios which were used as example tasks were selected based on the routines identified in the prestudy as well as on the frequency they are mentioned in smart home information material and related literature, e.g.[6]. The scenario most often found in all sources of information is the *morning routine* which constituted the first scenario in our study. The basic task was provided to the participant in written form (in german), the english translation is provided below.

Tasks 1: Morning Activity

Please imagine that you want to program your smart home in a way that it performs the following functions: After you get up (and open the door of your bedroom) the heating in the bathroom is raised to 25 ° Celsius and 10 minutes later the coffee maker is activated in the kitchen. As the second scenario *leaving the home in the morning* was selected. This scenario was mentioned as relevant by the majority of participants of the prestudy as well as it is thematised in smart home literature and information material. However, regarding the time of day when it occurs it is closely connected to the first scenario. This weakness has been discussed in the preparation phase of the study. An alternative would have been the scenario *coming home in the evening (e.g. from work)* which is described in diverse sources of information with a similar frequency as leaving home. But because of a hardware related limitation of the OSS and EPS systems (only scenarios could be programmed the components for which were physically present) finally the leaving home scenario was chosen.

Task 2: Leaving Home, Alarming Function

You leave your home and by pressing a switch near the entrance door the heating should be lowered and all the lights have to be switched off. After additional five minutes an alarm function is activated. After having programmed the two task scenarios on one system the participants had to fill out a short questionnaire stating their subjective opinion on the system they just used. The questionnaire included some open questions as well as the semantic differential of the UEQ, shown in Fig. 5. After having finished the whole sequence of programming scenarios on all three systems the participants were asked for a final comparative assessment of all systems as well as for a reasoned preference rating.

4 Results

4.1 Prestudy

In the following the results of the prestudy are presented. As described above, in the first step the participants were asked about activities and routines they typically perform at home. Table 1 shows the activities mentioned, how many participants mentioned them and what percentage in regard to the whole sample this represents. The second question addressed household goods that are associated to the activities mentioned in response to question one. Table 2 shows the effect of not priming the participant (not informing them about the objective of the study) because they also mentioned non-electrical or non-computerized things such as beds, door handles or books. Afterwards, we asked the participants to describe activities that are regularly performed (daily routines). Table 3 shows the results. Before being asked the following questions the participants were informed about the objective of the study, i.e. the potential to support activities in

Table 1. Activities at home

Activity	No. of Participants	Percentage
Sleeping	10	55,6
Reading	9	50
Relaxing Working	8	44,4
Cooking Listening to music Watch TV	7	38,9
Using PC Eating	6	33,3
Hygiene	5	27,8
Tidying up / Cleaning Receive guests or friends	4	22,2
Drink coffee Making music	3	16,7
Smoking Watering flowers Drawing Maintaining technology Crafting	1	5,55

the home by computerized technology. This was necessary to understand the outcome we expected from them in the following tasks. The participants were asked to explain as precise and detailed as possible the routines they identify in their daily activities. Because smart home technology currently is mainly capable of controlling electric, digital and computerized devices, they should associate this kind of devices to the routines and leave out other things possibly mentioned in the previous phase of the study. Table 4 shows the results of the detailed analysis of the routine mentioned by all participants, the morning routine. The second most mentioned routine was coming home in the evening. The resulting connections with devices are shown in Table 5. The information regarding the goal of the study obviously influenced the answers, specifically those given in response to the question: *"In what contexts or for which task could you imagine that technology could support you?"*. The answers included many products or services already available on the market such as automated cleaning or vacuuming devices (mentioned by 7 participants) or lawn mowing robots (4). The same applies to some kind of intelligent refrigerator which informs when goods run short (3), or a bathroom mirror displaying news (3). However, also some extraordinary imaginations were expressed, such as a self filling washing machine or a mechanism supporting the automated packing, unpacking and cleaning of sporting clothes, or a system that supports relaxing. On the question what means of interaction or devices the participants could imagine to be able to control smart home functionality, the majority of them named the smart phone (13), followed by voice control and wall mounted touch screen (8), remote control (6), tablet PC (4) and standard computers or websites (3).

Table 2. Household goods most frequently used at home

Household good	No. of Participants	Percentage
Computer	14	77,8
Toothbrush TV-Set	4	22,2
Coffee Maker Bed Couch Smartphone Dishes Water cooker Sanitary facilities	3	16,7
Books Remote controls Stove	2	11,1
Dishwasher, Ipod Docking Station Doorhandle, Cupboard Stereo equipment, Lights, Accordion, Slippers Vacuum cleaner, Carpet, Mp3-player, DVD-player Couch-table, Lighter, Ashtray, Kitchen table, Lawn mover, Alarm clock, Refrigerator, Closet, Kitchen	1	5,55

Table 3. Daily Routines

Routine	No. of Participants	Percentage
In the morning / getting up	18	100
Leaving home / coming home Cleaning	8	44,4
Going to bed In the evening	6	33,3
Friends, Guests	5	27,8
Before a journey Cooking At Lunchtime Before going out Weekend (Sunday)	2	11,1
After booting up computer, In the afternoon, At dinnertime, Relaxing, Shopping, Before/after sports activities	1	5,55

Table 4. Morning Routines. In the first column the involved devices are listed. The first row includes a time scale. The numbers represent the time in minutes elapsed after starting the activity (in this case getting up). The numbers in the grid illustrate how many participants stated that they use a certain device in the respective time slot and in what sequence they did so.

Device	Minutes								
	0	15	30	45	60	75	90	105	120
Cellphone	10	6	2						
Refridgerator	2	5	5	3	5				
PC		4	2	3	2				
Alarm-Clock	10								
Coffee Maker	3	3	2						1
Stove	1	2	3	1	1				
Lights	6	1							
Water Boiler		3	2	1					
TV		2		1				1	
Heating	3	1							
Dishwasher	1		1		1	1			
Radio	1	2							
Stereo Equip.	2	1							
Game Console				1				1	
MP3 Player		1	1						

4.2 Main Study

The first data analysis was performed to investigate how long it took to perform the scenario programming tasks described in section 3.2. Table 6 shows the results. As it can be observed in the column labeled maximum, the majority of values is 10:00 minutes. The reason for this is that 10 minutes was predefined as an upper threshold. In the case of reaching this threshold the task was terminated and defined as uncompleted. What is surprising on the data shown in the Table 6 is that not only our system, which constitutes a prototype, has major drawbacks hindering users to complete a task within an acceptable period of time, but also the commercial systems exhibit such barriers. Therefore a closer look at the reasons for these problems is necessary. In Table 7 the problems that occurred during the programming of our prototype are enumerated. A major drawback of our prototype is that it was not taken seriously. This is eligible, because it (in contrast to the other two) didn't connect to any specific devices which were physically present. However, it is build upon a platform which currently is used

Table 5. Coming Home Routines

Device	Minutes								
	0	15	30	45	60	75	90	105	120
Lights	5						2		
Refridgerator	4	3	3						
PC	4	1	2	3	1		1		1
Stove	3	3	4						
Heating	2			1					
TV	2		1	1	1		1	1	1
Cellphone	1		2		1				
Blinds	1								
Coffee Maker	1								
Stereo Equip.	1	1							2
DVD					1				
Microwave		1	1						
Radio			1						
Game Console				1	1				
Dishwasher		1				1	2		
Toaster			1						
Washing Machine					1	1			
Water Boiler		1	1						

Table 6. Time measures of task completion

Task	Time			
	Minimum	Maximum	Mean	STD
EPS 1	07:14	10:00	09:41	00:43
EPS 2	02:40	10:00	05:34	02:33
OUR 1	02:20	10:00	05:19	02:06
OUR 2	00:45	07:05	02:22	01:24
OSS 1	02:44	10:00	06:08	02:23
OSS 2	02:10	08:50	04:16	01:51

for controlling 21 smart home installation within the project mentioned above. Therefore enabling the prototype to actually control a smart home system would be possible with only little efforts. However, because the participants were not used to an interface based on an alternative approach (card sorting metaphor), they were skeptic regarding its potential performance. The other problems are

Table 7. Usability problems identified on our prototype

Problem	No. of Participants
Unsure if it really works	7
Starting time unclear	5
Usage of 3 scales Bad overlay of rooms/devices Scaling of time line suboptimal	4
Symbols not recognized (heating, door, coffee maker)	3
Status of door (open/closed) not clear Symbol for all rooms missing	2
Connection room / device not clear How to place symbols that everything works in parallel Devices are difficult to place on the time line Persons unclear Missing save button Short tutorial would be necessary	1

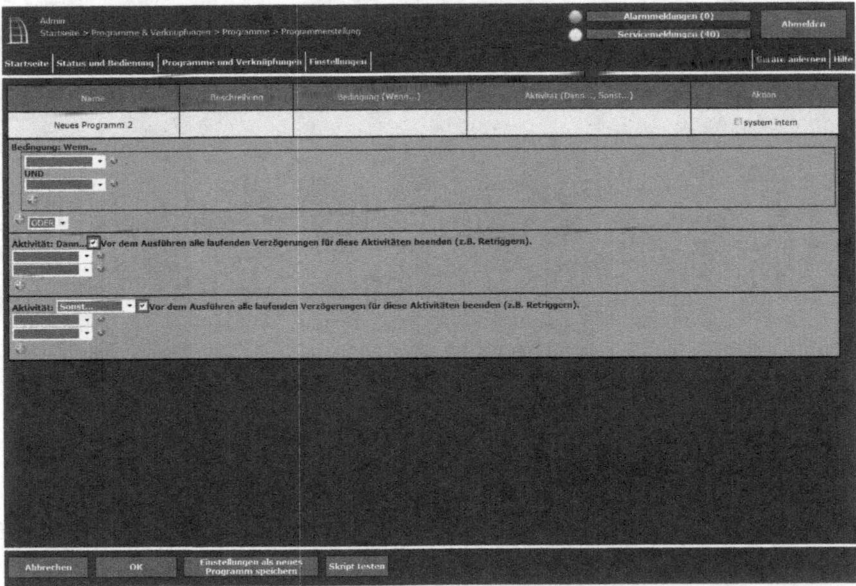

Fig. 3. Screenshot of the OSS System, contents courtesy of EQ-3 AG

comprehensible. In Fig. 3 a screenshot of the online sold system (OSS) is shown, Table 8 includes the list of the problems which occurred when the tasks had to be performed with this system. Major drawbacks of the OSS system have been that elements are sub-optimally designed and mislead users within the task sequence. Finally, usability problems found in the energy provider system, a screenshot of

Table 8. Usability Problems of the Online Sold System (OSS)

Problem	No. of Participants
Click on (non-functional) menu header	5
Unclear if alarm is correctly activated	4
Display flickers when keyboard is displayed Alarm in menu "system settings "	3
Red x seems to be an error message Instead *then* an *and* condition is programmed Trigger on change constitutes a bad description Preview covers background information Complex, because of many parameters Selection of devices not found	2
Only *if* found, *then* unclear Status door open/close unclear Minute entry does not work Adding a second device not clear Missing scroll bar in device overview Rooms not recognized	1

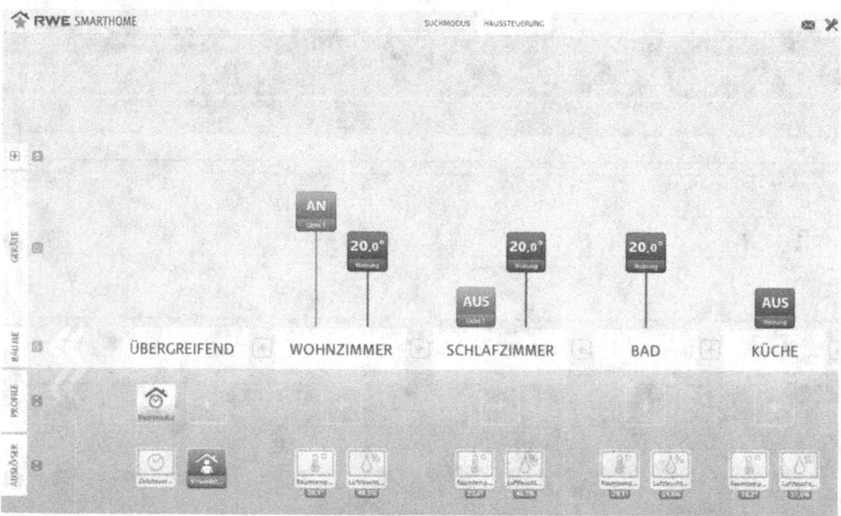

Fig. 4. Screenshot of the EPS system,contents courtesy of RWE AG

which is shown in Fig. 4, are analyzed. The problems that occurred during the evaluation of EPS are shown in Table 9. The problems within the usage of the EPS system were rather related to the complexity of the basic workflow than to problems on the elementary level (as it was the case in the two other systems). A detailed analysis and interpretation of the outcome of the study and a comparison of the three systems is given in section 5.

Table 9. Usability problems of the Energy Provider System (EPS)

Problem	No. of Participants
Uncertainty regarding correct profile (Scenario)	13
Combination of two devices	10
Changing temperature not clear Want to use trigger as device (8 heating, 1 door)	9
Programming more than one profile on error	6
Tries to use deactivated elements	4
Door open/closed unclear New rooms created Trigger difficult to find	2
Settings to small Dragging uncomfortable Doesn't find door contact Degree display is covered by fingers High complexity A reduced set of profiles would be enough Doesn't find heating Horizontal and vertical scrolling doesn't work	1

Table 10. General preference ranking of the three evaluated systems

System	Ranked 1st	Ranked 2nd	Ranked 3rd
Our System	6	4	7
OSS	10	4	3
EPS	1	9	7

4.3 Subjective Evaluation

In order to collect their subjective opinion on the three systems the participants were asked to fill out a semantic differential taken from the UEQ [31]. The results are shown in Fig. 5. The comparison shows that our system could outperform the other two in several positive aspects, for example, it is perceived as faster than the others, as better organized and more innovative. The OSS system is observed as more efficient than the other two, more secure and overall better than the others. The EPS system differentiates from the other two only on negative aspects, it is perceived as complicated and slow. On the level of user experience dimensions the comparative results are shown in Fig. 6. Both, our system and the OSS are seen as equally attractive, in relation to them the EPS system is judged as quite unattractive. In the dimensions perspicuity, dependability and efficiency the OSS system outperforms the other two. Our system achieves the best ratings in the dimensions stimulation and novelty. Finally, the participants

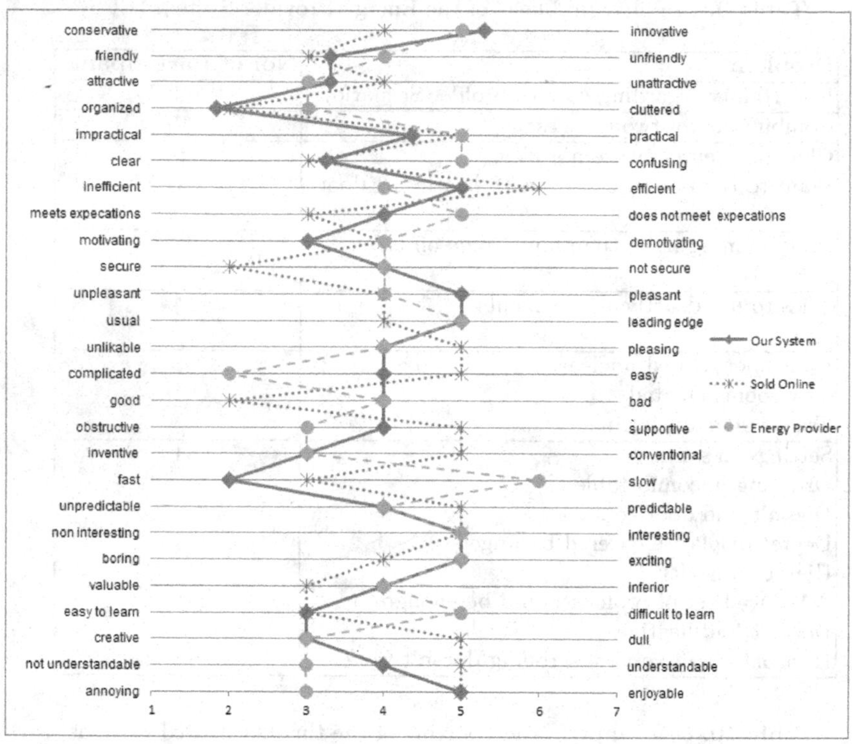

Fig. 5. Values of the evaluated systems on the basis of the UEQ items

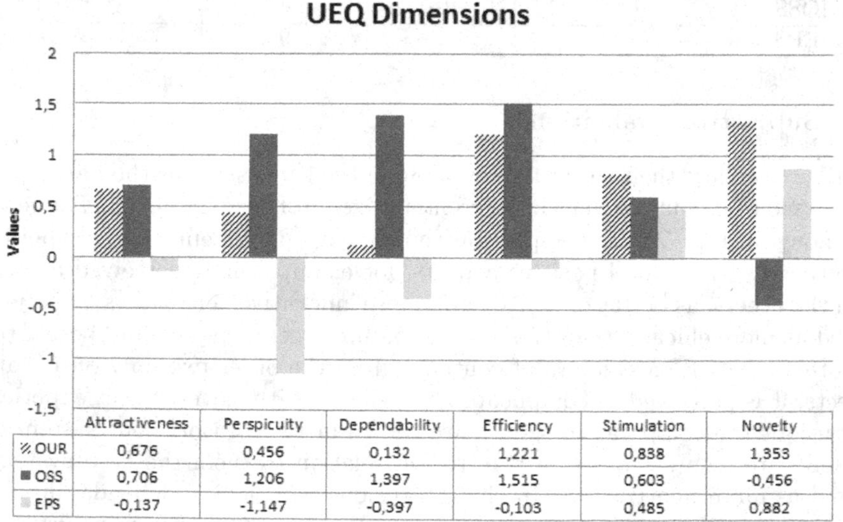

	Attractiveness	Perspicuity	Dependability	Efficiency	Stimulation	Novelty
OUR	0,676	0,456	0,132	1,221	0,838	1,353
OSS	0,706	1,206	1,397	1,515	0,603	-0,456
EPS	-0,137	-1,147	-0,397	-0,103	0,485	0,882

Fig. 6. Values of the evaluated systems in relation to the dimensions of UEQ

ranked the three systems on a general preference level. The results are shown in Table 10. The majority of participants ranked the OSS system on first position, followed by our system. Only one participant ranked the EPS system as first. Seven participants ranked the EPS as well as our system on third rank, only 3 did this for the OSS.

5 Discussion and Conclusion

This paper describes a case study evaluating usability aspects in regard to end user programming of smart home functionality. Because they constitute a specific functionality in this context, the focus was laid on scenarios. A two stage empirical evaluation approach was followed, whereas in the first stage the principal relevance of scenarios was evaluated. In the second stage a comparative usability test was performed within which a prototype application was evaluated and compared to two commercial smart home systems. The first stage of the study revealed that scenarios are potentially useful in the context of a smart home and based on the outcomes of the prestudy a digital prototype was developed to be able to simulate and evaluate realistic scenario programming tasks. The results show that the developed prototype could compete with the commercial systems and could even outperform them in dimensions such as novelty and stimulation. However, it has several drawbacks, which is not surprising because it constitutes an initial prototype. What was more surprising is that commercial smart home systems exhibit usability problems on the same level as our prototype. This result supports the assumption that still too little efforts are put into the enhancement of usability of interfaces in this domain, as it has been the case in the past with home appliances such as the VCR. Although only two low cost systems were evaluated it can be assumed that this kind of problems are present also in other systems sold on the market. The criticism often made in other domains, that consumers are misused as beta-testers, therefore seems to be applicable also in the context of smart home.

It has to be noted that the study has some limitations which influence the generalisability of the results. The first limitation is the composition of the participant samples. Whereas the participants of the prestudy are characterized by an acceptable heterogeneity in regard to their demographic characteristics, computing skills, professional background and marital status, all subjects having participated in the main study are students within the same age range. The reason of this has been that the main study had to be performed at the university and - although the study has been announced over diverse channels - people not being students obviously avoided the effort to come to the university to participate. But although the majority of participants (the computer science students) of the main study can be considered as computing experts insurmountable hurdles prevented them from finishing singular tasks. So it very improbable that laypersons would be able to achieve the goal of programming scenarios themselves with some of the state-of-the art systems available on the market. Another drawback of the study design is related to the hardware used. For efficiency reasons the three systems were installed on separate devices. Specifically the embedded PC had, compared to the other devices, an inferior handling and look and feel, which negatively influenced the subjective assessment of the EPS system. Despite of these shortcomings some of the results we could gain are promising as well as motivating to further develop the prototype and perform follow-up studies on the topic of scenario programming. One of the outcome is, that the online sold system (OSS) which is based on a visual representation of logic rules (if, then, else) was familiar to the majority of participants and specifically for the computer science students so straight forward that it was most trustable. This supports the results of other studies, e.g. described in [10]. The EPS system was characterized as optically very appealing, however, participants could not (in comparison to the OSS system) easily anticipate the next step(s) because the basic concepts (interface metaphor, workflow) were partly not comprehensible. Our system (prototype) was perceived as the most intuitive and direct one, however, as with the EPS system, several mechanisms were not familiar and therefore the participants had doubts that such a system would really work in the expected way. Specifically the social sciences students stated that they liked our approach more than the strict rule based approach represented by OSS. All these results constitute a good basis for a further development of the prototype and follow-up evaluations. We plan to integrate our prototype in the existing platform of the Casa Vecchia project within the next months. In the context of the project we have access to the target group of elderly people which could benefit from the alternative programming approach the prototype presented in this paper provides. This assumption will be evaluated in follow up studies. Elements other than electric devices, for example the conventional household goods such as beds, doors or books mentioned by the participants of the prestudy as well as the relevance of the additional *concepts* (people, rooms) we tried to integrate in our prototype have to be further investigated. In both studies performed these aspects were either ignored by the subjects or not understood. The reason for this could be, that existing smart home systems as well as electric or computerized devices do not offer individualization functions - although this kind of functionality is frequently described in the context of smart home advertisements. For example within the scenario of coming home the entertainment system in the smart home plays the preferred music. To our knowledge no system on the market is capable of differentiating between users in these situation, at least no system on a reasonable price level. Questions such as how a system can be programmed to support different users in poly-user environments (e.g. one and the same scenario for different persons preferring different music genres) or how it should react when not only one individual is involved (the whole family comes home, what music should be played?) could possibly be resolved with the programming metaphor we have in mind. To answer this and other questions further investigations are planned.

Acknowledgements. We thank the reviewers for their valuable comments. This paper is originated in the context of *Casa Vecchia* funded by the Austrian research promotion agency within the benefit program (Project.Nr. 825889).

References

1. Rode, J.A., Toye, E.F., Blackwell, A.F.: The domestic economy: A broader unit of analysis for end-user programming. In: Proceedings of CHI, pp. 1757–1760 (2005)
2. Chan, M., Esteve, D., Escriba, E., Campo, E.: A review of smart homes - Present state and future challenges. Computer Methods and Programs in Biomedicine 91(1), 55–81 (2008)
3. Saizmaa, T., Kim, H.C.: A Holistic Understanding of HCI Perspectives on Smart Home. Proceedings of NCM 2, 59–65 (2008)
4. Norman, D.A.: Living with complexity. MIT Press, Cambridge (2010)
5. Blackwell, A.F., Rode, J.A., Toye, E.F.: How do we program the home? Gender, attention investment, and the psychology of programming at home. Int. J. Human Comput. Stud. 67, 324–341 (2009)
6. Davidoff, S., Lee, M.K., Yiu, C., Zimmerman, J., Dey, A.K.: Principles of Smart Home Control. In: Dourish, P., Friday, A. (eds.) UbiComp 2006. LNCS, vol. 4206, pp. 19–34. Springer, Heidelberg (2006)
7. Brush, A.J., Lee, B., Mahajan, J., Agarwal, S., Saroiu, S., Dixon, C.: Home automation in the wild: Challenges and opportunities. In: Proc. of CHI, pp. 2115–2124 (2011)
8. Harper, R., et al.: The Connected Home: The Future of Domestic Life. Springer, London (2011)
9. Rode, J.A., Toye, E.F., Blackwell, A.F.: The fuzzy felt ethnography. Understanding the programming patterns of domestic appliances. Personal Ubiquitous Computing 8(3-4), 161–176 (2004)
10. Holloway, S., Julien, C.: The case for end-user programming of ubiquitous computing environments. In: Proceedings of the FSE/SDP Workshop on Future of Software Engineering Research (FoSER), pp. 167–172 (2010)
11. Mennicken, S., Huang, E.M.: Hacking the natural habitat: An in-the-wild study of smart homes, their development, and the people who live in them. In: Kay, J., Lukowicz, P., Tokuda, H., Olivier, P., Krüger, A. (eds.) Pervasive 2012. LNCS, vol. 7319, pp. 143–160. Springer, Heidelberg (2012)
12. Linskell, J.: Smart home technology and special needs reporting UK activity and sharing implemention experiences from Scotland, PervasiveHealth, pp. 287–291 (2011)
13. Das, S.K., Cook, D.J., Bhattacharya, A., Heierman, E.O., Lin, T.Y.: The role of prediction algorithm in the MavHome smart home architecture. IEEE Wireless Commun. 9(6), 77–84 (2002)
14. Rashidi, P., Cook, D.J.: Keeping the resident in the loop: Adapting the smart home to the user. IEEE Transac. on SMC, Part A 39(5), 949–959 (2009)
15. Augusto, J.C., Nugent, C.D.: Smart homes can be smarter. In: Augusto, J.C., Nugent, C.D. (eds.) Designing Smart Homes. LNCS (LNAI), vol. 4008, pp. 1–15. Springer, Heidelberg (2006)
16. Green, W., Gyi, D., Kalawsky, R., Atkins, D.: Capturing user requirements for an integrated home environment. In: Proceedings of NordiCHI, pp. 255–258 (2004)
17. Rodden, T., Benford, S.: The evolution of buildings and implications for the design of ubiquitous domestic environments. In: Proceedings of CHI, pp. 9–16 (2003)
18. Demiris, G., Hensel, B.K.: Technologies for an Aging Society: A Systematic Review of Smart Home Applications. IMIA Yearbook of Medical Informatics, 33–40 (2008)

19. Eckl, R., MacWilliams, A.: Smart home challenges and approaches to solve them: a practical industrial perspective. In: Tavangarian, D., Kirste, T., Timmermann, D., Lucke, U., Versick, D., et al. (eds.) IMC 2009. CCIS, vol. 53, pp. 119–130. Springer, Heidelberg (2009)
20. BMBF/VDE Innovationspartnerschaft AAL.: Interoperabilit/ät von AAL- Systemkomponenten. Teil 1: Stand der Technik. VDE Verlag, Berlin (2010)
21. Helal, S., Mann, W., El-Zabadani, H., King, J., Kaddoura, Y., Jansen, E.: Gator Tech Smart House: A Programmable Pervasive Space. IEEE Computer 38(3), 50–60 (2005)
22. Zimmermann, G., Vanderheiden, G., Gandy, M., Laskowski, S., Ma, M., Trewin, S., Walker, M.: Universal remote console standard - toward natural user interaction in ambient intelligence. In: Proceedings of CHI, pp. 1608–1609 (2004)
23. Nordman, B., Granderson, J., Cunningham, K.: Standardization of user interfaces for lighting controls. Computer Standards and Interfaces 34(2), 273–279 (2012)
24. Ko, A.J., Myers, B.A., Aung, H.: Six Learning Barriers in End-User Programming Systems. In: IEEE Symp. on VLHCC, pp. 199–206 (2005)
25. Saffer, D.: Designing Gestural Interfaces: Touchscreens and Interactive Devices. O'Reilly Media, Inc. (2008)
26. Sohn, T., Dey, A.K.: iCAP: An Informal Tool for Interactive Prototyping of Context-Aware Applications. In: Extended Abstracts of CHI, pp. 974–975 (2003)
27. Danado, J., Paterno, F.: A prototype for EUD in touch-based mobile devices. In: Visual Languages and Human-Centric Computing (VL/HCC), pp. 83–86 (2012)
28. Truong, K.N., Huang, E.M., Abowd, G.D.: CAMP: A magnetic poetry interface for end-user programming of capture applications for the home. In: Mynatt, E.D., Siio, I. (eds.) UbiComp 2004. LNCS, vol. 3205, pp. 143–160. Springer, Heidelberg (2004)
29. Röcker, C., Janse, M., Portolan, N., Streitz, N.A.: User Requirements for Intelligent Home Environments: A Scenario-Driven Approach and Empirical Cross-Cultural Study. In: Proceedings of sOc-EUSAI, pp. 111–116 (2005)
30. Shneiderman, B.: Leonardo's Laptop: Human Needs and the New Computing Technologies. MIT Press, Cambridge (2002)
31. Laugwitz, B., Held, T., Schrepp, M.: Construction and evaluation of a user experience questionnaire. In: Holzinger, A. (ed.) USAB 2008. LNCS, vol. 5298, pp. 63–76. Springer, Heidelberg (2008)

Understanding the Limitations of Eco-feedback: A One-Year Long-Term Study

Lucas Pereira, Filipe Quintal, Mary Barreto, and Nuno J. Nunes

Madeira Interactive Technologies Institute, Funchal, Portugal
{lucas.pereira,filipe.quintal,mary.barreto}@m-iti.org,
njn@uma.pt

Abstract. For the last couple of decades the world has been witnessing a change in habits of energy consumption in domestic environments, with electricity emerging as the main source of energy consumed. The effects of these changes in our eco-system are hard to assess, therefore encouraging researchers from different fields to conduct studies with the goal of understanding and improving perceptions and behaviors regarding household energy consumption. While several of these studies report success in increasing awareness, most of them are limited to short periods of time, thus resulting in a reduced knowledge of how householders will behave in the long-term. In this paper we attempt to reduce this gap presenting a long-term study on household electricity consumption. We deployed a real-time non-intrusive energy monitoring and eco-feedback system in 12 families during 52 weeks. Results show an increased awareness regarding electricity consumption despite a significant decrease in interactions with the eco-feedback system over time. We conclude that after one year of deployment of eco-feedback it was not possible to see any significant increase or decrease in the household consumption. Our results also confirm that consumption is tightly coupled with independent variables like the household size and the income-level of the families.

Keywords: Eco-feedback, Electric Energy Consumption, Sustainability.

1 Introduction

The notion of wellbeing based on personal ownership and mass consumption was largely identified as one of the factors leading to the growth of electricity consumption in the last years. As more people in developing countries have access to higher levels of comfort it is expected that the residential energy consumption will have a tremendous impact on the long-term effects in carbon emission and global warming.

Eco-feedback technology [1] is defined as the technology that provides feedback on individual or group behaviors with the goal of influencing future energy saving strategies. Eco-feedback has proven to be an effective way of promoting behavior change and considerable savings in electricity consumption [2]. However and despite the successful results reported, most research was conducted in short-term studies and therefore there is little evidence on the long-term effects of eco-feedback. Further research reported that after an initial period of exposure to eco-feedback the tendency

A. Holzinger and G. Pasi (Eds.): HCI-KDD 2013, LNCS 7947, pp. 237–255, 2013.

is towards a reduction in the attention provided to the feedback leading to behaviors relapse [3].

This paper contributes to the ongoing research efforts by presenting the results of the long-term deployment of a real time eco-feedback system in 12 households for the period of 52 weeks. We start by looking at the state of the art and defining our research goals for this study. In the next two sections we show how our eco-feedback system was designed and deployed, including an in-depth description of the sensing system and the process of selecting an adequate sample for the long-term study. We then present the most important results and discuss the key findings and implications of this work for forthcoming eco-feedback research. Finally we conclude and outline future work.

2 Related Work

In the last couple of decades considerable research was done in the field of eco-feedback technology. Fischer [2] reviewed approximately twenty studies and five compilations of publications between 1987 and 2008 exploring the effects of eco-feedback on electricity consumption and consumer reactions. The most notable findings reported that eco-feedback indeed resulted in energy savings between 5% and 12%, and that the greater changes in consumption would result from computerized eco-feedback like for instance in [4].

The literature is rich in interactive approaches to eco-feedback. For example in [5] the authors ran a three-month study in nine households where the eco-feedback was deployed in a clock-like ambient interface that would translate electricity consumption into graphical patterns. This study showed that people immediately became more aware of their energy consumption, and even developed the ability to associate the displayed patterns with the actual appliances used. In another example [6] the authors reported on a system that would give detailed eco-feedback information on individual appliances. Preliminary results on this experiment showed a reduction of 5% over the previous year when the eco-feedback was not available.

On the commercial side several models of eco-feedback systems reached the market in the last years. In an attempt to better understand the adoption and implications of commercial solutions, Miller and Buys [7] conducted a study with seven families that were using a commercial energy-and-water consumption meter and generated guidelines in how eco-feedback systems should be built and marketed. For instance, the cost of the system was a major issue for residents that were either engaged with the product or not. A second topic of discussion was the lack of product support advocated by the users, which immediately pointed out the problem of understanding the user experiences and perceptions around smart meters.

In an attempt to address this issue authors in [8] installed smart meters in 21 Belgian households between two and four weeks. They found out in accordance to other studies that despite an increased awareness there was no significant behavior change towards conservation of energy. Another important finding of this research reports that people had difficulties interpreting kilowatt-hour and that the corresponding conversion to monetary cost demonstrated irrelevant economic savings.

The emergence of smart handheld devices also presented new opportunities for researchers to test different eco-feedback systems. In [9] the members of 10 households were given access to a mobile power meter that would run either on a smartphone or a tablet. Findings suggested that the householders gained a deep understanding of their own consumption and that users found the feature of comparing their consumption with other community members useful. The mobility aspect was also important, as participants were able to access their consumption from virtually everywhere.

With the constant evolution of technology also came the possibility of providing disaggregated power consumption by individual appliance, division of the house or event daily activities like cooking dinner or doing the laundry. Recently in 10 Costanza and colleagues conducted a field study where they wanted to learn if users were able to easily leverage a connection between appliances and their day-to-day activities. This study lasted for two weeks and twelve participants were asked to use a system where they would be able to tag appliances to a time-series of their energy consumption. The results of the trial showed that the system was successful in engaging users and providing accurate consumption levels of some appliances that were consuming more than what they initially expected. Another important result was noticing that when tagging, users would refer to energy consumed by activity rather than just the tagged appliance.

Eco-feedback through persuasion was also attempted by some authors, for instance in [11] Gamberini and his pairs explored the possibility of encouraging electricity conservation practices through a mobile persuasive game. This eco-feedback game provided next-to-real-time, whole house and appliance based consumption information. Four families used the game during 4 months, and results showed that users kept playing it during the whole trial, despite the gradual reduction of accesses per day that was justified by the users as the result of getting more familiar with the application and what it had to offer.

All of the aforementioned studies clearly advocate for the short-term effectiveness of eco-feedback technology, in particular when considering disaggregated and interactive eco-feedback. However, we argue that these studies did not last enough to properly assess their long-term effectiveness. In fact, the gradual decrease of attention shown in the later study may indicate that once the novelty effect has passed users will go back (relapse) to their original behavior. This is defined in literature as response-relapse effect, where after a while the user behaviors (and hence consumption) will relapse to values prior to the intervention. This effect was reported by the authors' in [3] when investigating how the residents of a dormitory building would respond to different consumption information. The authors noticed that after a period when no feedback was provided the behaviors would approximate the ones before the study.

In this paper we argue that long-term studies of eco-feedback system are required in order to understand the lasting effects of this technology as a driver for behavior change. Here we explore some very important questions like: i) How is the eco-feedback system used after the novelty effect has passed? ii) How long does the novelty effect last? iii) How does the demographic data from the residents affect energy consumption?

This paper is a follow up to an initial short-term three-month study reported in [12]. In our first evaluation of this system we saw a reduction of 9% in consumption but also observed that users significantly decreased the interactions with the eco-feedback system after four weeks of deployment. Our initial results suggested that further research was needed in order to fully understand the potential and underlying issues with the long-term deployment of eco-feedback. In this second study we aimed to investigate further how the system was used after the novelty effect passed. We wanted to explore if there was a decrease in energy consumption as a result of eco-feedback intervention and also if further changes in the system could raise attention back to the eco-feedback. With a longer deployment we were also able to investigate other factors influencing behavior change, for instance demographic independent variables like family size and income.

3 System Design

In this section we briefly describe our eco-feedback research platform, which involves both the sensing infrastructure and the communication with the eco-feedback interface. For a throughout explanation of our framework please refer to [12] and [13].

3.1 Sensing Infrastructure

Our eco-feedback infrastructure is a low-cost, end-to-end custom made non-intrusive load monitoring system. Non-intrusive load monitoring (NILM) stands for a set of techniques for disaggregating electrical loads only by examining appliance specific power consumption signatures within the aggregated load data. NILM is an attractive method for energy disaggregation, as it can discern devices from the aggregated data acquired from a single point of measurement in the electric distribution system of the house [14].

Our NILM system consists of a netbook installed in the main power feed of each house (see Figure 1- left) covering the entire household consumption and thus removing the need to deploy multiple (intrusive) sensors. The netbook provides a low-cost end-to-end system: the audio input soundcard is used as the data acquisition module (two channels, one for current and another for voltage); the small display and the speakers provide the interactivity; the Wi-Fi card enables communication over the Internet; and the camera and built-in microphone serve as low-cost sensors for human activity sensing.

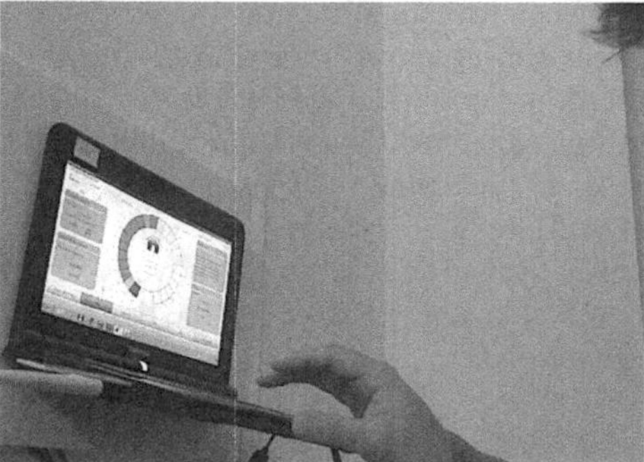

Fig. 1. System installed in the main power feed (left) and a householder interacting with the eco-feedback (right)

The current and voltage sample signals acquired with the soundcard are pre-processed and transformed into common power metrics (e.g. real and reactive power) that are representative of the energy consumption. These power values are used for event detection, event classification and, ultimately, the breakdown of consumption into individual appliances. In parallel, power consumption and power event data are stored in a local database to be used by any external application to provide eco-feedback to the householders.

3.2 Eco-feedback

The front-end eco-feedback component provides two different representations for the real time and historical consumption of the house. Our interactive visualization was implemented following the recommendations from previous studies in energy eco-feedback [15 - 17] that distinguish real-time and historical feedback. Real time feedback, which is said to be responsible for 5 to 15% of the changes in user behavior, displays the current energy consumption as well as major changes and trends in the current consumption. Conversely historical feedback refers to all the information collected (e.g. monthly values of energy consumption), and according to the literature can lead up to 10% of the users' behaviors towards future energy consumption by simply offering the possibility of reviewing and comparing data among different historical periods of time. To cope with these guidelines we have designed our eco-feedback user interface in a way that users could quickly switch between historic and real-time modes. Figure 2 shows how the information is organized in our user interface.

The center of the eco-feedback interface represents both real and historical consumption data using a wheel like graph. The left and right side of the interface present additional information, including weather, numerical consumption and comparison to previous periods. On the right hand side the interface provides notifications, suggestions and motivation hints. In the following we detail the interface for both the real-time and the historical views.

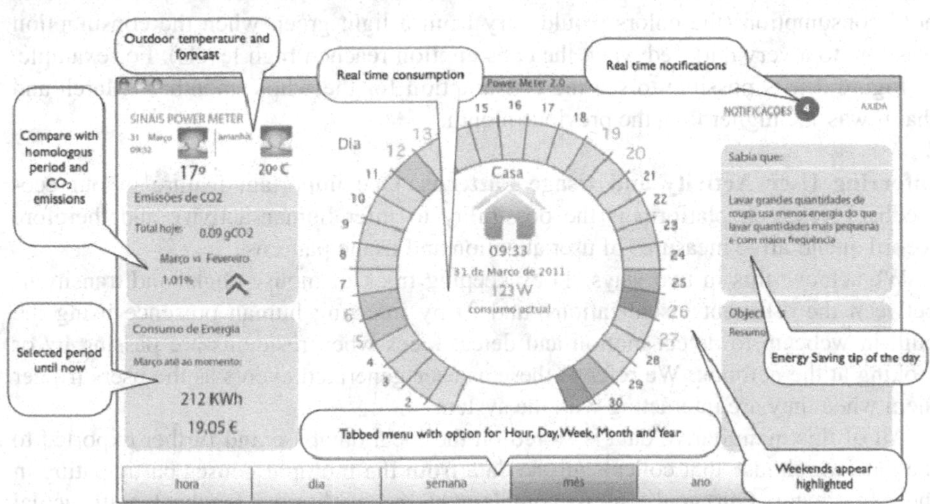

Fig. 2. Eco-feedback user interface showing the current month consumption

Real-Time Eco-feedback. The real time consumption as well as notifications and comparisons are always presented to the users regardless of the active view. The current consumption is presented in watts (in the center of the interface) and updated every second. The last hour view shows real time notifications, triggered every time the energy monitor detects a power change above a pre-defined threshold. The event notification is represented by a small dot that is added to the interface close to the time of occurrence, as shown in Figure 3. The size of the dot indicates the amount of power change, and clicking on it reveals the appliance that has the highest probability of triggering that event prompting the user to confirm, rectify or discard the guess.

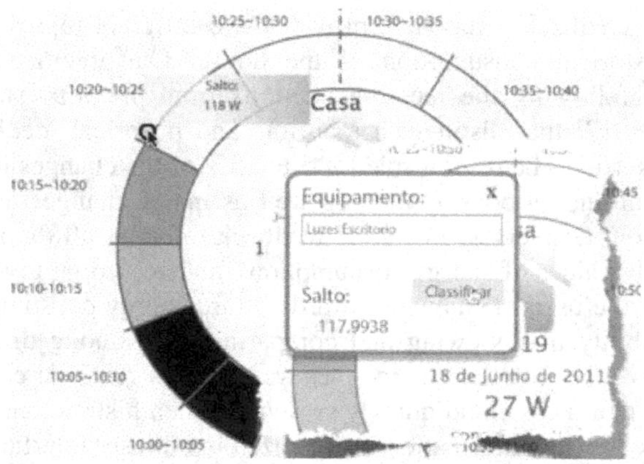

Fig. 3. Real-time notifications offering users the chance to label power changes

Historic Eco-feedback. The historic data (current day, week, month and year) is presented in two different modalities: the more traditional displays the quantities in numerical format (when hovering the mouse over a specific time period), while the less traditional consisted of a color-code that would change according to the household consumption (the colors would vary from a light green when the consumption was low to a very dark red when the consumption reached high levels). For example, in Figure 2 it is possible to see the consumption for the whole month of March and that it was 1% higher than the previous month.

Inferring User Activity and Usage Patterns. One important feature of our eco-feedback research platform is the possibility to infer human-activity and therefore record quantitative measures of user attention and usage patterns.

We achieve this in two ways: 1) by keeping track of mouse clicks and transitions between the different visualizations, and 2) by inferring human presence using the built-in webcam to detect motion and detect faces when residents are passing by or looking at the netbook. We refer to these as user-generated events as the users trigger them when they are interacting with the system.

All of this quantitative data is stored on the local database and further exported to the data warehouse that collects all the data from the multiple houses participating in the experiment. Our goal was to complement the qualitative feedback with actual measures of user activities and usage patterns.

4 Deployment

In this section we describe the process of deploying our eco-feedback research infrastructure for the field-testing. We start by explaining the initial sample selection, installation procedures and the nature of the collected data. We finalize with a description of our participants and how we reached the final numbers of 12 families using the system during 52 weeks.

4.1 Sample Selection

The sample selection for the first study was based on an extensive analysis of the consumption patterns of 46 000 household consumers in Funchal, a medium sized city of southern Europe (about 150 000 inhabitants).

For that purpose the local electricity company gave us access to the energy consumption data of all the consumers in the city for a period of two years. From that baseline data we divided the consumers in four levels according to their annual consumption in Euros. These four levels were then used to select a nearby neighborhood where we could easily find a sample that would be representative of the city.

The recruitment was done with personal visits to the selected buildings explaining the project and collecting additional demographic data from those that would volunteer to participate. In the end we were able install the system in 17 apartments as well as in six individual homes that volunteered to participate and five additional houses recruited from professors and students involved in the project.

The first version of our system was installed in June 2010, and was remotely updated to the new version in the last week of November. Users were informed of the update via informal phone calls or in the previous two weeks during the first round of interviews.

4.2 Data Collection

The deployment lasted until the end of 2011. In figure 4 we present the major milestones of the deployment including three interventions with the users: i) interviews in the third week of February 2011; ii) informal visits in the last two weeks of July 2011 and iii) a last round of interviews during all of December 2011. The system was removed from the houses during January and February 2012.

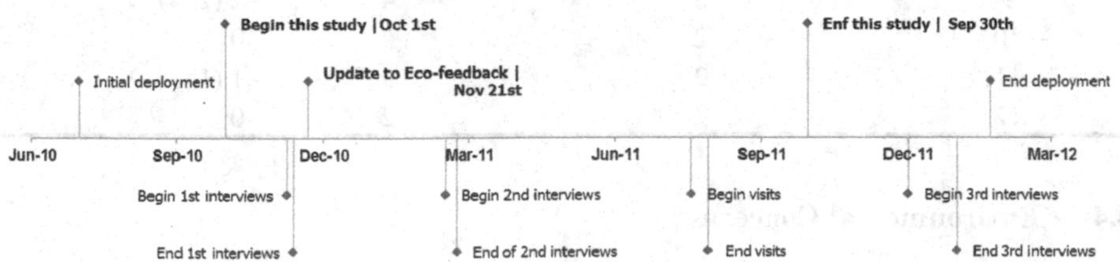

Fig. 4. Timeline chart showing the most relevant events of this eco-feedback deployment

During and prior to the study we collected both qualitative and quantitative data from the households. Quantitative data includes electric energy consumption samples (two per minute), power events (both on and off including the transient for active and reactive power) and user interactions with the eco-feedback system (including movement, face detection, mouse and menu selection). Qualitative data includes demographics (age, occupation, income, literacy, etc. for all family members), environmental concerns (from semi-structured interviews), and a detailed list of appliances in the house and reconstruction of family routines (from semi-structured interviews and diary studies).

4.3 Participants

From the 21 families that took part in the first study we ended with a final sample of 12 households. The main reasons for this high sample mortality (~ 43%) were mostly technical issues like stability of the Internet connection and long periods of absence of the families, including some moving out to a new house.

Our final analysis sample was the result of maximizing both the number of houses and the deployment time. Another requirement was that the final sample had to include data from both deployments in order to integrate the novelty effects introduced by the different versions of the eco-feedback system. This said we ended up choosing the period between the October 1st 2010 and September 30th 2011, as this represents exactly 52 consecutive weeks of data for 12 apartments all from the originally selected neighborhood. Table 1 summarizes the demographic data for our final sample participants.

Table 1. Demographics of the participating families

Family	Size (number of bedrooms)	People	Children (and ages)
1	3	4	2 (5, 10)
2	3	5	1 (10)
3	3	4	2 (8, 14)
4	3	3	1 (1)
5	3	4	2 (1, 7)
6	1	1	0
7	3	3	1 (5)
8	1	2	0
9	3	4	2 (2, 4)
10	2	2	0
11	2	3	1 (15)
12	3	2	0

4.4 Environmental Concerns

Since one important issue regarding the participants was their level of environment concerns we collected additional qualitative data from interviews with people from the households in the sample. When asked about environmental concerns nine out of the twelve families pointed global warming as a serious concern and six of these families considered that reducing their energy consumption would have a positive impact on the environment.

When asked about the adoption of sustainable actions eight families indicated that reducing personal costs and guaranteeing the wellbeing of future generations was their main motivations to reduce energy consumption. Despite this, when questioned about a particular number of sustainable behaviors less than half reported they had adopted these on a daily basis. The more frequently mentioned behaviors were: switching off the lights on empty rooms, washing full loads of clothes and acquiring energy efficient lights.

The complete list of actions reported by the families was over 40 actions related to energy, water and food consumption or conservation. These actions included not only individual measures as for example use public transportation, but also, social oriented activities such as carpool. The interaction with the participants, facilitated by the interviews and the overall study, allowed us to observe their increased level of awareness. The participants already performed actions related to saving energy previously to having the system. These levels of awareness were enhanced when exposed to the eco-feedback system. However, this was not evaluated through a scale of environmental concerns.

5 Evaluation

In this section we report the evaluation of the data collected during the long-term deployment of eco-feedback. We start by analyzing only the aggregated data, handling the sample as one group and not selecting any particular house. Then we rank our sample into three categories and investigate if there were significant changes in the consumption when considering background variables, such as the weather conditions, the household size and income-level of the families.

For this analysis we use the week as our standard period because it provides the most stable variance as it was expected because it corresponds more directly to a recurring family routine. For some specific cases where other variables were more appropriate day and month were also used as the aggregation time period.

5.1 Overall Power Consumption

We first looked at the average consumption of all the houses aggregated by week. The weekly average of power consumption ($n = 624$) was 62.45 kWh ($s = 27.49$ kWh). The high variance is explained by the considerable differences between households, for example three households had a weekly average of less than 40 kWh, while four houses had an average consumption between 80 and 110 kWh.

As a consequence we decided to rank our sample in three categories (low, average and high consumption), based on the 1/3 percentiles of the weekly consumption expressed in kWh. The following categories were defined: A (<= 42.3kWh), B (42.3kWh – 76.92 kWh) and C (> 76.92kWh) with four houses each and the average consumption ($n = 208$) was 36.45 kWh ($s = 7.76$ kWh) for category A, 56.30 kWh ($s = 11.59$ kWh) for category B and finally category C with 94.6 kWh ($s = 17.98$ kWh) respectively.

Changes in Consumption. To check for significant changes in consumption during the deployment of the eco-feedback we used a Wilcoxon signed-rank test, to compare repeated measures between consecutive months. For this analysis we used months instead of weeks because the test used is known to perform better for less than 20 values in each sample. Results showed no significant differences (for $p < 0.05$) in any of the categories.

In an attempt to get a better understanding of these results we individually asked the families about any changes in their consumption with most of them confirming that there were no real savings in the overall electricity bill. This was either because families found it difficult to reduce or even control their consumption levels as stated *"We didn't notice major changes. We already did a couple of things we would already disconnect some devices, toaster or radios. (...) The electricity is the hardest thing to control for us, the water seems easier."* (Family 11, Mother). Or due to the fact of current tax increases the local company put in practice as stated by this family *"Our consumption wasn't reduced. We compared with the bills and there was a tax increase, the consumption seems to be always the same for us, before and after having this device here."* (Family 5, Mother). However others noticed some decrease in consumption after having taken some measures: *"We changed all the lamps in the house to more energy efficient ones. I started to turn off the lights more often because I could see the impact of it so I had to do it. (…) Our bill wasn't reduced to 50% but it was reduced."* (Family 12)

Consumption and Weather Conditions. The deployment took place in Madeira Island known to have one of the mildest climates in the world with average temperatures ranging from 17°C in the Winter to 27° in the Summer. Still we wanted to understand how the climate might affect the electricity consumption. Therefore we compared the consumption between seasons as well as wintertime (WT) and daylight saving time (DST). For this analysis we used the daily consumption as the minimum unit of time.

The tests (for $p < 0.05$) have shown no significant differences in consumption between the seasons or between WT and DST for any of the categories suggesting that we should not expect a big variation of the energy consumption during the year. One possible explanation for this is, as previously mentioned, the low variation in the temperatures during the period of study - for the duration of the deployment ($n = 12$) the monthly average temperature was actually 19.8 ° C ($s = 2.89$ °C).

Consumption and Household Size. According to literature the number of people living in the household is the single most significant explanation for electricity consumption. The more people living in the house the more energy is used [18]. If we look at our consumption categories this is a direct conclusion. In fact the number of people in each household increases with each consumption category: category A has 9 people (7 adults and 2 children); category B 12 people (8 + 4); and category C 16 people (9 + 7). Therefore we have further investigated this topic by dividing the sample in categories according to the number of people in the house and looking for significant differences among these groups. We found four categories: 1 person (1 house), 2 people (3 houses), 3 people (3 houses) and 4 people (5 houses).

We tested consumption by household size using a Mann-Whitney U test and found a significant difference between 2 or 3 people (mean ranks were 21.86 and 51.14 respectively; $U = 121$; $Z = -5.935$; $p < 0.05$; $r = 0.699$) and 2 or 4 people (mean ranks were 27.86 and 60.88 respectively; $U = 337$; $Z = -5.623$, $p < 0.05$, $r = 0.57$). However no significant differences were found between 3 or 4 people. We haven't considered the single-family house due to being an isolated case in our sample.

To further analyze the effects of the number of people in the energy consumption we then categorized our sample by the number of children. We found three categories: 0 children (4 houses), 1 child (4 houses) and 2 children (4 houses). The same test shows significant differences between having none or one child (mean ranks were 26.96 and 63.22 respectively; $U = 118$, $Z = -6.743$, $p < 0.05$, $r = 0.74$) and zero or two children (means were 31.69 and 72.75 respectively; $U = 345$, $Z = -6.770$, $p < 0.05$, $r = 0.65$). No significant differences were found for having one or two children.

Our results confirm previous findings and general common sense that more people in the house result in more energy spent. Regardless one interesting finding worth investigating in the future is the fact that no significant differences appear when considering 3 or 4 persons or 1 or 2 children. One potential explanation is that after some point houses with more people will become more energy efficient, since the electricity usage per person tends to decrease.

Weekdays and Weekends Consumption. Finally we have also looked at average daily consumption and compared the weekdays and weekends. Table 2 summarizes the average consumption in each category for the given period.

Table 2. Weekday and weekend average consumption by consumptio category

	Weekdays (n=1044)	Weekend (n=416)
Category A	5.15 (s = 1.57)	5.39 (s = 1.72)
Category B	7.93 (s = 2.45)	8.40 (s = 2.67)
Category C	13.44 (s = 3.98)	13.78 (s = 3.82)

These results show that for all categories (and mostly B) there is a slightly higher consumption on weekends, which could be related to the fact that people tend to spend more time at home on weekends. Still (for $p < 0.05$) these differences are not significant in any of the categories.

5.2 Interaction with Eco-feedback

Similarly to the consumption analysis we will first look at the interaction events aggregated by week of study. The weekly average of mouse clicks ($n = 624$) was 8.93 ($s = 12.65$) and 23.92 ($s = 107.9$) for motion detection. After careful analysis of the data we found that the abnormal standard deviation for the motion events was related to the fact that one family also used the netbook to browse the web without closing the eco-feedback application. Hence our system detected a high number of non-intentional motion events (average motion for that house was 231.85 with a standard deviation of 304.64 for $n = 52$ weeks).

Therefore, after removing this house from the analysis, we ended up with a weekly motion detection average (n = 572) of 5.01 (s = 11.13) and 8.54 (s = 12.41) for mouse clicks. This difference between motion and clicks suggest that the notebook was only in the open position when users were in fact looking at the feedback and probably the main reason for that was its position behind the entrance door that would is some cases force the users to keep it closed. From this point forward we will be using only mouse clicks as the grouping variable for user interaction.

Long-Term Interactions. As mentioned in the introduction, one of the goals of this deployment was to achieve a better understanding on how the eco-feedback system was used after the novelty effect passed. Therefore we looked at the user-generated events (in this case the total number of mouse clicks) exactly when the novelty effect was introduced (deployment of the new user interface) until the last week of the study, as shown in Figure 5.

Our analysis shows an immediate increase of almost 25% in the user interaction right after installing the new interface (in weeks 8 and 9). These results confirm that as expected users react to new versions of the eco-feedback with an increased usage of the application. However our analysis also indicates that only three weeks after the new deployment the number of interactions dropped considerably until week 20 (a decrease of 45% when compared to the three weeks after the new deployment).

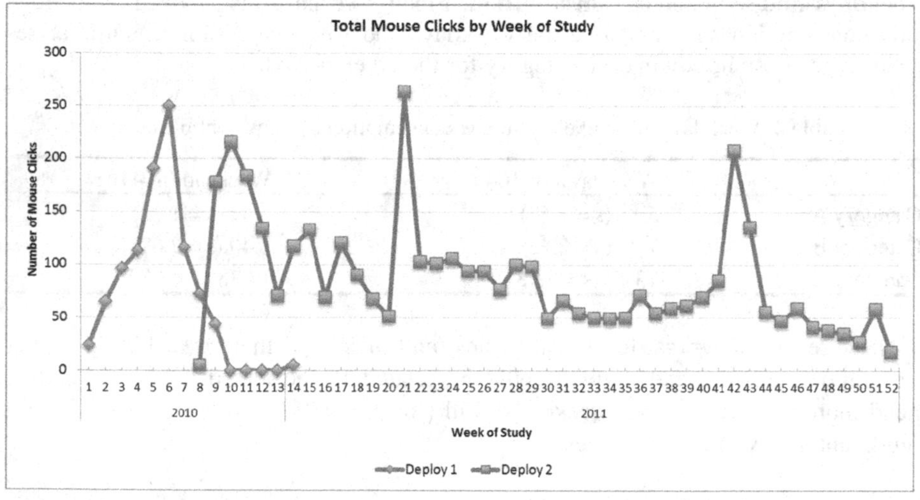

Fig. 5. Number of user interactions (mouse clicks) with the eco-feedback during 52 weeks

We clearly notice here the response relapse effect, which was significant if we consider that after 52 weeks the decrease in the number of interactions was almost 90% (at a drop rate of 2.2% per week). This decline was only interrupted by weeks 22 and the period between weeks 42 and 44 when we conducted interviews with the users, which also raised their awareness to the eco-feedback system.

Additionally, in the qualitative studies we asked users about this decrease of interest in the eco-feedback. Some families justified it with the lack of time in their routines, others felt like after a few weeks they already had a good perception of their consumption as shared by this family *"I wouldn't go there because most of the times I didn't have time to check it. I would just arrive home, get things done around here and go to sleep and start again the next day."* (Family 5, Mother). Their interaction with the system was reduced as a result of a more accurate picture of their consumption levels as stated by this family *"We would check our consumption more often initially until we got a rough idea or perception of what our consumption was but after that it would become less frequent."* (Family 12, Wife)

Another reason that was pointed for the lack of interest was the fact that the system became "yet just another electric device": *"And it became a habit to have it. I would check it whenever I would remember. I know already the power of each of device in the house, I already measured it (...) having this or that device working would not make me want to check the meter by itself."* (Family 7, father). Nevertheless other families kept using the system even if less frequently, as this father mentioned: *"We didn't ignore the device I would look at it everyday. What I noticed is that we achieve an average of consumption. And because we use the same devices all the time our attention to the system might decrease, we don't analyze it so carefully."* (Family 9, father)

5.3 Navigation in the Eco-feedback

We also wanted to understand which features of the system drew more attention to the users and the analysis showed that the most visited view was the current day consumption. This view had more accesses than all the other options together with an average (n = 11) of 179.27 (s = 63.84) mouse clicks. The second most used feature was the weekly consumption (average of 18 interactions, s = 13.15) while the least favorite was the year view (average of 6 interactions, s = 4.38). The total interactions with the different eco-feedback views are presented in figure 6.

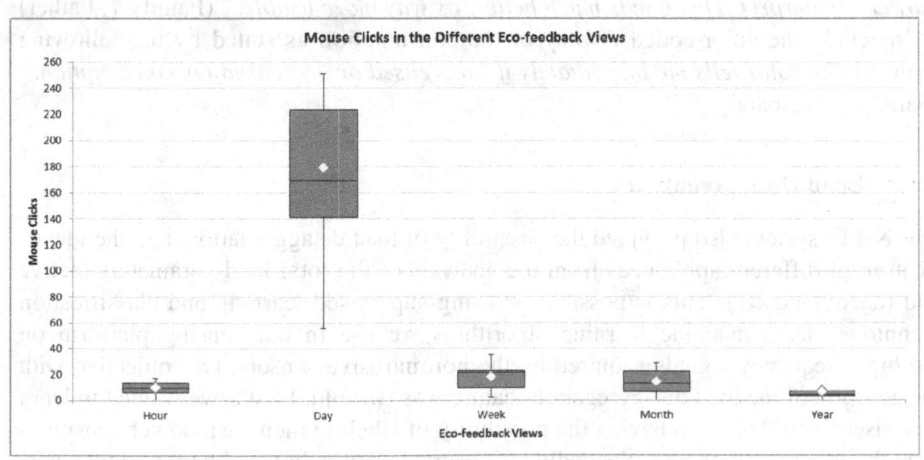

Fig. 6. Average interactions (mouse clicks) with the different eco-feedback screens

Another characteristic of the data that drew our attention was the fact that most interactions happened after 9PM peaking at 11PM with an average (n = 11) of 91.18 mouse clicks (s = 34.04). We believe this is a strong indicator that checking the consumption was something that users did at the end of the day. Most likely due to availability as referred by this family *"I would use it more at night when I was at home. I would see the consumption levels and if I saw something more than usual, I would assume that she had done something different or had used a device."* (Family 1, Husband) The average number of mouse clicks per hour of the day is shown in figure 7.

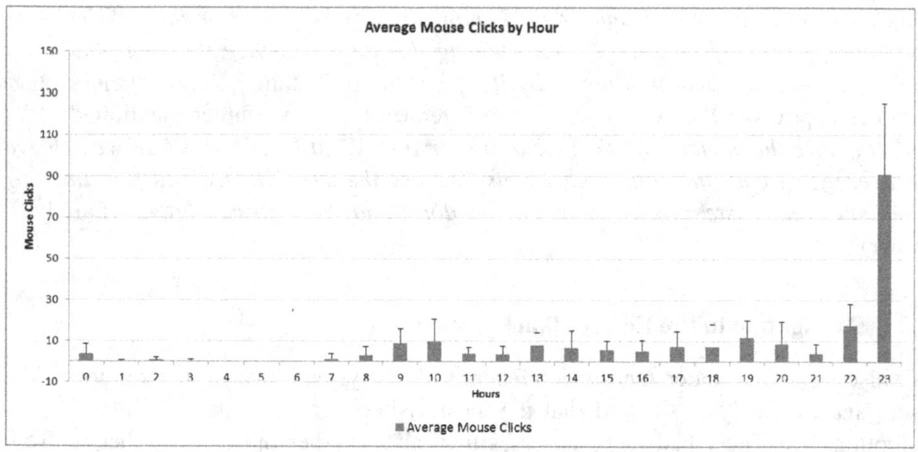

Fig. 7. Average interactions (mouse clicks) by hour of the day

When asked about the new user interface the tendency was to prefer the last one better, as shared by this family *"The other previous graph [column chart] was harder to interpret. This one is much better, its way more usable."* (Family 7, Father) - Especially the color-coded display of the consumption as stated by the following family *"The color tells me immediately if I increased or decreased my consumption."* (Family 9, Husband)

5.4 Load Disaggregation

Our NIML system also provided the possibility of load disaggregation, i.e., the identification of different appliances from the analysis of the total load parameters (active and reactive power). This is possible by using supervised learning and classification techniques from machine learning algorithms we use in our sensing platform on the high frequency signals acquired by the non-intrusive sensors. Our objective with the addition of the load disaggregation feature was twofold. Firstly we wanted to learn how users would react to having the possibility of labeling their own power consumption though power events. Secondly we wanted to see how adding a single new feature during the deployment would affect the interaction with the eco-feedback. The selection criterion for having this feature was being part of the top-5 most active families in terms of user-generated events, and it was installed during weeks 40 and 43 when we visited the families and helped them labeling some of the existing appliances.

Our expectations were that adding this new feature would work as another trigger to increase the interaction between the users and the eco-feedback system. However the results show that adding this feature did not result in a significant increase in the number of interactions during that period (the houses with the new feature had a 14.3 (s = 9.29) mouse clicks average while those without had a 10.5 (s = 16.4) mouse clicks average during the same period).

Also, and despite we have chosen to deploy the feature in the more active users, none of the members of the selected house managed to label power events on their own. We believe that one of the main reasons behind this was the high number of events that would show in the user interface making it hard to select and label the right appliance as described by this family member *"I think the most complicated thing to do is the consumption per device (…) it's complicated to manage such a large number of devices."* (Family 7, Husband)

Nevertheless, one important lesson learned for future designs was that not all the appliances seem to be of equal importance to the residents. This was especially seen when helping the users labeling their events as they kept asking questions about what they considered to high consumer appliances (e.g. oven and clothes washer / dryer).

6 Discussion and Implications

In this section with discuss the most relevant outcomes of the one-year (52 weeks) long-term study of eco-feedback. We start by summarizing the overall results of the study and then we discuss some of the weaknesses and possible implications of this work regarding future deployments of eco-feedback systems.

6.1 Results

Here we presented the results of the long-term deployment of a real-time eco-feedback solution during 52 consecutive weeks in a stable sample of 12 households. During this period families had access to their energy consumption with two versions of an eco-feedback interface that also gathered usage and interaction data.

Our findings show that after 52 weeks there was no significant reduction in energy consumption but also no increase. Our results contradict the literature that suggests a positive impact of eco-feedback on energy consumption. We argue that such conclusions could be based on typical short-term (2 or 3 week) studies, which are not long enough to capture the relapse behavior pattern after the novelty effect of the eco-feedback. We recognize that further research is needed to isolate the relationship between consumption and eco-feedback but when huge investments in smart grids and eco-feedback technologies are under way it would be important to deploy more long-term studies that investigate these results further. This is particularly relevant if we consider the latest results from the European Environment Agency (EEA) that show a 12.4% increase in the final energy consumption of households, with electricity emerging as the fastest growing source of energy between 1990 and 2010 [19].

We have also confirmed that energy consumption in households is tightly coupled with the number of residents and that large families tend to become more energy efficient when considering average consumption by household member. Another interesting finding was learning that the income-level of the family is another powerful explanatory variable of energy consumption, with low-income families being able to spend less energy than high-income with the same household size. This suggests that there is a minimum acceptable point of consumption that once reached it may become impossible to implement other consumption-reduction initiatives without negatively affecting the family needs and routines. This is what was noticed by one of the low-income families with 2 children: *"I think the changes were mostly on making us more aware of devices we used and habits we had. We had some bills that were a little expensive and we started to reduce some consumption (…) now I feel we have reached a constant value, we pay around the same amount each month and it won't cost more than this."* (Family 1, Wife) And also one of the average-income families, with a teenage daughter *"We try to reduce here and there but this is an apartment we can't walk here in the darkness, we need to turn on some lights."* (Family 11, Wife)

Our study also shows other results that are in line with most of the reviewed literature especially when considering increased awareness and better understanding on what appliances really consume as shared by the this family *"This helped us to know more about our consumption, and we did some changes around here (…) this device brought us a new kind of awareness but it didn't disturb our routine. We don't feel it disturbed us in any way. It was beneficial for us to have it."* (Family 1) Or even, by helping deconstruct some devices initial consumption levels' associated perceptions as stated by this family *"It helped us to see some devices were consuming more than we initially thought and it changed the way or time we used those devices, for example the iron or the kitchen hood."* (Family 9, Husband) And in some, rare cases, this better understanding lead to some routine changes: *"We have a very conscious way of consuming energy, we were careful before having it here. It helped us to see some devices were consuming more than we initially thought and it changed the way or time we used those devices (…) one of those was doing laundry and use the dishwasher only at night [to take advantage of the night tariff] and this changed our routine completely."* (Family 9, Husband)

On the long run users feel that there's nothing new to learn from the provided eco-feedback and therefore the number of interactions are reduced to marginal values which is a strong indicator that the eco-feedback provided needs to encourage users to learn more about their consumption as well as provide more tailored and personalized feedback especially after relevant changes happen (e.g. buying a new equipment or someone leaving the house for long periods).

6.2 Implications and Lesson Learned

One limitation of our study was the lack of a proper control group in order to better access the effectiveness of the eco-feedback as a way to promote energy reduction. This is particularly relevant because our results are contradicting many of the findings in the literature that rely on two or three week deployments of eco-feedback, which might be too optimistic about the potential of this technology. Furthermore, considering the long-term nature of this study it would be important to keep an updated profile of our participant families (e.g. holiday absences, some family member visiting for a long period, and a list of actual appliances in the house at every moment) as this would have allowed us to perform other kinds of comparative analysis like the consumption of similar houses or correlating user concerns with their actual consumption patterns. All of these possibilities involve significant costs in deploying and running the studies but are rightly justified in particular given the environmental and economic impacts of household energy consumption and the expectations with large-scale deployments of smart grids that could make eco-feedback widely available.

Physical Location and Security. Our eco-feedback system also presented some limitations that could have an impact in the results. The fact that our eco-feedback system was implemented using a netbook that acted both as the sensor and the visualization platform placed at the entrance of the household presented some limitations. The system was not easily accessible to all family members in particular children, as one of the mothers shared with us*: She didn't reach it (youngest daughter 7 years old)"*, (Family1, Mother). In addition the location of the netbook near the main power feed made it harder for family members to interact with the eco-feedback since it made them afraid of either dropping it in the floor or damaging the equipment since they considered it to be very fragile (the computer was stuck to the wall by only two adhesive velcro tapes). Finally some families also expressed concerns regarding the intrusiveness and safety of the system, even though it was properly and securely installed by a qualified electrician from the electrical company. For instance, some families did not allow their kids to come nearby or interact with fearing the risk of electric shock. In current deployments of our technology we are collecting data in the meters outside the houses and providing the eco-feedback using tablets and other mobile devices. Nevertheless our preliminary results are still consistent with the results presented here.

Appropriation of the Eco-feedback Technology. Finally, we have learned from the extensive interviews that family members tend to have naturally defined roles where some of them took over the task of checking and controlling the energy consumption and therefore, reducing the number of family members that would interact with the system. This made other family members feel they didn't need to worry or use the system, since someone (usually the husband or the person more comfortable around computers) was taking care of it as shared by two spouses: *"There are certain things I leave for him to do and other things I take care of myself. I was curious to use it and I would use it but not as often as him"* (Family 12, Wife) and *"He would check more because he would be more curious (husband) and me I would let him give me the report of it. He would summarize the information"* (Family 1, Wife)

7 Conclusion

This paper presents the results of a long-term deployment of eco-feedback technology in 12 apartment houses for 52 weeks in a southern European urban city. We collected both qualitative and quantitative data in order to assess the effectiveness of eco-feedback technology as a driver to promote energy conservation behaviors. Our results conflict the more promising expectations of eco-feedback based on short-term (two or three weeks) deployments reported in many HCI venues.

Despite the physical and methodological limitations of this study we have confirmed these results with a different deployment where the infrastructure is no longer placed inside the households removing the physical and security concerns with the eco-feedback device. We observed the same relapsing effects even when the eco-feedback is provided through a mobile device connected over the Internet to the non-intrusive sensor placed outside the house. After four weeks we observed the same decrease in attention and energy conservation behaviors.

We argue that in order to make eco-feedback technology effective further research is needed to understand what could lead users to retain attention over time in a way that promotes significant changes in their behavior capable of generating energy savings. We are also exploring other approaches like art-inspired eco-feedback [20] and social features like sharing energy consumption with in a community public display or social networks. In future work we also wish to further explore how the households perceive their consumption and how we can use family dynamic and routines to increase the effectiveness of eco-feedback technology.

In summary, our research highlights the importance of conducting long-term deployments of eco-feedback systems in order to understand the real potential and implications of this technology. Energy and resource consumption in general, are important application domains for persuasive technologies. However in order for this technology to have long-term impacts in domains like sustainability we need to overcome the novelty effect leading to response-relapse behaviors. Here we reported on such a study and presented some lessons learned that could lead to further research exploring new dimensions of eco-feedback.

Acknowledgements. This research was partially funded by the CMU | Portugal SINAIS project CMU-PT/HuMach/0004/2008 and the Portuguese Foundation for Science and Technology (FCT) doctoral grant SFRH/DB/77856/2011. Finally, we would also like to acknowledge all the participating families.

References

1. Froehlich, J., Findlater, L., Landay, J.: The design of eco-feedback technology. In: 28th International Conference on Human Factors in Computing Systems (2010)
2. Fischer, C.: Feedback on household electricity consumption: A tool for saving energy? Energy Efficiency 1, 79–104 (2008)
3. Peschiera, G., Taylor, J.E., Siegel, J.A.: Response-relapse patterns of building occupant electricity consumption following exposure to personal, contextualized and occupant peer network utilization data. Energy and Buildings 42, 1329–1336 (2010)
4. Ueno, T., Sano, F., Saeki, O., Tsuji, K.: Effectiveness of an energy-consumption information system on energy savings in residential houses based on monitored data. Applied Energy, 166–183 (2006)

5. Broms, L., et al.: Coffee maker patterns and the design of energy feedback artefacts. In: 9th ACM Conference on Designing Interactive Systems, pp. 93–102 (2010)
6. Spagnolli, A., et al.: Eco-Feedback on the Go: Motivating Energy Awareness. Computer 44(5) (2011)
7. Miller, W., Buys, L.: Householder experiences with resource monitoring technology in sustainable homes. In: 22nd Conference of the Computer-Human Interaction Special Interest Group of Australia ozCHI (2010)
8. Wallenborn, G., Orsini, M., Vanhaverbeke, J.: Household appropriation of electricity monitors. International Journal of Consumer Studies 35, 146–152 (2011)
9. Kjeldskov, J., Skov, M., Paay, J., Pathmanathan, R.: Using mobile phones to support sustainability: a field study of residential electricity consumption. In: 28th International Conference on Human Factors in Computing Systems, pp. 2347–2356 (2012)
10. Costanza, E., Ramchurn, S., Jennings, N.: Understanding domestic energy consumption through interactive visualizations: a field study. In: 14th ACM International Conference on Ubiquitous Computing (2012)
11. Gamberini, L., Spagnolli, A., Corradi, N., Jacucci, G., Tusa, G., Mikkola, T., Zamboni, L., Hoggan, E.: Tailoring feedback to users' actions in a persuasive game for household electricity conservation. In: Bang, M., Ragnemalm, E.L. (eds.) PERSUASIVE 2012. LNCS, vol. 7284, pp. 100–111. Springer, Heidelberg (2012)
12. Nunes, N.J., Pereira, L., Quintal, F., Berges, M.: Deploying and evaluating the effectiveness of energy eco-feedback through a low-cost NILM solution. In: 6th International Conference on Persuasive Technology (2011)
13. Pereira, L., Quintal, F., Nunes, N.J., Berges, M.: The design of a hardware-software platform for long-term energy eco-feedback research. In: 4th ACM SIGCHI Symposium on Engineering Interactive Computing Systems, vol. 1, pp. 221–230 (2012)
14. Hart, G.W.: Nonintrusive appliance load monitoring. IEEE 80, 1870–1891 (1992)
15. Mills, E.: Review and comparison of web- and disk-based tools for residential energy analysis. LBNL Paper 50950 (2002)
16. Motegi, N., Piette, M.A., Kinney, S., Herter, K.: Web-based energy information systems for energy management and demand response in commercial buildings. LBNL Paper 52510 (2003)
17. Darby, S.: The effectiveness of feedback on energy consumption. A review for DEFRA of the literature on metering, billing and direct displays. Environmental Change Institute, University of Oxford (2006)
18. Gram-Hanssen, K., Patersen, K.: Different everyday lives: Different patterns of electricity use. In: 2004 American Council for an Energy Efficient Economy Summer Study in Buildings (2004)
19. European Environment Agency. Final Energy Consumption by Source (February 2013), `http://www.eea.europa.eu/data-and-maps/indicators/final-energy-consumption-by-sector-5`
20. Nisi, V., Nicoletti, D., Nisi, R., Nunes, N.J.: Beyond Eco-feedback: using art and emotional attachment to express energy consumption. In: 8th ACM Conference on Creativity and Cognition (2011)

"...Language in Their Very Gesture" First Steps towards Calm Smart Home Input

John N.A. Brown[1,2], Bonifaz Kaufmann[1], Franz J. Huber[1], Karl-Heinz Pirolt[1], and Martin Hitz[1]

[1] Alpen-Adria Universität Klagenfurt
Universitätsstraße 65-67, 9020 Klagenfurt, Austria
{jna.brown,bonifaz.kaufmann,martin.hitz}@aau.at
{fhuber,kpirolt}@edu.aau.at
[2] Universität Politècnica de Catalunya
Neàpolis Building Rbla. Exposició, 59-69, 08800 Vilanova i la Geltrú, Spain

Abstract. Weiser and Brown made it clear when they predicted the advent of ubiquitous computing: the most important and challenging aspect of developing the all-encompassing technology of the early 21st Century is the need for computers that can accept and produce information in a manner based on the natural human ways of communicating. In our first steps towards a new paradigm for calm interaction, we propose a multimodal trigger for getting the attention of a passive smart home system, and we implement a gesture recognition application on a smart phone to demonstrate three key concepts: 1) the possibility that a common gesture of human communication could be used as part of that trigger, and; 2) that some commonly understood gestures exist and can be used immediately, and; 3) that the message communicated to the system can be extracted from secondary features of a deliberate human action. Demonstrating the concept, but not the final hardware or mounting strategy, 16 individuals performed a double clap with a smart phone mounted on their upper arm. The gesture was successfully recognized in 88% of our trials. Furthermore, when asked to try and deceive the system by performing any other action that might be similar, 75% of the participants were unable to register a false positive.

Keywords: Ubiquitous Computing, Calm Technology, Smart Home, Gestures, Gestural Interaction, Snark Circuit.

1 Introduction

Calm computing was introduced by Weiser and Brown as the most interesting and challenging element of what they dubbed the coming Era of Ubiquitous Computing [1]. While no one would seriously question whether or not computers have become ubiquitous, their claim that computers should "... fit the human environment instead of forcing humans to enter theirs..." has not been as universally accepted. Since calm computing was not built into our technology as it became ubiquitous, we must now find ways to add it post hoc and post haste. Fortunately, there is a ubiquitous device that is ready to meet our needs.

A. Holzinger and G. Pasi (Eds.): HCI-KDD 2013, LNCS 7947, pp. 256–264, 2013.

Smart phones are capable of both voice- and text-based telecommunication and they also provide their users with increased multimedia processing power far beyond the means of a previous generation's desktops and even mainframes, all in a pocket format. Furthermore, since such smart phones are equipped with high resolution cameras, GPS functionality, a digital compass, accelerometers and more, they are capable of providing a degree of additional information that is beyond the deliberate use or needs of most users.

Increasingly, developers are accessing the internal sensors through the API and putting the resultant technical data streams to use in a submerged manner: as input for software that interacts with the user in non-technical ways [2]. This robust exchange of data, graphics and sound has opened a new and very rich vein of interactive potential that is being pushed to its limits for technological and commercial purposes. We propose to put this technology to use in making computing more calm, in the sense described by Weiser and Brown [3], that is, to enable human computer interaction that is truly based on the natural human means of communication.

1.1 Interaction Unification of Distributed Smart Home Interfaces

We are working towards a new paradigm of conceptual unification of the distributed interfaces for environments based on ubiquitous computing and networks of embedded systems. The goal is to derive a conceptual unification rather than the simple spacial or technological concentration of hardware and software as has been attempted by Cohn et al. [4]. Nor do we intend to remove visible hardware as proposed by Streitz and Nixon [5] and Ishii et al. [6]. Our goal is to create the impression in the user that naturally multimodal behaviour will be understood by the dissembodied interpretive service that helps to convey their needs and desires to their smart home. In this way, the smart environment could become a space with a single holistic identity in the perception of the "user". This would be similar to the way in which our parents came to think of "the car" as they became less and less aware of the technological and mechanical components; the same way that the current generation ignores most of the functionality and technological underpinnings of their most personal computer, and think of it simply as "the phone".

This is fundamentally different from the conceptual framework underpinning the common development of smart homes, in that we do not intend to predict user intent, but rather to empower the user in the subtle and natural expression of their needs and preferences. A further fundamental difference is that we do not foresee smart homes as the exclusive domain of geriatric residences, embracing instead Weiser's prediction of ubiquitous computing.

1.2 Smart Home Controls

If it is accepted that using computers causes stress when the user feels that they are not in control, as per Riedl et al. [7], then it is a natural extension to assume that such stress would be an even greater threat in an immersive computing environment such as a smart home. Interviews and focus group sessions have shown that users prefer a centralized remote control to enable immediate interaction with a number of devices installed in a household [2]. While the concept of a control panel proved popular, as an interface, it is an artifact from the *Mainframe Era* of computing.

While some researchers, such as Chan et al. [8], foresee the coming of either wearable or implantable systems to complement domotic control with the provision of biomedical monitoring sensors, it will be some time before these features can become ubiquitous. They go on to stress that since smart homes promise to improve comfort, leisure and safety the interaction should be as natural as possible. If their proposed method of improvement is still developing technologically, our proposed method is built upon applying currently available technology in a novel manner; a manner based on the fact that human communication naturally involves complimenting words with gestures.

1.3 Using Gestures to Clarify Intent

> "In gestures we are able to see the imagistic form of the speaker's sentences. This imagistic form is not usually meant for public view, and the speaker him- or herself may be unaware of it…" [9].

All natural human interaction is multimodal. We constrain ourselves to a single modality only when required. When in a diving environment, scuba gear enables us to function without having to learn to breathe underwater, but formal communication is reduced to a single modality and becomes dependent on the use of strictly-defined and well-practiced gestures. When in a digital environment, the GUI interface enables us to function without having to learn machine language, but formal communication is reduced to a single modality and becomes dependent on the use of strictly defined and severely truncated words which have been removed from their usual ontological, cultural and environmental context.

We do not want to create a gestural recognition system based on the false paradigm of single modality interaction. Instead, our gestures will amplify spoken word interaction, observing the same phonological synchronicity rules that have been observed when gestures accompany speech in normal interaction [9]. This proves especially important in solving a problem common to speech interfaces: the noisy background.

A family having a conversation can confound a speech recognition system's interpretation of a command [2]. The use of speech or sound for interaction is a balancing act between the desire for immediate response and the expense of constant attentiveness – in terms of energy use, incessant processing and filtering, and false positive responses. Logically, this problem must be greatly reduced when a trigger is used to alert the system to attend to a spoken command that is about to be generated.

1.4 Triggering Attention

This experiment is the first step towards a new paradigm of smart home interaction which will provide a solution to the problems stated above. Our eventual goal is a system that will wait unobtrusively to be called into service. Ideally, the trigger should be one that is easy to perform intentionally but difficult to perform accidentally. The trigger should also allow the system to distinguish known users from one another and from strangers.

We propose that a passive system could become active when triggered by three roughly simultaneous commands delivered in different modalities. All three signals can be produced via the execution of a common human behavior for getting the attention of subordinates – the double clap paired with a spoken name. Brown et al. [10] have conducted an attempt towards the realization of a different aspect of this system, focusing on detecting and recognizing the audible signals of a double clap paired with a single-word voice-based command, in a noisy environment.

The first question addressed in our study is whether or not the separate components of a multimodal communicative technique could theoretically be used as separate parts of a trigger as discussed above. Secondly, we ask if some commonly understood gesture exists and can be used immediately, across cultural and linguistic barriers. Finally, we ask whether the signal communicated to the system could be extracted from secondary features of a deliberate human action.

We propose that the deliberate double clap can meet these conditions, and that it should be possible to automatically distinguish between common hand movements (such as waving or applause) and a deliberate double clap, as detected by the accelerometer in a smart phone. Smart phones have been used in many studies [11], [12], [13], [14], [15], but not with the intent that the signal should be generated incidentally during natural movement.

This also addresses the issue of previous exposure to technology (PET) by allowing the human to use a common human behavior rather than behavior based on technology [16]. In order to be certain that the phone's accelerometer is being used only incidentally, the entire device is mounted in an armband on the upper arm. This is not to say that an arm-mounted smartphone will be the final form of the device. Raso et al. [17] mounted a smartphone in that manner for unobtrusive measurement of shoulder rehabilitation exercises and we have followed their example simply to be certain that moving the device remains an incidental action – an unconscious side effect of the attempt to create the desired double clap gesture.

We have developed a smart phone application intended to recognize a double clap performed as a deliberate signal. In order to measure the functionality of the man-machine system (user, accelerometer and software), we have conducted a series of empirical trials.

2 Double Clap Recognition

For our smart phone application we used the LG Optimus 7 E900 and its built-in accelerometer. The application was implemented for the Windows Phone 7.5 framework. The accelerometer has 3-axes (X, Y, Z), and was set to 25Hz. It provides acceleration values normalized between -2.0 and +2.0. We used an armband to mount the device on the upper arm. We implemented a recognizer, which supports automatic segmentation, to capture the double clap gesture.

Performing a double clap is not only a commonplace gesture for getting someone's attention; it also has an easily-recognized pattern of accelerations and stops. In Fig. 1 the raw accelerometer data are displayed as a continuous function where all axes are separated.

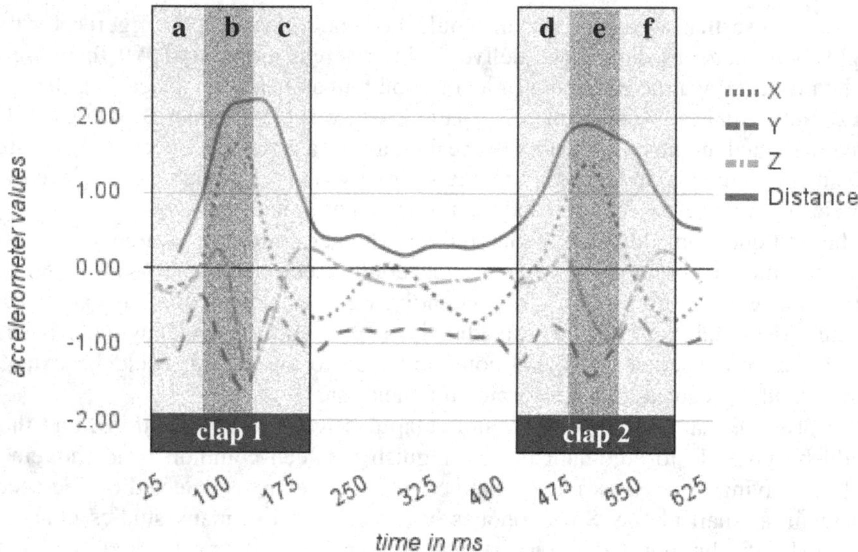

Fig. 1. Accelerometer data

The green line running above the others shows overall distance, as calculated using Equation 1. Significant regions are marked as follows. First, when the hands move towards one another, there is an increase in acceleration (Fig. 1a). Acceleration increases until it suddenly stops (b) and changes direction (c). Now the acceleration slows, readying for the beginning of the second clap (d-f).

$$D = \sqrt{(x_k - x_{k-1})^2 + (y_k - y_{k-1})^2 + (z_{k-} - z_{k-1})^2} \quad (1)$$

Equation 1 gives us the difference between two consecutive accelerometer values, telling us the increase or decrease of acceleration between the two points. After every interval, given by the refresh rate of 25ms, the Euclidean distance to the previous collected accelerometer data will be calculated. If the average of the last three continuous distance values reaches a certain threshold, this is recognized as the end of a double-clap sequence. Once a possible end has been found, the algorithm looks back at the last 20 data entries (500ms) and examines the average of the distance between two consecutive axis values, to find both, how often the axis crosses the zero line, and any sudden changes in direction. Depending on the results, the recorded values and values collected in the evaluation phase were compared to one another.

3 Experiment

We conducted our experiment in four stages, with the assistance of 16 right-handed participants (3 females) between the ages of 23 and 31 (m = 27, SD = 2.2). Participants were first given a survey regarding their familiarity with the double clap as a deliberate signal.

Secondly the participants performed the double clap six times with the device mounted on their upper arm as described above, and six times without, in accordance with the method used by Kühnel et al. [18].

Accelerometer data was then collected while participant performed the double clap 10 times, without other motions between the repetitions. An observer took note of their performance, recording false positives and false negatives following the procedure described by Sousa Santos et al. [19].

The third stage was a deliberate attempt by each participant to confound the system. They were encouraged to try to elicit false-positives with any common or uncommon movements that, it seemed to them, might be mistaken for the acceleration pattern of a double clap.

The final stage of our experiment was another survey, asking what the participants thought of the double clap as an interactive method and soliciting their opinions on the difficulty of learning and performing the action.

4 Results

The first of our qualitative excercises was a resounding success: All participants confirmed that they were familiar with the double clap as a deliberate action and could generate it themselves. 56.3% confirmed that they had used a double clap to get someone's attention and 62.5% confirmed both that they have witnessed the double clap used to garner attention and that they could imagine using a double clap to intitate interaction with their computer.

Our second exercise was answered quantitatively. Our method measured successful double claps in 88% of the total number of trials. To be more specific, half of our participants succeeded in all of their attempts and two-thirds of the participants succeeded in over 90% of their attempts.

We excludeed 2.5% of the overall number of double claps performed because they did not fall within the range of style, speed and/or noise level that was originally demonstrated.

Our third exercise resulted in the most entertaining portion of the data collection, as each participant tried to imagine and then execute some common or uncommon action which would be misread by our algorithm and mistaken for a double clap. Despite some wonderful performances, 75% of the participants were unable to deliberately generate a false positive. Two participants succeeded by vigorously shaking hands. Two others attempted to fool the system by doing the chicken dance. Only one of the dancers was successful.

In our final exercise, a post questionnaire, the participants told us how they felt about the ease or difficulty of using the double clap as a trigger for computer input.

Asked how easy or difficult it was to learn the double clap as demonstrated, 56.3% said that it was *very easy* and the rest described it as *easy*.

When we asked them how easy or difficult it was to perform the double clap, 56.3% reported that it was *very easy*, 37.5% described it as *easy*, and a single participant described the level of difficulty of the task as *normal*. No one rated learning or performing the task as either *difficult* or *very difficult*.

We wanted to know if participants would use a double clap gesture in order to interact with a computer. 50% responded with a simple *yes*, while 12.5% preferred *I think so*. One participant was *uncertain*, another chose *I don't think so* and 25% offered a simple *no*.

Those participants who did not like the idea of using a double clap as an interactive signal offered the following reasons:

"*The gesture is too loud, it gets too much attention*";

"*Snapping the fingers would be easier*", and;

"*Clapping is not so intuitive.*"

Two said that clapping is for "*getting the attention of animals*" and four said that they would be more likely to use the device if it were "*integrated into their clothes*" or "*in a watch*", but "*would not put it on every time*" that they wanted to use it.

5 Conclusions

As mentioned above, in this study, we sought to demonstrate three key concepts.

The first is the possibility that a common gesture of human communication could be used as part of a multimodal trigger for getting the attention of a passive smart home system. The Snark Circuit [10] is such a trigger, and it has been proposed that the common attention-getting action of clapping one's hands twice and calling out a name could provide three separate signals. The auditory components have already been tested [10], so we set out to test the possibility that the hand motions used to generate a double clap could be interpreted as computer input. We have shown that it can.

Secondly, we asked if some commonly understood gesture exists and can be used immediately, across cultural and linguistic barriers. 100% of our participants reporting familiarity with the use of a double clap to get attention and agreeing that it would be either *easy* or *very easy* to learn to perform it as demonstrated.

Finally, we asked whether the signal communicated to the system could be extracted from secondary features of a deliberate human action. To answer that question we developed a straightforward accelerometer-based smart phone gesture recognition application, which could recognise hand movements. More specifically, we developped it to distinguish between general hand movements and the movement pattern of a double clap, allowing the recognition of this unique movement pattern as a deliberate signal.

The recognizer does not use any machine learning or other statistical probability methods. Instead it is implemented with a basic template matching algorithm using a distance equation to identify an increase or decrease in acceleration. User tests with 16 participants showed an accuracy of 88%. What's more delibarate attempts to decieve the system and induce a false positive met with a 75% failure rate, despite the simplistic mathematical method used. It seems that template matching is sufficient for the recognition of this rather unique gesture.

These results provide evidence that the unconscious component(s) of natural human gestures can be used as deliberate components of multimodal commands, whether these components are redundencies for a simple trigger (as in the case of the Snark Circuit) or whether they are, instead, either additive or stand-alone commands for more complex interactions.

6 Future Work

Future work will be specifically directed at gesture recognition for either stand-alone purposes, or as a component of multimodal triggering as part of systems such as the Snark Circuit system.

With the double clap in use as described above, new gestures will have to be developed, or, more appropriately, old and common gestures will be sought out in order to match them with multimodal, multi-channel triggers for other activities.

The environments in which these systems can be applied will be further developed as well, including the testing of a *generative grammar*-based smart home ontology constructed from a set of *activities of daily living*-based use cases.

Further future work will include adaptation of our gesture recogniser by mounting a smartphone-linked accelerometer into a ring-based input device such as the one proposed by Brown et al. [20]. The additional articulation and small movements allowed thereby, when combined with the more robust recognition engine discussed above may allow for very subtle and complex gestural interaction, and bring us closer to the almost unconscious computer input promised by calm technology.

If we are to design towards that future, then we must accept that the technology will change and focus our skills on creating early versions of a welcoming and encalming environment in which humans interact easily and multimodaly with their surroundings, giving no more thought to operating systems or system requirements than they might devote to perceptual psychology or the anatomy of the vocal chords when talking with friends.

Acknowledgements. This work was supported in part by the Erasmus Mundus Joint Doctorate in Interactive and Cognitive Environments, which is funded by the EACEA Agency of the European Commission under EMJD ICE FPA n 2010-0012.

References

1. Weiser, M., Brown, J.S.: The Coming Age of Calm Technology. In: Denning, P.J., Metcalfe, R.M. (eds.) Beyond Calculation: The Next Fifty Years of Computing, pp. 75–85. Copernicus, New York (1997)
2. Koskela, T., Väänänen-Vainio-Mattila, K.: Evolution towards smart home environments: empirical evaluation of three user interfaces. Personal and Ubiquitous Computing (8), 234–240 (2004)
3. Weiser, M.: The Computer for the Twenty-First Century. Scientific American 265(3), 94–104 (1991)
4. Cohn, G., Morris, D., Patel, S.N., Tan, D.S.: Your noise is my command: Sensing gestures using the body as an antenna. In: Proceedings of the 2011 Annual Conference on Human Factors in Computing Systems, pp. 791–800. ACM (2011)
5. Streitz, N., Nixon, P.: The disappearing computer. Communications of the ACM 48(3), 32–35 (2005)

6. Ishii, H., Ullmer, B.: Tangible bits: towards seamless interfaces between people, bits and atoms. In: Proceedings of the SIGCHI Conference on Human Factors in Computing Systems, pp. 234–241. ACM, New York (1997)
7. Riedl, R., Kindermann, H., Auinger, A., Javor, A.: Technostress from a Neurobiological Perspective. Business & Information Systems Engineering, 1–9 (2012)
8. Chan, M., Estève, D., Escriba, C., Campo, E.: A review of smart homes - Present state and future challenges. Computer Methods and Programs in Biomedicine 9(I), 55–81 (2008)
9. McNeill, D.: Hand and mind: What gestures reveal about thought. University of Chicago Press, Chicago (1992)
10. Brown, J.N.A., Kaufmann, B., Bacher, F., Sourisse, C., Hitz, M.: "Oh I Say, Jeeves!': A Calm Approach to Smart Home Input (under review)
11. Kühnel, C., Westermann, T., Hemmert, F., Kratz, S., Müller, A., Möller, S.: I'm home: Defining and evaluating a gesture set for smart home control. International Journal of Human-Computer Studies 69(11), 693–704 (2011)
12. Prekopcsák, Z.: Accelerometer based real-time gesture recognition. In: Proceedings of International Student Conference on Electrical Engineering, Prague, pp. 1–5 (2008)
13. Joselli, M., Clua, E.: gRmobile: A framework for touch and accelerometer gesture recognition for mobile games. In: Proceedings of Brazilian Symposium on Games and Digital Entertainment, Rio de Janeiro, pp. 141–150 (2009)
14. Akl, A., Valaee, S.: Accelerometer-based gesture recognition via dynamic-time warping, affinity propagation, & compressive sensing. In: IEEE International Conference on Acoustics Speech and Signal Processing (ICASSP), Dallas, pp. 2270–2273 (2010)
15. Kratz, S., Rohs, M.: A $3 gesture recognizer: simple gesture recognition for devices equipped with 3D acceleration sensors. In: Proceedings of the 15th International Conference on Intelligent User Interfaces, IUI 2010, pp. 341–344. ACM, New York (2010)
16. Holzinger, A., Searle, G., Wernbacher, M.: The effect of Previous Exposure to Technology (PET) on Acceptance and its importance in Usability and Accessibility Engineering. Springer Universal Access in the Information Society International Journal 10(3), 245–260 (2011)
17. Raso, I., Hervas, R., Bravo, J.: m-Physio: Personalized Accelerometer-based Physical Rehabilitation Platform. In: Proceedings of UBICOMM 2010, Florence, pp. 416–421 (2010)
18. Kühnel, C., Westermann, T., Weiss, B., Möller, S.: Evaluating multimodal systems: a comparison of established questionnaires and interaction parameters. In: Proceedings of NordiCHI 2010, pp. 286–294. ACM, New York (2010)
19. Sousa Santos, B., Dias, P., Pimentel, A., Baggerman, J.-W., Ferreira, C., Silva, S., Madeira, J.: Head Mounted Display versus desktop for 3D Navigation in Virtual Reality: A User Study. Multimedia Tools and Applications 41, 161–181 (2008)
20. Brown, J.N.A., Albert, W.J., Croll, J.: A new input device: comparison to three commercially available mouses. Ergonomics 50(2), 208–227 (2007)

"Oh, I Say, Jeeves!" A Calm Approach to Smart Home Input

John N.A. Brown[1,2], Bonifaz Kaufmann[1], Florian Bacher[1], Christophe Sourisse[1], and Martin Hitz[1]

[1] Alpen-Adria Universität Klagenfurt
Universitätsstraße 65-67, 9020 Klagenfurt, Austria
{jna.brown,bonifaz.kaufmann,martin.hitz}@aau.at
{fbacher,csouriss}@edu.aau.at
[2] Universitàt Politècnica de Catalunya
Neàpolis Building Rbla. Exposició, 59-69, 08800 Vilanova i la Geltrú, Spain

Abstract. Now that we are in the era of Ubiquitous Computing, our input devices must evolve beyond the mainframe and PC paradigms of the last century. Previous studies have suggested establishing automatic speech recognition and other means of audio interaction for the control of embedded systems and mobile devices. One of the major challenges for this approach is the distinction between intentional and unintentional commands, especially in a noisy environment. We propose the *Snark Circuit*, based on the notion that a command received three times "must be true". Eventually, overlapping systems will recognize three triggers when a user claps twice (giving signals of sound and motion) and speaks the name of her computer. 20 participants took part in a study designed to test two of these three inputs: the sound of a double-clap and a spoken name. Vocal command recognition was successful in 92.6% of our trials in which a double clap was successfully performed.

Keywords: Smart Home, Ubiquitous Computing, Calm Technology, Audio Input, Multimodal Input, Snark Circuit.

1 Introduction

We are now living in Weiser and Brown's third age of computer acceptance into society, the Era of Ubiquitous Computing [1]. However, what they described as the most interesting and challenging element is missing: Calm Technology (CT). If interaction with submerged networks of embedded systems is truly to become "…as refreshing as taking a walk in the woods" [2], our methods of interaction have to evolve beyond the mainframe and PC input paradigms of the previous eras and meet us in the 21st Century. Weiser and Brown presented examples of CT-based output devices, but provided only a few examples of CT-based input devices [3].

1.1 Automatic Speech Recognition in Smart Environments

Research into smart homes has been going on for decades and detailed reviews of the literature have been conducted by Cook and Das [4] and by Chan et al. [5]. The focus

A. Holzinger and G. Pasi (Eds.): HCI-KDD 2013, LNCS 7947, pp. 265–274, 2013.

of these studies is often on Ambient Assistive Living (AAL) for the elderly or for people with special needs [6], but the entry threshold for AAL is dropping with the advent of innovative design and technology integration [7]. This is changing the nature of smart environments, especially as technological advances allow display and control to change from single-user to multi-user [8].

The control of networked and embedded systems through the use of automatic speech recognition has long been a feature of science fiction and fantasy interfaces. Attempts to translate the idea to real life have met only modest success. The *Sweet Home Project* in France is an attempt to design a new smart home interaction system based on audio technology [9]. They did not, however, conduct their experiments under normal, noisy conditions. Two unsolved questions in this realm have been whether or not to have a "live microphone" (constant sound detection) which means a constant drain on power and, worse than that, whether or not to have a "live processor" (which means a drain not only of electrical power, but of processing power, too).

We propose to evaluate whether or not a speech and sound recognition software similar to the one described above [9] can be made to work in an acoustically hostile environment, given the addition of a simple command protocol. This protocol is based on the triple-redundancy systems common to engineering [10], a truly user-centered perspective [11] and a hundred year old nonsense poem [12], in which the captain tells his crew: "I tell you three times, it must be true".

1.2 A New Paradigm – The Snark Circuit

We propose the *Snark Circuit* (Fig. 1), an early step towards a new paradigm of smart home interaction. Our goal is to provide a solution to the problem of using audio signals and voice commands in a noisy environment: a system that will wait unobtrusively to be called into service.

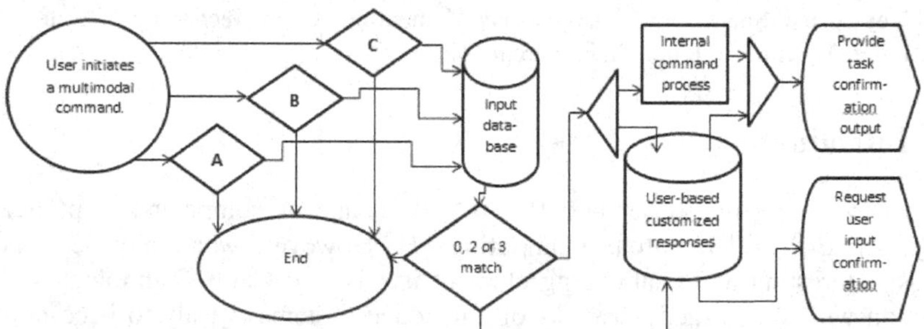

Fig. 1. The *Snark Circuit*: where any 3 recognized commands (A, B, C), detected within a small space of time, are compared to see if they hold the same meaning. If two of the recognized commands match, user confirmation is sought. If three match, the command is followed and task confirmation output is generated.

Fig. 1 shows the *Snark Circuit*, a "tell-me-three-times" command redundancy protocol or Triple Modular Redundancy [10] designed to fill the black box often assigned to filter noise from intentional command. Ideally, the trigger should be one that is easy to perform intentionally but difficult to perform accidentally. Conceptually, these parameters could be used to describe most naturally multimodal communication used by humans; the combination of voice with the "separate symbolic vehicle" that we call gesture [11]. These can be simple actions such as using the space between one's thumb and forefinger to illustrate the size of an object while also describing it verbally. An ambiguous gesture can be easily misunderstood. For example, waving one's hands loosely in the air may mean several different things; from cheering in formal sign language to cooling burnt fingers, from saying hello to saying goodbye. Other gestures are less likely to occur by accident.

Our chosen example is the double clap. Clapping an uncounted number of times may be common, but clapping twice is well understood to be a means of getting attention from either a group or an individual. As a gesture, clapping twice is quite unique, in that it involves limited inverted movements coming immediately one after the other, and it is clearly delimited by a rapid start and equally sudden stop. While many people are familiar with the decades-old technology of double-clap sound recognition used as an on/off switch for electrical devices, this is not what we are proposing. We are proposing that the sound and the movement of the double-clap both be used as independent signals which can make up two of the three inputs recognizable and useable in our triple redundancy. This introduces one aspect of Calm Technology [1] in that the user, intending to produce the noise of a double clap inadvertently produces the movement recognized as a separate signal. Inadvertent communication with a computerized system through natural human behavior is one of the key aspects of Calm Technology [13].

The trigger should also allow the system to distinguish known users from one another and from strangers. In "The Hunting of the Snark" [12] Lewis Carroll wrote: "I tell you three times, it must be true!" We propose that a passive system could become active when triggered by three roughly simultaneous commands of equivalent meaning, delivered in different modalities. All three signals can be produced via the execution of a common human behavior for getting the attention of subordinates – the motion and sound of a double clap paired with the sound of a spoken name.

2 User Study

The *Snark Circuit* relies on the detection of three different input signals. Brown et al. [14] have shown that the unique movement pattern of a double clap can be detected using the accelerometer in a standard smart phone. As mentioned above, in this study we have focused on the issues that were not addressed in the Sweet Home Project [9]; sound detection in noisy environment. To clarify this, Fig. 2 offers an illustration of a single-modality version of the *Snark Circuit*, detecting user intent with only two input signals (i.e. sound of clap and spoken name).

When the passively attentive system detects the sound of a double clap, it opens a three second window for recognizing the spoken word "Jeeves" (the name assigned to our system for this experiment). In future versions of this system, each user will assign an individually chosen name to the system.

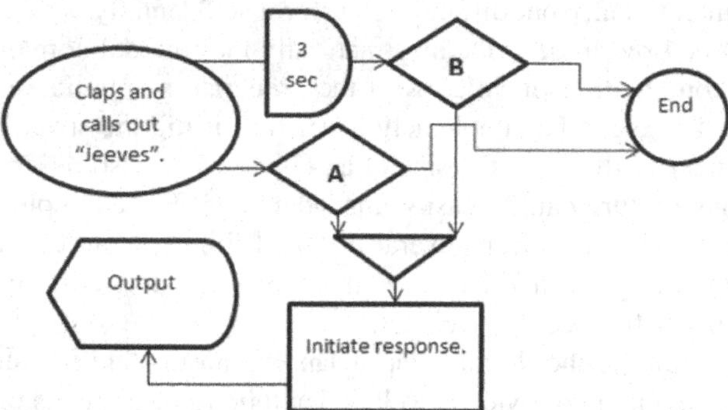

Fig. 2. The Simplified *Snark Circuit*: where two predefined commands (A – the sound of a double clap, and B – the sound of the spoken name "Jeeves"), detected in sequence, within three seconds, are recognized as a single multimodal command

2.1 Participants and Setup

10 men and 10 women, aged 19-28 (M=23, SD=2.635) from 8 different home countries, participated in our study. All participants said they were familiar with the use of a double-clap as a means of getting the attention of an individual or group.

The experiment was performed in a 3m x 3m office. Two microphones (*Sennheiser BF812* and *Sennheiser e8155* connected to the PC via a *Line6 UX1* audio interface) were hidden along the medial axis of the room, at roughly 0.5 meters from the walls, as shown in Fig. 3.

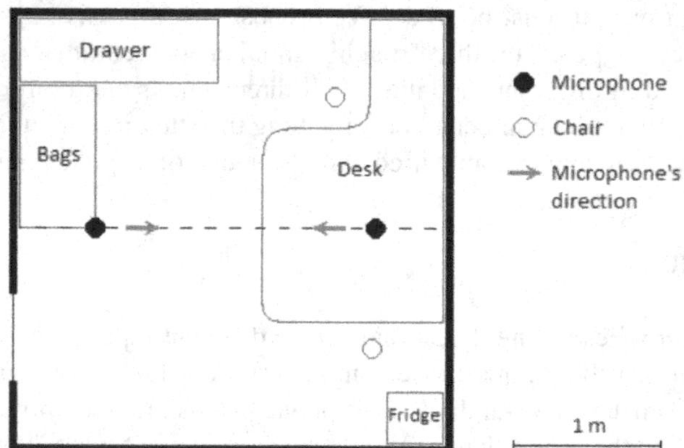

Fig. 3. Layout of the experimental setting

The number of microphones and their location reflects the work of Rouillard and Tarby [15]. The office is located in the administrative wing of a university, and so was surrounded by regular daily noises. On one side, there was an administration office and, on the other, a class room with computers. Noises of chairs moving or pieces of conversations could sometimes be heard, but the nature of our algorithm prevented that from becoming an issue. As will be explained in detail later, our system was constantly measuring noise levels, calculating the average over the ten most recent samples, and using that changing baseline as the comparative measure for recognizing a double clap. More precisely, a peak was defined as any noise over 120% of the background noise and two peaks occurring within 0.3-1.0 seconds were recognized as a double clap. This dynamic system is similar in nature to the interaction of humans who must adjust their volume as the volume around them changes.

2.2 Protocol

Ideally, the users of a future *Snark Circuit*-based system should each assign names to the system. Each user could assign a general, all-purpose name with or without additional specific names to deal with specific situations. This second option would allow each user to experience the illusion of easily distinguishable interactive personas with whom to interact when seeking to accomplish specific tasks in the environment. This could allow both the user and the system to more easily identify situational context when interacting.

Consider, for example, the household inhabited by an engineer, a nurse, and their four year-old daughter. The engineer dislikes play-acting and prefers to deal with the single manufacturer's default persona for their smart home, while her husband prefers to deal with a smart home persona that is more like an *aide-de-camp* or servant. Their daughter secretly believes that there are dozens of people living in the walls of the house, some of whom are strict and only speak to her in harsh voices about safety rules, while others invite her to play and answer when she calls.

At this stage of experimentation, we assigned a single name to the system for use by all participants: "Jeeves". We also used a single default automated response to successful attempts: "How may I help you?" If the system does not recognize that an attempt has happened, then it gives no response. For our current purposes, this was enough, and the exercise ended there.

When the full *Snark Circuit* is implemented, the protocol is based on triple redundancy rather than double. To clarify, three recognized commands with the same meaning would initiate the automated response for successful attempts: "How may I help you?" In the case where only two recognized commands have the same meaning, the system would query the user for clarification: "I'm sorry, were you trying to get my attention?" Finally, if there are no meaningful matches between recognized commands, no response is initiated.

2.3 Data Collection

Participants began by "getting the attention" of the system named Jeeves. This meant performing an audible double-clap, and following it with a correctly-pronounced utterance of the assigned name. When the system identifies a double-clap, it then switches into a listening mode and stays in this mode for a few seconds until it identifies a sound as matching a word in the database, or until the timeout is reached. In both cases, the system then goes back into waiting mode until it identifies another double clap.

Participants did not have to orient their claps or their voice towards any specific target in the office. Initially, the participant enters the room, closes the door, and then walks around the office for a few seconds. Double-claps and word utterance were performed just after the entrance into the office, and after having walked around for a few seconds. An observer made note of *false positive*s and *false negative*s.

2.4 Software

Like Rouillard and Tarby [15], the recognition software was implemented in C# with the Microsoft `System.Speech.Recognition` library. This library allows direct access to Window's speech recognition engine. In order to detect double claps, the software periodically calculates the average noise level, using the `AudioLevelUpdated` event of the created `speechRecognitionEngine` object. Every time the signal level exceeds 120% of the average noise level of the last ten samples, the software detects a peek. If there are exactly two noise peeks within 0.3s - 1s, the software classifies them as a double clap. Every time a double clap is detected the speech recognition engine is activated for 3 seconds. If the word "Jeeves" is uttered within this period, the engine recognizes it and gives a response.

3 Results

For each part of the command recognition (i.e. the double clap recognition and the word utterance recognition) we classified the input data into four categories:

- *true positive* (when the system is activated by a valid attempt),
- *true negative* (when the system is not activated by an invalid attempt),
- *false positive* (when the system is activated but no valid attempt took place), and
- *false negative* (when the system is not activated but a valid attempt took place).

As illustrated in Figure 4, successful double-claps were performed and recognized 71% of the time. 10% of our events were *true negatives*, 2% were *false positives* and 17% were *false negatives*.

Voice recognition was activated in 73% of our events (double-clap *true positives* + double-clap *false positives*). In other words, 73% of our original sample becomes 100% of the sample on which voice recognition is attempted. In this smaller pool, we found 83% *true positive*s, 7% *true negatives*, 3% *false positives*, and 7% *false negatives* (see Figure 4).

To calculate the overall system performance, we look at the number of times that the system performed correctly; identifying *positives* as *positives* and *negatives* as *negatives*. So we take all of the *true positives* and *true negatives* of voice recognition (i.e. 90% of the *true positive* clap recognition) and add it to the number of correctly-excluded double-claps (i.e. *true negative* clap recognition). This new total adds up to 76% of our original sample, giving us an overall system performance of 76%.

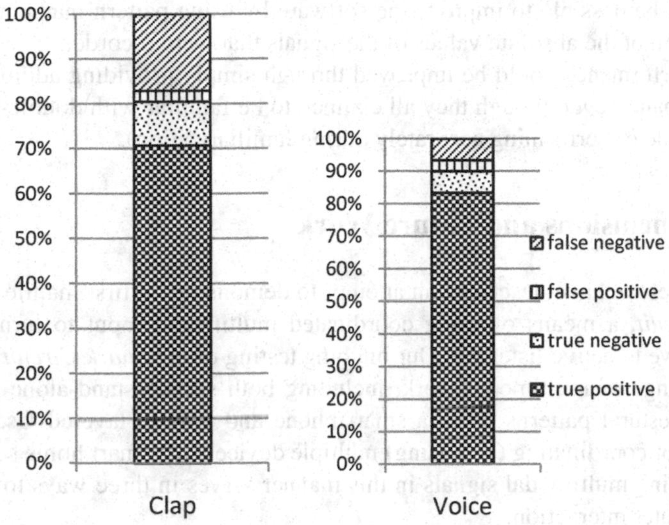

Fig. 4. Clap and voice recognition. Voice recognition was only activated in cases when clap detection was either true or false positive (i.e. 73% of the double claps).

4 Discussion

Performance constraints were set for both double claps and word utterances to decide if they could be classified as *true* or *false positives* or as *true* or *false negatives*, and an observer was in place to label each performance as either a valid or invalid attempt. Double claps should be audible, and a short silence should be easily heard between each clap. Word utterances should be audible too, and the word "Jeeves" should be pronounced clearly. According to the observer, some participants performed very quiet double-claps the software simply did not recognize, while others did not wait long enough between the two hand-claps for our generalized standard (*true negative,* 10%). Two percent of the claps detected were actually background noises that deceived the system (*false positive*).

On the matter of speech recognition, some participants pronounced the word "Jeeves" incorrectly, producing instead either "Yeeves" or "Cheese". These mispronunciations were recognized 3% of the time (*false positive*), and were rejected 7% of the time (*true negative*).

In this experiment, for the sake of expediency, default settings were used for the thresholds for recognizing double-claps and spoken words. Even the name "Jeeves" was used as a default. The interaction strategy proposed by this study is intended for customized environments and, as such, should include user-chosen names and user-modeled double-clap signals. Customizing the clap recognition to suit user abilities or preferences would certainly have improved the results.

It might be possible to improve the software by using pattern matching based on a combination of the absolute values of the signals that were recorded.

User performance could be improved through simply providing additional training for participants (even though they all claimed to be familiar with double-clapping and were capable of performing accurately during familiarization).

5 Conclusions and Future Work

In this paper we have described an attempt to demonstrate a first manifestation of the *Snark Circuit*, a means of using coordinated multimodal input to turn a computer from passive to active listening. Our ongoing testing of the *Snark Circuit* incorporates a larger range of multimodal work, including both simple stand-alone gestures and complex gestural patterns using a smart phone and a multi-layered Use Case-based ontology for coordinating (and using) multiple devices in a smart home setting.

Combining multimodal signals in this manner serves in three ways to improve human computer interaction.

1. It uses current state of the art tools to solve the problems that a single modality is not yet able to solve, such as ubiquitous interaction through speech despite a background full of confounding noises;
2. It leverages common processing and storage capacities to shift the majority of communicative effort from the human to the computer, and;
3. It increases the possibility of truly calm interaction as it was predicted by Weiser and Brown [1] roughly 20 years ago, through the simple expedient of allowing the user to communicate in a human-centered manner, with a computer that will treat it each user as an individual.

This last matter is of vital importance. Those of us living in the Era of Ubiquitous Computing must actively consider the consequences of the fact that we are all now surrounded by computerized systems that interact with us as though we were machines, capable of communicating constantly and consistently in a predetermined and generalized manner. We would do well to consider Mark Weiser's 18 year-old assertion that "...calm technology may be the most important design problem of the twenty-first century, and it is time to begin the dialogue"[16]. After all of these years, one might argue that it is well past time.

The first step of our future work will be to use a database of claps and commands recorded during this session to see if the commands and users can be recognized. If so, further testing will be conducted to see if the command recognizer can be generalized across participants or can even work with previously unknown users.

Concurrently, the recorded commands will be used to test a Use Case-based command ontology for smart homes. Once this testing has been completed in a laboratory setting, the ontology and multimodal commands will be tested in a real smart-home context in the Casa Vecchia smart homes [17].

The means of detecting location within an un-instrumented home proposed by Cohn et al. [18] will allow users to forego most of the navigation involved in centralized control. Future work should attempt to unify their *humantenna* toolset with a smart phone, a *Snark Circuit* for input unification, and the sort of underlying generative grammar and command ontology that will allow people a highly personalized experience of human computer interaction.

It is through small steps like these that we are approaching the Calm Technology that Weiser predicted as a condition of truly human-centered ubiquitous computing.

Acknowledgements. This work was supported in part by the Erasmus Mundus Joint Doctorate in Interactive and Cognitive Environments, which is funded by the EACEA Agency of the European Commission under EMJD ICE FPA n 2010-0012.

References

1. Weiser, M., Brown, J.S.: The Coming Age of Calm Technology. In: Denning, P.J., Metcalfe, R.M. (eds.) Beyond Calculation: The Next Fifty Years of Computing, pp. 75–85. Copernicus, New York (1997)
2. Weiser, M.: The Computer for the Twenty-First Century. Scientific American 265(3), 94–104 (1991)
3. Weiser, M.: Some Computer Science Issues in Ubiquitous Computing. Communications of the ACM 36(7), 75–84 (1993)
4. Cook, D.J., Das, S.K.: How smart are our environments? An updated look at the state of the art. Pervasive and Mobile Computing 3(2), 53–73 (2007)
5. Chan, M., Estève, D., Escriba, C., Campo, E.: A review of smart homes—Present state and future challenges. Computer Methods and Programs in Biomedicine 91(1), 55–81 (2008)
6. Holzinger, A., Ziefle, M., Röcker, C.: Human-Computer Interaction and Usability Engineering for Elderly (HCI4AGING): Introduction to the Special Thematic Session. In: Miesenberger, K., Klaus, J., Zagler, W., Karshmer, A. (eds.) ICCHP 2010, Part II. LNCS, vol. 6180, pp. 556–559. Springer, Heidelberg (2010)
7. Leitner, G., Fercher, A.J., Felfernig, A., Hitz, M.: Reducing the entry threshold of AAL systems: preliminary results from casa vecchia. In: Miesenberger, K., Karshmer, A., Penaz, P., Zagler, W. (eds.) ICCHP 2012, Part I. LNCS, vol. 7382, pp. 709–715. Springer, Heidelberg (2012)
8. Kaufmann, B., Kozeny, P., Schaller, S., Brown, J.N.A., Hitz, M.: May cause dizziness: applying the simulator sickness questionnaire to handheld projector interaction. In: Proceedings of the 26th Annual BCS Interaction Specialist Group Conference on People and Computers (BCS-HCI 2012), pp. 257–261. British Computer Society, Swinton (2012)
9. Vacher, M., Istrate, D., Portet, F., Joubert, T., Chevalier, T., Smidtas, S., Meillon, B., Lecouteux, B., Sehili, M., Chahuara, P., Méniard, S.: The sweet-home project: Audio technology in smart homes to improve well-being and reliance. In: 33rd Annual International IEEE EMBS Conference, Boston, Massachusetts, USA, pp. 5291–5294 (2011)
10. Kaschmitter, J.L., Shaeffer, D.L., Colella, N.J., McKnett, C.L., Coakley, P.G.: Operation of commercial R3000 processors in the Low Earth Orbit (LEO) space environment. IEEE Transactions on Nuclear Science 38(6), 1415–1420 (1991)
11. McNeill, D.: Hand and mind: What gestures reveal about thought. University of Chicago Press, Chicago (1992)

12. Carroll, L.: The hunting of the Snark. Macmillan, London (1876)
13. Brown, J.N.A.: Expert Talk for Time Machine Session: Designing Calm Technology as Refreshing as Taking a Walk in the Woods. In: Proceedings of IEEE International Conference on Multimedia and Expo (ICME 2012), pp. 423–423. IEEE, Melbourne (2012)
14. Brown, J.N.A., Kaufmann, B., Huber, F.J., Pirolt, K.-H., Hitz, M.: "...Language in their Very Gesture": First Steps Towards Calm Smart Home Interaction (under review)
15. Rouillard, J., Tarby, J.-C.: How to communicate smartly with your house? Int. J. Ad Hoc and Ubiquitous Computing 7(3), 155–162 (2011)
16. Weiser, M., Brown, J.S.: Designing calm technology. PowerGrid Journal 1(1), 75–85 (1996)
17. Leitner, G., Fercher, A.J.: Potenziale und Herausforderungen von AAL im ländlichen Raum. In: Demographischer Wandel - Assistenzsysteme aus der Forschung in den Markt (AAL 2011), pp. 10–15. VDE, Berlin (2011)
18. Cohn, G., Morris, D., Patel, S.N., Tan, D.S.: Your noise is my command: Sensing gestures using the body as an antenna. In: Proceedings of the SIGCHI Conference on Human Factors in Computing Systems (CHI 2011), pp. 791–800. ACM, New York (2011)

Optimizing Classroom Environment to Support Technology Enhanced Learning

Junfeng Yang[1,2], Ronghuai Huang[1], and Yanyan Li[1]

[1] R&D Center for Knowledge Engineering, Beijing Normal University, Beijing 100875 China
{yangjunfengphd,ronghuai.huang,liyy1114}@gmail.com

[2] Modern Educational Center, Hangzhou Normal University, Hangzhou 310015 China

Abstract. Researchers have found that classroom environment has close relationship to students' learning performance. When considering technology enriched classroom environment, researches are mainly on the psychological environment and the measurement of the environment. While as technology integrated in classroom, the physical classroom environment should be investigated to facilitate students' effective and engaged learning. First we carry out a survey on the current technology enriched classroom, after that we sample the Technology Involved Classroom (TIC) and Technology Uninvolved Classroom (TUC) to compare the differences between the two kinds of classroom; then we do the classroom observation and interview with teachers; finally based on the analysis of these data, we propose some solutions for optimizing the classroom environment to facilitate technology enriched learning in China.

Keywords: class environment, classroom environment, technology enriched classroom, flipped classroom, technology enhanced learning.

1 Introduction

Over the past four decades, the study of classroom environments has received increased attention by researchers, teachers, school administrators and administrators of school systems [1]. Research on the classroom environment has shown that the physical arrangement can affect the behavior of both students and teachers [2], and that a well-structured classroom tends to improve student academic and behavioral outcomes [3]. The nature of the learning environment is judged based on students' perceptual consensus about the educational, psychological, social, and physical aspects of the environment [4]. Generally, the physical, social and psychological aspects are the three dimensions of evaluating classroom environment, and there are direct associations between psychosocial environment and physical environment [5] [6]. Some well-validated and robust classroom environment instruments to measure students" perceptions are developed to measure the psychological environment in class, like Learning Environment Inventory (LEI) [7], Constructivist Learning Environment Scale (CLES) [8], What Is Happening In this Class? (WIHIC) questionnaire [9].

A. Holzinger and G. Pasi (Eds.): HCI-KDD 2013, LNCS 7947, pp. 275–284, 2013.

While as technology evolve dramatically, technology enriched learning environment can range from simple computer classrooms to extravagantly appointed classrooms equipped with computers, projectors, Internet access, and communications technology allowing for distance and real time access to a vast array of resources [10]. The use of computer and relevant digital devices has the potential to change physical and psychosocial classroom environments in either negative or positive ways. Many research have been done on the measurement of technology enriched classroom environment, and instruments like Constructivist Multimedia Learning Environment Survey (CMLES), New Classroom Environment Instrument (NCEI), and Technology-rich Outcomes-focused Learning Environment Inventory (TROFLEI), the Technology Integrated Classroom Inventory (TICI) are proposed and validated [11].

Although these researches and instruments could help to understand the physical and psychological classroom environment, they could not indicate how to construct and equip a classroom to facilitate effective and engaged learning and cultivate students' 21st survival skills. Especially in mainland China, there is few research concerning how to optimize today's classroom environment to match the needs of the new generation students from the perspective of effective teaching and learning. So in this research we try to carry out a survey on the current technology enriched classroom and then propose some solutions for optimizing the classroom environment to facilitate technology enriched learning.

2 Literature Review

In the age of information, both the physical classroom environment and the psychological classroom environment could be optimized through equipping "right" ICT and fusing "right" pedagogy.

In recent years, policy makers, institutions and researchers have realized the priority of classroom environment changing and they have initiated some projects on the improvements of classroom environment and the construction of future classroom. MIT initiated Technology Enhanced Active Learning (TEAL) project in 2000 to involve media-rich software for simulation and visualization in freshman physics carried out in a specially redesigned classroom to facilitate group interaction [12]. The student-centered activities for large enrollment undergraduate programs (SCALE-UP) project was initiated in North Carolina State University, with the aim to establish a highly collaborative, hands-on, computer-rich, interactive learning environment for large, introductory college courses [13] Kansas State in America initiated Technology Rich Classrooms project, and after the project Ault and Niileksela (2009) found that including technology in a classroom, training teachers how to use the technology, and providing support for technology use may change many aspects of learning [14]. Though these projects were able to demonstrate that the combination of newly designed classrooms and active learning approaches contributed to improving student learning achievements, but their research were lack of evidence for the findings because few of them isolated the relative effects of either space or pedagogy in research design.

According to Chinese scholar, the connotation of the classroom consists of three levels: (1) classroom is the physical environment (2) classroom is teaching activities, (3) classroom is integration of curriculum and teaching activities [15]. The classroom is not only a physical environment but also should provide support for carrying out various teaching and learning activities. From the late 90s, China started education information infrastructure construction in large-scale. After more than 10 years of construction, educational informatization has made remarkable achievements and the understanding of e-education has enhanced more than before [16]. Most teachings in class have transformed from the original "blackboard + chalk" mode to the "computer + projection" mode, but the teaching mode has not changed as we expected yet [17]. In some ways, the classroom and facilities have evolved dramatically, but in many ways they remain mired in the past. Wu (1998) indicated that the classrooms are mostly using the traditional seating layout [18]; Li (2006) expressed the functional advantages of the technology enriched classroom are not fully realized [19].

3 Research Method

This research involved a combination of a variety of methods. Whereas the classroom environment was surveyed with a questionnaire developed by the researchers, the class are observed with a classroom observation tool ICOT, the teachers was interviewed through an interview protocol.

3.1 Procedures

In this research we are trying to find the challenges in today's technology enriched learning environment, and then propose solutions to optimize the classroom environment. The procedure of this research can be divided into four steps as follows, as shown in Fig.1.

(1) Conduct a large scale survey on classroom environment from the perspective of teachers
(2) Use the sampling rules to select TIC and TUC. Technology Involved Class (TIC) refer to the class, in which internet is available and digital resources can be accessed conveniently, and digital technologies contribute to facilitate teaching and learning. Technology Uninvolved Class (TUC) refer to the traditional class in which internet is unavailable or digital resources could not be accessed conveniently, and technologies could not contribute to facilitate teaching and learning.
(3) Compare the differences between TIC and TUC.
(4) Go to classroom to observe the teaching practice in TIC and TUC, and after class to have an interview with teachers.

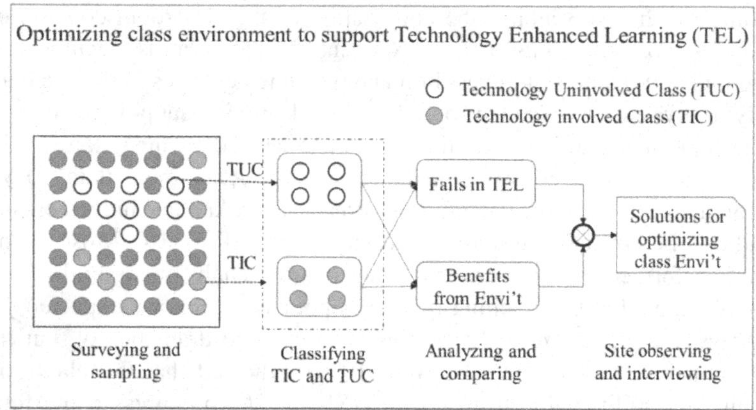

Fig. 1. Procedures of this research

3.2 Research Tools

Data was collected by ISTE Classroom Observation Tool (ICOT), and the Classroom Environment Questionnaire (CEQ) and the Focus Group Interview Protocol (FGIP) designed by ourselves.

The **ISTE Classroom Observation Tool** [20] is a computer-based rubric designed to help observers assess the nature and extent of technology integration in classroom, which is developed by International Society for Technology in Education (ISTE).

The **Classroom Environment Questionnaire** (CEQ) was developed based on the SMART classroom model proposed by Huang et. al. (2012) [21], as shown in Fig. 2.

Showing of learning and instructional content concern with the teaching and learning material's presenting capabilities in classroom. Not only should the learning contents be seen clearly, but also it should be suitable to learners 'cognitive characteristics. ***Managing of physical environment/instructional materials/students behavior*** represents diverse layouts and the convenience of management of the classroom. The equipment, systems, resources of classroom should be easy managed, including layout of the classroom, equipment, physical environment, electrical safety, network, etc. ***Accessing to digital resources*** represents convenience of digital resources and equipment accessing in the classroom, which includes resource selection, content distribution and access speed.

Real-time **interaction** ***and supporting technologies*** represents the ability to support the teaching/learning interaction and human-computer interaction of the classroom, which involves convenient operation, smooth interaction and interactive tracking. ***Tracking learning process/ environment*** represents tracking of the physical environment, instructional process and learning behavior in classroom.

According to the SMART classroom model, we developed the CEQ which consists of 65 questions, including the 11 questions about basic information and 54 questions on the dimensions of classroom environment. We use "content validity ratio" (CVR) to do the validity test of the questionnaire. Five experts (outstanding teachers and experts on subjects) are invited to give scores on the validity of the questionnaire. After collecting their scores and excluding some items that are not qualified, we finally use the **48** items with CVR over 0.7, as shown in Table 1.

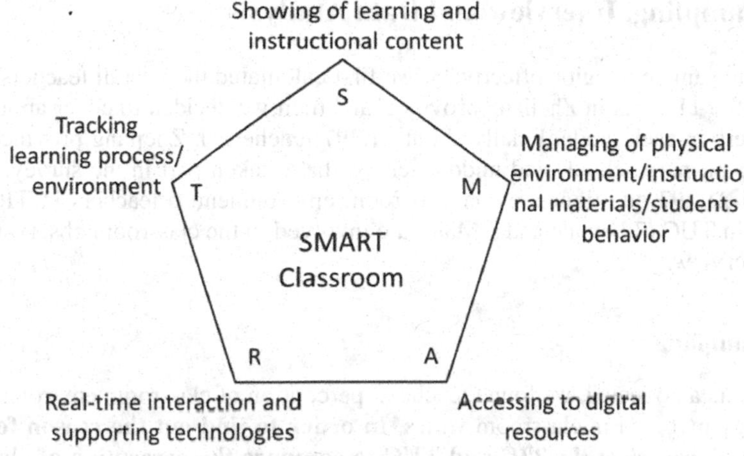

Fig. 2. SMART classroom model

Table 1. Dimensions and Items of CEQ

Dimensions	Items	Numbers of Items
Showing of learning and instructional content	Instructional showing, Learning showing, Audio effects	12
Managing of physical environment/instructional materials/students behavior	Physical environment, Instructional materials, students behavior and action	10
Accessing to digital resources	Internet, Instructional resources, Learning resources	9
Real-time interaction and supporting technologies	Instructor-students interaction, students-students interaction, Human-computer interaction	9
Tracking learning process/ environment	Instructional process, leaning behavior, other environmental factors	8

10 respondents are selected to fill in the same questionnaire again after a week to the test the reliability of the questionnaire, which results in the correlation coefficients ranging from 0.89 to 0.99.

Focus group interviews are a multi-faceted instrument that can be used alone, or in conjunction with other research methods allowing the researcher to delve more deeply into the study of a phenomenon and provide enhanced understanding to the research [22]. The FGIP consists of five parts which include showing content, managing environment, accessing resources, real-time interacting and tracking environment derived from the S.M.A.R.T. classroom model.

4 Sampling, Interview and Data Analysis

In order to sample a region effectively, we first calculated the overall teachers in each grade in the 11 cities in Zhejiang province, and then we decided to cover about 1/4 of all teachers in each grade. Finally about 21,397 teachers in Zhejiang province China including primary schools and middle schools have taken part in the survey, and we collected 21,397 questionnaires on classroom environment. 6 teachers in TIC and 6 teachers in TUC (7 Female and 5 Male) are involved in the classroom observation and focus interview.

4.1 Sampling

From the data collected, we found teachers' perception of classroom environment and technology involved in classroom varies. **In order to find out the reason for these differences, we select the TIC and TUC to compare the perception of classroom environment and the technology involved differences, which will enlighten us some solutions on optimizing classroom environment to support technology enhanced learning.** The sampling rules are: (1) Computer(s) and relevant digital devices are available in classroom; (2) Internet are available in classroom; (3) Digital resources are easy to access in classroom; (4) ICT are used to dispatch and collect learning materials frequently in class; (5) Students' works could be presented by using ICT in class frequently.

4.2 Comparison of TIC and TUC

Finally, 4046 out of 21,397 are selected as TIC (account for 18.9% of total), and 3376 are selected as TUC (account for 15.8% of total). From the comparison of TIC and TUC, we found:

(1) For classroom seating layout, 80.5% teachers in TIC express that the layout are conventional straight row layout, and 89.2% teachers in TUC express that. Teachers in TIC have adopted more U and O seating arrangements. 15.2% of teachers in TIC compared with 8.4% of teachers in TUC adopt U seating arrangement and 4.2% compared with 2.4% adopt O seating arrangement. Teachers in TIC change the classroom seating layout more often according to the pedagogy the use in class.
(2) For teaching console, 43.0% of teachers in TIC compared with 12.6% of teachers in TUC often change the place of teaching console and the classroom seating layout in order to carry out different teaching activities.42.7% of TIC compared with 59.2% of TUC would like to change the place of teaching console, but they con not because the console is fixed in front of the classroom.
(3) For showing content, 73.7% of teachers in TIC express that students could see clearly the showing content on the projection screen, while only 52.4% of teachers in TUC express this. 97.3% of teachers in TIC and 91.8% of teachers in TUC express PPT courseware could facilitate students' effective learning. For the reason why PPT courseware could not facilitate student's effective learning, most teachers express that "no time and no skill to do PPT" is the common reason.

31.1% of teachers in TUC express that the PPT is not good for student's digesting knowledge, while only 20.6% TIC think this.

(4) For technology enhanced interaction, we find TIC are more positive in "Students always learn collaboratively to finish the assignments in class", "Group students always learn together via interaction", "Students have more opportunities to discuss issues with the teacher", "Student have more opportunities to discuss with each other". These four questions are rating items using a five-point likert scale (5=Strongly Agree, 1=Strongly Disagree). 1.78, 1.81, 1.83, 1.84 are the four results in TIC; 2.39, 2.47, 2.65, 2.68 are the four results in TUC

(5) There are more senior teachers in TIC. Senior teachers take account 40.3% of teachers in TIC, but only 24.5% in TUC. In china, only a teacher have good pedagogy knowledge, domain knowledge and research ability could qualify himself to be a senior teacher. When consider the age of teaching and the degree or diploma, there is no significant difference.

Generally speaking, **the results are following**: (1) the seating layout is mainly conventional straight row layout and fixed; (2) the teaching console where the teaching computer and control system are located, is normally fixed in front of the classroom; (3) there are a larger proportion of teachers in TIC think students could see clearly the showing content on the projector screen than TUC; (4) Generally teachers' attitudes are positive to PPT courseware's effects on student's effective learning; (5) the deeper technology integration into classroom, the more collaborative learning strategy and digital technology are uses to facilitate interactions between teachers and students; (6) technology integration in classroom requires pedagogy knowledge, domain knowledge and research ability.

From the comparison, we find that the **physical classroom environment has a significant influence on teacher's teaching methods adoption,** which inspires us to think if the physical classroom environment could be improved to better facilitate teacher's teaching. So we went into classroom to do site observation in 6 TICs and 6 TUCs, and after that we carried out an interview with the teacher.

4.3 Site Observation and Interviews

In order to deeply understand the differences and to investigate the influences of physical environment on teacher's teaching, we first go into 6 TICs (2 Math, 2 English, 2 Chinese) and 6 TUCs (2 Math, 2 English, 2 Chinese) to observe the detail in-class behaviors, and then conduct interviews with the 12 teachers (4 Math, 4 English, 4 Chinese). All TIC are equipped with computers, projectors, wireless internet, interactive white board, Apple TV and other relevant digital technologies, while most TUC are traditional classroom with basic computers and projectors.

After the observation, focus group interview were carried out separately on 6 teachers in TIC and 6 teachers in TUC. The focus group interview protocol is based on the five dimensions of classroom environment. From the observation and interview, we find the following issues categorized into the five dimensions of classroom environment.

(1) **For showing content.** Most teachers , no matter in TIC or TUC express that because there is no curtain in classroom and the light from outside is so strong, some students could not see the content on the projector screen. When talking about the PPT usage, some teachers in TIC say they doubt whether students have enough time to take notes or digest knowledge before teachers change to the next slide, and some teachers in TUC are afraid of using PPT because it will distract student's attention.

(2) **For managing environment.** Almost all teachers express they are willing to adopt different teaching strategies to meet the teaching objects and students' needs, but they feel it a little difficult to conduct collaborative learning because of the conventional straight row layout, so they always want to change the seating layout to U shape. Teachers in TIC also express the inconvenience of the teaching console, which is evidenced from the observation that most teachers stay before the teaching console to manage computers for most time of the class. Teachers in TUC always complain the breakdown of computers and projectors.

(3) **For accessing resource.** Some teachers in TIC express they have built the website for sharing digital resources with students, and students could access to resources in class, which make it easier to adopt multiple teaching strategies, such as inquiry learning, collaborative learning, self-directed learning, etc. Students in TUC could not get access to digital resources. Teachers adopt more student-centered teaching method in TIC than in TUC, and students are more engaged in TIC than in TUC.

(4) **For real-time interaction.** Questions, discussion in peers, retell, role play, model, etc. are used to promote interactions between teachers and student. Students always show their learning outcome in TIC via airplay devices, while students seldom have opportunities to show the learning outcome in TUC. Interactive white board, interactive courseware and synchronous communication tools are used to promote communication between teachers and students in TIC.

(5) **For tracking environment.** Teachers both in TIC and TUC think it is necessary to record and analyze students learning behavior and teachers teaching behavior. From the interview, teachers have mentioned that the students' behavior should be recorded from the time students engaged, the time students take part in activities, the time students do practice, etc., and the teacher's behavior could be recorded and analyzed from the language in class, the teaching content, the activities conducted, the time using technology, etc.

The observation of TIC and TUC tell us some impressive results. In TIC, teachers often divided the class into several groups and conducted collaborative learning; while in TUC teachers always talk and students always listen and take notes. In TIC class, teacher use different kinds of technologies, but sometimes it seems the teachers is a little busy on technology; while in TUC class, teachers seldom use technology except for the projector for showing content.

5 Conclusion

The overall context for discussing our results reflects four important points based on the survey, observation and interview.

First, classrooms equipped with computers and projectors is the basic configuration of a technology rich classroom currently in mainland China, and some classrooms in top K-12 schools are equipped with different kinds of technology to facilitate teaching and learning, such as Apple TV, IPads, Interactive White Board, etc.

Second, from the schools participated in the research, we found technology may facilitate learning in case the technology enriched classroom was designed based on the pedagogy in association with "right" learning resources, "right" seating layout, place of teaching console and projector screen, etc.

Third, the five dimensions, such as showing content, managing environment, accessing resources, real-time interaction, tracking environment, can be taken into consideration in optimizing classroom environment.

Fourth, it is necessary for teachers to be aware of the potential risks for using slides. The teachers will perform better in technology enriched classroom if they have a fully understanding of the new generation students' learning needs, and have more technological knowledge and pedagogical knowledge. This results is coincided with Mishra and Koehler's TPACK model [23].

Acknowledgments. This research work is supported by Beijing Digital School (BDS) program the "The Survey on Students' Online Life Style and Strategy Study" (MAA10001), and the 2012 Zhejiang educational project "the Study on Promoting the Faculty Development under the TPACK framework (Y201223270)".

References

1. Aldridge, J.M., Dorman, J.P., Fraser, B.J.: Use of Multitrait-Multimethod Modelling to Validate Actual and Preferred Forms of theTechnology-Rich Outcomes-Focused Learning Environment Inventory (Troflei). Australian Journal of Educational & Developmental Psychology 4, 110–125 (2004)
2. Savage, T.V.: Teaching self-control through management and discipline. Allyn and Bacon, Boston (1999)
3. MacAulay, D.J.: Classroom environment: A literature review. Educational Psychology 10(3), 239–253 (1990)
4. Dunn, R.J., Harris, L.G.: Organizational dimensions of climate and the impact on school achievement. Journal of Instructional Psychology 25, 100–115 (1998)
5. Zandvliet, D.B., Fraser, B.J.: Physical and psychosocial environments associated with networked classrooms. Learning Environments Research 8(1), 1–17 (2005)
6. Zandvliet, D.B., Straker, L.M.: Physical and psychosocial aspects of the learning environment in information technology rich classrooms. Ergonomics 44(9), 838–857 (2001)
7. Fraser, B.J., Anderson, G.J., Walberg, H.J.: Assessment of learning environments: Manual for Learning Environment Inventory (LEI) and My Class Inventory (MCI) (3rd version). Western Australian Institute of Technology, Perth (1982)
8. Taylor, P.C., Fraser, B.J., Fisher, D.L.: Monitoring constructivist classroom learning environments. International Journal of Educational Research 27, 293–302 (1997)

9. Fraser, B.J., McRobbie, C.J., Fisher, D.L.: Development, validation and use of personal and class forms of a new classroom environment instrument. Paper Presented at The Annual Meeting of the American Educational Research Association, New York (1996)
10. Ott, J.: The new millennium. Information Systems Security 8(4), 3–5 (2000)
11. Wu, W., Chang, H.P., Guo, C.J.: The development of an instrument for a technology-integrated science learning environment. International Journal of Science and Mathematics Education 7(1), 207–233 (2009)
12. Dori, Y., Belcher, J.: How does technology-enabled active learning affect undergraduate students' understanding of electromagnetism concepts? The Journal of the Learning Sciences 14, 243–279 (2005)
13. Beichner, R., Saul, J., Abbott, D., Morse, J., Deardorff, D., Allain, R., et al.: Student-Centered Activities for Large Enrollment Undergraduate Programs (SCALE-UP) project. In: Redish, E., Cooney, P. (eds.) Research-Based Reform of University Physics, pp. 1–42. American Association of Physics Teachers, College Park (2007)
14. Ault, M., Niileksela, C.: Technology Rich Classrooms: Effect of the Kansas Model Contact author: Jana Craig Hare, MSEd 1122 West Campus Road, 239JRP Lawrence, KS 66045 (2009)
15. Wang, J.: Introductory remarks on classroom research. Educational Research 6, 80–84 (2003)
16. Wang, J.: The review on educational information and the informationization education. E-education Research 15(9), 5–10 (2011)
17. Huang, R.: ICT in Education Promotes Current Educational Change: Challenges and Opportunities. Chinese Educational Technology (1), 36–40 (2011)
18. Wu, K.: Educational Sociology, pp. 345–348. People's Education Press, Beijing (1998)
19. Li, X.: The problems and countermeasures on the applications in multimedia network classroom. China Modern Education Equipment (3), 25–27 (2006)
20. ISTE. ISTE Classroom Observation Tool (2012), http://www.iste.org/learn/research-and-evaluation/icot (retrieved from January 31, 2013)
21. Huang, R., Hu, Y., Yang, J., Xiao, G.: The Functions of Smart Classroom in Smart Learning Age (in Chinese). Open Education Research 18(2), 22–27 (2012)
22. Sinagub, J.M., Vaughn, S., Schumm, J.S.: Focus group interviews in education and psychology. Sage Publications, Incorporated (1996)
23. Mishra, P., Koehler, M.: Technological pedagogical content knowledge: A framework for teacher knowledge. The Teachers College Record 108(6), 1017–1054 (2006)

A Smart Problem Solving Environment

Nguyen-Thinh Le and Niels Pinkwart

Department of Informatics
Clausthal University of Technology
Germany
{nguyen-thinh.le,niels.pinkwart}@tu-clausthal.de

Abstract. Researchers of constructivist learning suggest that students should rather learn to solve real-world problems than artificial problems. This paper proposes a smart constructivist learning environment which provides real-world problems collected from crowd-sourcing problem-solution exchange platforms. In addition, this learning environment helps students solve real-world problems by retrieving relevant information on the Internet and by generating appropriate questions automatically. This learning environment is smart from three points of view. First, the problems to be solved by students are real-world problems. Second, the learning environment extracts relevant information available on the Internet to support problem solving. Third, the environment generates questions which help students to think about the problem to be solved.

Keywords: constructivist learning, information extraction, question generation.

1 Introduction

A smart learning environment may provide adaptive support in many forms, including curriculum sequencing or navigation [1], student-centered e-learning settings [2], or intelligent support for problem solving [3]. For the latter class, a smart learning environment should be able to provide students with appropriate problems and intervene in the process of problem solving when necessary. Researchers of constructivist learning suggest that students should learn with real-world problems, because real-world problems are motivating and require the student to exercise their cognitive and metacognitive strategies [4]. In the opposite, traditional learning and teaching approaches typically rely on artificial (teacher-made up) problems. Often though, students can then solve a problem which is provided in a class, but would not be able to apply learned concepts to solve real-world problems.

Hence, Jonassen [5] suggested that students should rather learn to acquire skills to solve real-world problems than to memorize concepts while applying them to artificial problems and proposed a model of constructivist learning environments. We adopt this model and propose a learning environment which supports students as they solve real-world problems collected from various crowd-sourcing problem-solution exchange platforms. For example, the platform *Stack overflow*[1] is a forum

[1] http://stackoverflow.com/

A. Holzinger and G. Pasi (Eds.): HCI-KDD 2013, LNCS 7947, pp. 285–292, 2013.

for programmers for posting programming problems and solutions; The platform *Wer-Weiss-Was*[2] provides a place for posting any possible problem and users who have appropriate competence are asked to solve a problem or to answer a question. These crowd-sourcing platforms can provide the learning environment to be developed with real-world problems. In order to coach and to scaffold the process of student's problem solving, the learning environment is intended to provide two cognitive tools: 1) an information extraction tool, and 2) a question generation tool.

The information extraction tool is required to provide students selectable information when necessary to support meaningful activity (e.g., students might need information to understand the problem or to formulate hypotheses about the problem space). The process of seeking information may distract learners from problem solving, especially if the information seeking process takes too long and if found information is not relevant for the problem being investigated. Therefore, the information extraction tool is designed to help the student to select relevant information. It can crawl relevant websites on the Internet and represent required information in a structured form.

Land [6] analyzed the cognitive requirements for learning with resource-rich environments and pointed out that the ability of identifying and refining questions, topics or information needs is necessary, because the process of formulating questions, identifying information needs, and locating relevant information resources forms the foundation for critical thinking skills necessary for learning with resource-rich environments. However, research has reported that students usually failed to ask questions that are focusing on the problem being investigated. For example, Lyons and colleagues reported that middle school children using the WWW for science inquiry failed to generate questions that were focused enough to be helpful [7]. For this reason, a question generation tool can potentially be helpful for students during the process of gathering relevant information. If a student is not able to come up with any question to investigate the problem to be solved, the learning environment should generate relevant questions for the student.

The constructivist learning environment to be developed is smart and a novel contribution due to three features. First, it deploys real-world problems for students to acquire problem solving skills. Second, even though information extraction is an established technology, it has rarely been deployed for enhancing the adaptive support for problem solving in smart learning environments. Third, while automatic question generation has also been researched widely, strategies of deploying question generation into educational systems are rarely found in literature [8]. This learning environment can be regarded as an open-ended learning environment which supports students acquire problem solving skills using information technology [6].

2 A Smart Constructivist Learning Environment

Currently, we are initiating a project which promotes the idea of learning by solving real-world problems. For this purpose, we develop a learning environment which

---

[2] http://www.wer-weiss-was.de/

collects real-world problems from crowd-sourcing platforms. Real-world problems occur almost every day, e.g., "in my area, it is snowing heavily. How can I bind a snow chain for my car?", "my bank offers me a credit of 100 000 Euro for a period of 10 years with an interest rate on 5%. Should I choose a fixed rate mortgage or a variable rate mortgage? Which one is better for me?", "I have a blood pressure of 170/86. Could you diagnose whether I have to use medicine?"

Two actors will play roles in this learning environment: instructors and students. The roles of instructors who are the expert of a specific learning domain include choosing the category of problems for their class and selecting real-world problems which are relevant for the learning topic being taught and at the right complexity level for their students. The challenge might here be how the platform should support instructors to choose appropriate problems, because if it takes too much time to search for relevant problems, instructors might give up and think of artificial problems. Through human instructors, real-world problems which are tailored to the level of their students can be selected. It is unlikely that the learning system might be able to select the right problem automatically for a given student model (this would require that a problem has a very detailed formal description, including complexity level).

Students can solve problems assigned by their instructors by themselves or collaboratively. They can use two cognitive tools during problem solving: information extraction for retrieving relevant information available on the Internet, and automatic question generation for helping students ask questions related to the problem to be solved. After attempting to solve these problems, students can submit their solution to the system. There, they can get in discussion with other students who are also interested in solving these problems. Let's name our learning environment SMART-SOLVER. In the following we illustrate how these tools can be deployed.

A university professor of a course *Banking and Investment* has collected the following problem from the SMART-SOLVER platform for his students:

"I want to buy a house and a bank for a loan of 100.000 Euro. The bank makes two offers for a yearly interest rate of 5%: 1) Fixed rate mortgage, 2) Variable rate mortgage. Which offer is better for me?"

John is a student of this course. He is asked to solve this problem using SMART-SOLVER. His problem solving scenario might be illustrated in Figure 1.

Peter is also a student of this course. However, he does not have an as good performance as John and is stuck. He does not know what kind of information or questions can be input into the information extraction tool. Therefore, he uses the question generation tool which proposes him several questions. The question generation uses the problem text as input and might generate the following questions which help Peter to understand basic concepts: "What is fixed rate mortgage?", "What is variable rate mortgage?" After receiving these questions, Peter might have a look into his course book or input these questions into the information extraction tool in order to look for definitions of these investment concepts.

John tries to solve the problem above using SMART-SOLVER. First, he uses the information extraction component to look for formulas for calculating the two mortgage options. He might have learned the concepts "fixed rate mortgage" and "variable rate mortgage" in his course. However, he might still have not understood these concepts; therefore John inputs the following questions into the information extraction tool:

- *"What is fixed rate mortgage?"*
- *"What is variable rate mortgage?"*

SMART-SOLVER searches on the Internet and shows several definitions of these concepts (not the whole websites). After studying the definitions, John may have understood the concepts and may want to calculate the two loan options. Again, he uses the information extraction tool in order to search for mathematical formulas. John might input the following questions into the tool:

- *"How is fixed rate mortgage calculated?"*
- *"How is variable rate mortgage calculated?"*

Using the formulas extracted from the Internet, John calculates the total interest amount for each loan option. He analyzes the advantage and disadvantage of each option by comparing the total amount of interest rate.

Fig. 1. A learning scenario using the information extraction tool

3 Architectural Approach

The architecture of the learning environment being proposed consists of five components: a user interface for students, a user interface for instructors, a database of real-world problems, an information extraction component, and a question generation component (Figure 2).

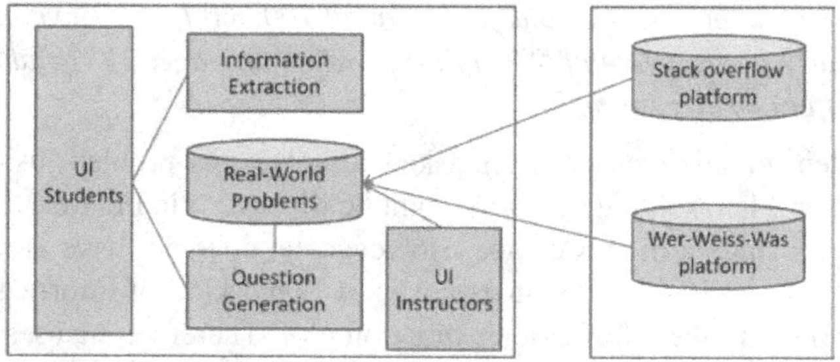

Fig. 2. The architecture of the smart learning environment

The database is connected with one or more crowd-sourcing platforms (e.g., Wer-Weiss-Was or Stack overflow) in order to retrieve real-world problems. The user interface for instructors is provided to support instructors in choosing appropriate problems for their students according to the level of their class. The user interface for students depends on the domain of studies. For each specific domain, a specific form for developing solutions should be supported, e.g., for the domain of law, the learning environment could provide tools for users to model an argumentation process as a graph. While attempting to solve problems, students can retrieve relevant information from the Internet by requesting the information extraction component, or they can ask the system to suggest a question via the question generation component. In the following, we will explain how these two components (information extraction and question generation) can be developed in order to make the learning environment in line with the constructivist learning approach.

3.1 Information Extraction

In order to extract relevant information on the Internet, we usually have to input some keywords into a search engine (e.g., Google or Bing). However, such search engines would find a huge amount of web pages, which contains these keywords, but do not necessarily provide relevant information for a task at hand.

Information extraction techniques can be used to automatically extract knowledge from text by converting unstructured text into relational structures. To achieve this aim, traditional information extraction systems have to rely on a significant amount of human involvement [9]. That is, a target relation which represents a knowledge structure is provided to the system as input along with hand-crafted extraction patterns or examples. If the user needs new knowledge (i.e., other relational structures) it is required to create new patterns or examples. This manual labor increases with the number of target relations. Moreover, the user is required to explicitly pre-specify each relation of interest. That is, classical information extraction systems are not scalable and portable across domains.

Recently, Etzioni and colleagues [10] proposed a so-called Open Information Extraction (OIE) paradigm that facilitates domain independent discovery of relations extracted from text and readily scales to the diversity and size of the Web corpus. The sole input to an OIE system is a corpus, and its output is a set of extracted relations. A system implementing this approach is thus able to extract relational tuples from text. The Open Information Extraction paradigm is promising for extracting relevant information on the Internet: TextRunner was run on a collection of 120 million web pages and extracted over 500 million tuples and achieved a precision of 75% on average [10]. Etzioni and colleagues suggested that Open Information Extraction can be deployed in three types of applications. The first application type includes question answering: the task is providing an answer to a user's factual question, e.g. "What kills bacteria?" Using the Open Information Extraction, answers to this question are collected across a huge amount of web pages on the Internet. The second application type is opinion mining which asks for opinion information about particular objects (e.g., products, political candidates) which is available in blog posts, reviews, or other texts. The third class of applications is fact checking which requires identifying claims that are in conflict with knowledge extracted from the Internet. The first type of applications using Open Information Extraction meets our requirement for developing an information extraction tool which helps students to submit questions for extracting relevant information.

3.2 Question Generation

Before students use the information gathering tool to retrieve relevant information for their problem, first they have to know what kind of information they need. Some of them may be stuck here. In this case, they can use the question generation tool which generates appropriate questions in the context of the problem being solved.

Graesser and Person [11] proposed 16 question categories for tutoring (verification, disjunctive, concept completion, example, feature specification, quantification, definition, comparison, interpretation, causal antecedent, causal consequence, goal orientation, instrumental/procedural, enablement, expectation, and judgmental) where the first 4 categories were classified as simple/shallow, 5-8 as intermediate and 9-16 as complex/deep questions. In order to help students who are stuck with a given problem statement, it may be useful to pose some simple or intermediate questions first. For example: "**What** is fixed rate mortgage?" (definition question), "**Does** a constant monthly rate include repayment and interest?" (verification question). According to Becker et al. [12], the process of question generation involves the following issues:

- Target concept identification: Which topics in the input sentence are important so that questions about these make sense?
- Question type determination: Which question types are relevant to the identified target concepts?
- Question formation: How can grammatically correct questions be constructed?

The first and the second issue are usually solved by most question generation systems by using different techniques in the field of natural language processing (NLP): parsing, simplifying sentence, anaphor resolution, semantic role labeling, and named entity recognizing. For the third issue, namely constructing questions in grammatically correct natural language expression, many question generation systems applied transformation-based approaches to generate well-formulated questions [13]. In principle, transformation-based question generation systems work through several steps: 1) delete the identified target concept, 2) a determined question key word is placed on the first position of the question, 3) convert the verb into a grammatically correct form considering auxiliary and model verbs. For example, the question generation system of Varga and Le [13] uses a set of transformation rules for question formation. For subject-verb-object clauses whose subject has been identified as a target concept, a "Which Verb Object" template is selected and matched against the clause. By matching the question word "Which" replaces the target concept in the selected clause. For key concepts that are in the object position of a subject-verb-object, the verb phrase is adjusted (i.e., auxiliary verb is used).

The second approach, which is also employed widely in several question generation systems, is template-based [14]. The template-based approach relies on the idea that a question template can capture a class of questions, which are context specific. For example, Chen et al. [14] developed the following templates: "*What would happen if* <X>?" for conditional text, "*When would* <X>*?*" and "*what happens* <*temporal-expression*>*?*" for temporal context, and "*Why* <*auxiliary-verb*> <*X*>*?*" for linguistic modality, where the place-holder <X> is mapped to semantic roles annotated by a semantic role labeler. These question templates can only be used for these specific entity relationships. For other kinds of entity relationships, new templates must be defined. Hence, the template-based question generation approach is mostly suitable for applications with a special purpose. However, to develop high-quality templates, a lot of human involvement is expected.

From a technical point of view, automatic question generation can be achieved using a variety of natural language processing techniques which have gained wide acceptance. Currently, high quality shallow questions can be generated from sentences. Deep questions, which capture causal structures, can also be modeled using current natural language processing techniques, if causal relations within the input text can be annotated adequately. However, successful deployment of question generation in educational systems is rarely found in literature. Currently, researchers are focusing more on the techniques of automatic question generation than on the strategies of deploying question generation into educational systems [8].

4 Discussion and Conclusion

We have proposed a vision and an architectural framework for a learning environment based on the constructivist learning approach. This learning environment is smart due to three characteristics: 1) this environment provides authentic and real-world problems, 2) the problem solving process performed by students are supported by exploration using the information gathering tool, and 3) the reflection and thinking process is supported by the question generation tool.

We are aware that some real-world problems might be overly complex especially for novice students. However, real-world problems can range from simple to highly complex – some of them might even be appropriate for students of elementary schools. In addition, since the learning environment being proposed provides cognitive tools (information extraction and question generation) which scaffold the process of problem solving, we think that using this learning environment, by solving real-world problems, students may improve their problem solving skills which can be used later in their daily life.

Numerous research questions can be identified in the course of developing this proposed learning environment. For instance, how should real-world problems be classified so that instructors can select appropriate problems easily? With respect to research on question generation with a focus on educational systems, several research questions need to be investigated, e.g., if the intent of a question is to facilitate learning, which question taxonomy (deep or shallow) should be deployed? Given a student model, which question type is appropriate to pose the next question to the student? Another area of deploying question generation in educational systems may be using model questions to help students improve the skill of creating questions, e.g., in the legal context. With respect to research on information extraction, several questions will arise, e.g., how should the problem solving process be designed so that students request appropriate information for solving an assigned problem? How much information should be retrieved for problem solving? We are sure, this list of research questions is not complete.

The contribution of this paper is twofold. First, it proposes a smart constructivist learning environment which enables students to solve real-world problems. Using this learning environment, students request the system for relevant information from the

Internet and the system can generate questions for reflection. Second, the paper identifies challenges which are relevant for deploying information extraction and question generation technologies for building the learning environment.

References

1. Yudelson, M., Brusilovsky, P.: NavEx: Providing Navigation Support for Adaptive Browsing of Annotated Code Examples. In: The 12th International Conference on AI in Education, pp. 710–717. IOS Press (2005)
2. Motschnig-Pitrik, R., Holzinger, A.: Student-Centered Teaching Meets New Media: Concept and Case Study. Journal of Edu. Technology & Society 5(4), 160–172 (2002)
3. Le, N.T., Menzel, W.: Using Weighted Constraints to Diagnose Errors in Logic Programming-The Case of an Ill-defined Domain. Journal of AI in Edu. 19, 381–400 (2009)
4. bRANSFORD, J.D., Sherwood, R.D., Hasselbring, T.S., Kinzer, C.K., Williams, S.M.: Anchored Instruction: Why We Need It and How Technology Can Help. In: Cognition, Education, Multimedia - Exploring Ideas in High Technology. Lawrence Erlbaum, NJ (1990)
5. Jonassen, D.H.: Designing Constructivist Learning Environments. In: Reigeluth, C.M. (ed.) Instructional Design Theories and Models: A New Paradigm of Instructional Theory, vol. 2, pp. 215–239. Lawrence Erlbaum (1999)
6. Land, S.M.: Cognitive Requirements for Learning with Open-ended Learning Environments. Educational Technology Research and Development 48(3), 61–78 (2000)
7. Lyons, D., Hoffman, J., Krajcik, J., Soloway, E.: An Investigation of the Use of the World Wide Web for On-line Inquiry in a Science Classroom. Presented at The Meeting of the National Association for Research in Science Teaching (1997)
8. Mostow, J., Chen, W.: Generating Instruction Automatically for the Reading Strategy of Self-questioning. In: Proceeding of the Conference on AI in Education, pp. 465–472 (2009)
9. Soderland, S.: Learning Information Extraction Rules for Semi-structured and Free-text. Machine Learning 34(1-3), 233–272 (1999)
10. Etzioni, O., Banko, M., Soderland, S., Weld, D.S.: Open Information Extraction From the Web. Communication ACM 51(12), 68–74 (2008)
11. Graesser, A.C., Person, N.K.: Question Asking during Tutoring. American Educational Research Journal 31(1), 104–137 (1994)
12. Becker, L., Nielsen, R.D., Okoye, I., Sumner, T., Ward, W.H.: What's Next? Target Concept Identification and Sequencing. In: Proceedings of the 3rd Workshop on Question Generation, held at the Conference on Intelligent Tutoring Systems, pp. 35–44 (2010)
13. Varga, A., Le, A.H.: A Question Generation System for the QGSTEC 2010 Task B. In: Proc. of the 3rd WS. on Question Generation, held at the ITS Conf., pp. 80–83 (2010)
14. Chen, W., Aist, G., Mostow, J.: Generating Questions Automatically From Informational Text. In: Proceedings of the 2nd Workshop on Question Generation, held at the Conference on AI in Education, pp. 17–24 (2009)

Collaboration Is Smart: Smart Learning Communities

Gabriele Frankl and Sofie Bitter

Alpen-Adria-Universität Klagenfurt, Austria
{gabriele.frankl,sofie.bitter}@aau.at

Abstract. Technological advances in the last decades have significantly influenced education. Smart Learning Environments (SLEs) could be one solution to meet the needs of the 21st century. In particular, we argue that smart collaboration is one fundamental need. This paper deals with the question what 'smart' is and why a SLE's design has to consider collaboration. Drawing on various theories, we argue that the community aspect plays a vital role in successful learning and problem solving. This paper outlines the benefits for the community and all parties involved (defined as a win-for-all or win^n-solution), as well as drivers that might influence collaboration. Design principles for SLEs, Smart Learning Communities (SLCs) and finally the conclusion close the paper.

Keywords: smart learning environment, smart learning communities, collaboration, social learning, win for all, design principles.

1 Introduction

In the last decades, we have been faced with tremendous technological advancements that have impacted greatly on the way people engage, interact and communicate with each other [1]. These developments in particular also affect teaching and learning as well as learning environments. Previous research studies [see references 1–5] in the context of smart learning environments (SLEs) have focused mainly on the technical development of SLEs. Additionally, only sporadic theoretical approaches to learning theory have been taken up by research studies. Certainly, learning theories do not provide a simple recipe for designing SLEs. Learning processes still happen in the brains of the learners, quite independently from technical supporting tools. Therefore, a profound and comprehensive consideration of the basics of learning processes is essential. This paper makes a contribution to this issue, with a particular focus on collaboration. However it is worth noting that the scope of this paper is limited, and therefore does not allow for an extensive picture. The following research questions are discussed:

- What does 'smart' mean in the context of (collaborative) learning environments and what does it mean for different individuals or groups?
- Why is it smart to collaborate and how can it be fostered through a SLE?

A. Holzinger and G. Pasi (Eds.): HCI-KDD 2013, LNCS 7947, pp. 293–302, 2013.

2 The Concept of 'Smart' Learning Environments

A consequence of all the technological changes in the last few years is that learning environments are increasingly being called 'smart' (even blackboards are called 'smart'). But what are actually "smart learning environments"? "Ubiquitous smart learning environments are always connected with WiFi, 3G and 4G and provide a learners' paradise where they can learn anywhere and anytime whatever they want to learn on the Net." [6]. Further, SLEs encourage multi-content on multi-devices [3]. Environments that are aware "of user behaviour in the learning process can be very helpful in providing the right content at the right time. The learning services that include the concept of such awareness and the capability of handling multi-media resources efficiently can be termed smart learning systems." [7] The definitions of 'smart' [e. g. 8] are manifold and do not provide a clear pathway that allows for a common understanding or an unambiguous definition.

We can cement our understanding of 'smart' learning environments as follows: "**S**ocial" (communication and interaction/connecting with others), "**M**otivating" (mutual benefits/reciprocity and enjoyment), "**A**utonomous" (self-paced and self-directed), "**R**eputation" (social appreciation, trust and competence) and "**T**echnology" (getting the maximum out of technology).

2.1 Smart – But for Whom?

Next, we would like to start by asking: smart for whom? For engineers, who develop SLEs and are eager to show what is possible from a technical point of view? For lecturers, who would like to decrease their (classroom) teaching time? For the individual learner, who wants to learn more comfortably with reduced effort? For the community of learners, who wish to get connected and share experiences and knowledge? Or for the society, that aims to have a strong and capable (but maybe not too critical) workforce? Our answer is simple: *Although the individual learner and the learning community should remain the focal point of interest, a SLE should be 'smart' for each stakeholder.* It should use up-to-date technology in a thorough didactical and pedagogical way. It should smartly support individual learning processes and enhance collaborative learning and learning groups. And (perhaps this is wishful thinking) SLEs should support societies to develop the necessary skills and knowledge to ensure sustainable solutions that show respect for all living beings and the environment. Although this is an extremely broad and comprehensive requirement that SLEs should fulfil, we have to keep in mind that if we want to build truly 'smart' solutions, we cannot neglect the fact that our world is one gigantic system.

2.2 'Smart' in (Learning) Theory

When we focus on the individual learner, some fundamentals of SLEs have been well known for decades, but very often ignored. Certainly, the complexity of human cognition and learning is rather high and until now there has been no single theory, which is able to cope with this complexity.

Bransford et al. [9] suggests four interconnected perspectives for implementing proper learning environments that seem particularly important: Firstly, there is the *learner-centred* component, which pays careful attention to the knowledge, skills, attitudes, and beliefs that learners bring to the educational setting. Secondly, the *knowledge-centred* component, reflecting the need to help students become knowledgeable, by learning in ways that lead to understanding and the ability to subsequently transfer this knowledge. The third component is *assessment-centred*, offering opportunities for feedback and revision aligned to the users' learning goals; and finally the *community-centred component*, which *embraces all the other components*.

In addition to Bransford et al. [9], the community and social learning aspect is an important part of other theories. Self-determination theory (SDT) by Edward Deci and Richard Ryan, e.g. [10, 11], addresses extrinsic forces and intrinsic motives and needs, as well as their interplay. One of its components – in addition to autonomy and competence - is '*relatedness*'; which is about *interacting and connecting with others:* "Conditions supporting the individual's experience of *autonomy, competence,* and *relatedness* are argued to foster the most volitional and high quality forms of motivation and engagement for activities, including enhanced performance, persistence, and creativity." [11]

Another useful theory in the context of SLEs is social learning theory. "The social learning theory explains human behaviour in terms of a continuous reciprocal interaction between cognitive, behavioural and environmental determinants" [12]. For SLEs, we argue that the environmental dimension is reflected in the community as well as on the technology as such.

2.3 Why Is It Smart to Collaborate?

It is of utmost importance for the well-being of the individual to have successful social interactions and to cooperate with others. The human brain is much more a social organ than a reasoning tool [13]. Hormones like dopamine and oxytocin are released through successful cooperation, which is also the basis for learning processes [14]. Further, our minds are susceptible to systematic errors [15], which can be counterbalanced by the group. Almost all learning, in particular the development of higher cognitive functions [16] or the acquisition of fundamentally (ontogenetically) new knowledge [17], is related to other people and embedded in social interaction. One more advantage of the group is heterogeneity. Plurality and the diversity of opinions are necessary to find workable solutions, for fundamental learning and for creativity. Knowledge sharing and knowledge creation are highly subject to social phenomena [18], which are inevitable for the exchange of *tacit knowledge or tacit knowing* [19, 20]. In contrast to codified or *explicit* knowledge, which is quite similar to information and thus can be formalized and stored in technical systems (e. g. SLEs), implicit or tacit knowing only exists in the brains of human beings. Usually people are not aware of it [21, 22], until it is requested, for example in a collaborative learning setting like a SLC.

Consequently, a SLE should provide explicit knowledge, and should use technology to support learning processes, for example through multi-media, simulations or serious games. However, when it comes to the acquisition of tacit knowledge, communities are necessary for exchange and interaction as tacit knowledge cannot be stored in any technical system. Hence, we propose that SLEs have to integrate both kinds of knowledge (see Fig.1). However, since tacit knowledge is achieved through exchange and interaction, communities have to be an integral part of SLEs. Nonetheless, tacit knowing can also be acquired through exercises and adequate technical support can encourage users to practice, i.e. e.g. real-world simulation or serious game-based learning.

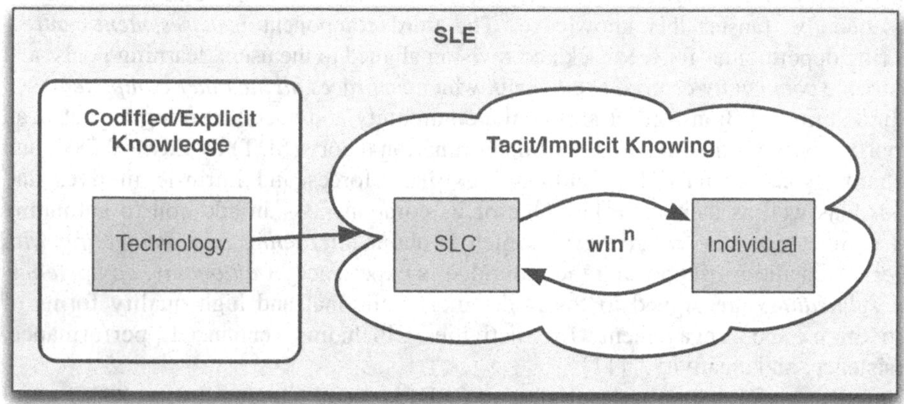

Fig. 1. Components of a SLE, integrating SLCs

When individuals share their valuable tacit knowledge with the SLC, which offers its growing knowledge-base to its members, a new virtuous circle and a win-for-all solution comes about (see Fig. 1).

Thus, interactivity is highly important for learners [23]. Even though interaction is not enough for online learning to be successful, it is considered as central to an educational experience [24], and facilitating interaction is key within a learning community [25]. Further, collaboration of competent individuals in a group or a team is necessary as today's problems are getting increasingly complex and therefore a single individual cannot deal with them alone anymore. A SLE focusing solely on the individual learner misses one important current demand – 'smart' in the 21st century means to collaborate successfully and sustainably. However, to a large extent there is a lack of knowledge on how we can collaborate successfully. Technology should be able to provide support in furthering this collaboration.

2.4 Why Do We Need SLEs?

One can argue, that humankind has been learning since the very beginning of time without SLEs. Why is it now necessary to invent something new? The answer is quite simple: A fundamental tenet of modern learning theory is that different kinds of learning goals require different approaches to instruction; new goals for education require changes in learning opportunities [9]. The complex problems we are facing today require changes in the way we educate people. One of these changes is the necessity to support groups and teams to learn successfully together, exchanging valuable tacit knowledge. A SLE should support collaboration processes and help people to overcome the potential obstacles to effective group processes and group dynamics when working together. If they manage to do so, SLEs could be one solution in meeting the demands of the 21st century.

3 Generating Benefits through the Community

A solution that is 'smart' for each stakeholder, as was claimed in point 2.1, needs to address the needs and wishes of each stakeholder and harmonize these different interests. Beyond that, a smart solution will meet needs of stakeholders they themselves were not aware of before. However, no matter how it is technically implemented, benefits for all stakeholders cannot be guaranteed, particularly not on a long-term basis in our rapidly changing world. Thus, we suggest a SLE, which enables a 'win for all'-solution [26, 27] within and through the community. Strictly speaking, such a SLE would be smart in the sense that it provides not only a SLE for individuals, but a framework that supports and fosters human interaction, information and knowledge sharing, cooperation and collaboration in order to realize a 'living' and emergent system. This can only come into being through 'smart' individuals working together in a 'smart' way. The mutual benefit generated through this community will be central to its vitality and success. Unfortunately, many learning environments are still not able to support or manage community building and knowledge sharing.

To make this social change happen, a shift in thinking will be necessary, which can be described as a *win-for-all, or win^n-solution*: A win^n-constellation means that each participant in this constellation (more precisely: all n participants) can only see him-/herself as a winner if all other participants win too and hence all participants themselves feel as winners [26]. Each individual using the SLE should feel responsible for his or her own learning process and progress, and, in addition, s/he has to support other students to reach their learning aims. A solution, where only one individual wins could be called win^1.

4 Drivers and Design-Principles for SLEs

Drawing on the theoretical concept of win^n, we argue that the main driver for collaboration and SLEs is ***usefulness and benefit***. People make use of tools that they regard as beneficial. Hence, the benefit for the individual is increased through a living community. Additionally, there has to be a benefit for this community (i.e., win^n). As a consequence, technical environments that do not benefit anyone or only selected users (in case the community is crucial for the overall success), should not be implemented at all. The initial point in designing a SLE is to ask for the benefits to each stakeholder, what new benefits would appear and for whom. Certainly, selected stakeholders should be incorporated into this process. The next challenging task is to harmonize the variety of interests of the various stakeholders. This is not easy, but necessary for a 'smart' solution. Where the condition for the main driver - that is providing a 'benefit'- is fulfilled, additional drivers can be striking.

A major benefit is the existing virtuous circle between the SLC and the individuals involved, as shown in Fig. 1. Taking a closer look on this circle it can be separated in two parts: the *central learning processes* and the *influencing processes* (see Fig. 2). Please note that Fig. 2 is work in progress and constitutes a first approach to tackle these highly complex and interwoven issues.

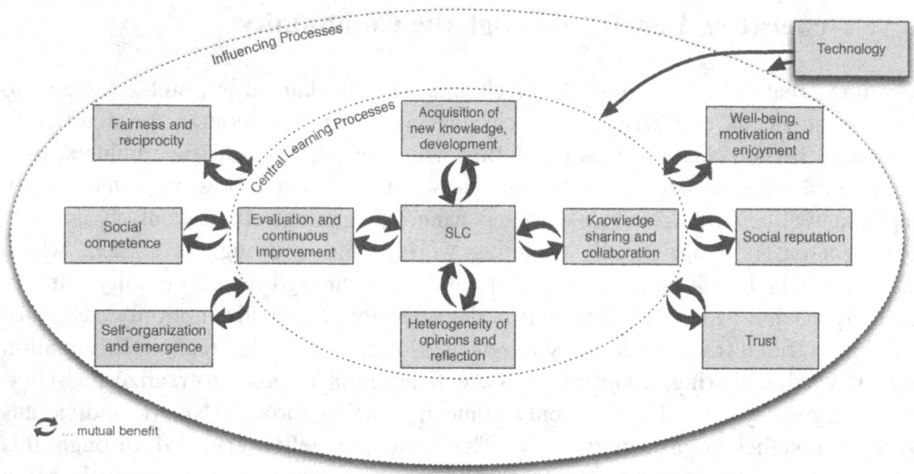

Fig. 2. Central and influencing processes of SLCs

Central learning processes deal with various smaller virtuous circles, impacting the win-for-all solution of a SLC (mutual benefits). As already mentioned, the individual needs the community to *acquire new knowledge* and thus to *further develop*. This in turn benefits the community, needing highly qualified members. Since experts are becoming increasingly specialized, *collaboration* and exchange of tacit knowledge are necessary to solve complex problems. Consequently, the quality of the community is augmented with successful *knowledge sharing*. Through the *heterogeneity of opinions* and various perspectives in the community, creativity and *reflection* are fostered, and one's own blind spots and mistakes are eliminated, again benefiting the individual and the community. Thus, there is an on-going *evaluation and continuous improvement* of individual and community knowledge, as well as knowledge areas.

The influencing processes include various virtuous circles as well: The *well-being* of individuals (as mentioned in 2.3), which is subject to the involvement of the community, has a positive impact on the community itself, fostering *motivation and enjoyment* to engage in the community. SLEs should be intrinsically motivating as well as motivating through collaboration. According to motivation theory people seek optimal stimulation and have a basic need for competence [28]. Consequently, through successful engagement, *social reputation* increases, nurturing individuals' motivation as well as trust in the community as a whole. Reputation is the outcome of what members promise and fulfil, hence, it reveals the participants' honesty and their concern about others and their needs [29]. The social appreciation of the community has shown to be an essential factor for individuals to contribute in knowledge sharing and supporting others to develop skills [30]. As several previous studies pinpoint (for example [31, 32]), *trust* is a vital component for knowledge sharing, as "knowledge sharing can be a demanding and uncertain process" [33] and learners have to feel "safe" to interact and share" [34]. This environment of trust enables the community to establish shared goals and values [27, 35], which have a positive impact on motivation. Hence, it could be argued that on the basis of trust, untypical solutions are found, as learners are more willing to engage in risky behaviour, supporting creativity as well as further developments. *Fairness and reciprocity* are critical and crucial motivating factors for human encounters. They are significant for the exchange of knowledge that is initiated by pro-social and altruistic behaviours [30, 36]. Reciprocity is defined as "voluntarily repaying a trusting move at a later point in time, although defaulting on such repayment is in the short-term self-interest of the reciprocator." [37]. We argue that reciprocity is necessary for the *mutual benefit* of collaboration in SLEs. The next circle is concerned with *social competence* of individuals. Participating within the community fosters social competences, which in turn leads to more successful interactions in the SLC. Having reached a critical amount of interactions, another important influencing factor can occur to the benefit for all: Social *self-organization,* which will keep the SLC alive and capable of acting, even with modified environmental conditions, and which is related to social *emergence*, which means that the SLC is more than the sum of its parts, namely the individuals involved. Surely, another influencing factor of a SLC is *technology.* Technology could lead to a virtuous circle by improving learning and sharing processes, which in turn improves technology; but this is not the norm.

Consequently, SLEs should provide an appropriate service and challenge the individual learner but at the same time offer interconnectedness with the learning community. The design of the SLE has to make this obvious and guarantee that the community fulfils each other's benefits. When designing an SLE, one should be aware, that there are mainly three different types of learning needs: Learning of single individuals (in interaction with technology); learning of well established work groups (task-oriented); and sharing between individuals, being loosely interconnected in a SLE (knowledge-oriented).

One requirement for both social-types of SLEs to be recognized as a comfortable and trustworthy environment is to be personalized and transparent. An example of a very simple way to do this is to set up a profile including picture(s) and some background information about the person. In order to foster learning processes, mutual support and collaboration, we propose that the SLE should request learners to define their tasks or learning goals and share them with the community. Additionally, learners should reveal topics on which they would like to receive support from the community but also issues where they are able to help other members. It would also be supporting that the SLE provides learners a scheduling function where they can post when they will have time and how much time they have to support others, since time is always short but required for the learning process. Learners should also report continuously to which degree they have already achieved their learning goals and whose support was helpful as well as what is still missing and where they would need (further) assistance. This process provides self-monitoring as well as self-reflection by the learners, which is essential for successful and sustainable learning and the improvement of self-education. Further, the design of a SLE should take into account the possibility that learners can give each other praise. This could be for example a medal being added to the profile-picture. This is one of the few exceptions where reward systems do not harm learning processes by destroying intrinsic motives. Ultimately, the architecture should initiate peer-discussion by suggesting peers with whom a discussion might be fruitful – this could be because of (dis-)similarities of knowledge, experience or interests. Finally, SLEs should foster different media qualities, the preparation of students for 'flipped classrooms' [23] and open spaces, crossing the borders of the Internet, because learning doesn't only happen in front of screens.

5 Conclusion

In this paper, we propose that 'smart' in the 21st century has to be associated with collaboration and sustainability. We draw on the concept of win^n, implying that *all* participants and stakeholders of a SLE should benefit mutually. Consequently, if the system does not provide sound benefits for its stakeholders, it should not be constructed. This might sound trivial, however, in practice a lot of systems were built lacking perceived usefulness. Needless to say, this is also an economic goal: although 'use' is quite a problematic word in the context of education, even inert knowledge, which is not used at any time, might be worthless for the individual, the team, the company or the society. Thus, the transfer of acquired experiences and knowledge in the SLE, i.e. the ability to extend what has been learned from one context to new contexts [38][39] and particularly to real world problems is an essential requirement for a SLE. The community can also support this transfer by defining various examples or scenarios on how to apply learned content in practice. Furthermore, the framework sets out some potential drivers for collaboration and design principles for SLEs, or SLCs. Certainly, there are many more relevant theories concerning smart learning and collaborative learning: social constructivist theory, social presence, social interdependence, situated learning, self-directed learning and self-regulation theory [34]. Additionally, (empirical) research in this case is needed, as this paper only offers an initial theoretical analysis of potential determinants and processes of SLEs and SLCs. The technological advances enabled us to connect globally with each other, providing us with an immense potential for synergetic, sustainable, creative or shorter: smart solutions. We have to take this opportunity.

References

1. Hirsch, B., Ng, J.W.P.: Education Beyond the Cloud: Anytime-anywhere learning in a smart campus environment. In: 6th International Conference on Internet Technology and Secured Transactions, Abu Dhabi, United Arab Emirates, December 11-14, pp. 11–14 (2011)
2. Burghardt, C., Reisse, C., Heider, T., Giersich, M., Kirste, T.: Implementing Scenarios in a Smart Learning Environment. In: 2008 Sixth Annual IEEE International Conference on Pervasive Computing and Communications (PerCom), pp. 377–382 (2008)
3. Kim, S., Yoon, Y.I.: A Model of Smart Learning System Based on Elastic Computing. In: Ninth International Conference on Software Engineering Research, Management and Applications, pp. 184–185. IEEE (2011)
4. Miyata, N., Morikawa, H., Ishida, T.: Open Smart Classroom: Extensible and Scalable Learning System in Smart Space Using Web Service Technology. IEEE Transactions on Knowledge and Data Engineering 21, 814–828 (2009)

5. Scott, K., Benlamri, R.: Context-Aware Services for Smart Learning Spaces. IEEE Transactions on Learning Technologies 3, 214–227 (2010)
6. Lee, J.R., Jung, Y.J., Park, S.R., Yu, J., Jin, D., Cho, K.: A Ubiquitous Smart Learning Platform for the 21st Smart Learners in an Advanced Science and Engineering Education. In: 15th International Conference on Network-Based Information Systems, pp. 733–738. IEEE (2012)
7. Kim, S., Song, S.-M., Yoon, Y.-I.: Smart learning services based on smart cloud computing. Sensors 11, 7835–7850 (2011)
8. Kim, T., Cho, J.Y., Lee, B.G.: Evolution to Smart Learning in Public Education A Case Study of Korean Public Education. In: Ley, T., Ruohonen, M., Laanpere, M., Tatnall, A. (eds.) OST 2012. IFIP AICT, vol. 395, pp. 170–178. Springer, Heidelberg (2013)
9. Bransford, J.D., Brown, A.L., Cocking, R.R.: How People Learn: Brain, Mind, Experience, and School. National Academy Press (2000)
10. Deci, E.L., Vansteenkiste, M.: Self-Determination Theory and basic need satisfaction: Understanding human development in positive psychology. Ricerche di Psicologia 27, 23–40 (2004)
11. N.A.: Self-Determination Theory. About the theory, `http://www.selfdeterminationtheory.org/theory`
12. Tu, C.-H.: On-line learning migration: from social learning theory to social presence theory in a CMC environment. Journal of Network and Computer Applications 23, 27–37 (2000)
13. Hüther, G.: Bedienungsanleitung für ein menschliches Gehirn. Vandenhoeck & Ruprecht GmbH & CoKG, Göttingen (2010)
14. Bauer, J.: Prinzip Menschlichkeit. Springer, Heidelberg (2008)
15. Kahneman, D.: Thinking, Fast and Slow. Farrar, Straus and Giroux, New York (2012)
16. Vygotsky, L.S.: Mind in Society. Harvard University Press, Cambridge (1978)
17. Miller, M.: Kollektive Lernprozesse. Suhrkamp, Frankfurt (1986)
18. Duguid, P.: The art of knowing: social and tacit dimensions of knowledge and the limits of the community of practice. The Information Society 21, 109–118 (2005)
19. Polanyi, M.: The Tacit Dimension. Peter Smith, Gloucester (1966)
20. Polanyi, M.: Knowing and being. Routledge & Kegan, London (1969)
21. Anderson, J.R.: Kognitive Psychologie. Springer, Heidelberg (1980)
22. Day, R.E.: Clearing up "Implicit Knowedge": Implications for Knowledge Management, Information Science, Psychology, and Social Epistemology. Journal of the American Society for Information Science and Technology 56, 630–635 (2005)
23. Khan, S.: Die Khan Academy. Die Revolution für die Schule von morgen. Riemann, München (2013)
24. Garrison, D.R., Cleveland-Innes, M.: Facilitating cognitive presence in online learning: Interaction is not enough. American Journal of Distance Education 19, 133–148 (2005)
25. Hill, J.R.: Learning Communities. Theoretical Foundations for Making Connections. Theoretical Foundations of Learning Environments, pp. 268–285. Routledge, N.Y. & London (2012)
26. Frankl, G.: win-n. win-win-Konstellationen im Wissensmanagement (2010)
27. Frankl, G.: Common Benefits and Goal Cooperativeness as Driving Forces for Knowledge Management. In: Proceedings of the 13th European Conference on Knowledge Management, Academic Conferences International (ACI), Reading, pp. 341–349 (2012)
28. Eccles, J.S., Wigfield, A.: Motivational Beliefs, Values, and Goals. Annual Review Psychology 53, 109–132 (2002)

29. Casaló, L.V., Cisneros, J., Flavián, C., Guinalíu, M.: Determinants of success in open source software networks. Industrial Management & Data Systems 109, 532–549 (2009)
30. Wasko, M.M., Faraj, S.: It is what one does: Why people participate and help others in electronic communities of practice. Journal of Strategic Information Systems 9, 155–173 (2000)
31. Hsu, M.-H., Ju, T.L., Yen, C.-H., Chang, C.-M.: Knowledge sharing behavior in virtual communities: The relationship between trust, self-efficacy, and outcome expectations. International Journal of Human-Computer Studies 65, 153–169 (2007)
32. Mooradian, T., Renzl, B., Matzler, K.: Who Trusts? Personality, Trust and Knowledge Sharing. Management Learning 37, 523–540 (2006)
33. Mooradian, T., Renzl, B., Matzler, K.: Who Trusts? Personality, Trust and Knowledge Sharing. Management Learning 37, 523–540 (2006)
34. Hill, J.R.: Learning Communities. Theoretical Foundations for Making Connections. Theoretical Foundations of Learning Environments, pp. 268–285. Routledge, N.Y (2012)
35. Guldberg, K., Pilkington, R.: A community of practice approach to the development of non-traditional learners through networked learning. Journal of Computer Assisted Learning 22, 159–171 (2006)
36. Wasko, M.M., Faraj, S.: Why should I share? Examining social capital and knowledge contribution in electronic networks of practice. MIS Quarterly 29, 35–57 (2005)
37. Gunnthorsdottir, A., Mccabe, K., Smith, V.: Using the Machiavellianism instrument to predict trustworthiness in a bargaining game. Journal of Economic Psychology 23, 49–66 (2002)
38. Thorndike, E.L., Woodworth, R.S.: The influence of improvement in one mental function upon the efficiency of other functions. Psychological Review 8, 247–261 (1901)
39. Byrnes, J.P.: Cognitive Development and Learning in Instructional Contexts. Allyn and Bacon, Boston (1996)

Smart Open-Ended Learning Environments That Support Learners Cognitive and Metacognitive Processes

Gautam Biswas, James R. Segedy, and John S. Kinnebrew

Dept. of EECS/ISIS, Vanderbilt University, Nashville, TN 37235. USA
{gautam.biswas,james.segedy,john.s.kinnebrew}@vanderbilt.edu

Abstract. Metacognition and self-regulation are important for effective learning; but novices often lack these skills. Betty's Brain, a Smart Open-Ended Learning Environment, helps students develop metacognitive strategies through adaptive scaffolding as they work on challenging tasks related to building causal models of science processes. In this paper, we combine our previous work on sequence mining methods to discover students' frequently-used behavior patterns with context-driven assessments of the effectiveness of these patterns. Post Hoc analysis provides the framework for systematic analysis of students' behaviors online to provide the adaptive scaffolding they need to develop appropriate learning strategies and become independent learners.

Keywords: open-ended learning environments, metacognition, measuring metacognition, scaffolding, sequence mining.

1 Introduction

Open-Ended Learning Environments (OELEs) are learner-centered [1]. They employ supporting technology, resources, and scaffolding to help students actively construct and use knowledge for complex problem-solving tasks [2,3]. Our research group has developed an OELE called Betty's Brain, where students learn about science topics by constructing a visual causal map to teach a virtual agent, Betty. The goal for students using Betty's Brain is to ensure that their agent, Betty, can use the causal map to correctly answer quiz questions posed to her by a Mentor agent, Mr. Davis [4].

In Betty's Brain, learners are responsible for managing both their cognitive skills and metacognitive strategies for learning the domain material, structuring it as a causal map, and testing Betty's understanding in order to obtain feedback on the quality of their map. This places high cognitive demands on learners [2], especially those who lack well-organized domain knowledge structures and effective self-regulation strategies [5].

An important implication of this is that OELEs require methods for measuring students' understanding of these important skills and strategies, while providing scaffolds that help them practice and learn these skills. Moreover, these systems must focus on cognition and metacognition in an integrated fashion. Supporting learners in developing their metacognitive knowledge may not be sufficient for achieving success in learning and problem solving, especially when learners lack the cognitive skills and background knowledge necessary for interpreting, understanding, and organizing critical aspects of the problem under study [2]. Learners with poor self-judgment abilities

A. Holzinger and G. Pasi (Eds.): HCI-KDD 2013, LNCS 7947, pp. 303–310, 2013.

often resort to suboptimal strategies in performing their tasks [6,7]. However, research studies have shown that proper scaffolding helps students improve their metacognitive awareness and develop effective metacognitive strategies [8].

In this paper, we present an extension of our previous work in analyzing students' behaviors on the system using sequential pattern mining methods [4]. Primarily, the extension involves interpreting and characterizing behavior patterns using metrics motivated by a cognitive/metacognitive model for managing one's own learning processes in OELEs. While the results in this paper represent a post hoc analysis of student behaviors, our goal is to use such results to assess students' cognitive and meta-cognitive processes in real-time as they work on their learning tasks[1].

2 Betty's Brain

Betty's Brain (Fig. 1) provides students with a learning context and a set of tools for pursuing authentic and complex learning tasks. These tasks are organized around three activities: (1) reading hypertext resources to learn the domain material, (2) building and refining a causal map, which represents the domain material, and (3) asking Betty to take a quiz. Students explicitly teach Betty by constructing a causal map that links concepts using causal relations, e.g., "absorbed heat energy increases global temperature." Students can check what Betty knows by asking her questions, e.g., "if garbage and landfills decrease, what effect does it have on polar sea ice?"

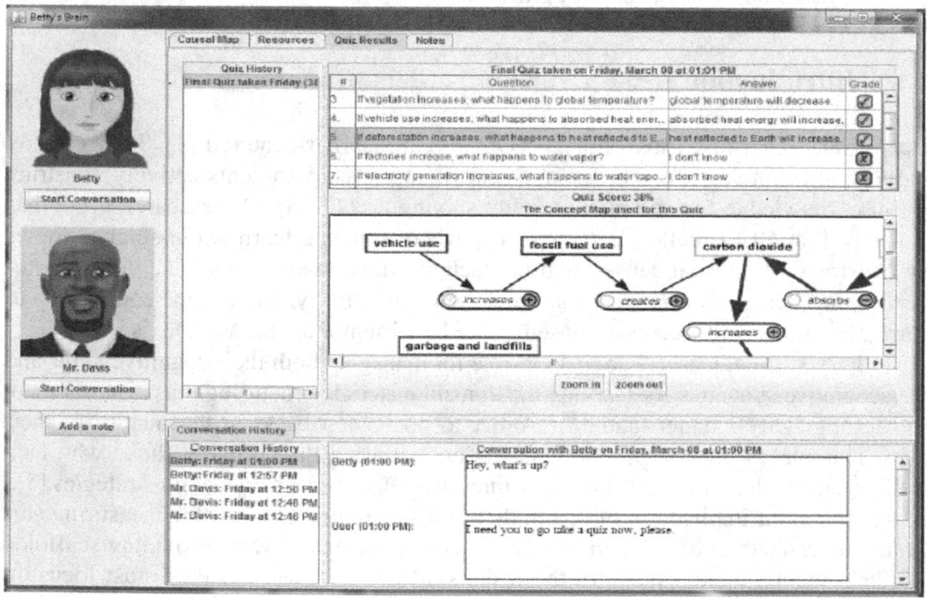

Fig. 1. Betty's Brain System showing Quiz Interface

[1] To download Betty's Brain, and get more details about the system, visit `www.teachableagents.org`

To answer questions, Betty uses qualitative reasoning that operates through chains of links [4]. The learner can further probe Betty's understanding by asking her to explain her answer. Betty illustrates her reasoning by explaining her solution and animating her explanation by highlighting concepts and links on the map as she mentions them.

Learners can assess Betty's (and, therefore, their own) progress in two ways. After Betty answers a question, learners can ask Mr. Davis, to evaluate the answer. Learners can also have Betty take a quiz on one or all of the sub-topics in the resources. Quiz questions are selected dynamically such that the number of correct answers is proportional to the completeness of the map. The remaining questions produce incorrect answers, and they direct the student's attention to incorrect and missing links.

After Betty takes a quiz, her results, including the causal map she used to answer the questions, appear on the screen as shown in Fig. 1. The quiz questions, Betty's answer, and the Mentor's assigned grade (i.e., correct, correct but incomplete, or incorrect) appear on the top of the window. Clicking on a question will highlight the causal links that Betty used to answer that question. To help students keep track of correct and incorrect links, the system allows students to annotate them with a green check-mark (correct), a red X (incorrect), or a gray question mark (not sure).

2.1 Measuring Cognition and Metacognition

To interpret students' learning behaviors on the system, we have developed a model that exploits the synergy between the cognitive and metacognitive processes needed for effective learning. Overall, this model includes four primary processes that students are expected to engage in while using Betty's Brain: (1) Goal Setting & Planning, (2) Knowledge Construction (KC), (3) Monitoring (Mon), and (4) Help Seeking. In this work, we focus on the KC and Mon process models.

Knowledge construction includes metacognitive strategies for (1) *information seeking*, i.e., determining when and how to locate needed information in the resources, and (2) *information structuring*, i.e., organizing one's developing understanding of the domain knowledge into structural components, e.g., causal links. In executing these metacognitive processes, learners rely on their understanding of the relevant cognitive processes linked to information seeking (finding causal relations while reading the resources) and structuring (converting the causal information to links that are added to appropriate places in the map).

Monitoring processes include (1) *model assessment* of all or a part of the causal model and then interpreting the derived information, and (2) *progress recording* to mark which parts of the causal model are correct, and which may need further refinement. Successful execution of metacognitive monitoring processes relies on students' abilities to execute cognitive processes for assessing the causal model (via questions, explanations, quizzes, and question evaluations) and recording progress (via note taking and annotating links with correctness information).

We have developed a set of data mining methods [7] for analyzing students' learning activity sequences and assessing their cognitive and metacognitive processes as they work in Betty's Brain. In addition, we have developed methods for measuring how student behaviors evolve during the course of the intervention depending on the type of feedback and support that they received from the Mentor agent. In particular, we are interested in studying whether students' suboptimal behaviors are replaced by better strategies during the intervention.

To assess student activities with respect to our cognitive/metacognitive model, we calculate four measures: *map edit effectiveness*, *map edit support*, *monitoring effectiveness*, and *monitoring support*. Map edit effectiveness is calculated as the per-cent-age of causal link edits that improve the quality of Betty's causal map. Map edit sup-port is defined as the percentage of causal map edits that are supported by previous activities. An edit is considered to be supported if the student previously accessed pages in the resources or had Betty explain quiz answers that involve the concepts connected by the edited causal link. Monitoring effectiveness is calculated as the percentage of quizzes and explanations that generate specific correctness information about one or more causal links. Finally, monitoring support is defined as the percent-age of causal link annotations that are supported by previous quizzes and explanations. Link annotations are supported (correct or incorrect) if the links being annotated were used in a previous quiz or explanation. For support metrics, a further constraint is added: an action can only support another action if both actions occur within the same time window; in this work, we employed a ten-minute window for support.

In order to calculate these measures and perform data mining of student behaviors, the system records many details of student learning activities and associated parameters in log files. For example, if a student accesses a page in the resources, this is logged as a Read action that includes relevant contextual information (e.g., the page accessed). In this work, we abstracted the log files into sequences of six categories of actions: (1) Read, (2) Link Edit, (3) Query, (4) Quiz, (5) Explanation, and (6) Link Annotation. Actions were further distinguished by context details, such as the correctness of a link edit.

3 Method

Our analysis used data from a recent classroom study with Betty's Brain in which students learned about the greenhouse effect and climate change. The study tested the effectiveness of two scaffolding modules designed to help students' understanding of cognitive and metacognitive processes important for success in Betty's Brain. The knowledge construction (KC) support module scaffolded students' understanding of how to identify causal relations in the resources, and the monitoring (Mon) support module scaffolded students understanding of how to use Betty's quizzes to identify correct and incorrect causal links on the causal map. Participants were divided into three treatment groups. The knowledge construction group (KC-G) used a version of Betty's Brain that included the KC support module and a causal-link tutorial that they could access at any time during learning. The KC module was activated when three out of a student's last five map edits were incorrect, at which point Mr. Davis would begin suggesting strategies for identifying causal links during reading. Should students continue to make incorrect map edits despite this feedback, the KC module activated the tutorial for identifying causal relations in short text passages. Students completed the tutorial session when they solved five consecutive problems correctly.

The monitoring group (Mon-G) used a version of Betty's Brain that was activated after the third time a student did not use evidence from quizzes and explanations to annotate links on the map. At this time, Mr. Davis began suggesting strategies for using quizzes and explanations to identify and keep track of which links were correct. Should students continue to use quizzes and explanations without annotating links correctly, the Mon module activated a tutorial that presented practice problems on link annotation. Students had to complete five problems correctly on the first try to complete the tutorial session. Finally, the control group (Con-G) used a version of Betty's Brain that included neither the tutorials nor the support modules.

Our experimental analysis used data collected from 52 students in four middle Tennessee science classrooms, taught by the same teacher. Learning was assessed using a pre-post test design. Each written test consisted of five questions that asked students to consider a given scenario and explain its causal impact on climate change. Scoring (max score = 16) was based on the causal relations that students used to explain their answers to the questions, which were then compared to the chain of causal relations used to derive the answer from the correct map.

Performance on the system was assessed by calculating a score for the causal map that students created to teach Betty. This score was computed as the number of correct links (the links in the student's map that appeared in the expert map) minus the number of incorrect links in the student's final map. We also used the log data collected from the system to derive students' behavior patterns, interpret them using our cognitive/metacognitive model, and study the temporal evolution of the observed KC and Mon strategies over the period of the intervention. Students spent four class periods using their respective versions of Betty's Brain with minimal intervention by the teacher and the researchers.

4 Results

Table 1 summarizes students' pre-post gains and final map scores by group. A repeated measures ANOVA performed on the data revealed a significant effect of time on test scores ($F = 28.66$, $p < 0.001$). Pairwise comparison of the three groups revealed that the Mon-G had marginally better learning gains than KC-G, which had better learning gains than the Con-G group. In particular, the Mon-G learning gains were significantly better than the Con-G gains at the 0.1 significance level ($p < .075$), indicating the intervention may have resulted in better understanding of the science content. The small sample size and the large variations in performance within groups made it difficult to achieve statistical significance in these results. However, one positive aspect of this finding is that while students in the Mon-G and KC-G spent an average of 10% and 17% of their time in the tutorials, respectively, they learned, on average, just as much, if not more, than the Con-G students.

Table 1. Pre-Post Test Gains and Map Scores

Measure	Con-G	KC-G	Mon-G
Pre-Post Test Gain	1.03 (1.99)	1.28 (2.33)	2.41 (1.92)
Map Score	8.87 (8.20)	9.55 (6.64)	9.53 (7.55)

To assess students' overall behaviors, we calculated the effectiveness and support measures in Table 2. The KC-G students had the highest scores on both map editing effectiveness and support, suggesting that the KC feedback helped students read more systematically while constructing their maps (however, only the map edit support showed a statistically-significant difference, KC-G > Con-G, $p = 0.02$, and the map edit effectiveness illustrated a trend, KC-G > Con- G, $p = 0.08$). However, the monitoring support did not seem to have the same effect on the Mon-G group. The Mon-G students did have the highest monitoring effectiveness, but it was not statistically significant. Further, the Con-G students had the highest monitoring support average ($p < 0.10$). It is not clear why the Mon or KC support and tutorials resulted in students performing less-supported monitoring activities.

Table 2. Effectiveness & Support Measures by Group

Measure	Con-G	KC-G	Mon-G
Map edit effectiveness	0.46 (0.13)	0.52 (0.07)	0.5 (0.12)
Map edit support	0..46 (0.26)	0.66 (0.18)	0.58 (0.22)
Monitoring effectiveness	0.3 (0.22)	0.32 (0.21)	0.4 (0.20)
Monitoring support	0.61 (0.30)	0.32 (0.4)	0.33 (0.32)

In order to investigate student learning behavior in more detail, we employed sequence mining analyses to identify the 143 different patterns of actions that were observed in the majority of students. Table 3 lists the 10 most frequent patterns that employed at least two actions and could be interpreted as a metacognitive strategy in our cognitive/metacognitive model. Each pattern is defined by two or more primary actions, and each action is qualified by one or more attributes. For example, [Add correct link] describes a student's addition of a causal link that was correct. The → symbol implies that the action to the left of the arrow preceded the action to the right of the arrow.

The average frequency is calculated as the average number of times per student a particular behavior pattern was used in each group. The last column represents our interpretation of the type of strategy a particular behavior represents. In this study, the type of strategy corresponding to a behavior was determined by the category of the cognitive process (KC or Mon) implied by the individual actions that made up the behavior. Therefore, some behaviors (e.g., pattern #3: [Quiz] → [Remove incorrect link]) span KC and Mon (KC+Mon) strategies.

The frequency numbers indicate that for almost all of the top 10 behaviors the CON-G showed a higher frequency of use than the two experimental groups. This may be partly attributed to the time the KC-G and Mon-G groups spent in tutorials, therefore, away from the primary map building task. However, an equally plausible reason is that the CON-G students used more trial-and-error approaches, spending less time systematically editing and checking the correctness of their maps. This is further supported by looking at the highest average frequency behaviors for each of the groups. Four of the top five behavior strategies for the Mon-G students are primarily Mon or KC+Mon related (patterns 1, 3, 5, and 7), involving quizzes, map editing, and explanations. KC-G students, on the other hand, more often employed KC strategies related to adding and removing links along with a couple of strategies that combine KC and Mon activities. The Con-G students seem to have employed KC and Mon strategies in about equal numbers.

Table 3. Comparison of Pattern Frequencies across Conditions

Pattern	Avg. Frequency			Model
	CON	KC	MON	Category
[Add incorrect link] → [Quiz]	11.20	7.35	8.24	KC+Mon
[Add incorrect link] → [Remove incorrect link]	6.00	12.65	3.71	KC
[Quiz] → [Remove incorrect link]	7.87	6.10	6.29	KC+Mon
[Add concept] → [Add correct link]	7.53	6.75	4.94	KC
[Quiz] → [Explanation]	8.40	3.80	5.35	Mon
[Remove incorrect link] → [Add incorrect link]	4.53	9.20	3.41	KC
[Add correct link] → [Quiz]	5.87	4.05	5.06	KC+Mon
[Remove incorrect link] → [Quiz]	5.93	4.45	4.12	KC+Mon
[Explanation] → [Explanation]	5.67	2.95	4.88	Mon
[Add incorrect link] → [Quiz] → [Remove same incorrect link]	5.27	3.75	3.82	KC+Mon

A particularly interesting example of a strategy is the pattern: [Add incorrect link (AIL)] → [Quiz (Q)] → [Remove same incorrect link (RIL)]. This could represent a strategy where a student first adds a link (which happens to be incorrect) and then takes a quiz to determine if the quiz score changes. Depending on the outcome (in this case, the score likely decreased), the student determines that the link added was incorrect, and, therefore, removes it. This may represent a trial-and-error strategy. To study this pattern further we developed two measures: (1) a measure of *cohesiveness* of the pattern, i.e., in what percentage of the AIL → Q → RIL patterns was the delete action supported by the quiz result; and (2) a *support* measure, i.e., in what percentage of the AIL → Q → RIL patterns was the addition of the link supported by recent actions. The MON group had higher cohesiveness (41.9 to 38.0 and 37.3 for the CON and KC groups) and support (27.7 to 20.3 and 187.7 for the CON and KC groups) measures, implying that they used this pattern in a more systematic way than the other two groups.

5 Discussion and Conclusions

The results presented in the previous section provide evidence that a combination of systematically-derived measures of cohesiveness and support can be used with data mining methods to better interpret students' learning behaviors and strategies. In our work on investigating cognitive and metacognitive processes in Betty's Brain, we had to carefully instrument the system to collect rich data on the students' activities and the context associated with those activities. Post hoc mining and analysis show that it is possible to interpret students' behaviors and that related context information allows us to better characterize and interpret these strategies. Our analyses in this study focused on students' knowledge construction and monitoring strategies. This study showed that the interventions produced changes in student behavior that were consistent with the scaffolding they were provided, implying that these metacognitive strategies can be effectively taught in computer-based learning environments.

Future work will involve refining the methods presented in this paper to discover and define strategies in a more systematic way. Further, we will extend our measurement framework to more tightly integrate theory-driven measures with data-driven mining for analyzing student cognition and metacognition during learning. Ultimately, we hope to find better ways of inferring students' intent (i.e., goals) from their observed behaviors and strategies while using the system..

Acknowledgments. This work has been supported by NSF-IIS Award #0904387 and IES Award #R305A120186.

References

1. Bransford, J., Brown, A., Cocking, R. (eds.): How people learn. National Academy Press, Washington, DC (2000)
2. Land, S., Hannafin, M.J., Oliver, K.: Student-Centered Learning Environments: Foundations, Assumptions and Design. In: Jonassen, D.H., Land, S. (eds.) Theoretical Foundations of Learning Environments, 2nd edn., pp. 3–26. Routledge, New York (2012)
3. Quintana, C., Shin, N., Norris, C., Soloway, E.: Learner-centered design: Reflections on the past and directions for the future. In: The Cambridge Handbook of the Learning Sciences, pp. 119–134 (2006)
4. Leelawong, K., Biswas, G.: Designing learning by teaching agents: The Betty's Brain system. International Journal of Artificial Intelligence in Education 18(3), 181–208 (2008)
5. Zimmerman, B.: Theories of self-regulated learning and academic achievement: An overview and analysis. In: Zimmerman, B., Schunk, D. (eds.) Self-Regulated Learning and Academic Achievement: Theoretical Perspectives. Erlbaum, Mahwah, pp. 1–37 (2001)
6. Azevedo, R., et al.: The Effectiveness of Pedagogical Agents' Prompting and Feedback in Facilitating Co-adapted Learning with MetaTutor. In: Cerri, S.A., Clancey, W.J., Papadourakis, G., Panourgia, K. (eds.) ITS 2012. LNCS, vol. 7315, pp. 212–221. Springer, Heidelberg (2012)
7. Kinnebrew, J.S., Biswas, G.: Identifying learning behaviors by contextualizing differential sequence mining with action features and performance evolution. In: Proceedings of the 5th International Conference on Educational Data Mining, Chania, Greece (2012)
8. Kramarski, B., Mevarech, Z.: Enhancing mathematical reasoning in the classroom: effects of cooperative learning and metacognitive training. AERA Journal 40(1), 281–310 (2003)

Curriculum Optimization by Correlation Analysis and Its Validation

Kohei Takada[1], Yuta Miyazawa[1], Yukiko Yamamoto[1], Yosuke Imada[1], Setsuo Tsuruta[1], and Rainer Knauf[2]

[1] Tokyo Denki University, Inzai, Japan
[2] Ilmenau University of Technology, Ilmenau, Germany

Abstract. The paper introduces a refined Educational Data Mining approach, which refrains from explicit learner modeling along with an evaluation concept. The technology is a "lazy" Data Mining technology, which models students' learning characteristics by considering real data instead of deriving ("guessing") their characteristics explicitly. It aims at mining course characteristics similarities of former students' study traces and utilizing them to optimize curricula of current students based to their performance traits revealed by their educational history. This (compared to a former publication) refined technology generates suggestions of personalized curricula. The technology is supplemented by an adaptation mechanism, which compares recent data with historical data to ensure that the similarity of mined characteristics follow the dynamic changes affecting curriculum (e.g., revision of course contents and materials, and changes in teachers, etc.). Finally, the paper derives some refinement ideas for the evaluation method.

Keywords: adaptive learning technologies, personalized curriculum mining, educational data mining.

1 Introduction

The way to compose personalized university curricula are different in different regions of the world. In some countries, students have a lot of opportunities to compose it according to their preferences. Especially newly entering students may be not able to cope with the jungle of opportunities, restrictions, prerequisites and other curriculum composition rules to compose a curriculum that meets all conditions, but also individual needs and preferences. Also, students may not know exactly in advance, which performance skills are challenged in the particular courses. Therefore, they cannot know whether or not they can perform really well in it. Moreover, students may be not consciously aware of their own performance traits and consider them to compose their curriculum to optimally make use of it.

The objective of this work is mining personalized optimal curricula not only based on grade point averages (GPA), but on the particular grade points (GP) reached in the particular courses. Since each particular course challenges certain

A. Holzinger and G. Pasi (Eds.): HCI-KDD 2013, LNCS 7947, pp. 311–318, 2013.

particular performance skills, the GP distribution on the various courses may characterize a student's performance traits. Thus, this information may be helpful to suggest best possible curricula complements to the courses taken so far, i.e. to a student's educational history. The approach refrains from explicit models, but uses a database of cases as the model and computes (positive and negative) correlations to a current case to derive suggestions for optimal curricula.

Proposals to characterize the skills challenged in university courses are usually explicit models in terms of describing challenged traits (such as the multiple intelligence model of Gardner [2]) and learning styles (such as the Felder-Silverman model [1]), to match with the teacher's style and material. Despite of their successful empirical evaluation such models cannot really been proven. But, such models are derived from educational data. So, why not using the data itself as the model instead of deriving explicit assumptions and guesses from it? The approach introduced here refrains from explicit models, but uses a database of cases as the model and computes (positive and negative) correlations to a current case to derive suggestions for optimal curricula. From an Artificial Intelligence (AI) point of view, this approach can be classified as Case Based Reasoning (CBR).

This may not be a contribution to reform the learning environment itself, but to utilize it in an optimal way, which is - from the viewpoint of learning success - a reformation as well.

The paper is organized as follows. Section two explains the concept of (positive and negative) performance correlations of a planned course to a course already taken. The correlation concept is expanded to a set of courses taken so far in section three. Section four explains the way to compose an upcoming semester by utilizing this concept. In section five, we discuss issues of data maintenance for computing such correlations. Section six introduces the performed evaluation method and results along with some refinement ideas.

2 Performance Correlation between Courses

The basic behind this approach is that people have profiles, which influences their success chance in learning. A learner's success is determined by many issues. A main issue is certainly the content to learn, which may challenge very theoretical and analytical towards practical creative matters. Another issue may be the organization form of learning, which rages from ex-cathedra teaching towards teamwork practices. Furthermore, the degree of matching between the teaching style of the teacher and the learning style of the learner matters. Another issue that influences the learning success are the kind of learning material. Additionally, social and cultural aspects such as the mutual relation between (a) teacher and learner, (b) in-between the learners, and (c) between the learner and people outside the learning process may influence the learner's performance.

There are various attempts to reveal all these issues and to derive a related learner modeling approach. Indeed, this is a useful research, because it reveals reasons for performing good or bad. Only by knowing these reasons learning scenarios can be improved with the objective to provide optimal conditions for each individual learner. Unfortunately, our former attempts to select and/or combine appropriate modeling approaches failed due to missing or impossible to obtain data.

However, learner profile items are inherent in the learners' educational data. Therefore, we shifted our approach to the idea of using the "lazy" model (a library of cases) instead of an explicit one. We believe that such individual profile parameters imply correlations between these degrees in different courses and thus, there are positive or negative correlations between the degrees of success in the courses. These correlations characterize implicitly a student's profile and can be derived by Educational Data Mining on (former) students' educational history.

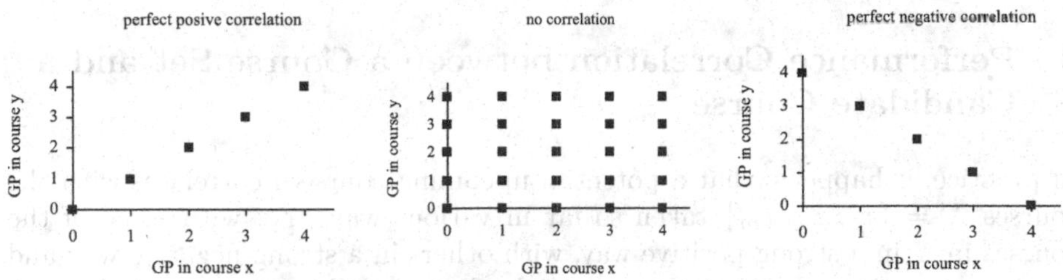

Fig. 1. Correlation types

In case of perfect positive correlation, this student will receive the same number of GP in course y than he/she achieved in course x (see left hand side of figure 1). In case of perfect negative correlation, this student will receive a number of grade points in course y which is the higher (lower), the lower (higher) the number of GP in course x was (see right hand side of figure 1). If there is no correlation at all between both courses, the challenged skills and the preferable learning style and learning material preferences are totally independent from each other (see center of figure 1).

To also include the circumstance that a correlation can change as a result of former learning experience, we consider only correlations $corr(x, y)$ of courses y taken at a later time than courses x.

Based on samples of k students $S = s_1, \ldots, s_k$, who took a course y after a course x, and each student s_i achieved g_i^x GPs in course x and g_i^y GPs in course y, the linear correlation coefficient can be computed as

$$corr(x, y) = \frac{\sum_{i=1}^{k}(g_i^x - \overline{g}^x)(g_i^y - \overline{g}^y)}{\sqrt{\sum_{i=1}^{k}(g_i^x - \overline{g}^x)^2}\sqrt{\sum_{i=1}^{k}(g_i^y - \overline{g}^y)^2}} \quad (1)$$

with $\overline{g}^z$ being the average number of GPs achieved by the k students $s_1, \ldots, s_k$ in course z.

Such correlations can be computed based on known educational histories of (former) students and applied to courses of the educational history of a current student to suggest optimal courses for upcoming semesters. Optimal means preferring courses with (1) high positive correlation with courses taken so far, in which the student achieved his/her best results and (2) high negative correlation with courses taken so far, in which the student achieved his/her worst results.

However, courses with no correlation should be spread in as well. They may represent learning scenarios that are not experienced so far by the considered learner. On the one hand, they bag the chance to perform better than so far. On the other hand, the is a risk to perform worse. We don't know a good trade-off between the chances and risks yet and leave this as a parameter in our curriculum composing technology. Of course, the proposed technology applies only to those candidate courses for the curriculum, for which all formal prerequisites are met by the considered student.

3 Performance Correlation between a Course Set and a Candidate Course

In practice, it happens, that a potential upcoming course y correlates with the courses $X = \{x_1, \ldots, x_m\}$ taken so far in various ways, i.e. with some of the courses in X in a strong positive way, with others in a strong negative way and maybe, with others not at all. Despite of the fact, that this may be an indication for not having sufficient data from (former) students, we have to cope with this situation.

Also, courses are weighted with their related number of units (in Japan) or credits (in USA and Canada as well as in Europe after the Bologna Reform has been completed in all participating countries). Since we apply our approach in a Japanese university, we use the term units here. Therefore, it is reasonable to weight the different correlations $corr(x_j, y)$ of a potential upcoming course y with the courses taken so far $X = \{x_1, \ldots, x_m\}$ with the number of units u_j of each $x_j \in X$ and to compute a weighted average correlation of the course y to the courses in X:

$$corr(\{x_1, \ldots, x_m\}, y) = \frac{\sum_{i=1}^{m} u_i * corr(x_i, y)}{\sum_{i=1}^{m} u_i} \tag{2}$$

- $\{x_1, \ldots, x_m\}$ is the set of courses the student, who looks for appropriate courses for his/her next semester, took so far, and m is the cardinality of this set, i.e. the number of these courses,
- y is a candidate course for the next semester, for which the students met all prerequisites and for which the correlation is subject of computing based on the students' samples in the database,
- k_j is the number of students in the database, who took course y after taking the course x_j ($xj \in \{x_1, \ldots, x_m\}$),
- $g_j^{x_i}$ is the number of grade points the j-th ($1 \leq j \leq k_i$) student in $\{s_1, \ldots, s_{k_i}\}$ of the data base received in the course x_i ($1 \leq i \leq m$),

– $\overline{g}^{x_i}$ is the average number of grade points achieved by the k_i students $\{s_1, \ldots, s_{k_i}\}$ in the database, who took the course x_i ($x_i \in \{x_1, \ldots, x_m\}$), and
– u_i is the number of units of course x_i ($1 \leq i \leq m$).

In this formula, we consider all courses taken by the student so far equally and independent from in which of the former semesters he/she took it for selecting courses for the upcoming semesters.

4 Composing a Curriculum for an Upcoming Semester

A curriculum for an upcoming semester is a subset of the courses $Y = \{y_1, \ldots, y_n\}$, which are possible to take in the next semester according to whether or not their prerequisites (and maybe individual other preferences such as vocational objectives) are met by the educational history $X = \{x_1, \ldots, x_m\}$.

As mentioned at the end of section 2, a reasonable relation between (a) the total number of units for courses that correlate very heavy with the educational history and (b) the total number of units for courses, which require traits and learning preferences, which do not correlate at all (and thus, may bring in new experiences) in the upcoming semester has to be determined. Let F ($0 \ll F < 1$) be a reasonable high fraction of highly correlating courses in the upcoming semester. Also, a reasonable total number U of units in the upcoming needs to be determined.

For each candidate course $y_j \in Y$, the correlation $corr(\{x_1, \ldots, x_m\}, y_j)$ to the educational history $X = \{x_1, \ldots, x_m\}$ is computed. Also, the educational history $X = \{x_1, \ldots, x_m\}$ has to be divided into (1) a sub-list $X^+ = \{x_1, \ldots, x_{m+}\}$ of courses, in which the considered student received the highest number of grade points and (2) a sub-list $X^- = \{x_1, \ldots, x_{m-}\}$ of courses, in which the student received the lowest number of grade points.

Additionally, the student is given the opportunity to add some of the 2nd-best performed courses of his/her educational history to X^+ and some of the 2nd worst performed courses to X^-. By doing so, we respect the fact, that a student may receive a "not best mark" by bad luck respectively a "not too bad" mark just by luck. Thus, we give the student the some space to add subjective feeling about being good or bad in the particular courses of his/her educational history.

Then, the set of candidate courses $Y = \{y_1, \ldots, y_n\}$ should be sorted towards a list $\overrightarrow{Y} = [y^1, \ldots, y^n]$ according to decreasing absolute values of (1) $corr(X^+, y)$, if $corr(X, y) > 0$, i.e. if y correlates positive to the (complete) educational history respectively (2) $corr(X^-, y)$, if $corr(X, y) < 0$, i.e. if y correlates negative to the (complete) educational history of the considered student. Of course, the correlations are computed based on the performance data (grade points) of former students, who too took y after all courses of X^+ respectively X^-.

Finally, the next semester is composed by subjects from the candidate set $Y = \{y_1, \ldots, y_p\}$ (1) the courses taken from the front end of this list until their total number of units reaches a value, which is as close as possible to $F * U$ and (2) the rest of the units are taken from the rear end of this list until the number of units taken from the rear reaches a value, which makes the total number of units in semester is as close as possible to U. Formally spoken, the Semester is composed as follows: $Sem := \{y^1, y^2, \ldots, y^k, y^l, y^{l+1}, \ldots, y^p\}$ with (1) $\sum_{i=1}^{k} u_i \approx F * U$ and (2) $\sum_{i=1}^{p} u_i \approx U$. More exactly, (1) means $|F * U - \sum_{i=1}^{k-1} u_i| \geq |F * U - \sum_{i=1}^{k} u_i|$ and $|F * U - \sum_{i=1}^{k+1} u_i| \geq |F * U - \sum_{i=1}^{k} u_i|$ and (2) means $|U - (\sum_{i=1}^{k} u_i + \sum_{i=l+1}^{p} u_i)| \geq |U - (\sum_{i=1}^{k} u_i + \sum_{i=l}^{p} u_i)|$.

Of course, such algorithmic result should be critically reviewed by both the student and an experienced teacher and manually adjusted.

5 Data Maintenance Issues

The correlation in the (former) students' data reflect all aspects, which may influence the learning success, which are (surly incompletely) drafted at the beginning of section 2. Thus, if any of parameter changes significantly, the former correlations may not reflect the true correlation after this change, i.e. after such a change the course should be considered as a new course, even if the content stays the same.

For example, in case a course teacher changes, the old data from the related former course need to be deleted from the database, which serves for computing the correlation, because at least the style of teaching the course and the kind of material changes by changing the teacher. In some cases, also the course content changes, too, and thus, the course profile changes even more, because the challenged traits may change additionally. Therefore, the correlation computing for between this course and any consecutive course has to begin from scratch.

Also, if a teacher revises a course and changes his/her way of teaching or the teaching material significantly, this course has to be handled in the same was as in the above case, i.e. deleted from the database. Deleting a course from the database does not mean, that the complete students' traces, which contain this course have to be deleted, of course. In fact, this would be a waste of the other useful information within these student's samples. It only means that this course does not count any more for correlation computation, i.e. it is excluded from that.

In case the same course is offered by multiple teachers, these courses have to be treated as different courses in the database, because different teachers may teach the same content in different ways.

In fact, each newly available data, also data of students, who did not complete their study yet, is a valuable source of information and has to be included into the database for making the correlation computation incrementally better and better.

6 Validation Approach

To validate the approach, we perform an experiment with available data of 186 sample students' traces. Since the composed upcoming semesters are compared with the real taken upcoming semesters of a subset of students, who took a very similar educational history, we set (1) the number of units U of the composed optimal next semester to the average number of units of the really taken upcoming semesters of those students (with the same educational history) and (2) the fraction of highly correlating courses F of the composed optimal next semester to the average fraction in the really taken upcoming semester.

Since any 2 students rarely have exactly the same educational history, we have to fine at least student clusters with "very similar" educational histories. For that purpose, we use the clustering algorithm K-means with $K = 10$ to form 10 clusters with about 15-20 students in each cluster, who have very similar educational histories in terms of courses taken so far and consider only those courses of their history, which all students of a cluster had had in common.

2nd Based on 1st Semester Experiment. First, for each cluster of students, who took the same curriculum (set of courses) $X = \{x_1, \ldots, x_n\}$ in common in their 1st semester, we compose an optimal curriculum for the 2nd semester based on our proposed technology.

For this purpose, we compute the correlation $corr(X, y_i)$ of all possible courses $Y = \{y_1, \ldots, y_n\}$ as described above and sort them towards $\overrightarrow{Y} = [y^1, \ldots, y^n]$ according decreasing values of $corr(X, y_i)$.[1]

To determine a reasonable value for F, we (1) compute the subset $Y^* \subseteq Y$ of really taken courses by the considered subset of students with the same common part of the history X, (2) determine the positions of the courses Y^* in $\overrightarrow{Y}$, (3) look for "the biggest gap" in-between any two position numbers, and (4) compute F by the number of units for the courses in front of this gap, divided by the total number of units for all courses in Y^*.

To determine U, we compute the average number of units that have really been taken in the considered students' second semester.

Based on $\overrightarrow{Y}$, F, and U, we compose an optimal 2nd semester with our technology as described in section 4. Then, we compute a "similarity" of each really taken 2nd semester by the students with the history X by their number of units, which are part of the optimal 2nd semester, related to (divided by) the total number of units in their really taken 2nd semester.

In case the real GPAs of the considered students in the 2nd semester increase with the similarity to the composed optimal 2nd semester, our technology is proven to be valid. This can easily be seen by drawing a diagram with the similarity to the composed optimum at its x-axes and the achieved GPS at the y-axes.

***i*-th Based on $(i-1)$-th Semester Experiments.** Next, we do the same experiment with each cluster of students, who took the same curriculum (set of courses) $\{x_1, \ldots, x_n\}$ in common in their 1st and 2nd semester.

[1] The top index in the sorted list is intended to clear up, that y_i is usually different from y^i.

Third, we do the same experiment with each cluster of students, who took the same curriculum (set of courses) $\{x_1, \ldots, x_n\}$ in common in their 1st, 2nd, and 3rd semester and so on until the subsets of students with identical educational histories become too small (at latest, when they have one element only).

7 Summary and Outlook

We introduced a refined technology to mine course characteristics similarities of former students' study traces and utilize them to optimize curricula of current students based to their performance traits revealed by their study achievements so far.

This way, our technology generates suggestions of personalized curricula. Furthermore, this technology is supplemented by an adaptation mechanism, which compares recent data with historical data to ensure that the similarity of mined characteristics follow the dynamic changes affecting curriculum (e.g., revision of course contents and materials, and changes in teachers, etc.).

Meanwhile, we collected sufficient data to mine such correlations (namely 186 students' traces) and start advising students optimal curricula based on this technology. Our next step is to perform the experiments as outlines in section six and analyze their results.

Furthermore, we investigate more refined correlation measures (than linear correlation), which are said to be more robust against outliers in the data base.

References

1. Felder, R.M., Silverman, L.K.: Learning and teaching styles in engineering education. Engineering Education 78(7), 674–681 (1988)
2. Gardner, H.: Frames of Mind: The Theory of Multiple Intelligences. Basic Books, New York (1993)
3. Knauf, R., Sakurai, Y., Takada, K., Dohi, S.: Personalized Curriculum Composition by Learner Profile Driven Data Mining. In: Proc. of the 2009 IEEE International Conference on Systems, Man, and Cybernetics (SMC 2009), San Antonio, TX, USA, pp. 2137–2142 (2009) ISBN: 978-1-4244- 2794-9
4. Knauf, R., Sakurai, Y., Tsuruta, S., Jantke, K.P.: Modeling Didactic Knowledge by Storyboarding. Journal of Educational Computing Research 42(4), 355–383 (2010) ISSN: 0735-6331 (Paper) 1541-4140 (Online)
5. Tsuruta, S., Knauf, R., Dohi, S., Kawabe, T., Sakurai, Y.: An Intelligent System for Modeling and Supporting Academic Educational Processes. In: Peña-Ayala, A. (ed.) Intelligent and Adaptive ELS. SIST, vol. 17, pp. 469–496. Springer, Heidelberg (2013)

The Concept of eTextbooks in K-12 Classes from the Perspective of Its Stakeholders

Guang Chen, Chaohua Gong, Junfeng Yang, Xiaoxuan Yang, and Ronghuai Huang*

Beijing Key Laboratory of Education Technology, Beijing Normal University, Beijing, China
{teastick,ronghuai.huang,gongchaohua,yangjunfengphd}@gmail.com,
yang_xiaoxuan@sina.cn

Abstract. With the emergence of eTextbooks initiatives on the rise, it is promising that eTextbooks support significant opportunities for improving the educational practices. However, there are both positive and negative reports of using eTextbooks in the academic setting. This study was aimed at exploring the concept of eTextbooks by confirming the opinions from stakeholders on the features and functions of eTextbooks in K-12 classes using a Delphi method. We conducted a three-round Delphi study with 56 respondents, including administrators, teachers, students, parents, and researchers from 14 organizations in Beijing. The findings identified 18 features and functions that covered a range of dimensions including structure and layout, interactive media, note-taking tools, assignment tools and management tools. Then, we tested eTextbooks in real classes at two primary schools initiatively. The results showed that eTextbooks could keep the instructional process running as smoothly as before.

Keywords: eTextbooks, concept, functions, stakeholder, Delphi study, K-12 classes.

1 Introduction and Literature Review

The prevalence of information technology is introducing a new trend in electronic publishing. Electronic publications are rapidly replacing printed materials. Many books are now available in electronic formats. With the tide of eBooks, eTextbooks have been under development in many countries. In Japan, the Ministry of Internal Affairs and Communications (MEXT) proposed the deployment of eTextbooks to all elementary and junior high school students by 2015, in the "Haraguchi-Vision", in late 2010[1]. In Korea, the "Education and Human Resources Development Ministry" and the "Korea Education and Research Information Service Korea" have been developing digital textbooks under the policy of "Government's Plan to Introduce Smart Education". Under this policy, eTextbooks are scheduled to be introduced into elementary and junior high school by 2014[2]. According to a report published in USA Today, the Obama Administration is advocating the goal of an eTextbook in every student's hand by 2017[3].

* Corresponding author.

A. Holzinger and G. Pasi (Eds.): HCI-KDD 2013, LNCS 7947, pp. 319–325, 2013.

So what is an eTextbook? Byun et al. defined eTextbooks from the aspect of multimedia: "a digital learning textbook that maximizes the convenience and effectiveness of learning by digitizing existing printed textbooks, to provide the advantages of both printed media and multimedia learning functions such as images, audiovisuals, animations, and 3D graphics as well as convenient functions such as search and navigation", which was also called, "digital textbooks"[4]. From the perspective of learning mode, Hsieh, et al. mentioned that the eTextbook is a kind of technology-mediated learning by nature[5]. The technology here is as mediation and communication for learning. Learning with eTextbooks is not independent activity but a community of practice that allows much communication among its participants. To meet the classroom teaching and learning requirement for Chinese K-12 schools, Chen et al. stated that eTextbook was a special kind of eBook developed according to curriculum standards, which meets the students' reading habits, facilitates organizing learning activities, and presents its contents in accordance with paper book style[6]. So how to give a clear definition for eTextbook still needed to refine.

Many researchers assert that eTextbooks have lots of advantages. Compared to paper textbooks, Sun & Flores asserted that what would really make eBooks viable for academic use is added functionality over printed versions. eTextbooks are portable and relatively easy to attain with complementary features, such as searching, hyper linking, highlighting, and note sharing[7]. Compared to eBooks, Hsieh, et al. suggested that the feature of eTextbooks could facilitate learning activities rather than just reading, which provide open access to students[5]. They asserted that eTextbooks could fit the characteristics of learning and the characteristics of the students would be understood.

However, there are still negative aspects of eTextbooks in instructional practice. According to Nicholas & Lewis, some of the disadvantages inherent with eTextbooks include the eyestrain from electronic displays[8]. From the aspect of user experience, Daniel et al. pointed out that the medium of eTextbook itself might not be as comfortable as a paper textbook experience for readers. The design of an eTextbook may need to make for a more constructive user experience[9]. Regarding to reading comprehension, some researchers argued that personal habits have developed early in life, which makes it difficult to adjust to new reading format for many people who grew up exclusively with printed texts[10]. Problems have emerged because learning with eTextbooks is beyond just reading with eBooks.

So, there are differences among opinions of eTextbooks from various researchers who have conducted studies in colleges and outside of classrooms, which cannot cover all the context of using eTextbooks. In this study, we tried to explore the concept of eTextbook from the perspective of the stakeholders using the Delphi method[11, 12], which will enable us to confirm what functions are needed for using eTextbooks in K-12 classes.

2 Research Design

2.1 Research Framework and Procedures

To explore the concept of eTextbooks in K-12 classrooms, we proposed the research framework. In the preliminary research period, we listed the advantages (functions) of eTextbooks through reviewing literature and identified respondents. In the first round of the Delphi study, we conducted a survey to examine what functions parents, teachers, students and administrators consider of great importance in using eTextbook in K-12 classrooms. For the second round of the Delphi study, we conducted face-to-face interview with 9 respondents. We summarized the required functions from the second-round of the study. In the third-round of the Delphi study, we interviewed 5 experts (researchers) from Beijing Normal University, so that we would refine the functions to be needed in real K-12 classes correctly. Then, we developed eTextbooks with specified functions and piloted the functions in real classed in order to confirm the required functions.

2.2 Participants

56 stakeholders were interviewed to determine the functions required for eTextbooks in K-12 classes, including 9 administrators, 19 teachers, 13 students, 10 parents, 5 researchers from 14 organizations in Beijing. To confirm the required functions, there were 9 teachers and 203 students in Grade 4 from two elementary schools in Beijing taking part in the research.

2.3 Data Collection

During the first round Delphi study, we proposed the interview outline with 3 open-ended questions:

(1)The advantages of eTextbooks: Compared to printed textbooks, what are the advantages of e-textbooks?

(2) Concept and features: What are your opinions on the concept of eTextbooks in K-12 classes? What kind of features from the checklist that should be considered when designing eTextbooks?

(3) Challenges: What are the challenges for eTextbooks in K-12 schools?

Defining consensus: Levels of agreement using the Likert scale was applicable to previous research, as quoted by McKenna [13] as 51% and by Williams & Webb [11] as 55%. Respondents' additional views and comments on items were also sought using both additional "tick boxes" and by providing space for written comments. This proved to be a valuable addition to the function checklist. The second and the third round were as the same as the first round.

After piloting the functions of eTextbook in K-12 classes, we collected the interviews from the 5 teachers and 10 students at random.

3 Results and Discussion

3.1 Results from the First-Round Delphi Study

Before conducting the first-round Delphi study, we listed the functions (advantages) of eTextbooks from the literature and the experiences of our research team, including structure, interactive rich-media, note-taking tools, assignment tools and management tools. Then we divided the draft functions checklist into 43 items, as shown in Fig. 1. We interviewed 51 respondents. All respondents were requested to rate the validity of each item on a 5-point Likert-type scale from 1 (not appropriate) to 5 (appropriate). The differences in opinions were discussed and resolved in a face-to-face interview with the relevant stakeholder groups later. The validity was evaluated by the same method.

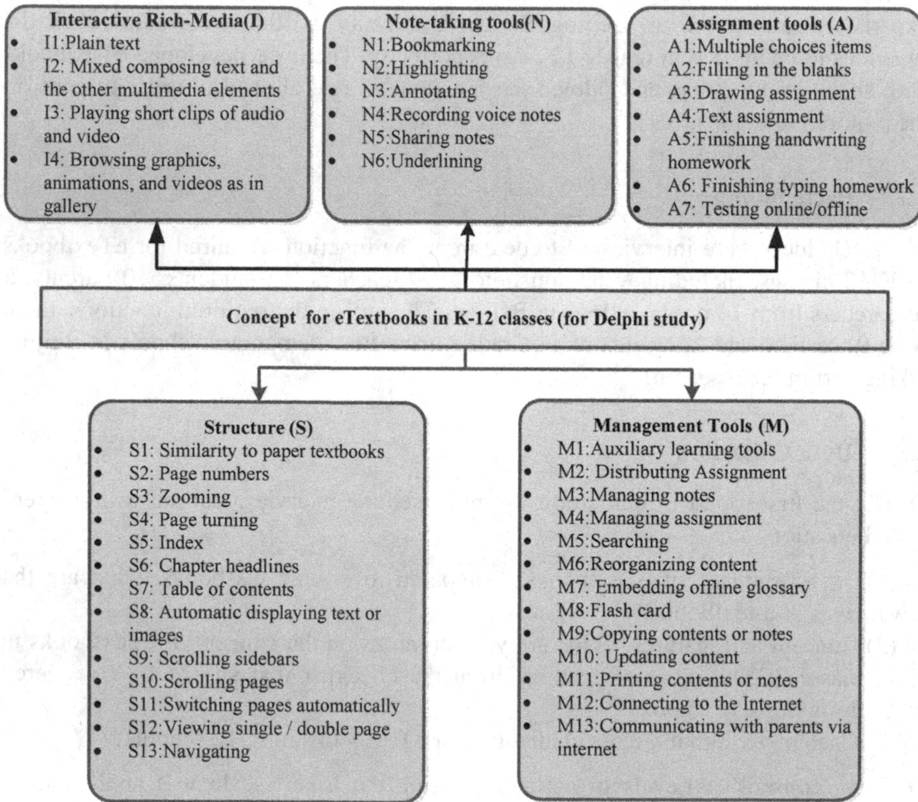

Fig. 1. Functions required for eTextbooks in K-12 Classes (from literature review)

Through analyzing the point of views from the stakeholders we found that they widely accepted the functions and features of eTextbooks to be divided into 5 dimensions (Mean>4.00, Min=3, Max=5). With regard to the items from each dimension, by calculating the mean from the function list, we deleted the items that were under 3.0. 24 items of 43 were remained. In all items, the median value was 4 or more and the difference between the minimum and maximum was 1.

Additionally, from respondents' opinions, we revised some expressions of the dimensions and items to make them more accurate and appropriate. For dimensions, we changed "Structure (S)" to "Structure and layout (S)", and "Interactive Rich Media" to "Interactive Media". For items, we changed "S1: Similarity to paper textbooks" to "S1: Similarity to paper books", "S4: Page turning" to "S4: Page flip effect", and "M5: Searching" to "M5: Searching in eTextbook and notes".

3.2 Results from the Second-Round Delphi Study

Using the results of 5 dimensions including 24 items from the first-round interviews, we conducted a face-to-face interview with the stakeholders. Nine (9) respondents (3 administrators, 3 teachers, and 3 parents) revised the function list according to readability and functions for supporting teaching and learning, and deleted 8 controversial items. Hence, after the second-round Delphi study, 16 items from the functions of eTextbooks that were required for classes remained.

3.3 Results from the Third-Round Delphi Study

We sent the revised version to 5 researchers who focused on eTextbook in education from Beijing Normal University. The validity test used "Content Validity Ratio" (CVR) method. 5 researchers were invited to give scores on the validity of the functions list. After collecting their scores, all the 16 items with CVR were over 0.8. The researchers suggested that 2 items (A8 and M14) should be added to the function list of eTextbooks. For item A8, researchers asserted that eTextbooks should enable students to submit their homework via a submitting system. Students' benefit from the system no matter in class or outside of class when they finish their homework and would like to submit them. For item M14, researchers pointed out that the function of synchronizing contents and notes was beneficial to both online and offline learning. Teachers and students would be able to synchronize various instructional materials simultaneously, which supports ubiquitous learning. The result from the third-round Delphi study was shown in Fig. 2.

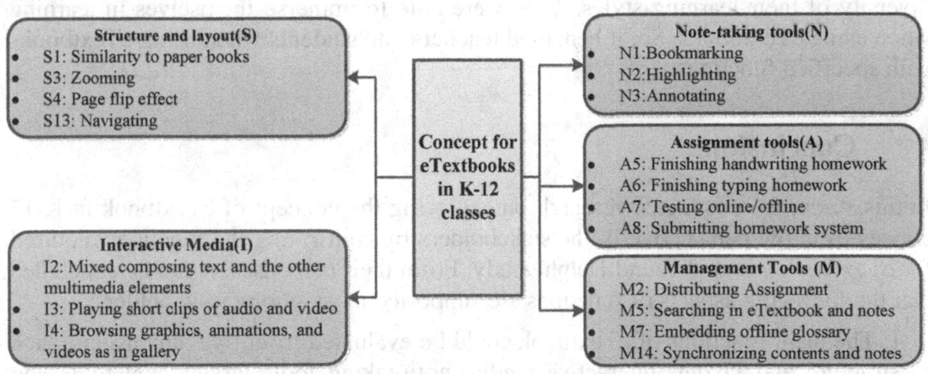

Fig. 2. Result from the third-round Delphi study

3.4 Piloting eTextbooks with Particular Features in Class

According to the results from third-round Delphi study, we designed eTextbooks with 18 required functions (as in Fig. 2). Then we conducted the pilot experiment to refine the functions with responds from teachers and students. Based on the piloting school conditions, we chose iPads as learning devices in classroom for each teacher and student; iBooks were used as eTextbooks reader software for the presentation of the learning contents; iTeach developed by the team of researchers was used as an interaction platform for teacher, students and contents. We observed and recorded fourteen classes taught by nine teachers to clarify the correspondence by considering the function and its practical usage. We collected the teachers' reflection documents after class and interviewed nine teachers for investigating their experiences.

Fig. 3. Using eTextbooks in real class

Teachers and students confirmed that the 18 functions had all played an important role in facilitating the teaching and learning performance in the classes. It was of great significance to design eTextbooks with the concept that had a focus on these functions. From teachers' perspectives, among the 5 dimensions, they mentioned that eTextbook should inherit the functions of paper textbooks that keep the classroom instruction running smoothly, including annotating, highlighting, etc. From students' opinions, the structure and layout of the eTextbooks should be able to match the diversity of their learning styles. They were able to immerse themselves in learning when using eTextbooks. So, it benefited teachers and students when using eTextbooks with specified functions.

4 Conclusion

In this study, we conducted research on exploring the concept of eTextbook in K-12 classes from the perspective of the stakeholders by confirming the functions required for eTextbook with a 3-round Delphi study. From their responses, it can be concluded that the following aspects of functions are supported most among stakeholders:

1. The main functions of eTextbook could be evaluated from five dimensions, such as structure and layout, interactive media, note-taking tools, assignment tools and management tools.

2. The eTextbooks with similar structures and layout of paper textbooks would help students follow their reading habits for accepting them initiatively. The structure of paper textbook, such as table of contents, chapter headlines, index, page numbers, are considered to be of great importance; the unique feature of eBook such as zooming, page flip effect, and navigating should also be taken into consideration.

3. From the teachers' perspective, the management tools of eTextbooks are very important for utilizing them in normal classes and expanding their use, such as distributing assignment, searching in eTextbook and notes, embedding offline glossary, synchronizing contents and notes.

4. From the teachers' and parents' perspective, the assignment tools of eTextbooks are necessary for doing homework and communicating between parents and teachers, such as testing online/offline, finishing typing/handwriting homework assignments, submitting then via the built-in system.

5. From the students' perspective, the note-taking tools of eTextbooks enable them to take notes like paper textbooks. Additionally, the unique features of note-taking tools would enhance the students' note-taking abilities in and out of classes, including bookmarking, highlighting and annotating.

This is a tentative research on exploring the concept of eTextbook in K-12 classes and a Delphi study has its own limitations. So investigation on a larger scale of sample capacity will be needed to collect more evidences in order to confirm these findings. A series of further experiments should be carried out to examine the findings of this study.

References

1. IBTimes Staff Reporter: Japan to pilot digital textbooks in primary schools soon, http://www.ibtimes.co.uk/articles/65459/20100924/digital-textbooks-in-japan-primary-schools-electronic-textbooks-in-japan-primary-schools.htm
2. Taizan, Y., Bhang, S., Kurokami, H., Kwon, S.: A Comparison of Functions and the Effect of Digital Textbook in Japan and Korea. International Journal for Educational Media and Technology 6, 85–93 (2012)
3. Liu, Z.: Is it time for wider acceptance of e-textbooks? An examination of student reactions to e-textbooks. Chinese Journal of Library and Information Science 5, 76–87 (2012)
4. Byun, H., Cho, W., Kim, N., Ryu, J., Lee, G., Song, J.: A Study on the effectiveness measurement on Electronic Textbook, Korean Education & Research Information Service. Research Report CR 2006-38, Republic of Korea (2006)
5. Hsieh, C.-C., Chen, J.W., Luo, D., Lu, C.-C., Huang, Y.L.: Insights from the Technology System Method for the Development Architecture of e-Textbooks. Presented at the ICDS 2011, The Fifth International Conference on Digital Society (February 23, 2011)
6. Chen, G., Gong, C., Huang, R.: E-textbook: Definition, Functions and Key Technical Issues. Open Education Research 18, 28–32 (2012)
7. Sun, J., Flores, J., Tanguma, J.: E-Textbooks and Students' Learning Experiences. Decision Sciences Journal of Innovative Education 10, 63–77 (2012)
8. Nicholas, A.J., Lewis, J.K.: The Net Generation and E-Textbooks. International Journal of Cyber Ethics in Education 1 (2011)
9. Daniel, D.B., Woody, W.D.: E-textbooks at what cost? Performance and use of electronic v. print texts. Computers & Education 62, 18–23 (2013)
10. Kang, Y.-Y., Wang, M.-J.J., Lin, R.: Usability evaluation of E-books. Displays 30, 49–52 (2009)
11. Williams, P.L., Webb, C.: The Delphi technique: a methodological discussion. Journal of Advanced Nursing 19, 180–186 (1994)
12. Beretta, R.: A critical review of the Delphi technique. Nurse Researcher 3, 79–89 (1996)
13. McKenna, H.P.: The Delphi technique: a worthwhile research approach for nursing? Journal of Advanced Nursing 19, 1221–1225 (2006)

A Multi-dimensional Personalization Approach to Developing Adaptive Learning Systems

Tzu-Chi Yang[1], Gwo-Jen Hwang[2], Tosti H.C. Chiang[1], and Stephen J.H. Yang[1]

[1] Department of Computer Science and Information Engineering,
National Central University, No. 300, Jung-da Road, Chung-li, Taoyuan 320, Taiwan
{tcyang,tosti.Chiang,jhyang}@csie.ncu.edu.tw

[2] Graduate Institute of Digital Learning and Education,
National Taiwan University of Science and Technology,
43, Sec.4, Keelung Rd., Taipei, 106, Taiwan
gjhwang.academic@gmail.com

Abstract. In this study, a multi-dimensional personalization approach is proposed for developing adaptive learning systems by taking various personalized features into account, including learning styles and cognitive styles of student. In this innovative approach, learning materials were categorized into several types and associated as a learning content based on students' learning styles to provide personalized learning materials and presentation layouts. Furthermore, personalized user interfaces and navigation strategies were developed based on students' cognitive styles. To evaluate the performance of the proposed approach, an experiment was conducted on the learning activity on the learning activity of the "Computer Networks" course of a college in Taiwan. The experimental results showed that the students who learned with the system developed with the proposed approach revealed significantly better learning achievements than the students who learn with conventional adaptive learning system, showing that the proposed is effective and promising.

Keywords: adaptive learning, personalization, learning style, cognitive style.

1 Introduction

In the past decade, various personalization techniques have been proposed for developing adaptive hypermedia learning systems, and have demonstrated the benefit of such an approach [1], [2]. Many researchers have indicated the importance of taking personal preferences and learning habits into account [3]. Among those personal characteristics, learning styles which represent the way individuals perceive and process information have been recognized as being an important factor related to the presentation of learning materials [4]. On the other hand, cognitive styles have been recognized as being an essential characteristic of individuals' cognitive process. Researchers have tried to develop adaptive learning systems based on either learning styles or cognitive styles; nevertheless, seldom have both of them been taken into consideration, not to mention the other personalized factors [1], [5], [6].To cope with

A. Holzinger and G. Pasi (Eds.): HCI-KDD 2013, LNCS 7947, pp. 326–333, 2013.

this problem, this paper presents an adaptive learning system which is developed by taking students' preferences and characteristics, including learning styles and cognitive styles, into consideration. Moreover, an experiment has been conducted to show the effectiveness of the proposed approach.

2 Literature Review

2.1 Learning Styles and Cognitive Styles

Learning styles have been recognized as being an important factor for better understanding the model of learning and the learning preferences of students [7]. Keefe [8] defined an individual's learning style as a consistent way of functioning that reflects the underlying causes of learning behaviors, he also pointed out that learning style is a student characteristic indicating how a student learns and likes to learn. The Felder–Silverman Learning Style Model (FSLSM) developed by Felder and Silverman [9] have been recognized by many researchers as being a highly suitable model for developing adaptive learning systems. Kuljis and Lui [10] further compared several learning style models, and suggested that FSLSM is the most appropriate model with respect to the application in e-learning systems. Consequently, this study adopted FSLSM as one of the factors for developing the adaptive learning system.

On the other hand, cognitive style has been recognized as being a significant factor influencing students' information seeking and processing [10]. It has also been identified as an important factor impacting the effectiveness of user interfaces and the navigation strategies of learning systems [1]. Among various proposed cognitive styles, the field dependent (FD) and field independent (FI) styles proposed by Witkin, Moore, Goodenough and Cox [11] are the most frequently adopted. Therefore, in this study, FI/FD cognitive style is adopted as another factor for developing the adaptive learning system.

Accordingly, in this study, learning styles are used to provide personalized learning materials and presentation layouts, while cognitive styles are used to develop personalized user interfaces and navigation strategies.

2.2 Adaptive Learning Systems

An adaptive learning system aims to provide a personalized learning resource for students, especially learning content and user-preferred interfaces for processing their learning. Researchers have indicated the importance of providing personalized user interfaces to meet the learning habits of students [1]. It can be seen that the provision of personalization or adaptation modules, including personalized learning materials, navigation paths or user interfaces, has been recognized as an important issue for developing effective learning systems [12].

Several studies have been conducted to develop adaptive learning systems based on learning styles or cognitive styles [1], [3], [13]. However, few studies have considered multiple learning criteria, including learning styles, cognitive styles, and knowledge levels, for developing adaptive learning systems.

3 System Implementation

The proposed adaptive learning system (ALS) consists of four modules: the Learning content-Generating Module (LCGM), the Adaptive Presentation Module (APM), the Adaptive Content Module (ACM) and the Learning Module (LM). Fig. 1 shows the framework of the ALS modules.

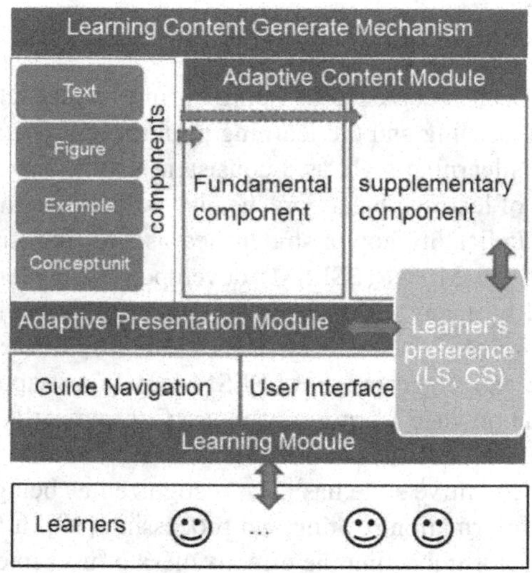

Fig. 1. The framework of ALS modules

3.1 Learning Content-Generating Module

Fig. 2 shows the concept of the learning content-generating module, which is used to extract contents from raw materials and generate chunks of information for composing personalized learning materials based on the presentation layout. Each subject unit contains a set of components, such as the ID of the unit, texts, photos, etc. The components of a subject unit are classified into the following six categories:

- Concept unit: containing the title, concept ID, abstract and representative icon of the course unit.
- Text components: the text content of the course unit.
- Example component: the illustrative examples related to the course content.
- Figure component: the pictures, photos and figures related to the course unit.
- Fundamental component: Fundamental components contain the primary contents of a course, including the title of each learning unit or concept, and the corresponding texts, figures, examples and exercises.
- Supplementary component: Supplementary components contain supplementary materials that are helpful to students in extending the learning scope or realizing the concepts to be learned.

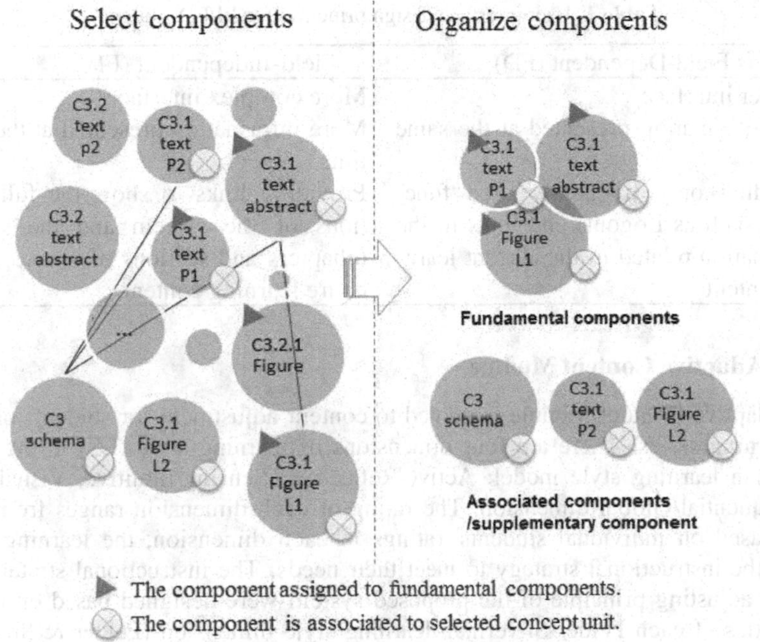

Fig. 2. The concept of LCGM

After selecting the appropriate components (learning materials), LCGM organizes the selected components based on individual students' learning styles and cognitive styles. The selection rule of LCGM depends on the metadata of each component. For example, concept C3.1 has six components (Fig. 2), three of them assigned to fundamental component, the others selected as associated / supplementary components. The organized learning content is then presented to individual students based on the presentation layout framework (process with APM), which consists of the following areas: The system reservation area, curriculum navigation area, learning content area, supplementary material area and the guided navigation area. For example, for the FD students, "next stage" and "previous stage" buttons are provided to guide the students to learn the course materials in an appropriate sequence.

3.2 Adaptive Presentation Module

The adaptive presentation module (APM) consists of two parts: the layout strategy based on student cognitive styles and the instructional strategy based on their learning styles. The layout strategy focuses on adjusting the presentation layout for individual students based on their cognitive styles. The FD students prefer structured information, promotion and authority navigation; on the other hand, the FI students like to organize information by themselves. From the perspective of the course lesson navigation, the user interface for the FD students is designed to show less information at the same time to avoid distracting them, which is called "Simpler interface" in this study; on the contrary, the interface for the FI students presents more information to help them make a comprehensive survey of the learning content, which is called "more complex interface," as shown in Table 1.

Table 1. User interface design principle for FI/FD students

Field-Dependent (FD)	Field-Independent (FI)
Simpler interface	More complex interface
Less information presented at the same time	More information presented at the same time
Providing only frequently used functions (such as Logout) and links to the information related to the current learning content	Providing links to show the full functions of the system and the schema (chapters and sections of course) of the entire learning content

3.3 Adaptive Content Module

The Adaptive Content Module is related to content adjustment for students of different learning styles. There are four dimensions of learning style (LS) in the Felder-Silverman learning style model: Active/Reflective, Sensing/Intuitive, Visual/Verbal and Sequential/Global dimension. The rating of each dimension ranges from -11 to +11. Based on individual students' ratings in each dimension, the learning system adapts the instructional strategy to meet their needs. The instructional strategies and content adjusting principle of the proposed system were designed based on the characteristics of each Felder-Silverman learning style dimension (Felder & Silverman, 1988). The adjusting principles of ACM which refer to FSLSM are constructed by rule-based decision tree. For example, visual-style learners tend to learn better from visualized materials, such as pictures, diagrams and films, while verbal-style learners prefer text materials. The Sequential-style learner receives learning materials in a logical order, while Global-styles learner is able to browse through the entire chapter to get an overview before learning.

3.4 Learning Module

The LM provides students with the learning content and user interface generated based on their cognitive styles and learning styles. As mentioned above, Sequential students tend to gain understanding in linear steps, with each step following logically from the previous one, and tend to follow logical stepwise paths when finding solutions. Global students tend to solve problems quickly once they have grasped the big picture, and tend to learn in large jumps without seeing connections. Fig. 3 shows a learning module for an FI student with [SEQ/GLO: 4] (tend to Global), and a learning module for another FI student with [SEQ/GLO: 8] (more tend to Global).

Fig. 4 shows another illustrative example to show the similarities and differences between the learning modules generated for FD students with verbal and visual learning styles. It can be seen that the learning content has been adjusted to meet the students' learning styles. Moreover, the user interface in Fig. 4 (for FD students) is much simpler than that in Fig. 3 (for FI students), showing part of the adjustments made for the students with different cognitive styles. The user interface for FI students (Fig. 3) included the course schema in the left panel and a navigation button on the top of the screen, while that for the FD students only had the title of current course unit. From literature, most FD students were likely to be affected by contexts.

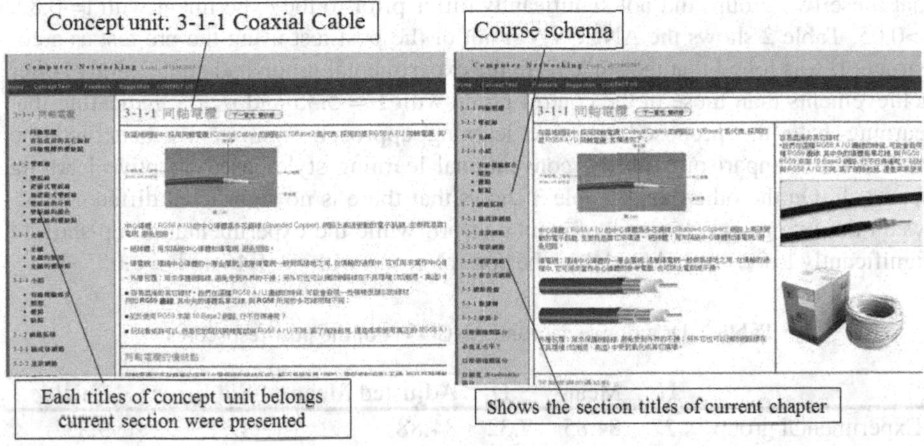

Fig. 3. Illustrative example of a learning module

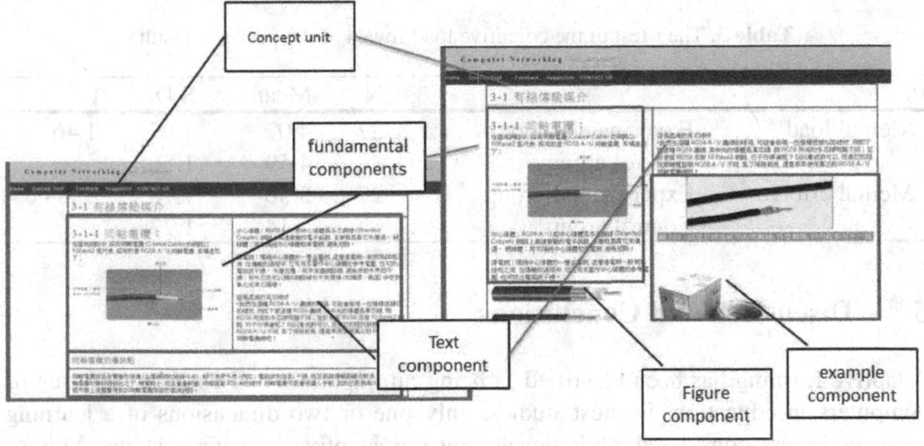

Fig. 4. Learning modules for FD students with verbal and visual learning styles

4 Experiment and Evaluation

To evaluate the performance of the proposed approach, an experiment was conducted on the learning activity of the "Computer Networks" course of a college in Taiwan. The participants were randomly divided into a control group (n = 27) and an experimental group (n = 27). The students in the experimental group were arranged to learn with the adaptive learning system; On the other hand, the students in the control group learned with a conventional adaptive e-learning approach.

To evaluate the effectiveness of the proposed approach, a pre-test, a post-test, and the measures of cognitive load was employed in the experiment. The result implying that these two groups did not significantly differ prior to the experiment with t=-0.82, p>0.05. Table 2 shows the ANCOVA result of the post-test using the pre-test as a co-variate. It was found that the students in the experimental group had significantly better achievements than those in the control group with F = 5.35 and p<.05, indicating that learning with the proposed adaptive learning approach significantly benefited the students in comparison with the conventional learning style-based adaptive learning approach. On the other hand, Table 3 shows that there is no significant difference between the two groups in terms of mental effort, while the experimental group showed significantly lower mental load than the control group with t = 1.46 and p < .05.

Table 2. Descriptive data and ANCOVA of the post-test scores

	N	Mean	S.D	Adjusted Mean	Std.Error.	*F value*
Experimental group	27	84.85	7.32	84.88	4.76	5.35*
Control group	27	80.33	7.22	80.30	4.76	

*p<.05

Table 3. The t-test of the cognitive load levels of the post-test result

		N	Mean	S.D	t
Mental load	Experimental group	27	4.07	1.07	1.46*
	Control group		4.59	1.50	
Mental effort	Experimental group	27	5.56	1.78	-0.84
	Control group		5.14	1.77	

5 Discussion and Conclusions

Adaptive learning has been identified as being an important and challenging issue of computers in education. In most studies, only one or two dimensions of a learning style model are considered while developing the adaptive learning systems. Moreover, in most systems, only a fixed type of user interface is provided. In this paper, we propose an adaptive learning system developed by using both learning styles and cognitive styles to adapt the user interface and learning content for individual students; moreover, the full dimensions of a learning style model have been taken into account. The experimental results showed that the proposed system could improve the learning achievements of the students. Moreover, it was found that the students' mental load was significantly decreased and their belief of learning gains was increased.

From the experimental results, it can be seen that the proposed approach is promising. The developed system can be applied to other applications by replacing the learning components with new ones. Furthermore, it is expected that the learning portfolios of students can be analyzed and more constructive suggestions can be given to teachers and researchers accordingly.

References

1. Mampadi, F., Chen, S.Y.H., Ghinea, G., Chen, M.P.: Design of adaptive hypermedia learning systems: A cognitive style approach. Computers & Education 56(4), 1003–1011 (2011)
2. Nielsen, L., Heffernan, C., Lin, Y., Yu, J.: The Daktari: An interactive, multi-media tool for knowledge transfer among poor livestock keepers in Kenya. Computers & Education 54, 1241–1247 (2010)
3. Tseng, J.C.R., Chu, H.C., Hwang, G.J., Tsai, C.C.: Development of an adaptive learning system with two sources of personalization information. Computers & Education 51(2), 776–786 (2008)
4. Papanikolaou, K.A., Grigoriadou, M., Magoulas, G.D., Kornilakis, H.: Towards new forms of knowledge communication: the adaptive dimension of a web-based learning environment. Computers & Education 39, 333–360 (2002)
5. Hsieh, S.W., Jang, Y.R., Hwang, G.J., Chen, N.S.: Effects of teaching and learning styles on students' reflection levels for ubiquitous learning. Computers & Education 57(1), 1194–1201 (2011)
6. Hwang, G.J., Tsai, P.S., Tsai, C.C., Tseng, J.C.R.: A novel approach for assisting teachers in analyzing student web-searching behaviors. Computers & Education 51(2), 926–938 (2008)
7. Filippidis, S.K., Tsoukalas, L.A.: On the use of adaptive instructional images based on the sequential-global dimension of the Felder-Silverman learning style theory. Interactive Learning Environments 17(2), 135–150 (2009)
8. Keefe, J.W.: Learning style: Cognitive and thinking skills. National Association of Secondary School Principals, Reston (1991)
9. Felder, R.M., Silverman, L.K.: Learning styles and teaching styles in engineering education. Engineering Education 78, 674–681 (1988)
10. Kuljis, J., Liu, F.: A comparison of learning style theories on the suitability for elearning. In: Hamza, M.H. (ed.) Proceedings of the IASTED Conference on Web-Technologies, Applications, and Services, pp. 191–197. ACTA Press (2005)
11. Witkin, H.A., Moore, C.A., Goodenough, D.R., Cox, P.W.: Field-dependent and field-independent cognitive styles and their educational implications. Review of Educational Research 47(1), 1–64 (1977)
12. van Seters, J.R., Ossevoort, M.A., Tramper, J., Goedhart, M.J.: The influence of student characteristics on the use of adaptive e-learning material. Computers & Education 58, 942–952 (2012)
13. Hwang, G.J., Yin, P.Y., Wang, T.T., Tseng, J.C.R., Hwang, G.H.: An enhanced genetic approach to optimizing auto-reply accuracy of an e-learning system. Computers & Education 51(1), 337–353 (2008)

Extending the AAT Tool with a User-Friendly and Powerful Mechanism to Retrieve Complex Information from Educational Log Data*

Stephen Kladich, Cindy Ives, Nancy Parker, and Sabine Graf

Athabasca University, Canada
sjkladich@hotmail.com, {cindyi,nancyp,sabineg}@athabascau.ca

Abstract. In online learning, educators and course designers traditionally have difficulty understanding how educational material is being utilized by learners in a learning management system (LMS). However, LMSs collect a great deal of data about how learners interact with the system and with learning materials/activities. Extracting this data manually requires skills that are outside the domain of educators and course designers, hence there is a need for specialized tools which provide easy access to these data. The Academic Analytics Tool (AAT) is designed to allow users to investigate elements of effective course designs and teaching strategies across courses by extracting and analysing data stored in the database of an LMS. In this paper, we present an extension to AAT, namely a user-friendly and powerful mechanism to retrieve complex information without requiring users to have background in computer science. This mechanism allows educators and learning designers to get answers to complex questions in an easy understandable format.

Keywords: Learning Analytics, Learning Management System, Log Data.

1 Introduction

Learning management systems (LMSs) collect and store a great deal of data about learners and learning materials/activities along with how learners interact and utilize these materials/activities [1]. As learners access more and more learning material/activities online, and as LMSs collect more and more data about learners and their interactions, analysing and understanding that data has become critical for many reasons. For instance, this data can inform educators and learning designers about how students learn and how they use the provided materials/activities, leading to valuable insight into student behaviours and enabling educators and learning designers to better adapt learning content and teaching strategies to student needs. Such insights may lead to better understanding of the effectiveness of course designs and teaching strategies, and thus to better student satisfaction and increased student performance from improved learning designs and teaching strategies.

* The authors acknowledge the support of Athabasca University and NSERC.

A. Holzinger and G. Pasi (Eds.): HCI-KDD 2013, LNCS 7947, pp. 334–341, 2013.

The concept of analysing learning system data is not new. Rudimentary analytics have been employed since the beginning of computer-based training [2], for example to add a degree of personalization to the learner's experience. However, when dealing with aggregated data of a high number of users in a large instructional setting, more advanced techniques and tools are required to effectively understand learners' interactions with the course materials and resources of the LMS, particularly when analyzing such data across entire programs of study. The large amount of data generated is often stored in complex log files or databases, and therefore, specialized techniques are required to process the data and derive meaningful information. These techniques are often not within the skill-set of instructional designers and educators as they are highly technical or computational in nature and require specific skills and tools (such as knowledge of database schema/SQL). Therefore, such educational log data are typically not accessed by learning designers and educators, and thus, little feedback is available for them on whether their teaching strategies and learning designs are actually helping students. Siemens [3] writes that the field of learning analytics (and its related research) must make such activities relevant to educators and administrators – this is only possible if tools are developed that specifically aim at this group to help them understand, evaluate and act on the data that learning systems generate.

Given this skills gap, researchers have started to design and build solutions to assist educators in understanding the data in their learning systems. Tools such as GLASS [4], LOCO-Analyst [5], and CosyLMSAnalytics [6] have been developed to meet this need, however they all have either complex user interfaces, are specific to only one LMS, only extract data without providing even basic analytical functionality, or require other specialized software to be installed.

This paper focuses on the Academic Analytics Tool (AAT) [7]; a browser-based application that can access and report on the data generated by any LMS. While a general description of the tool is provided by Graf et al. [7], the aim of this paper is to introduce an extension of the tool which increases the user-friendliness of AAT and at the same time provides users with advanced access to complex information from educational log data.

The paper is structured as follows: Section 2 provides a brief overview of AAT. Section 3 presents the proposed extension of the tool and Section 4 concludes the paper and discusses future work.

2 Overview of AAT

AAT allows users to access and analyse educational log data from any LMS. It uses input data from one or several databases of an LMS, extracts and analyses the data that are specified by users, stores the results within the Academic Analytics database, and allows exporting the results as CVS and HTML files to be further used by visualization or advanced statistical tools. In the following paragraphs, the main design decisions and building blocks of the tool are described to provide an overview of AAT.

In order for users to be able to specify what data they are interested in accessing and analysing, one of the main building blocks of AAT is the notion of *concepts,* which refer to logical constructs of interest to the user (such as a course, discussion forum, quiz etc.). Most of these concepts are *learning objects* (e.g., discussion forum, quiz, resource, etc.) and therefore, have a pedagogical type which defines their educational purpose (e.g., a quiz could be for self-assessment or graded).

While concepts only allow users to specify the area in which they are interested, *patterns* allow them to specify in more detail what data they exactly are looking for. Patterns are similar to reports or queries and can be simple lists of concepts (e.g., a list of students who have posted more than 5 times in a forum) or an analysis/calculation on concepts (e.g., the average amount of time students spent on quizzes).

In order to make the tool applicable for different LMSs, *templates* have been introduced, which can be seen as the interface between the tool and the databases. While patterns specify what data should be extracted from a database, templates specify where (i.e., what tables and columns) the respective data resides within the database of a particular learning system, considering the version of the system (e.g., Moodle 2.2). Each LMS version/instance has its own template, and the template is configurable to encompass any alterations or modifications to the out-of-the-box schema. Therefore, the tool can be used for an LMS even if its database schema has been modified to meet the needs of the institution.

In the context of AAT, a *database* is the underlying repository for an LMS. AAT can connect to different databases as well as to multiple databases at the same time.

Another central entity to AAT is the *dataset*. A dataset specifies one or more courses that the user is interested in. The dataset defines on which data the patterns are executed (e.g., all courses in Nursing, the course "Introduction to Biology", etc.).

In order to allow a user to run patterns repeatedly over time on the same or different datasets and databases, *profiles* have been introduced. A profile can be seen as an experiment for extracting and analysing particular data. In a profile, a user specifies which LMS is used (through selecting a template), which courses and time spans should be investigated (through selecting the dataset), and which data the user is interested in (through selecting patterns). AAT guides the user through this specification process. Once the profile is created, it can be used to extract and analyse the specified data.

Another important feature of AAT is *user management*, including user roles and the authentication process. All users authenticate to the tool via built-in login or via CAS [10]. The authentication method is set via the application's configuration file. There are two principle types of users in AAT: Administrators and Standard Users. Administrators can add or edit templates and are assumed to have an understanding of the LMS database's schema. Administrators also define the database connections the tool uses to access LMS data, including the authentication credentials. Standard Users (e.g., educators, course designers) can use the tool to select an LMS to connect to, select a dataset, create/edit/store/delete patterns, and create/edit/store/delete profiles. Standard users are not expected to have any knowledge of database schema or SQL.

3 Pattern Creation Mechanism

While the previous section provided a brief overview of AAT, this section presents an advanced pattern creation mechanism that extends AAT to provide users with easy-to-use functionality to extract and analyse complex information from educational log data. The mechanism includes three components to improve the user-friendliness and the capacity of the tool to enable the user to ask complex questions (and therefore build complex patterns) while still remaining easy to use for users without computer science background. These components include: (1) an ontology for easy use of the tool, (2) a refined approach for pattern chaining, and (3) further options and a wizard-style user interface for pattern creation. In the following subsections, these three components are presented in more detail.

3.1 AAT Ontology

Ontologies in the information systems context are utilized to represent the knowledge about a system (i.e., its parts, relationship of parts, activities, etc.), to share that knowledge, and house a vocabulary about that system [11]. Such an ontology forms the reference point for design and implementation considerations for AAT. The AAT system domain knowledge is represented by the AAT Ontology illustrated in Fig. 1. In this figure only the most relevant attributes are included as examples since a complete visualization of all the attributes is space-prohibitive. There are three key aspects to the AAT Ontology: Concepts, Databases and Templates.

Concepts represent the entities that are considered in the system to get information about and perform analysis. Each concept can have one or more attributes that contain data about this concept. In this ontology, there are five main types of concepts: User, Course, Learning Object, Message and Log. The User concept consists of three subtypes which represent common roles in an LMS: Student, Teacher, and Course Creator. Similarly, the Learning Object concept has several subtypes that encompass the types of learning objects typically used by LMSs: assignments, quizzes, forums, books, resources, wikis, blogs, surveys, and lessons. The list of learning objects, users and concepts can be easily extended if needed, making the ontology extensible.

A *Database* is the repository that includes data about the learners and their behaviour/performance (i.e., a database of an LMS) and is mapped by the template to facilitate pattern creation.

A *Template* is composed of Concept Mappings, one for each concept that is supported by the LMS. A Concept Mapping not only maps the relationship between the concept and the concept's location in the LMS database, but also the relationship between concepts. Specifically, the concept mapping includes four types of information. First, it provides information on where a concept and its attributes are stored in an LMS database, including the concrete information in terms of tables and columns. Second, it also includes information about the relationships between concept attributes and the database schema, which is critical when a concept is physically stored in multiple database tables. For example in Moodle, data about quizzes is stored in several tables, so the concept mapping for the concept quiz provides information about how these quiz tables are related to each other (e.g., which question belongs to which quiz) and how information in these tables can be found. Third, a concept mapping stores the relationships between the respective concept and other concepts, as well as indicates the unique id of each concept. For example, information about how the concept *Student* is related to other concepts such as *Discussion Forum*, *Quiz*, etc. is stored (e.g., to find out which student has posted messages in a discussion forum). Fourth, a concept mapping stores labels for the concept and its attributes. Such labels are names for the concept and its attributes that can be easily understood by users of the tool without knowledge about the database schema and how tables and columns are called in the database. These labels are then used throughout the system as names for concepts and their attributes (e.g., in the pattern creation process) so that users are able to employ terminology they are comfortable with. These labels are LMS/template specific, so that different terminology can be used for different learning systems. This is an important feature to make the system user-friendly, as different vendors use their own terminology, to which users of that LMS become accustomed.

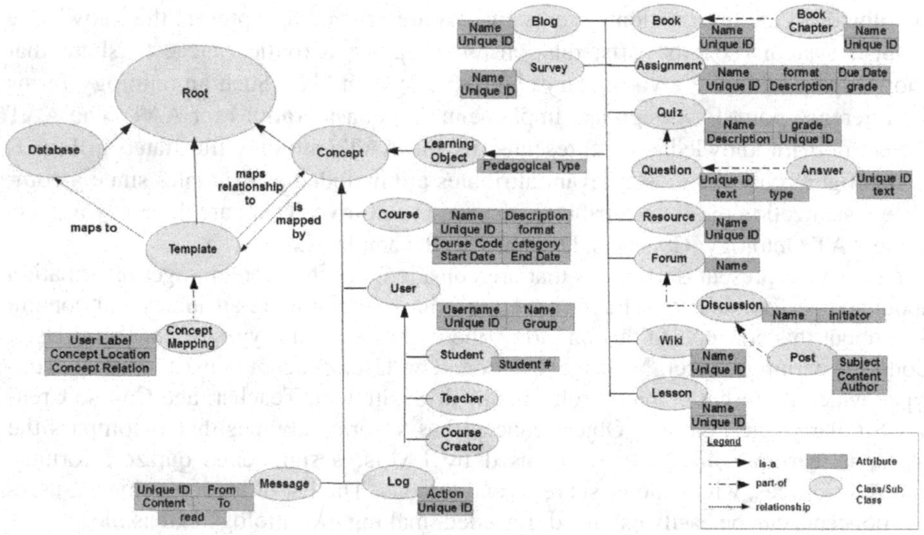

Fig. 1. AAT Ontology

3.2 Pattern Chaining

Pattern chaining facilitates the creation of complex patterns through utilizing simpler patterns as inputs. This way, complex patterns can be easily created. A user can start with a simple pattern (e.g., a list of students' grades on assignments) and then refine the pattern more and more, deepening his/her investigation about learners' behaviours/actions or the quality of learning material. This functionality allows a user to incrementally and progressively build increasingly complex patterns. For example, if a user wishes to identify difficult quiz questions, he/she can build a pattern that extracts data about the average performance of students on questions within quizzes. On top of this pattern, the user can create another pattern that outputs all quiz questions where the average performance of students is lower than, for example, 70%.

There are two types of pattern chaining. First, one pattern can be used as input for another pattern, and therefore restrict the result set of the base pattern. Second, two existing patterns can be merged, leading to a combination of the result sets of these two patterns.

In order to make it possible to chain patterns (for both types), particular additional data as well as meta-data need to be stored for each pattern. Such additional data include identifies of the table(s) from which the data have been retrieved. Such identifiers are indicated in the ontology as the unique id of a concept. For example, if we retrieve a list of all assignment grades that have been achieved by students, then additional data such as the assignment id has to be stored (even if the user is not interested in this id). Furthermore, meta-data are needed that indicate from which location the respective data have been retrieved. This information can be easily retrieved from the AAT ontology, in particular the concept mappings.

3.3 Advanced and User-Friendly Pattern Creation Interface

In this section, an easy to use wizard-style interface for creating patterns is presented. From technical point of view, the end result of a pattern is a SQL statement that is used to retrieve the data the user specified. This SQL statement is built from the template as it is the result of joins defined in a template's concept mappings. Patterns can be reused across different LMS instances.

The interface of the pattern creation wizard is mainly based on questions that users can answer in order to "tell" the system what data they want to access and how they want to analyse them. As a first step, users can decide if they want to: (1) create a new pattern from scratch, (2) use a pattern as input for building another pattern, (3) merge two patterns, or (4) perform an analysis on an existing pattern.

To create a new pattern from scratch, five steps are needed. First, the user can *select the concepts* he/she wishes to investigate. These concepts are (as per the AAT Ontology) linked to the concept mappings for a particular LMS' template. If an LMS does not support that concept (i.e., there is no concept mapping for it) then that concept is not available for selection. Second, the user can *select the concept attributes* he/she wishes to investigate (Fig. 2). These attributes are presented based on the selected concepts. In both the first and second steps, the text (i.e., names of concepts and attributes) comes from the ontological labels for each concept and attribute rather than directly from the database and therefore, is easy to understand for users without knowledge about the database structure. Third, the user can define filters for each of the attributes. The filter type is based on the attribute's data type (as per the ontology). Thus, a user can filter a string attribute by exact match or by a sub-string; an integer can be filtered by an exact match to a number or by a lower and/or upper bound (or both). The user also specifies if he/she wishes to apply all the filters (logical AND) to the pattern or not (logical OR). Fourth, the user can define the sort order of the result set and how each attribute will be sorted (ascending or descending). Fifth, the user can save the pattern, with the option to share the pattern with other users.

In each step, a Pattern Result pane and SQL pane is shown (as in Fig. 2), presenting the user with the tabular result of the query (limited to the top 10 rows) and the exact SQL statement that will be run against the source database to obtain the result set. The SQL pane is more for advanced users/administrators and can be hidden via the application configuration file.

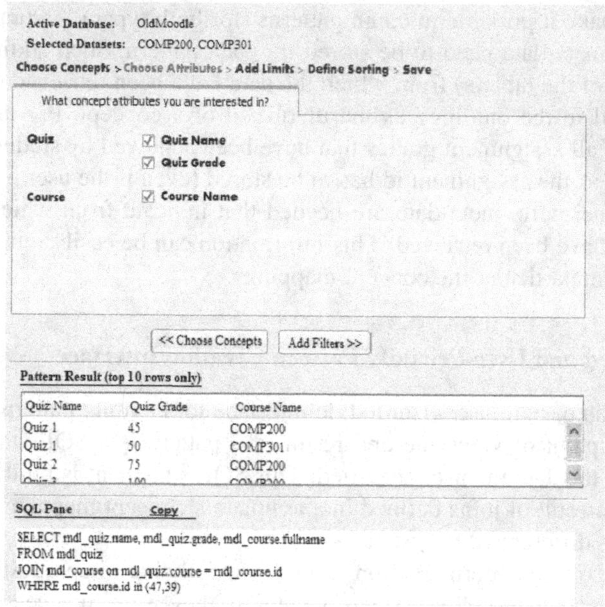

Fig. 2. The Pattern Creation Interface: Selecting Concept Attributes

In order to use a pattern as input or to merge two patterns, the user first selects the base pattern or the two patterns to be merged. After that, the same wizard as for creating a new pattern is used to go through the five steps of selecting concepts, concept attributes, filters, and the sort order, as well as storing the newly created pattern.

To perform an analysis (or calculation) on an existing pattern, the user can select the base pattern, the type of analysis (i.e., counting, calculating the sum or average, and presenting the minimum or maximum) and the concept attributes on which the respective analysis should be performed. Such analyses can either be performed for one attribute, resulting in a single value (e.g., the number of forum postings in a course), or for one attribute per concept, resulting in an additional column of the result set of the base pattern (e.g., the average number of postings per student). This new column can then be named by the user.

4 Conclusions and Future Work

AAT is an innovative tool that enriches smart learning environments by reducing the complexity to access educational log data and therefore enabling educators and course designers to perform simple and complex analytical queries on students' behaviour across courses and programs. In this paper, a user-friendly and powerful mechanism is introduced that extends the AAT tool to be more intuitive but at the same time, advances the functionalities to retrieve complex information from a learning system's log data. In order to do so, an extensible ontology has been built, a refined approach for pattern chaining has been developed and a wizard-style user interface with further options for pattern creation has been implemented.

By extending the functionality and user-friendliness of AAT, the tool becomes more relevant for its target group, namely course designers and educators. On one hand, this user group typically does not have background in computer science or knowledge about the database schema of a learning system. Therefore, advanced user-friendliness in form of a quick and easy way to access data is required. On the other hand, users want to perform comprehensive analyses on the log data in order to get meaningful insights into how students learn, how well teaching strategies work, how much a particular course design supports students, etc. While increased functionality and flexibility often comes at the cost of usability, the proposed mechanism addresses both of these needs and provides users with a user-friendly and powerful way of accessing and analysing educational log data. By offering such access and possibilities for analysing data, the tool provides a smart environment that facilitates course designers' learning about the effectiveness of their courses as well as educators' learning about student behaviours that impact on learning outcomes.

Future work will deal with advanced visualization of data, adding statistical functionality (e.g., regression, correlation), and conducting a study where learning designers and educators evaluate AAT with respect to its usability and usefulness. Another direction of future work will be to investigate the potential of AAT to increase learners' awareness of their learning processes through providing them with access to their behaviour and performance data.

References

1. Aljohani, N.R., Davis, H.C.: Significance of Learning Analytics in Enhancing the Mobile and Pervasive Learning Environments. In: International Conference on Next Generation Mobile Applications, Services and Technologies (NGMAST), pp. 70–74 (2012)
2. Ross, S.M., Anand, P.G.: A computer-based strategy for personalizing verbal problems in teaching mathematics. Educational Tech. Research & Development 35(3), 151–162 (1987)
3. Siemens, G.: Learning analytics: envisioning a research discipline and a domain of practice. In: International Conference on Learning Analytics and Knowledge (LAK), pp. 4–8. ACM, New York (2012)
4. Leony, D., Pardo, A., Valentín, L., Castro, D., Kloos, C.D.: GLASS: a learning analytics visualization tool. In: International Conference on Learning Analytics and Knowledge (LAK), pp. 162–163. ACM, New York (2012)
5. Liaqat, A., Hatala, M., Gasevic, D., Jovanovic, J.: A Qualitative Evaluation of Evolution of a Learning Analytics Tool. Computers & Education 58(1), 470–489 (2012)
6. Retalis, S., Papasalouros, A., Psaromiligkos, Y., Siscos, S., Kargidis, T.: Towards Networked Learning Analytics – A concept and a tool. In: International Conference on Networked Learning, Lancaster (2006)
7. Graf, S., Ives, C., Rahman, N., Ferri, A.: AAT: a Tool for Accessing and Analysing Students' Behaviour Data in Learning Systems. In: International Conference on Learning Analytics and Knowledge (LAK), pp. 174–179. ACM, New York (2011)
8. Fowler, M.: Patterns of enterprise application architecture. Addison-Wesley, Boston (2002)
9. PHP Manual–Database Extensions, `http://www.php.net/manual/en/refs.database.php`
10. CAS, `http://www.jasig.org/cas`
11. Chandrasekaran, B., Josephson, J.R., Benjamins, V.R.: What are ontologies, and why do we need them? Intelligent Systems and their Applications 14(1), 20–26 (1999)

Automating the E-learning Personalization

Fathi Essalmi[1], Leila Jemni Ben Ayed[1], Mohamed Jemni[1], Kinshuk[2], and Sabine Graf[2]

[1] The Research Laboratory of Technologies of Information and Communication & Electrical Engineering (LaTICE), Higher School of Sciences and Technologies of Tunis (ESSTT), University of Tunis, Tunisia

[2] School of Computing and Information Systems, Athabasca University, Canada

fathi.essalmi@isg.rnu.tn, leila.jemni@fsegt.rnu.tn, mohamed.jemni@fst.rnu.tn, kinshuk@athabascau.ca, sabineg@athabascau.ca

Abstract. Personalization of E-learning is considered as a solution for exploiting the richness of individual differences and the different capabilities for knowledge communication. In particular, to apply a predefined personalization strategy for personalizing a course, some learners' characteristics have to be considered. Furthermore, different ways for the course representation have to be considered too. This paper studies solutions to the question: How to automate the E-learning personalization according to an appropriate strategy? This study finds an answer to this original question by integrating the automatic evaluation, selection and application of personalization strategy. In addition, this automation is supported by learning object metadata and an ontology which links these metadata with possible learners characteristics.

Keywords: Personalization strategy, E-learning, Evaluation of personalization parameters.

1 Introduction

This research originated from the recognition of the need of a complete solution to the central question: How to automate the E-learning personalization according to an appropriate strategy? Through a comprehensive literature review, [1] identified 16 personalization parameters[1] and 23 personalization systems implementing 11 personalization strategies[2]. The different reported strategies express different personalization needs. For example, PERSO uses Case Based Reasoning (CBR) approach to determine which course to propose to the students based on their levels of

---

[1] A personalization parameter includes a set of complementary learners' characteristics. For example, the learner's level of knowledge (includes the learners' characteristics beginner, intermediate and advanced), motivation level (includes the learners' characteristics low and high motivation) and the active/reflective dimension of the Felder-Silverman learning style model are personalization parameters.

[2] A personalization strategy includes a set of personalization parameters.

A. Holzinger and G. Pasi (Eds.): HCI-KDD 2013, LNCS 7947, pp. 342–349, 2013.

knowledge and their media preferences [2]. MetaLinks, an authoring tool and web server for adaptive hyperbooks, has been used to build a geology hyperbook [3]. MetaLinks uses three personalization parameters, namely learner's level of knowledge, learning goals and media preferences. AHA! [4] uses stable presentations, adaptive link (icon) annotations and adaptive link destinations for personalizing E-learning. The personalization strategy of AHA includes the parameters Felder–Silverman learning style, media preference and navigation preference. Milosevic, Brkovic, and Bjekic [5] used *Kolb's learning cycle* for tailoring lessons. Their work also incorporated the *learner motivation* as a personalization parameter, which is used to determine the complexity and the semantic quantity of learning objects. Others personalization systems implementing personalization strategies are reported in the literature. PASER [6] has been developed for course planning according to *learners' goals* and their *level of knowledge*, using a domain ontology which describes a hierarchy of the artificial intelligence area. Protus [7] considers the *Felder-Silverman Learning Styles Model* and the *learner's level of knowledge* to recommend relevant links and activities for learners. [8] uses Web mining techniques to deliver appropriate content to learners according to their interests and needs. The number of theoretical and possible personalization strategies, that can be used for personalization, is very high (>50000) [1]. This high number expresses a richness of the E-learning personalization domain that could be exploited by automating the E-learning personalization according to appropriate strategy. Personalizing all courses according to only one predefined personalization strategy would not fit the specificities of courses [9] and teachers' preferences [1]. Therefore, we need to select and apply the appropriate personalization strategy for each course.

This paper answers to the research question: How to automate the E-learning personalization according to an appropriate strategy? This central question could be divided into three sub-questions. (1) How to automate the selection of the appropriate personalization strategy? (2) How to automate the design of personalized learning scenarios? and, (3) How to integrate the solutions to the above mentioned two sub-questions?

Some parts of the central question have already been solved. In particular, the first sub-question (how to automate the selection of an appropriate personalization strategy?) has been studied in [2], where an approach has been presented for the automatic evaluation of personalization strategies. Metrics evaluating personalization strategies are supported by an Ontology representing and managing the Semantic Relations between Values of Data elements and Learners' characteristics (OSRVDL). The second sub-question (how to automate the design of personalized learning scenarios?) has also been partially studied. In particular, a manual (not automated) solution to this second sub-question has been presented in [1] where design and experiment of an architecture for the personalization have been proposed in two complementary levels: the E-Learning Personalization level 1 (ELP1), and the E-Learning Personalization level 2 (ELP2). ELP1 allows for the personalization of learning contents and structure of the course according to a given (specified within ELP2) personalization strategy. ELP2 allows for defining the personalization strategy flexibly. This level of personalization enables teachers to select the learning scenario and to specify manually the personalization strategy (to be applied on the selected learning scenario) by choosing a subset of personalization parameters. To achieve the integrated solution of the central question, the following gaps need to be studied. The answer to the second sub-question (how to automate the design of personalized learning scenarios) is not yet fully resolved (automated). This paper integrates the ontology OSRVDL with ELP1+ELP2 to automatically apply the selected personalization strategy. This ontology allows for automatically generating the learning objects appropriate to learners' characteristics included in the selected personalization strategy. Another gap still remaining is the third sub-question, namely how to integrate the automatic evaluation, selection and application of personalization strategy. This paper provides answer to this question by integrating metrics evaluating personalization strategies with ELP1+ELP2+OSRVDL.

The next section of the paper presents the approach for automatic design of personalized learning scenarios and evaluation of personalization strategies. Section 3 presents an integrated framework for automating the E-learning personalization according to the appropriate strategy. Finally, section 4 concludes the paper with a summary of the work, its limitations and potential future research directions.

2 Automating the Design of Personalized Learning Scenarios and Evaluation of Personalization Strategies

This section presents two processes which are needed to achieve the central question. Then, section 3 presents an integration of these processes in the whole architecture. The first process is concerned with the automatic design of personalized learning scenarios. This process raises its importance from the need to generate appropriate learning scenario by considering the personalization strategy and the learner profile. The second process is needed to evaluate personalization strategies and help teachers in selecting the appropriate one.

The first process (automatic design of personalized learning scenarios) uses the ontology OSRVDL [9, 10] which includes 76 semantic relations between metadata elements and learners characteristics. The richness of the ontology and its extensibility is the basis for an extensible and generic process. This process links learning objects with the appropriate learners characteristics based on OSRVDL. Then, appropriate and non-appropriate learning objects can be used for personalizing E-learning courses. For example, in an adaptive navigational support, appropriate learning objects could be marked with green icons and non-appropriate learning objects could be marked with red icons. If no information is available for the adaptation decision for some learning objects, the adaptation is considered as neutral for those learning objects. This process is based on metadata (which is commonly used for the reuse of learning objects), course, and OSRVDL. For example, if we assume that: (1) there is a semantic relation between the data element *4.1 Format* [11] associated with the value *image* and the *learner media preference* of *graphic*; and, (2) *a course* contains a learning object O1 which is described by the data element *4.1 Format* associated with the value *image*, we can conclude that the learning object O1 is appropriate for the *media preference graphic* (see Figure 1). This process can be used for operationalizing the personalization of courses. In addition, this approach can also be used for the analyses of the metadata describing the learning objects in order to evaluate personalization strategies.

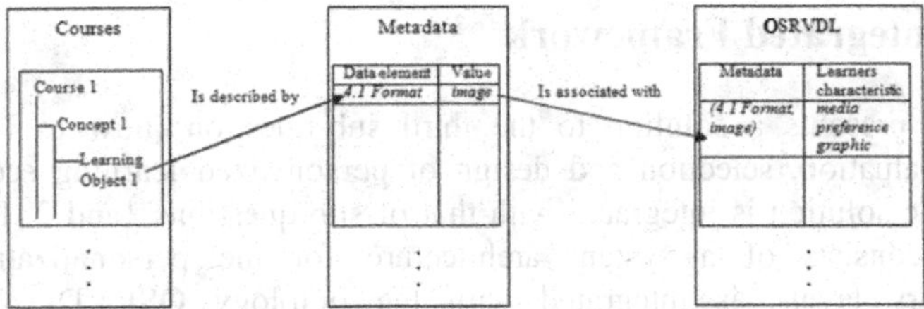

Fig. 1. Appropriate learning object

The second process (evaluation of personalization strategies) benefits from the result of the first process. The generic approach for automatic selection of appropriate learning objects can be used for earlier evaluation of personalization parameters (before starting the learning process and determining the learners' characteristics). This is because of two reasons. The first one is: it is possible to study automatically the feasibility and the easiness of personalizing a given course according to a personalization parameter. For example, when a given course contains learning objects appropriate for each learner's characteristic included in a personalization parameter, the parameter is considered as useful for personalizing the given course. However, if the given course does not contain learning objects appropriate for the learners' characteristics included in another personalization parameter, this parameter is considered as non-useful for personalizing the course. The second reason is the feasibility of comparing personalization parameters. Figure 2 presents *the structure of a course* and *a matrix of appropriate learning objects* used for the evaluation of personalization parameters. This matrix contains the personalization parameters and their divergent characteristics in the columns. The rows of the matrix contain the courses and the concepts included in them. Each cell contains the learning objects presenting a specific concept according to a specific characteristic. The last lines of the matrix can include metrics evaluating personalization parameters. For example, one of these metrics calculates, for each personalization parameter, the number of cells which include a learning objects divided by the number of cells. This rate increases when there are more learners' characteristics considered by learning objects. This metric allows for comparing personalization parameters.

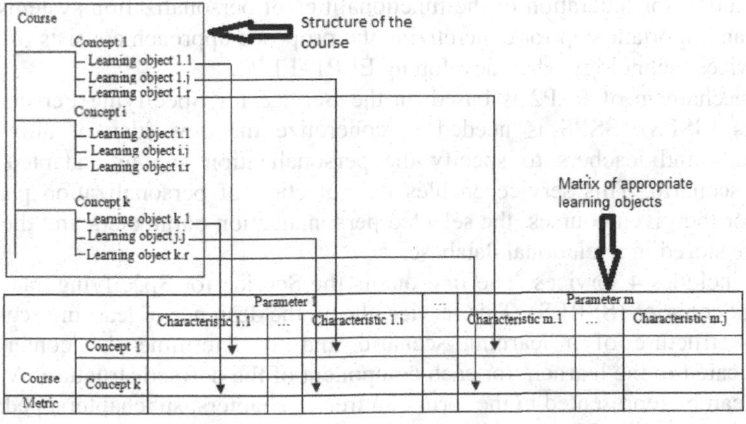

Fig. 2. Early evaluation of personalization parameters

3 An Integrated Framework

This section presents a solution to the third sub-question (how to integrate the automatic evaluation, selection and design of personalized learning scenarios). At first, the basic solution is integrated with that of sub-questions 1and 2. ELP1+ELP2 [1], which consists of a system architecture for the personalization at two complementary levels, is integrated with the ontology OSRVDL, the service evaluating personalization strategy and the service for automating the design of personalized learning scenarios.

ELP1+ELP2 is built by integrating components which focus on the personalization level 1 (ELP1) and the personalization level 2 (ELP2). Furthermore, ELP1 must apply the personalization strategy specified by the teacher in ELP2. ELP1+ELP2 is a new vision of personalization that offers a solution towards some fundamental limitations of E-learning personalization systems. The main advantages of ELP1+ELP2 include the ability of teachers to select the most suitable personalization parameters for their learning scenarios and the possibility of applying more than one personalization parameter according to the specifics of the learning scenarios. The personalization systems available in the literature offer important functionalities for determining the learner characteristics according to a predefined subset of the personalization parameters. The federation of these functionalities and their combination allows for generation of other personalization strategies. However, the personalization systems are developed with different programming languages and tested/used in different contexts. This makes the combination of the functions offered by these systems rather difficult. In this context, the Web services technology offers a powerful solution for the interoperability between multiple applications. In fact, a service can be considered as a distant function which is executed when it is called. In this way, when using services, developers are not interested in the implementation (algorithm, structure, programming language) and the platform of the service. Developers want to only call the service when they need it. Therefore, Web service is an emergent solution for integration of applications. Besides, the personalization systems are tested on different Web servers. This also advocates use of Web services technology for the integration of these personalization systems. Web services technology also offers a major solution for federation of the functionalities of personalization systems. In this context, an important step for concretizing the proposed approach consists of utilizing Web services technology when developing ELP1+ELP2.

The mechanism of ELP2 is based on the Service for Specifying Personalization Strategies (SSPS). SSPS is needed to concretize the new idea of allowing the pedagogues and teachers to specify the personalization strategy adapted for the learning scenario. This service enables the selection of personalization parameters (SPP). For the given courses, the selected personalization parameters and their list of values are stored in a relational database.

ELP1 includes 4 services. The first one is the Service for Specifying and Reusing Learning Scenarios (SSRLS). This service allows the designer of learning scenarios to define a structure of a learning scenario and to determine the content to be communicated to the learners for each component of the defined structure. A learning scenario can be represented in the form of a tree of chapters, subchapters, pedagogical activities, and so on. The second service is the Services for Determining Learners'

Characteristics (SDLC). The aim of SDLC is to federate the set of services for determining the learners' characteristics where each of them is associated with a personalization parameter. The third service is the Service for Applying Personalization Strategies (SAPS). SAPS allows for the application of the personalization strategy specified in SSPS by combining the learner profile with the learning scenarios. Besides, SAPS is responsible for building the learner profile by gathering the output of the selected services for approximating the required learner characteristics. The fourth service is the Service for Learner Navigational Support (SLNS). SLNS allows for the illustration of the learning content in the form of adaptive navigational support. SLNS displays the structure of learning scenarios designed with SSRLS in an adaptive way.

The integrated framework is presented in the figure 3. ELP1+ELP2 is enhanced by integrating the ontology OSRVDL, the service evaluating personalization strategy and the service for automating the design of personalized learning scenarios.

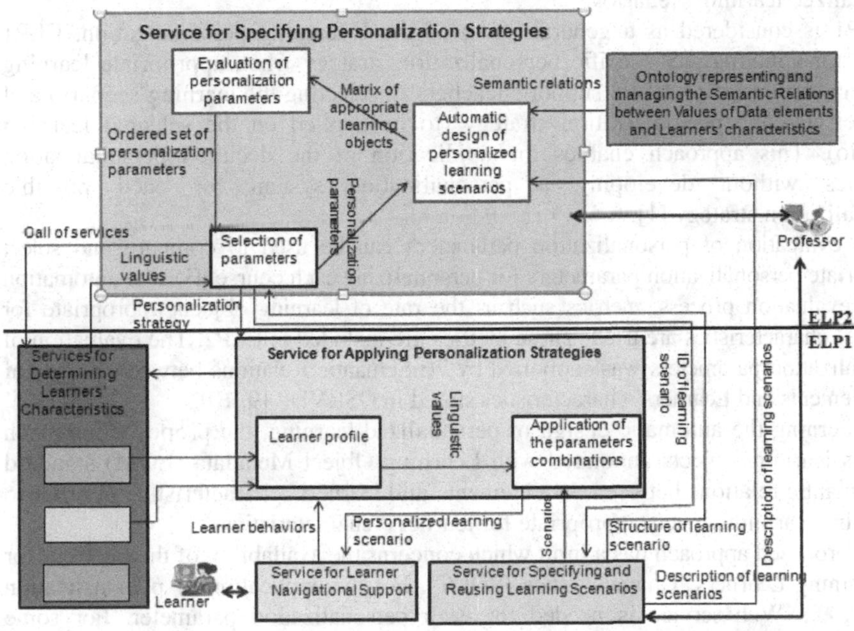

Fig. 3. An integrated framework

The automatic design of personalized learning scenarios plays two roles. The first one is to prepare the matrix of appropriate learning objects as presented in the figure 2. This matrix is used by the component *Evaluation of personalization parameters*. After the selection of the appropriate personalization parameters based on their evaluation, the automatic design of personalized learning scenarios allows for having learning scenarios appropriate for the selected personalization strategy and the learner profile. This is done by considering only those columns of the matrix which include the selected personalization parameters.

4 Conclusion, Limitation and Potential Future Research Directions

There is a rich set of personalization strategies which could help for the success of E-learning. These personalization strategies need to be evaluated to select the appropriate one for each course. Furthermore, personalized learning scenarios need to be designed based on the selected personalization strategy. These processes (evaluation of personalization strategy and design of personalized learning scenarios) need to be automated and integrated in order to reduce the efforts and times of course personalization.

This paper presented a solution to the central question: How to automate the E-learning personalization according to an appropriate strategy? An integrated framework is presented for the personalization of E-learning at two levels (ELP1 and ELP2), evaluation of personalization parameters, and automatic design of personalized learning scenarios.

ELP1 is considered as a generalization of the E-learning personalization. ELP1 allows for applying any specific personalization strategy when appropriate learning scenarios are designed. ELP2 supports teachers in selecting the learning scenario and in specifying the personalization strategy (to be applied on the selected learning scenario). This approach enables the application of the declared personalization strategies without developing a personalization system for each possible personalization strategy [1].

The evaluation of personalization parameters can be used to compare and select appropriate personalization parameters for personalizing each course. For the automation of the evaluation process, metrics such as the rate of learning objects appropriate for learners' characteristics are used. These metrics are included in ELP2. The evaluation of personalization parameters was supported by 76 Semantic Relations between Values of Data elements and Learners' characteristics stored in OSRVDL [9, 10].

Concerning the automatic design of personalized learning, the proposed approach exploits learning objects annotated with Learning Object Metadata (LOM) standard and semantic relations between data elements and learners' characteristics in order to determine learning objects appropriate for learners' characteristics.

The proposed approach has a limit which concerns the availability of the Services for Determining Learners' Characteristics (SDLC). For the application of personalization strategies, a Web service is needed for each personalization parameter. For some personalization parameters, Web services are implemented and used for the evaluation of the proposed approach. Other personalization parameters are reported in the literature without publication of Web service (or software components) for determining learners' characteristics. The absence of published Web service for each personalization parameter is a constraint towards an easy specification of personalization strategies. It might be interesting to collaborate with the research structures working on these parameters for the capitalization of the developed components (for determining learners' characteristics) by their implementation and publication as Web services. In this way, each component could be used/called by several personalization systems.

Beside the future works for reducing the limit of the proposed approach, there are other potential future works concerning ELP1+ELP2 and OSRVDL that deserve some consideration.

Concerning ELP1+ELP2, ELP3 should be studied as an additional layer of the E-learning personalization. ELP3 symbolizes the E-learning systems which support the personalization by educational institutes as personalization logistics according to the personalization needs and environments.

Concerning OSRVDL, future directions of this research should deal with extending OSRVDL for describing the Web services implementing the personalization parameters (including the URL of the Web service, available functions, organizations, researchers working on the personalization parameters, and so on). This extension could facilitate the reuse of the personalization parameters. Furthermore, OSRVDL should be extended by considering additional data elements, learners' characteristics and semantic relations between them.

References

1. Essalmi, F., Jemni Ben Ayed, L., Jemni, M., Kinshuk, Graf, S.: A fully personalization strategy of E-learning scenarios. Computers in Human Behavior 26(4), 581–591 (2010)
2. Chorfi, H., Jemni, M.: PERSO: A System to customize e-training. In: 5 th International Conference on New Educational Environments, Lucerne, Switzerland, (May 26-28, 2003)
3. Murray, T.: MetaLinks: Authoring and affordances for conceptual and narrative flow in adaptive hyperbooks. International Journal of Artificial Intelligence in Education 13, 199–233 (2003)
4. Stash, N., Cristea, A., de Bra, P.: Adaptation to Learning Styles in ELearning: Approach evaluation. In: Reeves, T., Yamashita, S. (eds.) Proceedings of World Conference on E-Learning in Corporate, Government, Healthcare, and Higher Education, pp. 284–291. AACE, Chesapeake (2006)
5. Milosevic, D., Brkovic, M., Bjekic, D.: Designing lesson content in adaptive learning environments. International Journal of Emerging Technologies in Learning 1(2) (2006)
6. Kontopoulos, E., Vrakas, D., Kokkoras, F., Bassiliades, N., Vlahavas, I.: An ontology based planning system for e-course generation. Expert Systems with Applications 35, 398–406 (2008)
7. Klasnja-Milicevic, A., Vesin, B., Ivanovic, M., Budimac, Z.: E-Learning personalization based on hybrid recommendation strategy and learning style identification. Computers & Education 56, 885–899 (2011)
8. Khribi, M.K., Jemni, M., Nasraoui, O.: Toward a Hybrid Recommender System for E-Learning Personalization Based on Web Usage Mining Techniques and Information Retrieval. In: Proceedings of World Conference on E-Learning in Corporate, Government, Healthcare, and Higher Education (E-learn 2007), vol. 7(1), pp. 6136–6145 (2007)
9. Essalmi, F., Jemni Ben Ayed, L., Jemni, M.: An ontology based approach for selection of appropriate E-learning personalization strategy, DULP Workshop. In: The 10th IEEE Int. Conf. on Advanced Learning Technologies, Sousse, Tunisia, pp. 724–725 (2010)
10. Essalmi, F., Jemni Ben Ayed, L., Jemni, M., Kinshuk, Graf, S.: Selection of appropriate E-learning personalization strategies from ontological perspectives. Special Issue on the Design Centered and Personalized Learning in Liquid and Ubiquitous Learning Places. Interaction Design and Architecture(s) Journal 9-10, 65–84 (2010)
11. IEEE Inc. Draft Standard for Learning Object Metadata (2002)

Teaching Computational Thinking Skills in C3STEM with Traffic Simulation

Anton Dukeman, Faruk Caglar, Shashank Shekhar, John Kinnebrew, Gautam Biswas, Doug Fisher, and Aniruddha Gokhale

Department of Electrical Engineering and Computer Science
Vanderbilt University, Nashville, TN, USA
{anton.dukeman,faruk.caglar,shashank.shekhar,john.s.kinnebrew, gautam.biswas,douglas.h.fisher,a.gokhale}@vanderbilt.edu

Abstract. Computational thinking (CT) skills applied to Science, Technology, Engineering, and Mathematics (STEM) are critical assets for success in the 21st century workplace. Unfortunately, many K-12 students lack advanced training in these areas. C3STEM seeks to provide a framework for teaching these skills using the traffic domain as a familiar example to develop analysis and problem solving skills. C3STEM is a smart learning environment that helps students learn STEM topics in the context of analyzing traffic flow, starting with vehicle kinematics and basic driver behavior. Students then collaborate to produce a large city-wide traffic simulation with an expert tool. They are able to test specific hypotheses about improving traffic in local areas and produce results to defend their suggestions for the wider community.

Keywords: Computational Thinking, Smart Learning Environments, Simulation, Visual Programming.

1 Introduction

There is an increasing awareness that the United States is not doing well in K-12 STEM education. In a science proficiency assessment that included 400,000 students from 57 countries, the U.S. ranked 25th of 30 developed nations, with 25% of its students at or below proficiency (the largest percentage among the 30 developed nations) [1]. Improved STEM education in high schools will better prepare students for college education in the STEM disciplines, an important requirement for the 21st century workforce [2]. Developing paradigms that combine STEM learning and problem solving along with preparation for future learning is critical for our nation's future [3].

One way to revitalize STEM education is to make learning engaging and ubiquitous through real-world problem solving that extends beyond the classroom and into the community. In addition, cyber-enabled educational infrastructure that seamlessly integrates personalized and collaborative learning will further advance engagement and participation in STEM education. In this paper,

A. Holzinger and G. Pasi (Eds.): HCI-KDD 2013, LNCS 7947, pp. 350–357, 2013.

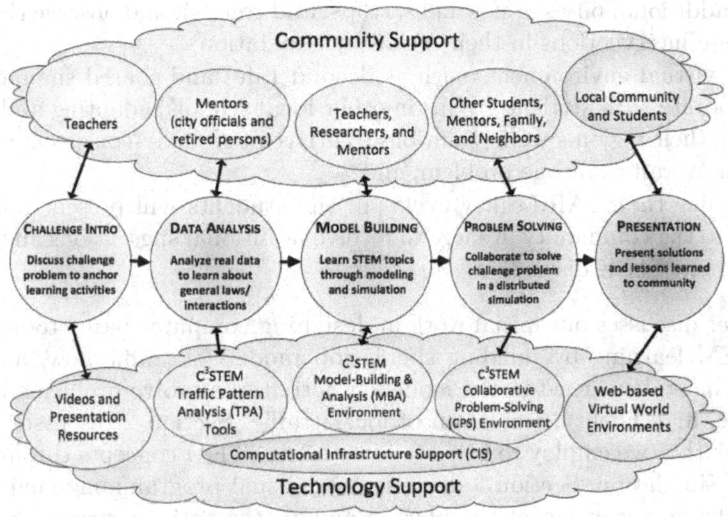

Fig. 1. C3STEM Workflow

we lay the initial groundwork for an innovative community-situated, challenge-based, collaborative learning environment (C3STEM) that harnesses computational thinking, modeling, simulation, and problem-solving to support ubiquitous STEM learning.

Through these combined learning activities and larger-scale collaborations across schools, there will be concrete curriculum-related science and mathematics lessons that the activities will exercise and remediate. Figure 1 illustrates the workflow for participants in C3STEM. The proposed system will support a challenge-based curriculum that integrates STEM concepts, model building, problem solving, and collaboration. Through C3STEM projects high school students will collaboratively address problems of traffic congestion and safety in urban and suburban environments. The traffic domain is attractive because it is a rich source of STEM-related problems, data and simulations are readily available for building our pedagogical tools, and we have local expertise in modeling transportation systems. In particular, small classroom-based student groups will

1. identify traffic patterns by analyzing actual traffic data collected in the form of streaming video by city and state traffic departments;
2. interact with traffic engineers, city planners, and the C3STEM research team to understand how to analyze the traffic data, model traffic patterns, and develop and analyze their solutions;
3. develop agent-based models that align with the observed patterns (e.g., traffic congestion along selected thoroughfares at different times of day and the effects of stoplights and interstate on/off-ramps);
4. design and analyze interventions using their models (e.g., revising traffic light cycles, suggesting carpooling, or extending a public transportation system

with additional buses or new hubs, stops, and routes), and observe the effects of these interventions in their region by simulation;
5. use a virtual environment (such as Second Life) and shared simulations to collaborate with student groups in other local schools, adapting and coordinating their region-specific solutions to arrive at globally-consistent solutions for an overall challenge problem; and
6. much like the STAR.Legacy cycle [4], the students will present their solutions to the community at large to receive additional suggestions and further community-wide discussion.

This paper discusses our initial work in designing computer-based tools to support STEM learning by building simulation models of traffic flow, and using the fundamentals learned in the modeling activities to solve problems in a collaborative, high-fidelity simulation of local traffic. Section 2 discusses the CT framework that we employ to help students learn STEM concepts through modeling and simulation. Section 3 discusses our visual programming environment that emphasizes learning of the physics and mathematics concepts. Section 4 then discusses our initial design efforts to construct the collaborative traffic simulation system for problem solving. Finally, section 5 presents conclusions and future work.

2 Background: CT and the CTSiM Environment

C3STEM employs a CT framework to promote effective STEM learning and preparation for future learning. Many of the epistemic and representational practices central to the development of expertise in STEM disciplines are also primary components of CT. Wing has described computational thinking as a general analytic approach to problem solving, designing systems, and understanding human behaviors [5, 6]. CT draws upon concepts that are fundamental to computer science, including practices such as problem representation, abstraction, decomposition, simulation, verification, and prediction. These practices, in turn, are also central to modeling, reasoning and problem solving in a large number of scientific and mathematical disciplines [7].

Although the phrase "Computational Thinking" was introduced by Wing in 2006, earlier research in the domain of educational technology also focused on similar themes, e.g., identifying and leveraging the synergies between computational modeling and programming on one hand, and developing scientific expertise in K-12 students on the other. For example, Perkins and Simmons showed that novice misconceptions in math, science and programming exhibit similar patterns in that conceptual difficulties in each of these domains have both domain-specific roots (e.g., challenging concepts) and domain general roots (e.g., difficulties pertaining to conducting inquiry, problem solving, and epistemological knowledge) [8]. Complementary work by Harel and Papert argued that programming is reflexive with other domains, i.e., learning programming in concert with concepts from another domain can be easier than learning each separately [9]. Along similar lines, several other researchers have shown that programming and computational modeling can serve as effective vehicles for learning challenging science and math concepts [10–12]. Games and simulation can also be effective at encouraging learning. Work at MIT has led to several programs designed to engage students while teaching subjects such as electromagnetism and the American Revolution [13]. Some groups have even taken commercial games not originally meant for education and developed modules conforming to core curriculum requirements. An example is SimCityEDU, for which lesson plans ranging from math to social interaction have been developed [14].

To engage students in computational thinking and STEM learning, we employ the Computational Thinking in Simulation and Modeling (CTSiM) learning environment [15, 16]. CTSiM provides an agent-based, visual programming interface for constructing computational models and allows students to execute their models as simulations and compare their models' behaviors with that of an expert model. In CTSiM, students design, build, simulate, and verify computational models through four interrelated sets of activities:

1. *Conceptualization*: Initially, students conceptualize the science phenomena and mathematical relationships by structuring their model in terms of the types of agents involved, their properties, behaviors, and interactions.
2. *Construction*: With the visual programming interface, students build a computational model by composing and parameterizing available actions/commands with values and properties in the form of visual primitives. Students select primitives from a library and arrange them spatially, using a drag-and-drop interface to generate their computational model, which defines the behavior of individual agents in the simulation.
3. *Enactment*: A microworld, in the form of a multi-agent simulation using NetLogo [17], allows students to simulate and visualize the agent-level behaviors defined by the student in the computational model [18].
4. *Envisionment*: Students go beyond simply simulating their own models by designing simulation experiments to analyze, refine, and validate the behavior of their model with comparison to a simulation of an expert model.

3 Building Traffic Models in CTSiM

Traffic modeling and simulation provides an excellent domain for learning a variety of STEM concepts in a computational thinking framework. There are many different levels of abstraction available from the lowest level involving basic kinematics equations to the interactions between adjacent neighborhoods with several levels in between. In this section, we present the design of initial CTSiM units for the micro-level traffic modeling activities in C3STEM.

3.1 Position, Velocity, and Acceleration

To understand traffic at a low level requires knowing the relationship between position, velocity, and acceleration, so this is where students start. Students are asked to build a simple one dimensional model of a single car on a straight road with zero initial velocity speeding up and then cruising at a specified velocity. Because students are required to learn the relationship between position, velocity, and acceleration in the context of Newton's laws of motion they will develop systematic mathematical equations using the laws to compute the required values. Visual primitives are provided for students to compute the car's position and velocity using the car's current acceleration, velocity, and position. A library of mathematical functions allow students to model acceleration profiles, such as constant, linear, and square root. Students can us the visual environment to build and run these models and they can simultaneously observe the movement of the vehicle on a road as well as plots of position, velocity, and acceleration over time to facilitate their understanding of the concepts.

After completing the first module, students will move to modeling deceleration. Students can model a scenario where a single car moving with an initial velocity comes to a stop at a stop sign and then accelerates to its initial velocity on the road. If the new functions are used correctly, then the computed velocities will be correct for the entire trajectory, and the car will follow a deceleration profile, stop at the stop sign, and accelerate again.

After learning the basics of kinematics of motion, students move on to study driver behavior and interactions. The second module (enactment world shown in Fig. 2) introduces the parameters, such as *stopping sight distance*, the distance from a stop sign or another stationary object that the driver activates the stop

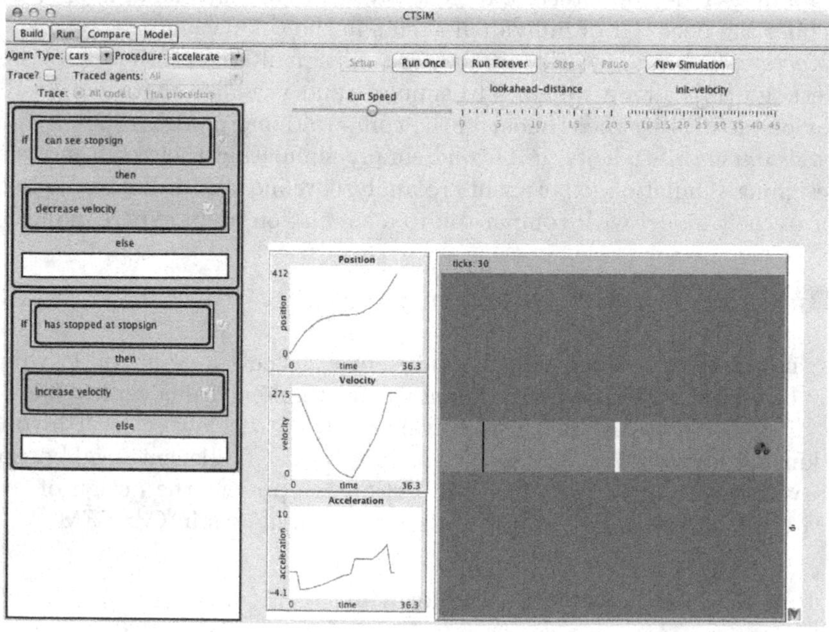

Fig. 2. Enactment world for stop sign model

behavior [19]. Students will set two parameters: *initial velocity* and *stopping sight distance* (with sliders visible in the upper portion of Fig. 2) to build two functions, accelerate and go. As any experienced driver knows, the faster the car's velocity, the earlier the driver has to initiate the deceleration function. Depending on the initial parameters, the acceleration plot (bottom plot in Fig. 2) may have a steep, uncomfortable drop rather than a smooth deceleration to stop.

3.2 Scaling Up: Modeling Simple Traffic Flows

In this model, students will model multiple cars moving along roads with intersections that have stop signs and stop lights [20]. The road configurations will be provided to the students, and they may represent a traffic network in a section of the city. Like before, cars may enter the environment with an initial velocity or start from rest at different positions. The students' models will scale up to include multiple cars that operate together. As cars approach the stop sign or the end of the queue of cars waiting at the stop sign, they will have to slow down, and then take turns going through the intersection. Some cars may be able to go at the same time, such as two cars traveling in opposite directions on the same road or two cars on perpendicular roads both turning right. Because this module builds on previous ones, students will focus on the modeling issues associated with multiple interacting queues, such as waiting at a stop lights and signs, and following cars. Lane changing functions will be implemented as part of driver behavior, and students will study wave patterns, such as the propagation effects when a car suddenly slows down. In addition to basic physics, this module will introduce students to parameters, such as *average vehicle velocity* and *throughput*, allowing students to learn higher level mathematical analyses and methods for optimizing parameter values given constraints. A subsequent module may introduce *gap acceptance time*, and the *average turn time* at an intersection [21].

4 From Micro to Macro: Using SUMO

Another component of C3STEM is collaborative problem solving. Students will collaborate with other groups at their school and also groups at other schools to develop models of traffic flow for their entire city. Each group will model traffic flow in neighborhoods and connected city streets and highways near their school. Initially, they will interact with the researchers (the authors) and city traffic engineers to understand the traffic flow patterns and related parameters locally and globally (i.e., across the entire city). This study will help them formulate specific solutions to reduce traffic congestion.

For city-wide traffic simulation and analysis, students will interact with a simulation of traffic in the region around their school using web-based tools that display results of a back-end simulation run in the Simulation of Urban MObility (SUMO) environment, a continuous simulation for large road networks [22]. A web interface using Google Maps will be used as the interface with the simulation server as it is a familiar interface for students. The map interface is easily recognizable and supports a drag-and-drop interface for manipulating the simulation parameters; cars, stop lights, and stop signs can be easily manipulated on the map.

5 Conclusions and Future Work

Computational thinking skills are an essential part of STEM-related careers, and the United States lacks qualified applicants in these disciplines. C3STEM seeks to improve students' STEM learning and problem solving skills using the traffic domain as a motivating example. Students start with the basic low-level physics simulations involving position, velocity, and acceleration in CTSiM. They then move on to model driver behaviors, such as gap acceptance time and stopping sight distance. After mastering the micro-level physics concepts, they move up to SUMO, to analyze and solve more complex analytic problems. At the macro-level in SUMO, students can manipulate intersections in many ways such as changing stop light timing or adding stop signs. Solving complex problems analytically requires students to think at multiple levels of abstraction and draw conclusions across the different layers. We will run experimental studies at two high schools in Chattanooga, Tennessee in the U.S. Students will be given a pretest to measure each student's understanding of physics principles and CT skills. After the program, students will be given a similar posttest to determine the efficacy of the program at promoting both CT and STEM skills in the students.

Some future work involves enhancing the collaboration aspect. Good collaboration requires good communication and this will be provided through a persistent online world. In this environment, students will be able to interact through text and video. They will be able to see the simulation as it is running and share ideas about how to improve it. Students working on adjacent neighborhoods will have to communicate often as they are the most likely to be affected by each other's changes, however, neighborhoods that are far apart may have to communicate as well and will be able to do so. We will also create additional CTSiM modules to simulate other aspects of driver behavior to students, such as lane changing and highway merging.

References

1. Fleischman, H.L., Hopstock, P.J., Pelczar, M.P., Shelley, B.E.: Highlights from PISA 2009: Performance of US 15-Year-Old Students in Reading, Mathematics, and Science Literacy in an International Context. NCES 2011-004. National Center for Education Statistics (2010)
2. Karoly, L., Panis, C.: The 21st Century at Work: Forces Shaping the Future Workforce and Workplace in the United States. RAND Labor and Population Technical Report 0-8330-3492-8 (2004)
3. Board, N.S.: Preparing the Next Generation of STEM Innovators: Identifying and Developing our Nation's Human Capital. Report NSB 10-33 (2010)

4. Schwartz, D.L., Brophy, S., Lin, X., Bransford, J.D.: Software for Managing Complex Learning: Examples from an Educational Psychology Course. Educational Technology Research and Development 47, 39–59 (1999)
5. Wing, J.M.: Computational Thinking. Communications of the ACM 49, 33–35 (2006)
6. Wing, J.M.: Computational Thinking: What and Why? Link Magazine (2010)
7. Council, N.R.: Taking Science to School: Learning and Teaching Science in Grades K-8. National Academy Press (2007)
8. Perkins, D.N., Simmons, R.: Patterns of Misunderstanding: An integrative model for science, math, and programming. Review of Educational Research 58, 303–326 (1988)
9. Harel, I.E., Papert, S.E.: Constructionism. Ablex Publishing (1991)
10. Guzdial, M.: Software-Realized Scaffolding to Facilitate Programming for Science Learning. Interactive Learning Environments 4, 1–44 (1994)
11. Sherin, B.L.: A Comparison of Programming Languages and Algebraic Notation as Expressive Languages for Physics. International Journal of Computers for Mathematical Learning 6, 1–61 (2001)
12. Blikstein, P., Wilensky, U.: An Atom is Known by the Company it Keeps: A constructionist learning environment for materials science using agent-based modeling. International Journal of Computers for Mathematical Learning 14, 81–119 (2009)
13. Jenkins, H., Klopfer, E., Squire, K., Tan, P.: Entering the Education Arcade. Computers in Entertainment (CIE) 1, 8 (2003)
14. GlassLab: About SimCityEDU (2013), http://www.simcityedu.org/about/
15. Basu, S., Dickes, A., Kinnebrew, J.S., Sengupta, P., Biswas, G.: CTSiM: A Computational Thinking Environment for Learning Science through Simulation and Modeling. In: The 5th International Conference on Computer Supported Education (2013)
16. Sengupta, P., Kinnebrew, J.S., Basu, S., Biswas, G., Clark, D.: Integrating Computational Thinking with K-12 Science Education using Agent-based Computation: A Theoretical Framework. In: Education and Information Technologies (2013)
17. Wilensky, U.: NetLogo. Center for Connected Learning and Computer-Based Modeling. Northwestern University, Evanston (1999), http://ccl.northwestern.edu/netlogo/
18. Papert, S.: Mindstorms: Children, Computers, and Powerful ideas. Basic Books, Inc. (1980)
19. Fambro, D.B., Fitzpatrick, K., Koppa, R.J.: Determination of Stopping Sight Distances, vol. 400. Transportation Research Board (1997)
20. Horowitz, A.J.: Revised Queueing Model of Delay at All-Way Stop-Controlled Intersections. Transportation Research Record 1398, 49 (1993)
21. Pollatschek, M.A., Polus, A., Livneh, M.: A Decision Model for Gap Acceptance and Capacity at Intersections. Transportation Research Part B: Methodological 36, 649–663 (2002)
22. Behrisch, M., Bieker, L., Erdmann, J., Krajzewicz, D.: SUMO - Simulation of Urban MObility: An Overview. In: SIMUL 2011, The Third International Conference on Advances in System Simulation, Barcelona, Spain, pp. 63–68 (2011)

Learning Analytics to Support the Use of Virtual Worlds in the Classroom

Michael D. Kickmeier-Rust and Dietrich Albert

Cognitive Science Section, Knowledge Management Institute
Graz University of Technology
Inffeldgasse 13, 8010 Graz, Austria
{michael.kickmeier-rust,dietrich.albert}@tugraz.at

Abstract. Virtual worlds in education, intelligent tutorial systems, and learning analytics – all these are current buzz words in recent educational research. In this paper we introduce ProNIFA, a tool to support theory-grounded learning analytics developed in the context of the European project Next-Tell. Concretely we describe a log file analysis and presentation module to enable teachers making effectively use of educational scenarios in virtual worlds such as OpenSimulator or Second Life.

Keywords: Learning analytics, CbKST, virtual worlds, OpernSimulator, data visualization.

1 Introduction

Although the hype over open, freely accessible virtual worlds is abating a little bit, there is still a significant amount of interest from teachers, particularly technology-minded ones, to adopt the rich possibilities of virtual worlds for their classroom work. In particular a strong beneficial aspect is seen in using existing communities and worlds to illustrate, to demonstrate, and to experience historical facts and events or other cultures. There is also a strong community for foreign language learning. Examples of how to find such "educationally meaningful" places offers the *SecondLife* destination guide (http://secondlife.com/destinations) ordered by topics or for *OpenSimulator* the *3D Learning Experience Services* (http://3dles.com/en).

But what are virtual worlds or virtual environments? Virtual worlds are persistent, computer-simulated, graphical environments in which individuals can appear through avatars, i.e., artificial representations of themselves. Within these worlds, people can interact with the virtual environment and with others, regardless of their physical locations, through a variety of integrated communication channels ranging from text-based chat to video communications. Virtual worlds, which were first developed in the late 1970s, are viewed as a subset of virtual reality applications that have moved through several stages of development [1].

While virtual worlds are common in (multiplayer) online games and role-playing games (such as *World of Warcraft*), virtual environments are rapidly emerging as a

A. Holzinger and G. Pasi (Eds.): HCI-KDD 2013, LNCS 7947, pp. 358–365, 2013.

complementary means to the physical world for communicating, collaborating, and organizing activities in a variety of fields, ranging from education to management. Of special relevance to the educational sector is that virtual worlds offer a rich potential for collaboration, exploration, and creativity due to characteristics such as immersion, avatar-based interaction, multi-modal collaboration, or the feeling of presence. For example, virtual environments have been used for activities such as brainstorming or iteratively and interactively creating (from objects such as clothing or buildings to interactive art and music shows to simulations of natural disasters), while simultaneously or asynchronously sharing the act of creation with other users. Many of the objects in the virtual world contain scripts that run animations of the object, play media files (such as sound or video) or otherwise enable the user to do or experience something new or perhaps even impossible. Another aspect is the fact that virtual worlds are not constrained by the real world physics (Figures 1). While this may appear to be self-evident, it is worth consideration when imagining and planning what one can do in these virtual environments; it can be just emptiness populated with objects or physically impossible representations such as being depicted as a cancer cell avatar within a human body. Undoubtedly, the *holodeck* of the 1970s is becoming reality. Therefore, virtual worlds and avatar-supported interaction seem to be a convincingly natural playground with an unbelievable diversity of tools to learn.

During the past few years, virtual worlds and virtual teams have received an increasing amount of attention by educational researchers. The results, however, are still unclear. The work of [2], for example, yielded that distinct characteristics of teaming in virtual worlds such as physical distance, device dependence, structural dynamism, or natural and cultural diversity may reduce (learning) performance. On the other hand, in a larger scale study [3] found that aspects like virtual proximity, communication modalities, and task coordination can significantly support performance. Further enabling factors coming from the research are trust, support, encouragement, freedom, challenge, goal clarity, motivation, commitment, sufficient resources and time. Further research addressed also several distinct aspects of virtual environments and teams, for example, the effects of avatar reference frames and realities, multimodality, simulation fidelity, immersion, etc. (e.g., [4, 5]). Summarizing such findings, virtual worlds may (and likely will) serve as a promising basis for educational solutions supporting and facilitating new forms of learning. This holds true for conventional schooling but also for distance education measures.

Fig. 1. Examples for virtual worlds (OpenSimulator and Second Life)

In the past years, *OpenSimulator* (http://opensimulator.org) has been increasing in popularity, particularly for serious purposes (such as education). *OpenSimulator* is an open source multi-platform, multi-user 3D application server that enables individuals and firms across the globe to customize their virtual worlds based on their technology preferences (Figure 1). The project is powered by the efforts of the community members, who devote their time and energy to the development processes. The project has a global reach and the community hosts a very diverse group of actors: independent users, freelance developers, non-profit organizations (e.g., universities), and commercial players.

2 Bringing Virtual Worlds into the Classroom

A learning scenario for *OpenSimulator,* we developed recently, is an English learning adventure. The idea is that an entire class logs into the virtual world and forms small teams. The teams then are supposed to solve a mysterious riddle: *Why is the town they find abandoned? Where are all the people?* There are only a view characters left (e.g., a drunkard or a priest) who can provide the teams with foretelling and throughout the world various hints are hidden. Accomplishing this scavenger hunt, the teams must read and listen to English texts and must understand the (often complex) meaning and must identify the main points. A highly motivational, competitive element is a reward for the team who solves the riddle first. From a pedagogical perspective, the scenario is designed around the so-called *CEFR* skills, a common specification of second language competencies (cf. http://www.cambridgeesol.org/about/standards/cefr.html).

In educational settings, usually activities and test results are stored with scores or qualitative descriptors in an overview and are included separately in a series of results or outcomes alongside other activities. This is a strongly behavior/activity-oriented approach, which most often cannot life up to the demands of 21st century education. Also educational policies in Europe presently are moving from a focus on knowledge to a focus on competency, which is reflected in revisions on curricula in the various countries. For example, in Norway, the learning goals catalogue now covers five broad areas: communication; language learning; culture, society and literature - each of which comprises sets of competencies (e.g., "the ability to read and understand the main content of texts on familiar topics"). Equally, in Austria there is a revision of the curricula in progress heading towards competence-based schooling.

In alignment with this increasing competency and ability focus of modern teaching, the virtual learning scenarios are a perfect teaching context, because they combine experiential, active, constructivist learning, with the need to directly apply the competencies in a meaningful setting, not least receiving direct feedback (e.g., by progressing with the scavenger hunt or by feedback of virtual educational entities). In that sense, a playful use of virtual worlds enables designing instructionally brilliant lesson, grounding for example on the important "First Principles of Education", as stated by famous M. David Merrill. The principles are (1) demonstration, (2) application, (3) activation of prior knowledge, (4) integration of new information into the mental and physical world of the learner, and (4) an orientation to meaningful tasks.

The problem with a broad virtual scenario (as the adventure described above) is the massive amount of relevant educational data being produced and the inability of teachers to monitor, record, aggregate the information in a formative sense in or to generate a fair and correct model of a learner's activities and competencies – without the support of smart software solution enabling such level of "learning analytics". In other words, imagine an entire school class with 25 students; all are entering a large virtual world and disperse quickly all over this world. The teacher has almost no chance to monitor what is going on in the world, who is active or inactive, who is communicating to whom in which manner, etc.

One option is log file data. Usually, virtual environments such as *OpenSimulator* provide detailed log files for specific sessions. Unfortunately, the amount of information, stored in such log files, is massive and it takes software to analyze the log files – in an educationally meaningful way. This is the prerequisite that the activities and the performance in the virtual worlds can really contribute to a formative assessment and thus a tailored support of students.

3 Learning Analytics with ProNIFA

ProNIFA is a tool to support teachers in the assessment process that has been developed in the context of the European Next-Tell project (www.next-tell.eu). The name stands for *probabilistic non-invasive formative assessment* and, in essence, establishes a handy user interface for related data aggregation and analysis services and functions. Conceptually, the functions are based on *Competence-based Knowledge Space Theory* (CbKST), originally established by Jean-Paul Doignon and Jean-Claude Falmagne [6, 7], which is a well elaborated set-theoretic framework for addressing the relations among problems (e.g., test items). It provides a basis for structuring a domain of knowledge and for representing the knowledge based on *prerequisite relations*. While the original idea considered performance only (the behavior; for example, solving a test item), extensions of the approach introduced a separation of observable performance and latent, unobservable competencies, which determine the performance [8, 9].

CbKST assumes a finite set of more or less atomic competencies (in the sense of some well-defined, small scale descriptions of some sort of aptitude, ability, knowledge, or skill) and a prerequisite relation between those competencies. A prerequisite relation states that competency *a* (e.g., to multiply two positive integers) is a prerequisite to acquire another competency *b* (e.g., to divide two positive integers). If a person has competency *b*, we can assume that the person also has competency *a*. To account for the fact that more than one set of competences can be a prerequisite for another competency (e.g., competency *a* or *b* are a prerequisite for acquiring competency *c*), *prerequisite functions* have been introduced, relying on and/or-type relations. A person's *competence state* is described by a subset of competencies. Due to the prerequisite relations between the competencies, not all subsets are admissible competence states.

By utilizing *interpretation and representation functions* the latent competencies are mapped to a set of tasks (or test items) covering a given domain. By this means, mastering a task correctly is linked to a set of necessary competencies and, in addition, not mastering a task is linked to a set of lacking competencies. This assignment induces a *performance structure*, which is the collection of all possible *performance states*. Recent versions of the conceptual framework are based on a *probabilistic mapping* of competencies and performance indicators, accounting for making lucky guesses or careless errors. This means, mastering a task correctly provides the evidence for certain competencies and competence states with a certain probability.

ProNIFA retrieves performance data (e.g., the results of a test or the activities in a virtual environment) and updates the probabilities of the competencies and competence states in a domain. When a task is mastered, all associated competencies are increased in their probability, vice versa, failing in a task decreases the probabilities of the associated competencies. A distinct feature in the context of formative assessment is the multi-source approach. ProNIFA allows for connecting the analysis features to a broad range of sources of evidence. This refers to direct interfaces (for example to *Google Docs*) and it refers to connecting, automatically or manually, to certain log files. Using this level of connectivity, multiple sources can be merged and can contribute to a holistic analysis of learners' achievements and activity levels. The interpretation of the sources of evidence occurs depending on a-priori specified and defined conditions, heuristics, and rules, which associate sets of available and lacking competencies to achievements exhibited in the sources of evidence. The idea is to define certain conditions or states in a given environment (regardless if it is a Moodle test or a status of a problem solving process in a learning game). Examples for such conditions may be the direction, pace, and altitude a learner is flying with a space ship in an adventure game or a combination of correctly and incorrectly ticked multiple choice tasks in a regular online school test. The specification of such state can occur in multiple forms, ranging from simply listing test items and the correctness of the items to complex heuristics such as the degree to which an activity reduced the 'distance' to the solution in a problem solving process. The next step is to assign a set of competencies that can be assumed to be available and also lacking when a certain state occurs. This assumption can be weighted with the strength of the probability updates. In essence, this approach equals the conceptual framework of *micro adaptivity* as, for example, described by [10]. Figure 2 is a screenshot of ProNIFA analyzed data from a *Second Life* activity.

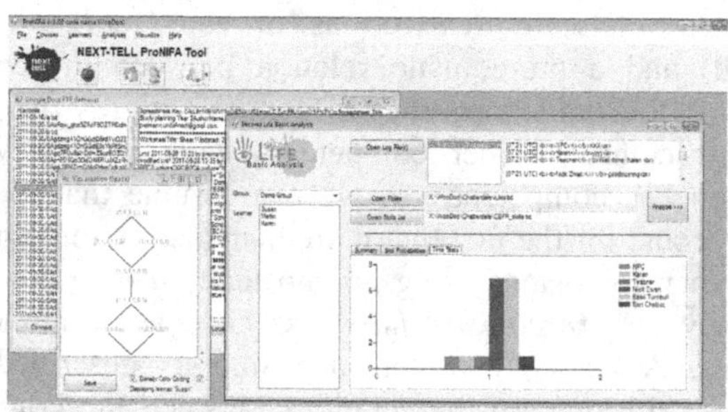

Fig. 2. Screenshot of ProNIFA

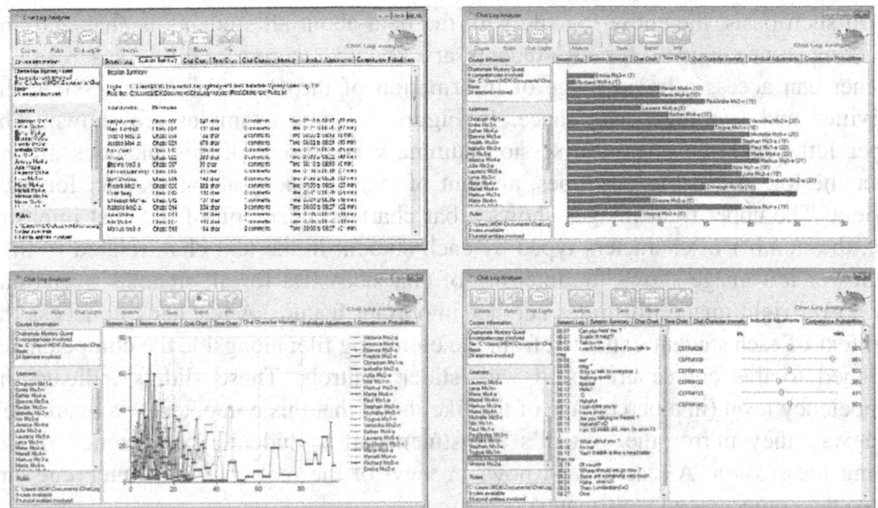

Fig. 3. Screenshots of the log file analysis module

3.1 Working with Log Files

Related to aforementioned English scenario for *OpenSimulator*, we developed a chat log analysis module (cf. Figure 3). The module allows using a course with a set of assigned students and a set of involved competencies (in this particular case 6 of aforementioned CEFR skills); in a second step the teacher may apply a set of rules to interpret the log files. The possibilities range from simple counting of certain events up to using scripting code to identify competencies. In the following box an example is given; this rule defines a time-based quest. The students have to listen to what the drunkard says; his talk finishes with "…, now go there!". The actual target is a box with a hidden letter in a hotel. If a student, indicated by "<NAME>", has understood the speech and manages to get there within 8 minutes, the system takes this as an indicator that the student has competence number 2 and consequently increases the probability in the competence model for this student by the value 0.2. In the same manner competencies could be decreased in the probability as well.

```
[Time1]
Type=1
WhoS=Drunkard
WhoE=Hotel Hint
WhatS=<NAME>, now go there!
WhatE=<NAME> arrived
Whom=Mario Wolf; Maria Wolf
Up=002
UpVal=0,2
Down=
DownVal=
```

On such basis, the entire log file with the data about all students is analyzed and the competence models for all students are updated incrementally. As a result, a teacher can access a broad range of information of the *OpenSimulator* session, the activities and learning performance. In Figure 3, some examples are shown. The upper left panel shows how a session summary is presented; this includes among other the log in and log out times, amount of chat contacts, activities, etc. for each student. The upper right images shows a bar chart visualization of the chat intensity (i.e., the number of characters typed by each student in the text chat, related to that, the lower left image shows the intensity of chat activities for all students over time. The lower right image illustrates another important feature. A teacher can access the chat text of each student extracted from the entire log file; alongside the competencies assigned to this course are listed with slider controls. These sliders indicate the competency level (in a percentage of the likelihood that this competency is available), in a way they mirror the system's adjustments of a student's competency model during the session. A teacher can now, in view of the real chat text, intervene and adjust the competency levels manually.

As mentioned above, ProNIFA operates on the basis of probability distributions over competence structures. In simple words this means that there is an order set of meaningful states a student can be in, ranging from having none of the competency to having all of them. However, due to the underlying prerequisite relations not all states are possible. For example, it is highly unlikely that a student cannot understand a written text in a foreign language and, at the same time, has the ability to understand the same text spoken by a native speaker. Since we are applying a probability distribution, the probabilities to be in one of the states sum up to 1. A teacher can access these information in form of so-called *Hasse diagrams* (an example is shown in the lower left part of Figure 2). The competence states are arrange according their structure and the assigned probabilities are displayed in form of color coding – the more salient a state appears the higher is its probability (ProNIFA allows custom color themes).

4 Conclusion

There is no doubt that learning scenarios for and in virtual worlds will be a part of classroom education. The advantages are convincing: immersion and fun, collaboration and interaction, exploration and active competence construction. The downside is that a reasonable and effective implementation of activities in the virtual world requires smart software solutions and sound approach to learning analytics to record, aggregate, analyze, visualize and store the data. ProNIFA is a tool that supports exactly those needs. Moreover, ProNIFA has a scientific, psycho-pedagogical framework in the background enabling a competence and learning performance oriented analysis of data.

Presently we are conducting classroom, studies to elucidate the usefulness of the existing features and to collect real-world demands for such learning analytics system – also beyond the focus on log files (as described here).

Acknowledgements. This project is supported by the European Community (EC) under the Information Society Technology priority of the 7th Framework Programme for R&D under contract no 258114 NEXT-TELL. This document does not represent the opinion of the EC and the EC is not responsible for any use that might be made of its content.

References

1. Dionisio, J., Burns, W., Gilbert, R.: 3D virtual worlds and the Metaverse: Current status and futre possibilities (2011) (unpublished manuscript)
2. Gibbs, C.B., Gibson, J.L.: Unpacking the concept of virtuality: The effects of geographic dispersion, electronic dependence, dynamic structure, and national diversity on team innovation. Administrative. Science Quarterly 51, 451–495 (2006)
3. Kratzer, J., Leenders, R.T., Engelen, J.M.: Managing creative team performance in virtual environments: an empirical study in 44 R&D teams. Technovation 26, 42–49 (2006)
4. Larach, D., Cabra, J.: Creative problem solving in Second Life: An action research study. Creativity and Innovation Management 19, 167–179 (2010)
5. Ward, T.B., Sonneborn, M.S.: Creative expression in virtual worlds: Imitation, imagination, and individualized collaboration. Psychology of Popular Media Culture 1, 32–47 (2011)
6. Falmagne, J., Cosyn, E., Doignon, J., Thiéry, N.: The Assessment of Knowledge. Theory and in Practice (2003), `http://www.scribd.com/doc/3155044/Science-Behind-ALEKS`
7. Doignon, J., Falmagne, J.: Knowledge Spaces. Springer, Berlin (1999)
8. Albert, D., Lukas, J. (eds.): Knowledge spaces: Theories, empirical research, and applications. Lawrence Erlbaum Associates, Mahwah (1999)
9. Korossy, K.: Modelling knowledge as competence and performance. In: Albert, D., Lukas, J. (eds.) Knowledge Spaces: Theories, Empirical Research, and Applications, pp. 103–132. Lawrence Erlbaum Associates, Mahwah (1999)
10. Kickmeier-Rust, M.D., Albert, D.: Micro adaptivity: Protecting immersion in didactically adaptive digital educational games. Journal of Computer Assisted Learning 26, 95–105 (2011)

Evaluation of Optimized Visualization of LiDAR Point Clouds, Based on Visual Perception

Sašo Pečnik, Domen Mongus, and Borut Žalik

Faculty for Electrical Engineering and Computer Science, University of Maribor, Slovenia
{saso.pecnik,domen.mongus,zalik}@uni-mb.si

Abstract. This paper presents a visual perception evaluation of efficient visualization for terrain data obtained by LiDAR technology. Firstly, we briefly summarize a proposed hierarchical data structure and discuss its advantages. Then two level-of-detail rendering algorithms are presented. The experimental results are then provided regarding the performance and rendering qualities for both approaches. The evaluation of the results is finally discussed in regard to the visual and spatial perceptions of human observers.

Keywords: LiDAR, level-of-detail, visual perception, human visual system, image quality assessments.

1 Introduction

Over the last decade with on-going advances, Light Detection and Ranging (LiDAR) has become the leading technology for obtaining spatial data [1–3]. LiDAR is a laser-based remote sensing technology that measures the time delay between the transmission of the laser pulse and the detection of its reflection when determining the range of a distant object [4]. LiDAR systems can be mounted on aircraft, and are capable of acquiring surface data quickly with high accuracy and great density, as shown in Fig. 1. The obtained points are georeferenced according to the position of the aircraft by using the Differential Global Positioning System (DGPS), and the orientation of the aircraft by using the Inertial Measurement Unit (IMU) [5]. In this way, LiDAR data consist of three-dimensional, georeferenced point clouds of the terrain surface. LiDAR achieves higher precision and is less dependent on the weather and light conditions because of the use of a short wavelength laser light [6]. Furthermore, LiDAR is able to penetrate through vegetation and due to its ability to separate individual reflections, it is able to record a terrain under the vegetation [6].

Airborne LiDAR technology provides huge quantities of spatial data, but the processing and interactive visualization of such massive point clouds can be very difficult. Regrettably, the more data we want to visualize, the longer it takes to render, thus inducing a noticeable lag of the system. Humans are extremely sensitive to visual lags, and as a result the users' efficiency reduce when they have to work with such systems [7]. Therefore an efficient rendering technique is needed for interactively visualizing of LiDAR data. Several techniques have been developed in an effort to

A. Holzinger and G. Pasi (Eds.): HCI-KDD 2013, LNCS 7947, pp. 366–385, 2013.

reduce system lag like: visibility culling [8], double buffering [9], antialiasing [10], level-of-detail [11], and hierarchical subdivision [12]. All these techniques are based on a graphic workload reduction. Only the Level-Of-Detail (LOD) technique can balance the system's load in real-time, to the point that no considerable frame drops are noticeable at any time. In essence, LOD attempts to trade spatial fidelity against temporal fidelity: detailed geometry requires more time to display but contains more details, and vice versa. It is a good compromise for real-time interactive graphic systems, since it is less perceptible to human vision than temporal delays during visual updating [13].

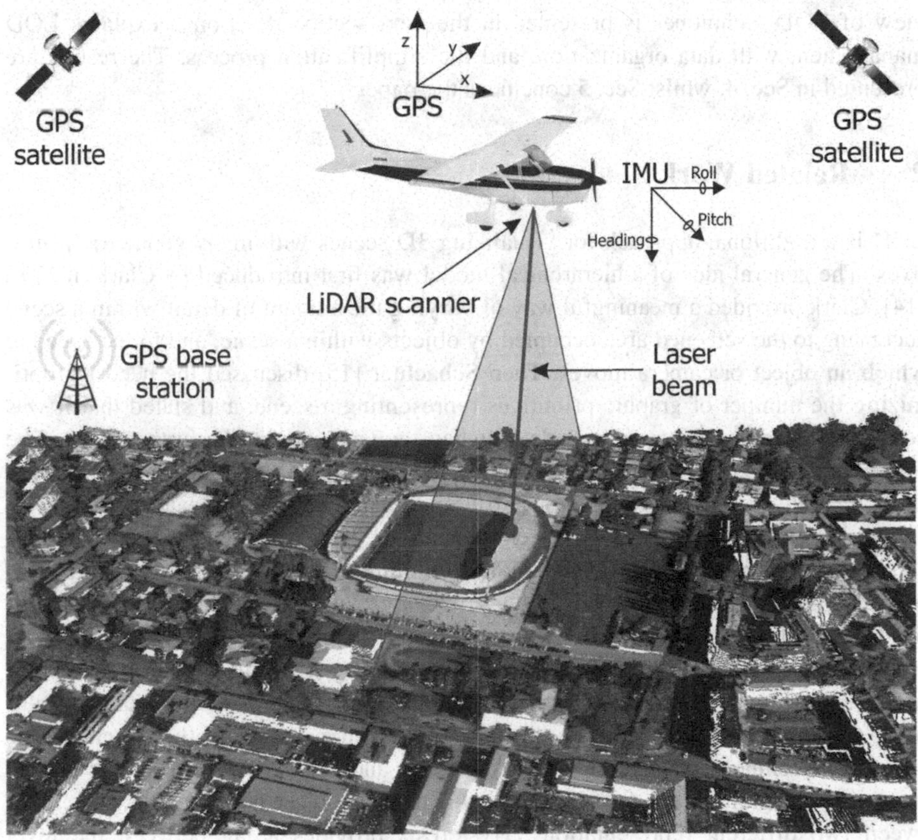

Fig. 1. Gathering the LiDAR data

LOD is realized by rendering a detailed geometry when the object is close to the viewer and coarser approximation when the object is distant or small. In this way, the graphic workload is significantly reduced and image quality is preserved without affecting the fidelity of the display. However, LOD with its advantages regarding performance enhancements also comes with some significant drawbacks. The most problematic is the popping effect, which happens when the graphic system switches between different detail levels, and flicker is noticed. This visual artifact occurs due to an improper model for selecting an optimal detail level. In order to avoid this effect, a smooth visual transition is needed in the form of a continuous model for measuring the degree of visual details that a user can perceive. This information is then used by the model to select the optimal LOD, in order to avoid noticeable changes in visualization. For this it is essential to find the relation between the computer's capabilities and human perception.

This paper presents an evaluation based on the visual perception of optimized LOD visualization regarding LiDAR data. This paper is organized as follows: a brief overview of LOD techniques is presented in the next section. Section 3 explains LOD management with data organization, and the simplification process. The results are presented in Sec. 4, whilst Sec. 5 concludes the paper.

2 Related Work

LOD is a traditional approach for visualizing 3D scenes with many geometric primitives. The general idea of a hierarchical model was first introduced by Clark in 1976 [14]. Clark provided a meaningful way of varying the amount of detail within a scene according to the screened area occupied by objects within a scene, and to the speed at which an object or camera moves. Then Schachter [15] discussed the need for optimizing the number of graphic primitives representing a scene, and stated that it was common to display objects in less detail when they appeared to be further away. The first applications using such LOD techniques for optimizing graphic workloads were flight simulators. In these applications different detail levels were created by hand.

Several Discrete LOD (DLOD) techniques have been proposed over past decades [16–19]. It is typical for DLOD techniques to simplify a 3D object into a number of objects with different detail levels. Then proper LOD of the object is selected according to the calculated distance from the viewpoint. Falby et al. [20] used the DLOD technique for managing massive terrain datasets using a terrain-paging algorithm to manage the swapping of visible terrain tiles. They used three different dataset resolutions and displayed the terrain within a resolution, depending on the distance. The Hysteresis technique was presented by Astheimer and Pöche [21] to overcome DLOD weaknesses regarding the flickering effect when an object constantly switches between two different representations. Hysteresis provides a lag between the level switches, thus the object switches to a lower level at a different distance then to the higher level.

Most LOD techniques for terrain optimization in real-time use hierarchical space subdivision for scene representation. A method for recursively subdividing terrain into quarters using a quadtree was presented by Lindstrom at al. [22]. They used a hierarchy in which each depth of the tree corresponded to a certain detail level. Therefore, it was easy to combine different detail levels into representation of the terrain. Their system also considered the roughness of the terrain and used lower details where the topography was smoother. Similar methods have also been developed using Delaunay triangulation for continuous terrain LOD [23, 24]. The resulting hierarchies in these methods are not trees but directed acyclic graphs, which make it difficult to combine different detail levels.

The field of progressive meshes as a technique of dynamic LOD was intensively promoted by Hoppe [25–28]. He presented an efficient data structure with a simplification procedure that preserves the geometry of the mesh and its overall appearance [25, 26]. This methods supported geomorphs, progressive transmission, compression, and selective refinement. In his further work, he presented a view-dependent framework for selectively refining arbitrary progressive meshes [27, 28]. He defined a criteria based on the view, surface orientation, and screen-space geometric error for incrementally adapting the approximation of progressive meshes. In this way, it is possible to apply progressive meshes on terrain data.

Visualization of LiDAR point clouds is a hard challenge. Recently, Kovač and Žalik [29] developed a two-pass point-based rendering technique that uses elliptical weighed average filtering for solving problems relating to aliasing. A web-based LiDAR visualization with point-based rendering was proposed by Kuder and Žalik [30]. They used an efficient data organization within a data structure that enabled quick handling of range and LOD queries.

More sophisticated approaches are based on the Human Vision System (HVS) [31–34], where perceptual metrics like spatial frequency and visual acuity are used to determine visible differences between images. The Just Noticeable-Difference (JND) approach was presented by Cheng et al. [35]. They used a perceptual analysis for improving the results of geometric measures and so identified redundant data that cannot be identified using conventional geometric metrics. A volume-rendering algorithm that follows the user's gaze and smoothly varies the display resolution has been developed by Levoy and Whitaker [36]. Gaze-contingency uses models of human spatial perception and can be applied to geographic data representation.

3 LOD Management

The first, and probably most important decision, having direct impact on display quality and performance corresponds to the choice of rendering primitives. Due to the nature of the data we use points as display primitives for rendering instead triangles or quads.

As already mentioned, the main problem working with LiDAR datasets is the amount of data. A typical LiDAR data set contains several tens of millions of points. Today's graphic cards are incapable of storing such amounts of geometric primitives. Consequently, they cannot show a whole dataset using real-time interaction. Therefore, LOD management is needed. In addition, presenting three-dimensional spatial data on a two-dimensional display weakens human perception of 3D data. Therefore, it is necessary to find a LOD balance between technical and perceptual abilities. LOD is always implemented with an efficient data organization that allows for fast simplification of the scene. In the continuation we describe our data organization and the simplification process.

3.1 LiDAR Data Organization

Data collected by LiDAR technology are stored in ASPRS LAS [37], a public binary file format for exchanging spatial data. LAS files store remotely sensed point clouds without any topological information amongst individual points. Airborne LiDAR data are typically saved into LAS files as flight swathes. Because of this, the data are in the form of stripes. Usually it takes a number of flights to collect enough surface data for the desired resolution.

The basis for the fast and effective visualization of point clouds is hierarchical space partitioning, where data are divided into smaller segments. Although we are dealing with 3D point cloud data, a quadtree data structure has been applied, since LiDAR data describing terrain is considered as 2.5D data. A quadtree is a 2D data structure, where the root covers the whole area and has 4 children that halved space into 4 equal quadrants. Due to the shape of the point cloud, the spatial hierarchy can be ineffective; since space is always halved and the stripe form is preserved throughout all levels of the tree. Therefore, the root of the tree is divided in such a way as to approximate a square shape. In this way, our structure can have one or more children in the root. Obviously, with such a root division our structure is no longer a traditional quadtree anymore (see Fig. 2).

In our case, space partitioning is constructed during the pre-processing phase. The construction of the tree is accomplished over three steps:

- During the initialization step the root and the corresponding quadtrees are created. The root bounding volume is divided into rectangles along the longer dimension. The number of divisions is determined as the ratio between the length and width of the root area.
- Actual space partitioning is obtained during construction of the quadtrees. The construction is conducted by sending, point after point through the corresponding quadtree nodes, depending on their positions in space. The points are inserted until the predetermined threshold has been met regarding points per quadtree leaf. When this happens, four new quadtree nodes are created by dividing this leaf and rearranging the points from previous leaves into new ones.
- When the whole tree has been constructed, the points within individual quadtree leaves are randomly sampled to avoid re-sampling during the rendering phase. By using OpenGL's Vertex Buffer Objects (VBO) [38], the geometry of the points is stored within the graphic card memory. This allows for maintaining only point indexes and VBO references in the leaves of the quadtree, which are located within the main memory. Those LiDAR point attributes that are not needed for visualization remain in the source file and can be accessed on request via the point indexes. The points, which are displayed, are selected by random sampling.

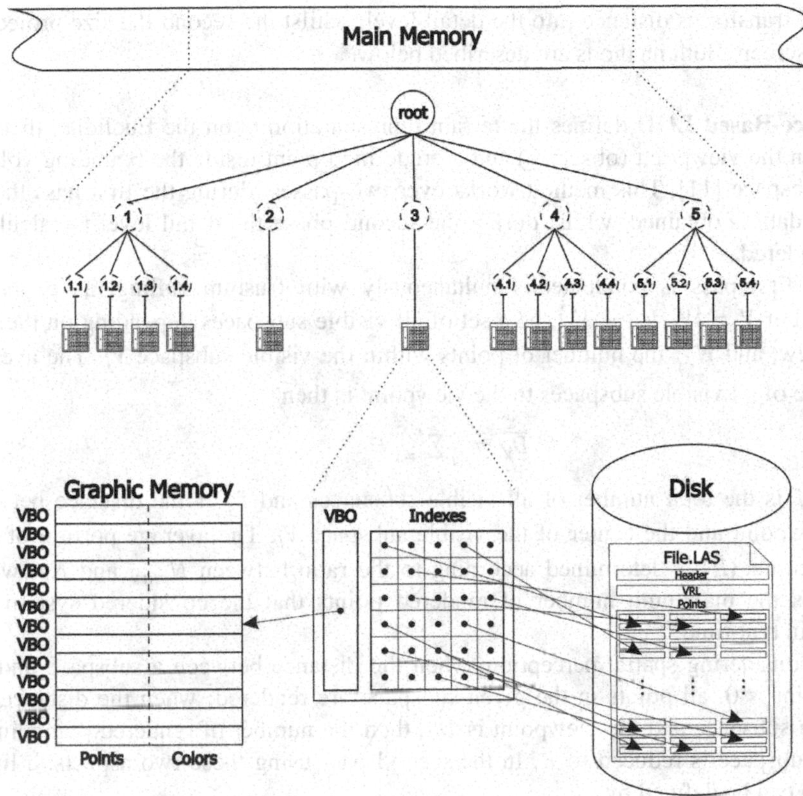

Fig. 2. Data organization

3.2 Simplification of the Scene

In order to achieve faster visualization, reducing the number of the display primitives is needed, where only the points lying outside the view can be removed without loss of image quality. A frustum culling technique on the space subdivision hierarchy needs to be applied for this. Frustum culling excludes tree nodes from rendering if they are outside the viewing frustum. This process must be performed each time the view changes. The described technique is effective and provides a significant reduction of display primitives. However, it has some drawbacks if insufficient points are reduced, especially when the view covers the entire scene. Therefore, points being located inside the viewing frustum have to be removed, too. For this purpose, a LOD technique is applied to simplify the scene.

Our design of data organization ensures a simple LOD implementation, where only the number of points is calculated for those quadtree nodes that need to be rendered. This is possible because of using VBOs. However, two different LOD selection factors have been developed that adapt the abilities of the computer system. The first method transforms distance into the detail level, whilst the second the size projection on the screen. Both methods are described below.

Distance-Based LOD defines the terrain representation upon the Euclidian distance between the viewpoint (observer) and a predefined point inside the bounding volume of a subspace [11]. This method works over two-passes: during the first pass the required data is obtained, whilst during the second phase the detail level is calculated and rendered.

The first pass is conducted simultaneously with frustum calling, as explained above. Let $V = \{V_1, V_2, \dots, V_J\}$, be a set of all visible subspaces depending on the current view, and N_{V_j} the number of points within the visible subspace V_j. The average distance of all visible subspaces to the viewpoint is then

$$\overline{D_V} = \frac{1}{J}\sum_{j=1}^{J} D_j, \tag{1}$$

where J is the total number of all visible subspaces and D_j is the distance between the viewpoint and the center of the visible subspace V_j. The average portion of rendered points $(\bar{R})$ is determined according to the ratio between N_{max} and N_V, where N_{max} is the maximum number of rendered points that the considered system can render in real-time.

By considering spatial perception, when the distance between a subspace and the viewpoint is 0, all points in the given subspace are rendered; when the distance between a subspace and the viewpoint is $\overline{D_V}$, then the number of rendered points in the given subspace is reduced to $\bar{R}$. In the second pass using these two aspects, a linear relation can be defined by

$$R_j = \frac{D_j \cdot (\bar{R}-1)}{\overline{D_V}} + 1, \tag{2}$$

where the percentage of points rendered (R_j) in subspace V_j is calculated based on its distance to the viewpoint (D_j).

Nevertheless, there are several issues related to the domain of the variable $R_j \in [-\infty, \infty]$. First, the negative percentage of points cannot be drawn. Moreover, there are often no subspaces at distance 0, and thus no subspaces are drawn in full details. Another issue occurs when subspaces are too far and are not drawn, but must be visible. In order to avoid this, the optimal number of rendered points within a subspace V_j can then be calculated by a piecewise function of distance D_j

$$LOD_D(D_j) = \begin{cases} N_{V_j}, & R_j > \frac{85}{100} \\ \frac{5}{100} \cdot N_{V_j}, & R_j < \frac{5}{100} \\ R_j \cdot N_{V_j}, & otherwise \end{cases}, \tag{3}$$

In this way, we render all points within the subspaces where the render factor is higher than 85% and in the other at least five percent of the contained points, as shown in Fig. 3. Using such an organization and definition of LOD, we are able to smooth transitions between different detail levels. In this way, we are able to remove the popping effect.

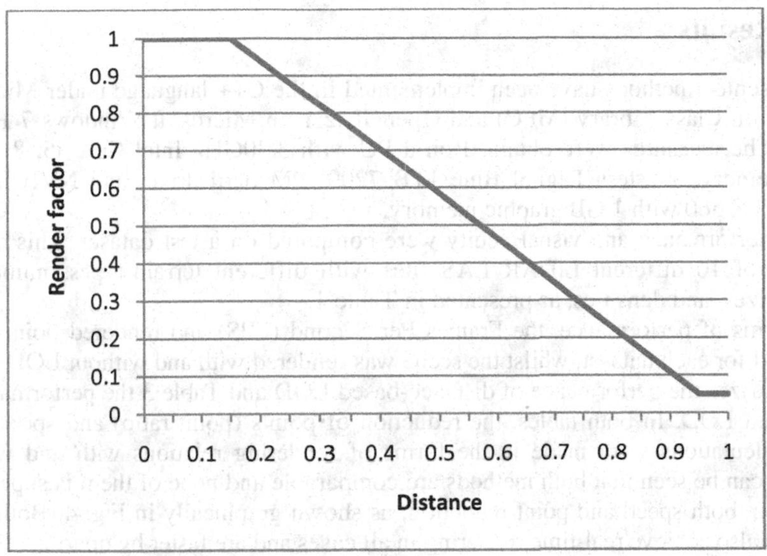

Fig. 3. The relation between detail level and distance

Size-Based LOD defines the terrain representation based on the projected size in screen coordinates [11]. This method is also performed in two passes like distance-based LOD.

During the first pass, the projected screen area for all visible subspaces is defined by

$$A_V = \sum_{j=1}^{J} A_j, \tag{4}$$

where A_V is the area of all subspaces on the screen in pixels, A_j is the area for each individual subspace V_j. Then the maximum density of the points projected on the screen is calculated by:

$$\rho = \frac{A_V}{N_{max}}. \tag{5}$$

In this way, the upper limit is set so that the system cannot produce any lag.

The area A_j of subspace V_j is transformed into a number of points that will be rendered by the following equation:

$$LOD_S(A_j) = \frac{A_j}{N_{V_j} \cdot \rho} \tag{6}$$

The calculation of LOD_S where the number of points is defined with the area occupied on the screen is computationally more expensive. However, size-based LOD is more accurate and can calculate a more accurate detail level.

4 Results

The presented methods have been implemented in the C++ language under Microsoft Foundation Class Library (MFC) and OpenGL 2.1 on Microsoft Windows 7 Professional. The measures were obtained on a PC with 3.30GHz Intel Core i5, 8 GB of main memory, Western Digital Blue 1TB 7200RPM hard drive, and NVIDIA GeForce GTX 560 with 1 GB graphic memory.

The performance and visual acuity were compared on a test dataset. This dataset consists of 10 different LiDAR LAS files with different terrain-types, number of points, sizes, and densities, as presented in Table 1.

In terms of performance, the Frames Per Second (FPS) and rendered points were measured for each dataset, whilst the scene was rendered with and without LOD. Table 2 summarizes the performance of distance-based LOD and Table 3 the performance of size-based LOD. In both tables, the reduction of points (point ratio) and speed (FPS ratio) calculations were made in the forms of rendering relations with and without LOD. It can be seen that both methods are comparable and none of them is superior to another in both speed and point reduction, as shown graphically in Fig. 4. Both LOD methods also achieve real-time rendering in all cases and are faster by up to 2.5 times.

Table 1. Description of test LiDAR LAS files

Dataset	*Terrain type*	*No. points*	*Size (m × m)*	*Density*
File 1	Flat rural	6190800	781 × 880	9.01
File 2	Moderate rural	16438927	866 × 3704	5.13
File 3	Steep forested	13677384	500 × 750	36.47
File 4	Flat rural	19426701	1000 × 1500	12.95
File 5	Mountains	34000000	2593 × 3385	3.87
File 6	Flat urban	21147263	3294 × 1501	4.28
File 7	Flat forested	17949909	1000 × 895	20.07
File 8	Flat urban	13861142	1957 × 2114	3.35
File 9	Moderate urban	37933412	4012 × 3025	3.13
File 10	Moderate urban	20657230	683 × 1383	21.87

Table 2. Distance-based LOD performance

Dataset	*Without LOD*		*Distance-based LOD*			
	Points	FPS	Points	FPS	Points ratio	FPS ratio
File 1	6190800	58	5051868	62	0.82	1.07
File 2	15328038	29	5611506	50	0.37	1.72
File 3	11884312	35	4828047	53	0.41	1.51
File 4	18194184	26	4956370	52	0.27	2.00
File 5	27355270	17	5323969	40	0.19	2.35
File 6	20208650	23	5188756	42	0.26	1.83
File 7	11667829	38	4992648	45	0.43	1.47
File 8	11446218	38	4966357	55	0.43	1.45
File 9	18714995	25	5864291	47	0.31	1.88
File 10	19765258	23	5674141	42	0.29	1.83

Table 3. Size-based LOD performance

Dataset	*Without LOD*		*Size-based LOD*			
	Points	FPS	Points	FPS	Points ratio	FPS ratio
File 1	6190800	58	5266644	60	0.85	1.03
File 2	15328038	29	4241778	53	0.28	1.83
File 3	11884312	35	5083744	52	0.43	1.49
File 4	18194184	26	5932876	48	0.33	1.85
File 5	27355270	17	5718530	41	0.21	2.41
File 6	20208650	23	4459356	45	0.22	1.96
File 7	11667829	38	3901293	60	0.33	1.58
File 8	11446218	38	5144027	53	0.45	1.39
File 9	18714995	25	4305302	55	0.23	2.20
File 10	19765258	23	5294721	44	0.27	1.91

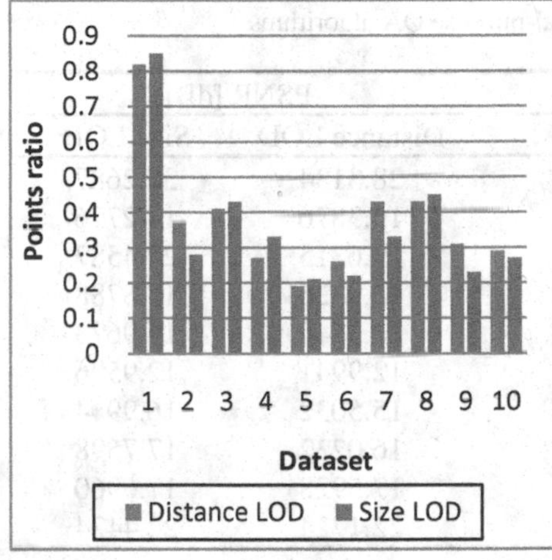

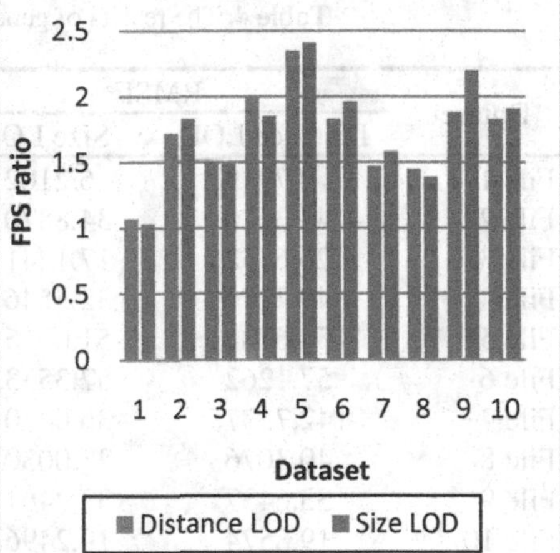

Fig. 4. The comparison between distance-based LOD and size-based LOD in terms of point reduction (left) and speed (right)

In terms of image quality, both LOD methods were compared using seven different Quality Assessment (QA) algorithms:

- **Root-Mean-Square Error (RMSE)** one of the more commonly used signal fidelity measures [39];
- **Peak Signal-to-Noise Ratio (PSNR)** another widely used metric of quality [40];
- **Structural Similarity (SSIM)** index for measuring the similarity between two images [41];
- **Visual Difference Predictor (VDP)** a metric for predicting whatever differences between two images are visible to the human observation [42];
- **Visual Information Fidelity (VIF)** an image information measure that quantifies the information relation between reference image and distorted image [43];

- **Naturalness Image Quality Evaluator (NIQE)** an image quality analyzer based on statistical regularities observed in natural images [44] and
- **Blind Image Quality Index (BIQI)** another image quality assessment that works on the statistics of local Discrete Cosine Transformation (DCT) coefficients [45];

The presented QA algorithms can be grouped into three categories: general-purpose algorithms, HVS algorithms, and blind algorithms.

The general-purpose and HVS algorithms are full-reference metrics [41], based on measuring the error between the distorted (with LOD) and the reference image (without LOD). The general-purpose algorithms RMSE and PSNR results are presented in Table 4. Distance LOD refers to the error between the reference image and the image rendered using distance-based LOD, whilst Size LOD is the error between the reference image and the image rendered using size-based LOD. The range of RMSE metric is from 0 to infinity, with 0 being a perfect score, whilst the PSNR metric has inverse scoring (i.e. the higher the number the better the score).

Table 4. The results of general-purpose QA algorithms

Dataset	RMSE		PSNR [dB]	
	Distance LOD	Size LOD	Distance LOD	Size LOD
File 1	9.7919	6.2102	28.3134	32.2687
File 2	43.3670	34.8870	15.3876	17.2776
File 3	26.6052	17.1241	19.6315	23.4587
File 4	40.7296	32.5546	15.9326	17.8788
File 5	54.6781	51.0715	13.3746	13.9673
File 6	57.1262	57.3533	12.9941	12.9596
File 7	42.7937	36.0430	15.5032	16.9944
File 8	40.3076	33.0030	16.0232	17.7598
File 9	33.6437	33.2461	17.5928	17.6960
File 10	19.8574	19.2496	22.1724	22.4424

It is observed that the size-based LOD has better values than the distance-based LOD thus indicating higher consistency, as shown in Fig. 5. The size-based LOD has lower RMSE and higher PSNR values for all files except for insignificant difference in file 6. However, these methods are too general and inconsistent, according to human eye perception, for obtaining more adequate information about perceived details.

The VDP, SSIM, and VIF Image Quality Assessment (IQA) algorithms were adapted to objectively evaluate the rendered image quality in terms of HVS, and are shown in Table 5. These algorithms are also full-reference metrics [41] like general-purpose algorithms.

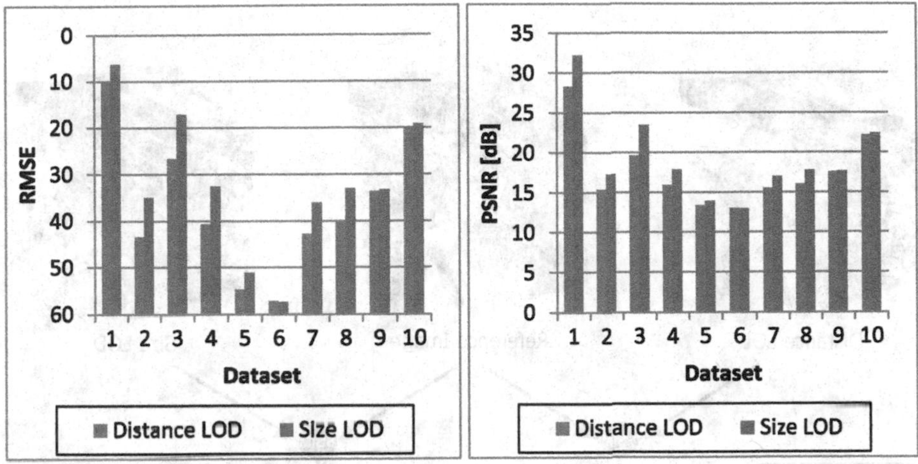

Fig. 5. The comparison between distance-based LOD and size-based LOD with general-purpose QA: RMSE (left) and PSNR (right)

Table 5. The results of the HVS methods

Dataset	SSIM		VDP [%]		VIF	
	Distance LOD	Size LOD	Distance LOD	Size LOD	Distance LOD	Size LOD
File 1	0.9733	0.9875	87.97	85.76	0.4021	0.7595
File 2	0.9174	0.9528	71.09	69.89	0.1863	0.6065
File 3	0.9038	0.9474	39.93	30.35	0.2362	0.5741
File 4	0.8738	0.8996	69.02	44.67	0.1349	0.2675
File 5	0.7050	0.7296	63.03	25.30	0.0489	0.0725
File 6	0.8645	0.8567	60.15	63.33	0.0997	0.1282
File 7	0.8388	0.8727	68.42	57.40	0.1363	0.5156
File 8	0.9190	0.9488	64.83	59.83	0.1430	0.3116
File 9	0.9394	0.9322	56.96	53.92	0.2177	0.4934
File 10	0.9581	0.9627	58.98	52.40	0.3050	0.4090

Fig. 6 shows the LOD evaluation with the SSIM metric [41]. It can be seen from the SSIM index map (darker pixels' greater visual differences) that most of the differences either occurred within dense point regions that represent vegetation or at sharp edges, for both methods. Moreover, size-based LOD shows fewer differences in the front regions (see Fig. 6 and Fig. 12) that appear larger on the screen and have more details. The SSIM index is a value between -1 and 1, where SSIM=1 for images that are perceptually equivalent. In general, size-based LOD has fewer differences (higher SSIM value), as can be seen in Fig. 8. However, a lower SSIM index was achieved in the cases of file 6 and 9. The reason for this exception can be related to the outliers contained in those files. File 6 was selected in order to further compare the visual acuity of the LOD methods with noisy data.

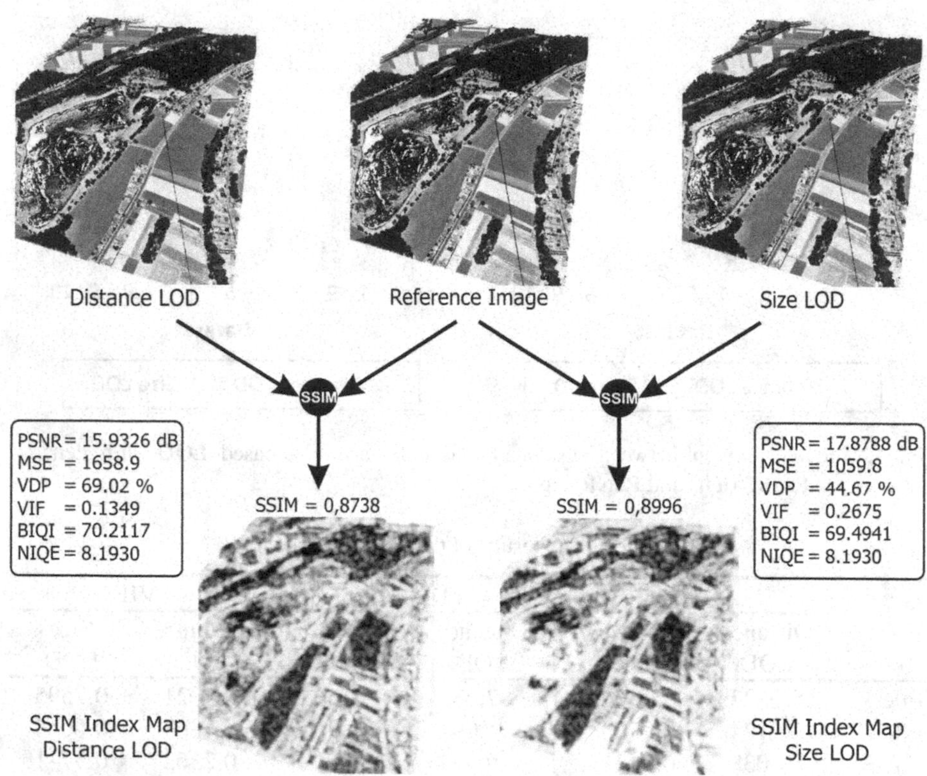

Fig. 6. LOD evaluation of dataset 4 with SSIM algorithm in terms of HVS

VDP is an algorithm for describing the human visual response. The goal of the algorithm is to determine the degree to which physical differences (i.e. incorrect luminance and chrominance) become visible to the human observer [42]. The results of the evaluation from the VDP metric show that the probability of difference detection within the compared images is high, but the size-based LOD has lesser probabilities than the distance-based LOD, except for file 6, as shown in Fig. 9. The VDP metric gives comparable results with the SSIM metric (see Table 5). The exception is file 9 where the size-based LOD provided a better result under the VDP metric than the distance-based LOD, in comparison by using SSIM metric. It can also be observed that the detection map of the VDP metric gave the same graphical result as SSIM metric (compare Fig. 6 and Fig. 7). Both VDP and SSIM graphical results show that the size-based LOD produces lees differences (e.g. more visual acuity and more details) in the front regions, as can be seen in Fig. 12.

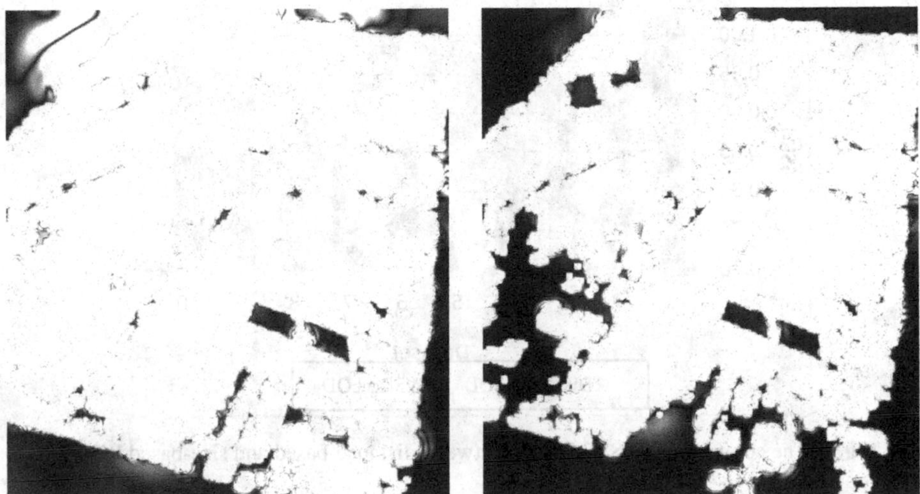

Fig. 7. The VDP detection map result of dataset 4, for distance-based LOD (left), and for size-based LOD (right)

VIF is a full-reference IQA, which is derived from a statistical model for natural scenes, a model for image distortion, and a HVS model [43]. It is a number that quantifies the information fidelity of a tested image that the HVS can extract. VIF is a very effective metric; it can capture the effects of linear contrast enhancements on images and compute any improvement in visual quality. A VIF value greater than 1 specifies improvement, whilst values of less than 1 suggest a loss of visual quality. The results, as shown graphically in Fig. 10, indicate that all LOD-rendered images lost visual quality, which is obvious when reducing the number of primitives to be rendered. In all datasets VIF metric yielded better results for size-based LOD than the distance-based LOD. Furthermore, the results were significantly better, except for files 5 and 6: The reasons for this are considered in the continuation.

Table 6. The results of measures using blind algorithms

Dataset	NIQE		BIQI	
	Distance LOD	Size LOD	Distance LOD	Size LOD
File 1	10.9488	11.0849	18.7885	18.7126
File 2	9.9463	10.0650	75.8597	75.8024
File 3	6.3427	6.3208	44.2735	32.1958
File 4	8.1410	8.1930	70.2117	69.4941
File 5	10.8441	10.1716	75.8859	75.8859
File 6	10.5231	9.9334	55.1908	56.5241
File 7	10.1567	10.5150	54.4409	52.8999
File 8	17.8624	18.0699	75.3882	75.3306
File 9	11.3614	11.2506	74.9341	74.7894
File 10	8.5702	7.7800	22.1538	21.1642

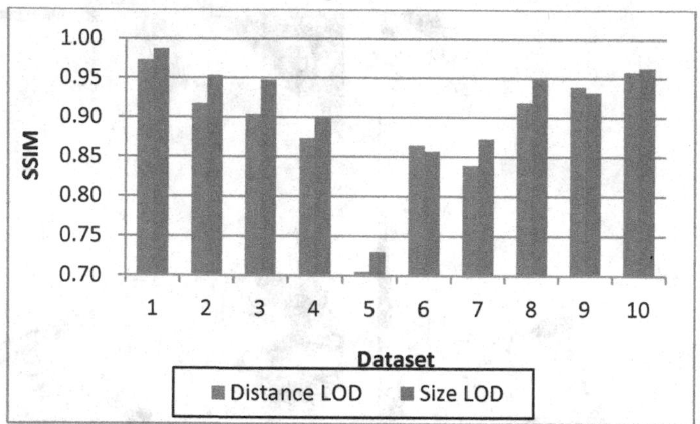

Fig. 8. The comparison of SSIM index between distance-based and size-based LODs

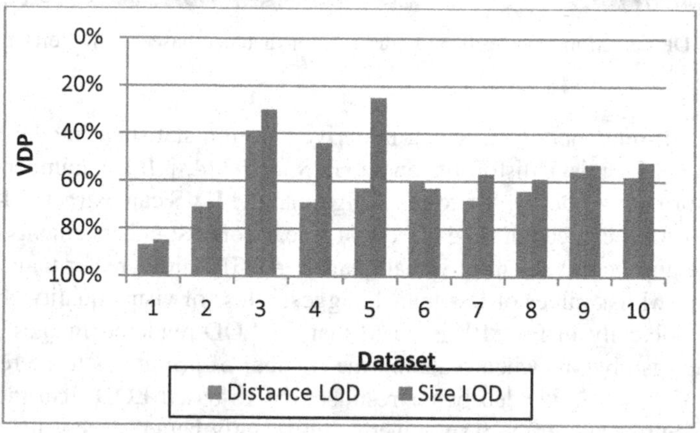

Fig. 9. Graphical comparison of VDP metric between both LOD

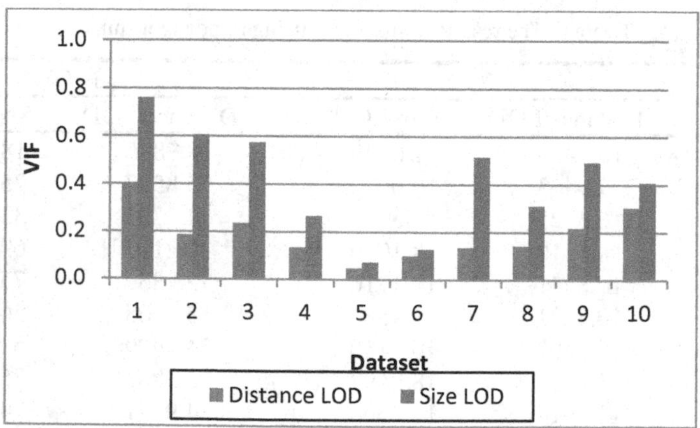

Fig. 10. The comparison between both LODs using VIF

The third category are No-Reference QA algorithms or blind algorithms, where the algorithm has only the distorted image and has to estimate the quality [44, 45]. Both methods, NIQE and BIQI, are not correlated with human judgment of visual quality. The results are shown in Table 6. From the obtained BIQI metric the size-based LOD had better results (smaller values) than the distance-based LOD, with the exception of file 6 (see Fig. 11). This was unlike the NIQE metric where the values were more diverse and no relevant information could be extracted, which is a consequence of the measurable deviations from statistical regularities that can be observed from these methods.

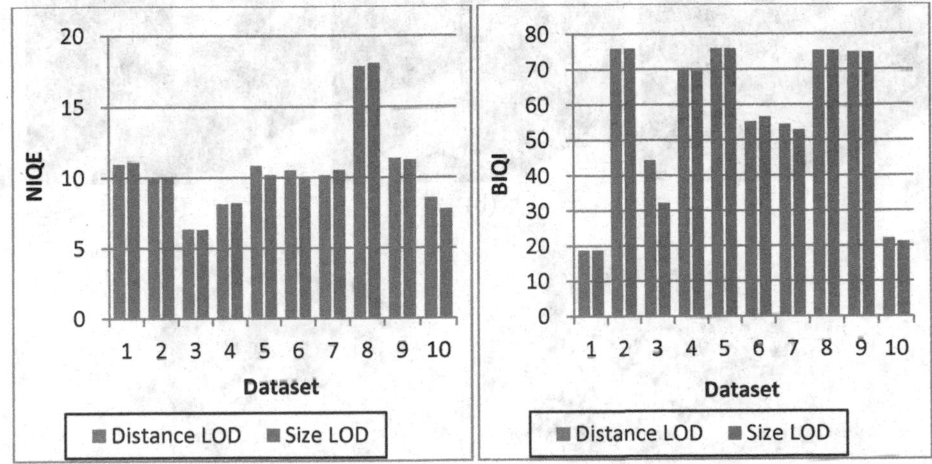

Fig. 11. The comparison between distance-based LOD and size-based LOD with no-reference QA: NIQE (left) and BIQI (right)

The summation is that particularly good results are obtained with the RMSE, PSNR, and BIQI methods, which are not correlated to HVS. However, these methods are not authentic enough for our evaluation. File 6 showed poor results in almost all metrics, which can be related to the specific data contained within the dataset. The data in file 6 described a city and thus contained a considerable amount of noise. This noisy data appears as highly elevated points known as high outliners. Subspaces describing such large differences in z-coordinates occupy relatively large areas on the screen even if the subspaces are somewhere in the middle or in the background. In this way, such subspaces are rendered with high detail levels but they have almost no relevant data, whilst other subspaces with relevant data are rendered with lower detail levels. The SSIM index had again problems due to the noise presented in file 9. In the case of file 5, it can be observed that all HVS matrices had the best results. File 5, consisting of bare mountains, had the biggest points reduction ratio from all the datasets, and therefore, it could be displayed with almost imperceptible differences.

The results show significant variances across each dataset, because each metric covers a particular aspect. Important information about the structure of the objects in the visual scene was tested using the SSIM metric, whilst the VDP metric was used to obtain the important visual information within the reduced LOD picture. The VIF metric was used to measure the perceptually relevant information that could be extracted from a picture. With the evaluation of HVS metrics, it could be concluded that the size-based LOD better represents data in terms of human vision.

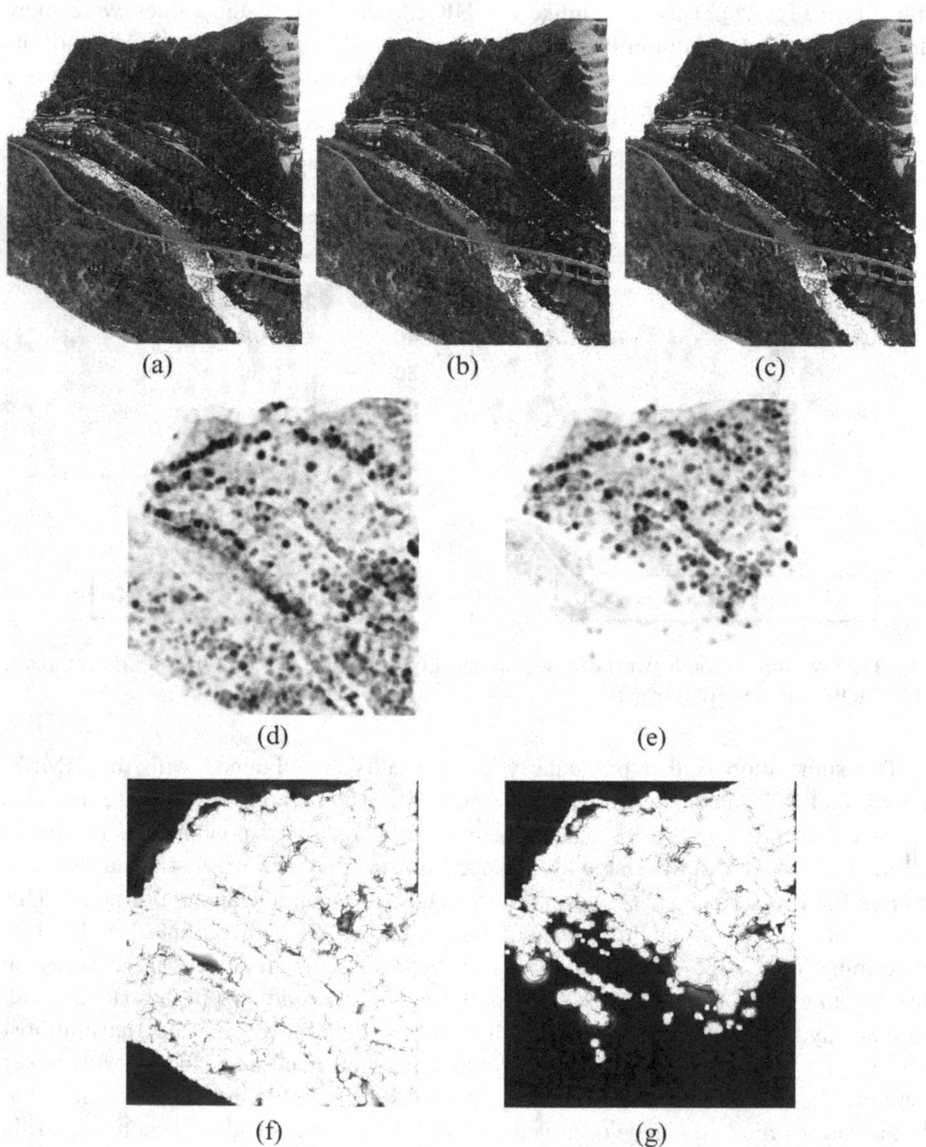

Fig. 12. LOD evaluation of dataset 3: distance-based LOD (a), without LOD (b), size-based LOD (c), SSIM index map of distance-based LOD (d), SSIM index map of size-based LOD (e), VDP result of distance-based LOD (f), and VDP result of size-based LOD (g)

5 Conclusion

In this paper, two different LOD methods were developed and evaluated in terms of visual perception and human judgment. We have presented an efficient method for the visualization of LiDAR data. Within this context, LOD hierarchy has been introduced that enables the elimination of invisible parts of a scene, as well as the allocation and fast rendering of visible bounding volumes. Two different LOD criteria were created, for this purpose. The first method calculates LOD depending on distance, whilst the second depends on the projected size. Both methods are constructed to remove all visual artifacts like 'popping'. The results show that the presented approach is capable of rendering large amounts of LiDAR data in real-time.

Little is known about how to select optimal LOD, especially in terms of human vision and spatial perception. Bearing this in mind, we evaluated both LOD methods using 7 different QA methods. We believe from the results, that the size-based LOD is more suitable for representing terrain data gathered by LiDAR airborne scanners.

Acknowledgements. This work was supported by the Slovenian Research Agency (grants L2-3650 and P2-0041), the European Union, European Regional Fund, within the scope of the framework of the Operational Programme for Strengthening Regional Development Potentials for the Period 2007-2013, contracts No. 3211-10-000467 (KC CLASS) and within the framework of the operation entitled "Centre of Open innovation and ResEarch UM" co-funded by the European Regional Development Fund and conducted within the framework of the Operational Programme for Strengthening Regional Development Potentials for the period 2007-2013, development priority 1: "Competitiveness of companies and research excellence", priority axis 1.1: "Encouraging competitive potential of enterprises and research excellence".

References

1. Briese, C., Norbert, P., Dorninger, P.: Applications of the robust interpolation for DTM determination. International Archives of Photogrammetry Remote Sensing and Spatial Information Sciences 34, 55–61 (2002)
2. Lienert, B.R., Sharma, S.K., Porter, J.N.: Real Time Analysis and Display of Scanning Lidar Scattering Data. Marine Geodesy 22, 259–265 (1999)
3. Mongus, D., Žalik, B.: Parameter-free ground filtering of LiDAR data for automatic DTM generation. ISPRS Journal of Photogrammetry and Remote Sensing 67, 1–12 (2012)
4. Wehr, A., Lohr, U.: Airborne laser scanning—an introduction and overview. ISPRS Journal of Photogrammetry and Remote Sensing 54, 68–82 (1999)
5. Mongus, D., Žalik, B.: Efficient method for lossless LIDAR data compression. International Journal of Remote Sensing 32, 2507–2518 (2011)
6. Persson, Å., Söderman, U., Töpel, J., Ahlberg, S.: Visualization and analysis of full-waveform airborne laser scanner data. International Archives of Photogrammetry, Remote Sensing and Spatial Information Sciences 36, 103–108 (2005)
7. Gregory, R.L.: Eye and Brain: The psychology of seeing. Princeton University Press (1997)

8. Bartz, D., Meißner, M., Hüttner, T.: OpenGL-assisted occlusion culling for large polygonal models. Computers & Graphics 23, 667–679 (1999)
9. Scher Zagier, E.J.: A human's eye view: motion blur and frameless rendering. Crossroads 3, 8–12 (1997)
10. Liu, N.L.N., Jin, H.J.H., Rockwood, A.P.: Antialiasing by Gaussian integration. IEEE Computer Graphics and Applications 16, 58–63 (1996)
11. Luebke, D., Reddy, M., Cohen, J.D., Varshney, A., Watson, B., Huebner, R.: Level of detail for 3D graphics. Morgan Kaufmann (2002)
12. Bittner, J., Wonka, P., Wimmer, M.: Visibility preprocessing for urban scenes using line space subdivision. In: Proceedings of Ninth Pacific Conference on Computer Graphics and Applications, pp. 276–284 (2001)
13. Smets, G.J.F., Overbeeke, K.J.: Trade-off between resolution and interactivity in spatial task performance. IEEE Computer Graphics and Applications 15, 46–51 (1995)
14. Clark, J.H.: Hierarchical geometric models for visible surface algorithms. Communications of the ACM 19, 547–554 (1976)
15. Schachter, B.J.: Computer Image Generation for Flight Simulation. IEEE Computer Graphics and Applications 1, 29–68 (1981)
16. Garland, M., Heckbert, P.: Simplification using Quadric Error Metrics. In: Proceedings of SIGGRAPH, pp. 209–216 (1997)
17. Klein, R., Liebich, G., Strasser, W.: Mesh reduction with error control. In: Proceedings of Visualization 1996, pp. 311–318 (1996)
18. Erikson, C., Manocha, D., Baxter III, W.V.: HLODs for faster display of large static and dynamic environments. In: Proceedings of the 2001 Symposium on Interactive 3D Graphics, pp. 111–120. ACM, New York (2001)
19. Cohen, J., Varshney, A., Manocha, D., Turk, G., Weber, H., Agarwal, P., Brooks, F., Wright, W.: Simplification envelopes. In: Proceedings of the 23rd Annual Conference on Computer Graphics and Interactive Techniques, pp. 119–128. ACM, New York (1996)
20. Falby, J.S., Zyda, M.J., Pratt, D.R., Mackey, R.L.: NPSNET: Hierarchical data structures for real-time three-dimensional visual simulation. Computers & Graphics 17, 65–69 (1993)
21. Astheimer, P., Pöche, M.-L.: Level-of-detail generation and its application in virtual reality. In: Proceedings of the Conference on Virtual Reality Software and Technology, pp. 299–309. World Scientific Publishing Co., Inc., River Edge (1994)
22. Lindstrom, P., Koller, D., Ribarsky, W., Hodges, L.F., Faust, N., Turner, G.A.: Real-time, continuous level of detail rendering of height fields. In: Proceedings of the 23rd Annual Conference on Computer Graphics and Interactive Techniques, pp. 109–118. ACM, New York (1996)
23. Ferguson, R.L., Economy, R., Kelly, W.A., Ramos, P.P.: Continuous terrain level of detail for visual simulation. In: Proceedings of IMAGE V Conference, pp. 144–151 (1990)
24. Scarlatos, L.L.: A refined triangulation hierarchy for multiple levels of terrain detail. In: Proceedings of IMAGE V Conference, pp. 114–122 (1990)
25. Hoppe, H.: Progressive meshes. In: Proceedings of the 23rd Annual Conference on Computer Graphics and Interactive Techniques, pp. 99–108. ACM, New York (1996)
26. Hoppe, H.: Efficient implementation of progressive meshes. Computers & Graphics 22, 27–36 (1998)
27. Hoppe, H.: View-dependent refinement of progressive meshes. In: Proceedings of ACM SIGGRAPH, pp. 189–198 (1997)
28. Hoppe, H.: Smooth view-dependent level-of-detail control and its application to terrain rendering. In: Proceedings of Visualization 1998, pp. 35–42 (1998)

29. Kovač, B., Žalik, B.: Visualization of LIDAR datasets using point-based rendering technique. Computers & Geosciences 36, 1443–1450 (2010)
30. Kuder, M., Žalik, B.: Web-Based LiDAR Visualization with Point-Based Rendering. In: 2011 Seventh International Conference on Signal Image Technology & Internet-Based Systems, pp. 38–45. IEEE (2011)
31. O'Sullivan, C., Howlett, S., Morvan, Y., McDonnell, R., O'Conor, K.: Perceptually adaptive graphics. State of the Art Report EUROGRAPHICS 4 (2004)
32. McNamara, A.: Visual Perception in Realistic Image Synthesis. Computer Graphics Forum 20, 211–224 (2001)
33. Watson, B., Walker, N., Hodges, L.F.: Supra-threshold control of peripheral LOD. ACM Transactions on Graphics (TOG) 23, 750–759 (2004)
34. Scoggins, R.K., Moorhead, R.J., Machiraju, R.: Enabling level-of-detail matching for exterior scene synthesis. In: Proceedings of the Conference on Visualization 2000, pp. 171–178. IEEE Computer Society Press, Los Alamitos (2000)
35. Cheng, I., Shen, R., Yang, X., Boulanger, P.: Perceptual Analysis of Level-of-Detail: The JND Approach. In: Eighth IEEE International Symposium on Multimedia (ISM 2006), pp. 533–540. IEEE (2006)
36. Levoy, M., Whitaker, R.: Gaze-directed volume rendering. ACM SIGGRAPH Computer Graphics 24, 217–223 (1990)
37. Samberg, A.: An Implemetation of the ASPRS LAS Standard. In: Proceedings of the ISPRS Workshop Laser Scanning 2007 and SilviLaser 2007, vol. 36, pp. 363–372 (2007)
38. Cozzi, P., Riccio, C.: OpenGL Insights. CRC Press (2012)
39. Applegate, R.A., Ballentine, C., Gross, H., Sarver, E.J., Sarver, C.A.: Visual acuity as a function of Zernike mode and level of root mean square error. Optometry and Vision Science 80, 97–105 (2003)
40. Chandler, D.M., Hemami, S.S.: VSNR: a wavelet-based visual signal-to-noise ratio for natural images. IEEE Transactions on Image Processing 16, 2284–2298 (2007)
41. Wang, Z., Bovik, A.C., Sheikh, H.R., Simoncelli, E.P.: Image quality assessment: from error visibility to structural similarity. IEEE Transactions on Image Processing 13, 600–612 (2004)
42. Bradley, A.P.: A wavelet visible difference predictor. IEEE Transactions on Image Processing 8, 717–730 (1999)
43. Sheikh, H.R., Bovik, A.C.: Image information and visual quality. IEEE Transactions on Image Processing 15, 430–444 (2006)
44. Mittal, A., Soundararajan, R., Bovik, A.: Making a Completely Blind Image Quality Analyzer. IEEE Signal Processing Letters, 1 (2012)
45. Moorthy, A.K., Bovik, A.C.: A two-stage framework for blind image quality assessment. IEEE Signal Processing Letters 17, 513–516 (2010)

Visualising the Attributes of Biological Cells, Based on Human Perception

Denis Horvat[1], Borut Žalik[1], Marjan Slak Rupnik[2], and Domen Mongus[1]

[1] Faculty of Electrical Engineering and Computer Science
[2] Faculty of Medicine,
University of Maribor, Slovenia
{denis.horvat,zalik,marjan.rupnik,domen.mongus}@uni-mb.si

Abstract. This paper presents a new automatic colouring technique for grey-scale images that extends the CSL model from the visual perception point of view. Colour-coding is based on the attributes of contained objects, such as their area or radius of the bounding circle. Their extraction is achieved using advanced concepts of connected operators from mathematical morphology, whilst CIELab LCH colour-space is considered for their visualisation. A comparison between the proposed attribute-based visualisation (ABV) model and the CSL model was performed during a test-case on biological-cells. Whilst both models were superior to the original grey-scale image representation, we showed that the ABV model significantly increased the clarity of the visualisation in comparison to the CSL model, as it produced smoother transitions from low to high attribute values and avoided creating visual boundaries between regions of similar attributes.

Keywords: HCI, visual human perception, connected operators, mathematical morphology.

1 Introduction

The use of colours within HCI systems is primarily motivated by their higher dynamic range compared to grey-scales [1]. By using appropriate colour-scales, the number of just-noticeable differences can be considerably increased [2]. However, the usage of colours that is motivated purely by aesthetic reasons is not always the right choice, especially in science and engineering. Although such colour-mappings may increase the attractiveness of the visualisation, they often cause confusion when users try to understand their meaning [3]. The use of colours that do not increase the information perceived by the user should, therefore, be avoided. For these reasons, it is important to understand the information and select an appropriate colour-scale in order to achieve a clear representation of the data.

In biomedical data visualisation colours play an increasingly important role. By increasing the number of just-noticeable differences, they help the detection, diagnosis, and management of many ailments, whilst the increased dimensionality of the data representation allows for visualisation of information obtained

A. Holzinger and G. Pasi (Eds.): HCI-KDD 2013, LNCS 7947, pp. 386–399, 2013.

by supplementary sensors. A realistically hued colour map, the construction of which is based on Hounsfield units (HU), was proposed by Silverstein [4] to represent a density measurement of peritoneal fluid. This colour map is then applied on 2D CT (computed tomography) slices to create volume visualisation. Colours are also effectively used in Doppler colour flow imaging [5,6], where the blood-flow direction is displayed as red or blue, whilst variations of hues from dull to bright represent the velocities. Additional information, such as the turbulence of the flow can be displayed with green and hue/saturated combinations. Since complex spectral velocity visualisations are avoided in this way, such colour-flow visualisation makes the data more readable and easier to interpret by an inexperienced user. As the blood-flow can be used to quantify the inflammatory response to a burn, the same colour pallet was extended by Pape [7] to show the prediction of burn healing potential. The occurrence of colour can also be found in ultrasound image interpretations, where so-called B-colour [8] was proposed for emphasising the meniscal tear. More recently, magnetic resonance imaging (MRI) for brain-structure studies has been extended to functional magnetic resonance imaging (fMRI) [9] that can measure brain activity based on excess blood-supply. This brain activation is graphically colour-coded based on the strength of the activation. Although these are just some of many examples where the dimensionality of biomedical data visualisation has been increased by the use of the colours, they show the effectiveness of such information presentation. Nevertheless, these methods only consider the information obtained from supplementary sensors, whilst an automated colouring method capable of extracting information about structures contained in biomedical images, has not as yet been considered.

This paper presents an automatic colouring technique for the grey-scale images of biological cells. The main purpose of this method is to increase the information presented to the user by considering several characteristics of objects present within a grey-scale image. Conceptually, this work is an extension of the so-called MSLS segmentation scheme [10] and the CSL model [11] from the human perception point of view. An efficient image-preprocessing framework, provided by these morphological methods allows for automated extraction of objects. Their characteristics (e.g. area, volume, or radius) can then be used for image colouring that provides the user with a better understanding of the content.

The structure of this paper is as follows: Section 2 gives a short background of the used image preprocessing techniques that extract attributes used for image colouring. Section 3 discusses the use of colour-scales and gives the definition of the proposed visualization model. The results are presented in Section 4, and Section 5 concludes the paper.

2 Extraction of Visualization Attributes

By providing quantitative algebraic descriptions of geometrical structures and shapes, mathematical morphology (MM) has long been an effective tool for pattern recognition on grey-scale images [12]. Over recent years, connected operators have become increasingly popular [10,11,13,14]. These morphological operators act on connected sets of pixels with constant intensities called flat-zones, rather that on individual pixels. They can merge flat-zones, but they cannot break them. Because of this, they cannot introduce new edges or modify the positions of the existing ones. Consequentially, they either remove a feature from an image or they leave it perfectly preserved. A particularly interesting type of connected operators are attribute-filters [15,16]. These filters do not use structuring elements. Instead, they decide which flat-zones to merge using a specific attribute (e.g. area, volume, or bounding volume) of a connected set of flat-zones called a connected component. For this purpose, an image $f : E \rightarrow Z$ is viewed as a mapping function that maps a pixel $n \in E$ from a space $E \subset \mathbb{Z}^d$ (usually $d = 2$) to a space $Z \subset \mathbb{Z}$. In order to define an attribute-filter $\gamma(f)$, f is decomposed into a set $T = \{t_h\}$ of thresholded sets t_h, each given by [17]:

$$t_h = \{n \mid f[n] \geq h\}. \tag{1}$$

Let a function $C_h[n] \in t_h$ return a connected set of pixels from t_h (i.e. connected component) that contains a pixel n or $\emptyset$ otherwise, and $\Lambda(C_h[n])$ a function that estimates its attribute. An opening $\gamma_\lambda^\Lambda(f)$ that filters f according to Λ with a criterion λ, is defined as [18]:

$$\gamma_\lambda^\Lambda(f)[n] = \bigvee \{h \mid \Lambda(C_h[n]) > \lambda\}, \tag{2}$$

where $\bigvee$ is the maximum. In short, $\gamma_\lambda^\Lambda(f)$ assigns to each pixel n the highest value h at which it still belongs to a peak-connected component $C_h[n]$ that satisfies a criterion $\Lambda(C_h[n]) > \lambda$. Note that Λ is considered in this paper as an increasing attribute given by the following relationship:

$$C_{h_A}[n] \subseteq C_{h_B}[n] \rightarrow \Lambda(C_{h_A}[n]) \leq \Lambda(C_{h_B}[n]). \tag{3}$$

Amongst the increasing attributes, the area of the connected components $A(C_h[n])$ is especially well-studied. In particular, area opening $\gamma_a^A(f)$, introduced by Vincent [18], has recently come back into focus by the introduction of so-called differential area profiles (DAPs) [19]. DAP is essentially a top-hat scale-space computed from granulometry based on $\gamma_a^A(f)$. A granulometry is an ordered set of filters that progressively removes image details by filtering them at an increasing scale, whilst a top-hat scale-space is obtained by comparing the differences between filtered images. Consider a granulometry obtained by an ordered set of area thresholds $\mathbf{a} = \{a_i\}$, where $i \in [0, ..., I]$, $a_0 = 0$ and $a_{i-1} < a_i$. DAP $\Delta_\mathbf{a}^A$ is defined as:

$$\Delta_\mathbf{a}^A(f) = \{\gamma_{a_{i-1}}^A(f) - \gamma_{a_i}^A(f) \mid i \in [1, ...I]\}. \tag{4}$$

In other words, $\Delta_\mathbf{a}^A$ assigns to each pixel a $I-1$ long vector containing information about the reductions achieved by each of the corresponding area openings $\gamma_{a_i}^A$ at scale a_i. $\Delta_{\mathbf{a}}^A$ is called a positive response vector, whilst its negative equivalent can be obtained by an ordered set of area closings $\phi_{\mathbf{a}}^A$ as:

$$\Theta_{\mathbf{a}}^A(f) = \{-(\phi_{a_{i-1}}^A(f) - \phi_{a_i}^A(f)) \mid i \in [1, ...I]\}. \quad (5)$$

On this basis, the so-called MSLS mapping scheme, proposed for classical morphological filter [20], has been extended by Ouzounis [10]. At each pixel n, the MSLS mapping scheme estimates two characteristics: $r(f)$ is the maximal response and $s(f)$ is the scale at which it is obtained. By considering $\Delta_{\mathbf{a}}^A$, it is given as follows ($\bigwedge$ is minimum):

$$r_{\Delta_{\mathbf{a}}^A}(f)[n] = \bigvee \Delta_{\mathbf{a}}^A(f)[n], \quad (6)$$

$$s_{\Delta_{\mathbf{a}}^A}(f)[n] = \bigwedge i \mid (\gamma_{a_{i-1}}^A(f) - \gamma_{a_i}^A(f))[n] = r_{\Delta_{\mathbf{a}}^A}(f)[n]. \quad (7)$$

The same definitions are used in order to obtain $r_{\Theta_{\mathbf{a}}^A}(f)$ and $s_{\Theta_{\mathbf{a}}^A}(f)$ from $\Theta_{\mathbf{a}}^A(f)$, whilst the final output is obtained by:

$$r(f)[n] = \bigvee r_{\Delta_{\mathbf{a}}^A}(f)[n], r_{\Theta_{\mathbf{a}}^A}(f)[n], \quad (8)$$

$$s(f)[n] = \begin{cases} s_{\Delta_{\mathbf{a}}^A}(f)[n] & \textbf{if } r_{\Delta_{\mathbf{a}}^A}(f)[n] > r_{\Theta_{\mathbf{a}}^A}(f)[n] \\ s_{\Theta_{\mathbf{a}}^A}(f)[n] & \textbf{if } r_{\Delta_{\mathbf{a}}^A}(f)[n] < r_{\Theta_{\mathbf{a}}^A}(f)[n] \\ 0 & \textbf{if } r_{\Delta_{\mathbf{a}}^A}(f)[n] = r_{\Theta_{\mathbf{a}}^A}(f)[n] \end{cases} . \quad (9)$$

This MSLS scheme has further been optimised for visualization by the so-called CSL model introduced by Pesaresi [11]. In addition to $r(f)$ and $s(f)$, the CSL model maps the intensity level $l(f)$ of the pixel n before the iteration of the maximal response filter. That is:

$$l(s)[n] = \begin{cases} \gamma_{a_{s(f)[n]-1}}^A[n] & \textbf{if } r_{\Delta_{\mathbf{a}}^A}(f)[n] > r_{\Theta_{\mathbf{a}}^A}(f)[n] \\ \phi_{a_{s(f)[n]-1}}^A[n] & \textbf{if } r_{\Delta_{\mathbf{a}}^A}(f)[n] < r_{\Theta_{\mathbf{a}}^A}(f)[n] \\ f[n] & \textbf{if } r_{\Delta_{\mathbf{a}}^A}(f)[n] = r_{\Theta_{\mathbf{a}}^A}(f)[n] \end{cases} . \quad (10)$$

Note that extraction of DAPs can be in linear time using a max-tree structure [21], where only three passes over the image are required. A pseudo-code of computationally efficient implementation of DAPs is given in [10].

3 Applying the Colour Scales

By enhancing visual dimensionality, colours are very important in human-computer interaction. Namely, they allow better perception of the image content by visualising information obtained during the preprocessing step. Colours are determined as single points within the coordinate system named the colour-space (or colour-model [2]). Two categories of colour spaces are known:

- **Device-dependent** colour spaces are defined in such a way as to be easily displayed on a physical device such as a computer monitor. The best-known device-dependent colour space is the RGB cube [1,2].
- **Device-independent** colour spaces, on the other hand, are not limited by the number of colours that a device can display. A good example is the CIE colour space [2,22,23] that is able to represent all the colours visible by a human eye.

A colour-scale is defined as a parametric curve within a colour space [1,24]. An appropriate colour-scale can greatly enhance data representation, but if used incorrectly may cause confusion and misinterpretation [3]. In order to avoid problems relating to colour-scales, the following properties are desirable when considering representing a sequence of numerical values $\{v_1 \leq \ldots \leq v_n\}$ by colours $\{c_1 \ldots c_n\}$ [24,25,26]:

- **Order** preservation is a property that requires visually ordered colour-codes for ordered sets of values. This property can be seen in the temperature colour-scale, where gradual transition from blue (used to display cold or low values) to red (used for hot or large values) is achieved.
- **Uniformity and representative distance** demands that colours satisfy the separation principle [25], which states that clearly-separated values should be separated from colours clearly perceived as different, whilst near values have to be presented by colours that are perceived as similar. The colours encoding two numerical values should correspond to the distance between them.
- **Boundaries** should not be created by a colour-scale if there are no significant differences between the numerical data.

Colour-scales can then be divided into two types of representation [24,27]. The first is the univariate representation that is essentially a mapping function assigning 3D (or 4D) colour-coordinates to a 1D scalar value. Such colour-scales can either be continuous or discontinuous according to the definition of the mapping function. These univariate colour-scales are popular during biomedical visualisation in those cases where intensity values exceed the range $[0, 255]$ (e.g. CT images), and their representation using grey-scales leads to information loss [4]. On the other hand, multivariate colour-scales are able to colour-code two or more independent input variables. Since the attributes obtained by the previously explained image preprocessing step describe significantly different characteristics of the objects (e.g. intensity, maximal intensity difference, and attributes), multivariate colour-scales are more appropriate. However, selecting an appropriate colour-scale becomes significantly more complex in this case since intuitive correlations between attributes and colours have to be established [28]. The RGB model, for example, produces unsatisfactory results when each particular attribute is used for independent control of a particular channel, since two objects with the same attribute value may be of different colours. It was for this reason that the HSV colour-space was proposed in the CSL model [11].

3.1 Using HSV for Attribute Visualisation

HSV is a cone-shaped multivariate colour space that describes a colour by its hue, saturation, and value. The *hue* component is represented by an angle around the central vertical axis and it determines the colour from the rainbow-scale spectrum [2]. *Saturation* is defined as the distance from the centre axis and determines the amount of grey present in the colour, whilst *value* is determined as the distance along the vertical centre axis, and indicates the intensity of the presented light.

The main motivation behind the HSV model is to achieve more perceptually-intuitive colour representation in comparison with the traditional Cartesian RGB model. However, this model is still device-dependent because it is defined directly in terms of the displayed RGB drive signals [3]. A transformation from HSV colour coordinates to the device-dependent RGB coordinates is therefore straightforward.

According to the CLS model [11], $s(f)$ (obtained by (9)) provides the value of the *hue*. For this purpose, $s(f)$ is normalised and scaled within the range $[0, 360]$. In practical terms, all the connected components that share similar attributes are of similar colours, and the whole colour-spectrum is used. The maximum response $r(f)$, on the other hand, is normalised to $[0, 1]$ and then used as *saturation*. Consequently, those connected components that have high-intensity relative to their neighbouring backgrounds should contain less grey. Finally, $l(s)$ is applied to *value*, thus ensuring that the colour level is the same for all pixels belonging to the same most contrasted connected component.

Despite being quite popular, applying HSV has some significant disadvantages [2,24]. As it usually follows the scale of colour within the visible spectrum, it does not appear as linear [2,24,28]. Because of this, an observer may see no perceptually initiated ordering between the colours. In addition, the varying hues are defined by a circle. Consequentially, the hue starts and ends with the same colour (usually red), meaning that the more extreme attributes will be coloured red or very similar (Fig 1). Although this can be solved by limiting the value of the hue, a significant range of colours is eliminated from the visualisation in this way. Another problem is imposed by the colour-scale itself as yellow is used to visualise mid-range values. However, yellow is naturally perceived as a very striking colour and mid-range values are emphasised in this way. This colour-scale also contains the perceptual discontinuities that lead to perception of boundaries even when they are absent within the visualised data. Some of these problems (but not all) can be solved by the CIELab colour space, where attributes can be coloured in a linear fashion and perceptual discontinuities are avoided.

3.2 Using CIELab LCH for Attribute Visualisation

CIELab is a variation of the device-independent colour model named CIE [2,22,23]. CIE is defined as describing the whole human visual gamut using only positive values called tristimulus colour coordinates [2]. CIE is perceptually non-linear [22], meaning that the linear changes in the components do not result in

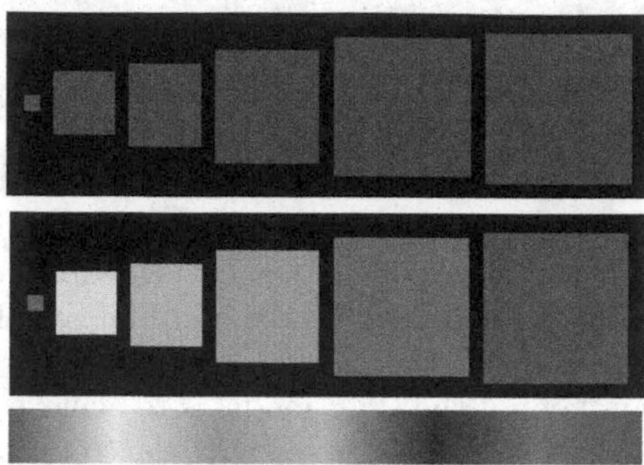

Fig. 1. Area-based image representation using the CLS model with a HSV colour-scale, where the smallest and largest objects are of the same colour

a linear change in the perceived colour [2,3]. CIELab was created in order to overcome this problem using three components called L, $a*$, $b*$. Component L represents the lightness and ranges from 0 to 100 for darker and lighter colors, respectively. On the other hand, $a*$ and $b*$ are defined on the principle that a color cannot be both red and green, or yellow and blue at the same time. Component $a*$, therefore, describes the transition from the green at one extreme to the red at the other. Component $b*$ gives the same transition, but for the yellow and the blue. The ranges for these two values are dynamic and depend on L. Because of this, L, $a*$, and $b*$, as represented by Cartesian coordinates, are usually mapped into a more meaningful model called LCH. LCH uses polar coordinates that match more closely with the human visual experience [22]. LCH represents lightness, chroma, and hue and is in that sense similar to the previously described HSV model. Hue is again represented by degrees from the range $[0-360]$, whilst chroma and lightness are from the range $[0-100]$. Similar to CIELab, the lightness represents the transition from dark to bright and the chroma represents colourfulness.

3.3 Attribute-Based Visualisation Model

In this section, a new attribute-based visualisation (AVB) model is proposed that extends the previously discussed CSL model from the human visual perception point of view. Three important modifications are applied for this purpose:

- first, MSLS labelling schema is avoided in order to emphasise the visualisation of particular objects of interest,
- the level, defined by the CLS model as the pixel intensity before the interaction of the maximal-response filter is changed to be equal to the original pixel intensity, in order to preserve the information from the original image, and
- finally, CIELab LCH is applied instead of HSV in order to increase the efficiency of the data representation.

Accordingly, $ABV(f) : f \to (r(f), s(f), l(f))$ maps the characteristic values of a grey-scale image f at pixel n as:

$$r(f)[n] = \bigvee \Delta_{\mathbf{a}}^{A}(f)[n], \tag{11}$$

$$s(f)[n] = \bigwedge i \mid (\gamma_{a_{i-1}}^{A}(f) - \gamma_{a_i}^{A}(f))[n] = r_{\Delta_{\mathbf{a}}^{A}}(f)[n], \tag{12}$$

$$l(f)[n] = f[n]. \tag{13}$$

CIELab LCH colour-scale is then proposed to be applied on ABV values. The extracted maximal responses $r(f)$ are normalized to the range $[0 - 100]$ and assigned to chroma, corresponding attribute values $s(f)$ normalised within the range $[0, 360]$ are represented by hue, and pixel intensity levels $l(f)$ are normalised within the range $[0, 100]$ to define the lightness. Since CIELab LCH is device-independent, colours that cannot be displayed have to be clipped to the nearest displayable value. Nevertheless, the obtained transitions of colours are more linear in comparison with the HSV colour space, as confirmed by the results.

4 Results

The efficiency of the proposed ABV model was studied in a test-case of biological-cells. Freshly-prepared pancreatic tissue slices were loaded with membrane labelling di-4-anneps dye for 20 minutes at room temperature. Excitation was performed with an argon laser using 476 nm wavelength, and emission collected between 500-650 nm using a HyD hybrid detector. Images were then obtained using a Leica TCS SP5 II invert confocal microscope. Two of the five images included in the test-set are shown in Fig. 2.

The DAPs of the test-images were estimated based on area A and radius R attributes, i.e. $\Delta_{\boldsymbol{a}}^{A}$ and $\Delta_{\boldsymbol{r}}^{R}$. Area-zones of interest were heuristically defined by $\boldsymbol{a} = \{0, 290, 440, 590, \ldots, 2690\}$, whilst $\boldsymbol{r} = \{0, 18, 29, 40, \ldots, 117\}$ was used to

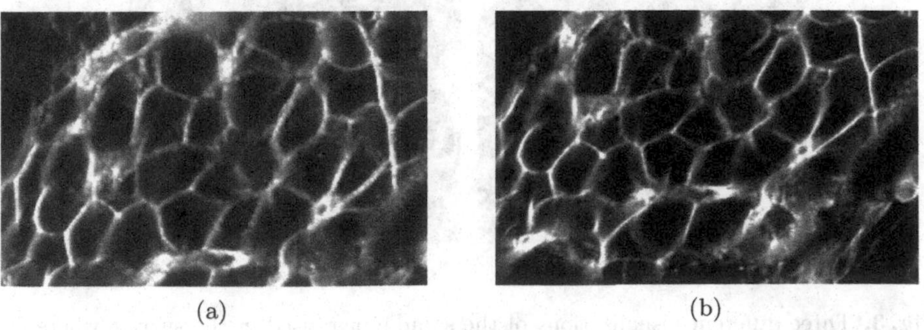

(a) (b)

Fig. 2. Grey-scale images of biological cells

obtain radius-zones. In all the cases, the first attribute-zones were ignored due to their subjectivity to noise. Since the cells appeared darker than the fluorescent dye loaded in the bordering membrane (see Fig. 2), their attributes were obtained from the bottom-hat scale-space. In order to emphasise this notion, $\Delta_{\boldsymbol{a}}^{A}(-f)$ and $\Delta_{\boldsymbol{r}}^{R}(-f)$ were considered when applying the ABV models. Thus, comparison between the original image representation, CSL model, and the proposed ABV model was achieved based on 5 different test images, where 10 coloured images were obtained (5 for each of the two attributes) by each of the two models.

In order to evaluate the visualization methods, a group of 14 test subjects was selected during a survey, which was achieved over two sessions (for each attribute). All the test subjects were familiar with the contents of the images prior to this survey, where 9 of them were professionals working with images of biological cells on a daily basis. A set of 143 unique cells were selected from 5 test images and the test subjects were then asked to classify them as small or large. In the case of coloured images, the meaning of colours was not explained to the test subjects. Three of the examples used in the survey are shown in Fig. 3.

Each of the test images was shown to the test subjects on a different sheet of paper in a carefully chosen order, where two successive images were always different and visualised using a different model. The test subjects were instructed to write S or L in white transparent circles for which they thought to represent

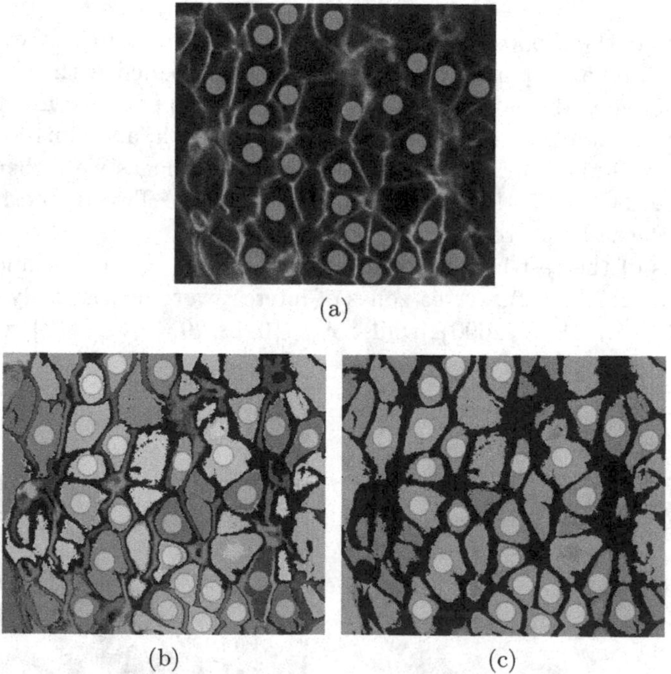

Fig. 3. Three different visualisations of the same image used in the survey, where (a) is the original representation, (b) is coloured by CLS, and (c) by ABV

small or large cells, respectively. The time and accuracy of the classification was measured for each individual image. The latter was defined as the percentage of correctly classified cells as smaller (or larger) than the average cell-size. Fig. 4 shows the results obtained from the survey.

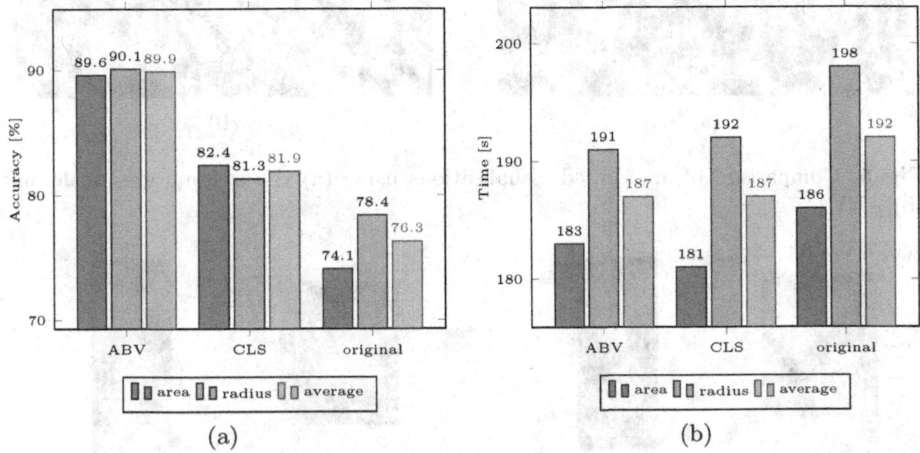

Fig. 4. Results obtained from the survey for (a) accuracy and (b) the average time spent on each image

As seen in Fig. 4b, the average time spent by the test subject did not depend significantly on the type of the visualisation, whilst the accuracy of the classification (shown in Fig. 4a) was significantly lower in the case of the original image representation in comparison with both colouring models. The first reason for this is that when the cells are visualised using CSL and ABV, their textures appear flat because the saturation (defined by the maximal response contained in DAP) is equal for all the pixels belonging to a single cell. They are also of similar colour as determined by the corresponding attribute value of the maximal response, which is assigned to the hue. As shown in Fig. 5, this greatly simplifies the cell identification as boundaries are sharp and easy to distinguish, whilst this is not always the case with original grey-scale images.

When considering the results of the original grey-scale image visualisation (see Fig. 4a), a greater error rate was noticeable when classification was based on the area than on the radius attribute. Namely, the test subjects relatively often misclassified long thin cells as large. On the other hand, they were able to intuitively relate the colour of the cell with the corresponding attribute, and avoid this error when CLS and ABV models were used. The example shown in Fig. 6 caused considerable problems to test subjects as it demonstrates two cells with the same radius (Fig. 6b) but significantly different areas (Fig. 6a). Consequentially, their colour-codes are similar in the former and different in the latter cases. However, they were often wrongfully classified within the same group during grey-scale image representation.

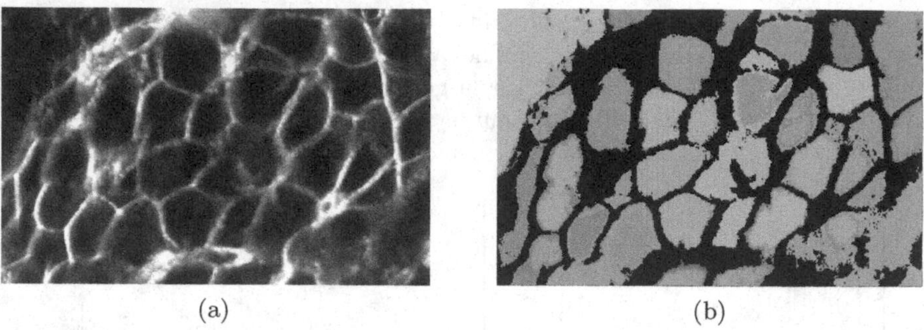

Fig. 5. Comparison of area-based visualisations using (a) the original grey-scale and (b) ABV

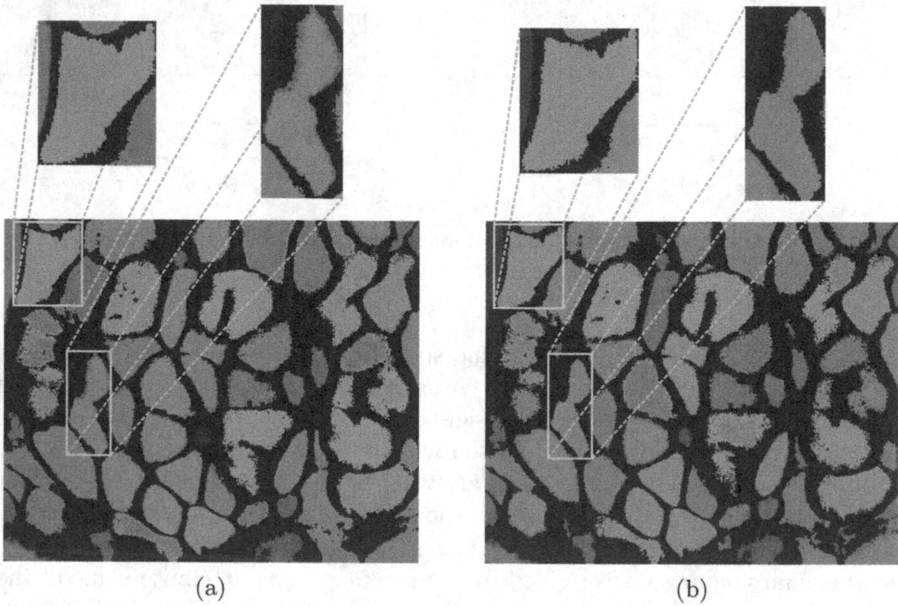

Fig. 6. Comparison of the visualisations using the ABV, based on the (a) area and (b) radius attributes

Another important aspect of image colouring can be observed when comparing the error rates obtained by CLS and ABV models. The latter appears to be advantageous as the used CIELab LCH colour-space is more linear in comparison to HSV. Since CIELab LCH dose not create obvious visual boundaries between gradually increasing attribute values, the test subjects correctly related the colour differences to differences in attribute values. The great majority of errors obtained in the case of ABV model were noted during classification of cells with attributes close to the threshold value. On the other hand, the error rate was higher in the case of the CLS model since the non-linearity in colour transition misled the test subjects when classifying two cells with similar attributes (radius or area), in different groups. Two examples where the test subjects often made this mistake can be seen in Fig. 7. It could arguably be claimed that cell 1 in Fig. 7a is more similar to cell 3, if only the colours were observed. However, the actual area difference between cell 1 and 2 is 372, whilst the difference between cells 1 and 3 is equal to 858. On the other hand, test subjects did not made this error when visualisation was based on the ABV model. As seen in Fig. 7b, colours of cells 1 and 2 are much more similar than the colours of cells 1 and 3. The same drawback for the CSL model can observed when comparing the areas of cells 3, 4, and 5. The test subjects often classified cell 4 as large because of the resemblance to the colour of cell 3 when visualisation was based on the CLS model. However, cell 5 is from the same class as cell 4, which is small. This is represented significantly better by ABV model in Fig. 7b.

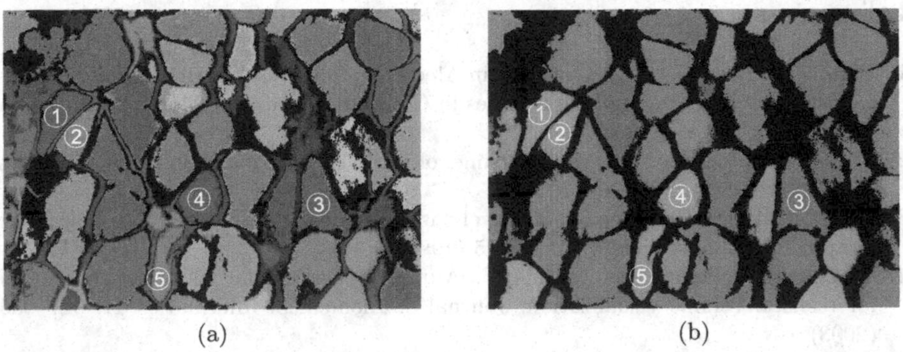

Fig. 7. Example image of area visualisation in the survey where errors were caused by CIE (a) in comparison with ABV (b)

5 Conclusion

This paper proposed a new colouring technique for grey-scale images named the ABV model. Similar to the concurrent CLS model, this method is capable of colour-coding objects present within a grey-scale images, based on their attributes. The extraction of attributes is achieved through a morphological concept known as DAPs, where the ABV model only consider negative response vectors whilst applying CIELab LCH colour-space for the visualising the biological cells. In comparison with their original grey-scale representations, the visualisation significantly improves. Both models increase visual distinction between the cell's interior and membrane and allow for straightforward visual recognition of those cells with similar attributes. However, the ABV model appears to be better in comparison with the CLS. Since it avoids boundaries during the transition from low to high attribute values, a better perception of the relationships between cell-attributes is achieved. Finally, different relationships between objects can be established according to the attributes used for the visualisation. As shown by the results, this type of visual grouping can significantly improve the human-computer perception of visualized objects.

Acknowledgments. This work was supported by the Slovenian Research Agency under grants 1000-08-310105, L2-3650, and P2-0041. The paper was produced within the framework of the operation entitled Centre of Open innovation and ResEarch UM. The operation is co-funded by the European Regional Development Fund and conducted within the framework of the Operational Programme for Strengthening Regional Development Potentials for the period 2007-2013, development priority 1: Competitiveness of companies and research excellence, priority axis 1.1: Encouraging competitive potential of enterprises and research excellence.

References

1. Preim, B., Bartz, D.: Visualization in Medicine: Theory, Algorithms, and Applications. The Morgan Kaufmann Series in Computer Graphics. Morgan Kaufmann (2007)
2. Silva, S., Santos, B., Madeira, J.: Using color in visualization:a survey. Computers & Graphics 35(2), 320–333 (2011)
3. MacDonald, L.: Using color effectively in computer graphics. IEEE Computer Graphics & Applications 19(4), 20–35 (1999)
4. Silverstein, J., Parsad, M., Tsirline, V.: Automatic perceptual color map generation for realistic volume visualization. Journal of Biomedical Informatics 41, 927–935 (2008)
5. Zeller, J., Griewing, B., Morgenstern, C., Walker, M., Kessler, C.: Color flow doppler versus power doppler imaging in the examination of vertebral arteries. European Journal of Ultrasound 5, 133–139 (1997)
6. Williamson, T., Harris, A.: Color doppler ultrasound imaging of the eye and orbit. Survey of Ompthalmology 40, 255–267 (1996)
7. Pape, S., Baker, R., Wilson, D., Hoeksema, H., Jeng, J., Spence, R., Monstrey, S.: Burn wound healing time assessed by laser doppler imaging (ldi). part1: Derivation of a dedicated colour code for image interpretation. Burns 32, 187–194 (2012)
8. Huang, Z., Long, W., Xie, G., Kwan, O., DeMaria, A.: Comparison of gray-scale and b-color ultrasound images in evaluating left ventricular systolic function in coronary artery disease. American Heart Journal 123, 395–402 (1992)
9. Rissman, V.C., Wagner, J., Imaging, A.: the human medial temporal lobe with high-resolution fmri. Neuron, 298–308 (2009)
10. Ouzounis, G.K., Pesaresi, M., Soille, P.: Differential area profiles: Decomposition properties and efficient computation. IEEE Transactions on Pattern Analysis and Machine Intelligence 32(8), 1533–1548 (2012)
11. Pesaresi, M., Ouzounis, G.K., Gueuguen, L.: A new compact representation of morphological profiles: report on first massive vhr image processing at the jrc. In: Algorithms and Technologies for Multispectral, Hyperspectral, and Ultraspectral Imagery XVIII, vol. 8390, p. 839025 (May 2012)
12. Shih, F.Y.: Image processing and mathematical morphology: fundamentals and applications. CRC Press, Boca Raton (2009)

13. Salembier, P., Wilkinson, M.H.: Connected operators: A review of region-based morphological image processing techniques. IEEE Signal Processing Magazine 136(6), 136–157 (2009)
14. Hernández, J., Marcotegui, B.: Shape ultimate attribute opening. Image and Vision Computing 29(8), 533–545 (2011)
15. Wilkinson, M.: Attribute-space connectivity and connected filters. Image and Vision Computing 20(2), 120–126 (2007)
16. Westenberg, M., Roerdink, J., Wilkinson, M.: Volumetric attribute filtering and interactive visualization using the max-tree representation. IEEE Transactions Image Processing 16(12), 2943–2952 (2007)
17. Maragos, P., Ziff, R.: Threshold superposition in morphological image analysis systems. IEEE Transactions on Pattern Analysis and Machine Intelligence 12(5), 498–504 (1990)
18. Vincent, L.: Morphological grayscale reconstruction in image analysis: Applications and efficient algorithms. IEEE Transactions on Image Processing 2(2), 176–201 (1993)
19. Ouzounis, G., Soille, P.: Differential area profiles. In: 20th International Conference on Pattern Recognition (ICPR), pp. 4085–4088 (2010)
20. Pesaresi, M., Benediktsson, J.A.: A new approach for the morphological segmentation of high-resolution satellite imagery. IEEE Transactions on Geoscience and Remote Sensing 39(2), 309–320 (2001)
21. Salembier, P., Oliveras, A., Member, A.O., Garrido, L.: Anti-extensive connected operators for image and sequence processing. IEEE Transactions on Image Processing 7(4), 555–570 (1998)
22. Ford, A., Roberts, A.: Colour space conversions. Technical report (1998)
23. Robertson, P.: Visualizing color gamuts: a user interface for the effective use of perceptual color spaces in data displays. IEEE Computer Graphics and Applications 8, 50–64 (1988)
24. Rheingans, P.: Task-based color scale design. In: Applied Image and Pattern Recognition, pp. 35–43 (1999)
25. Trumbo, B.E.: A theory for coloring biavariate statistical maps. The American Statistician 35, 220–226 (1981)
26. Levkowitz, H., Herman, G.: Color scales for image data. IEEE Computer Graphics and Applications 12, 72–80 (1992)
27. Silva, S., Madeira, J., Santos, B.: There is more to color scales than meets the eye: A review on the use of color in visualization. Information Visualization 4, 943–950 (2007)
28. Bergman, L.: A rule-based tool for assisting colormap selection. In: Proceedings Visualization 1995, pp. 118–125 (1995)

Interactive Visual Transformation for Symbolic Representation of Time-Oriented Data

Tim Lammarsch[1], Wolfgang Aigner[1], Alessio Bertone[2], Markus Bögl[1], Theresia Gschwandtner[1], Silvia Miksch[1], and Alexander Rind[1]

[1] Institute of Software Technology and Interactive Systems, Vienna University of Technology, Austria
{lammarsch,aigner,boegl,gschwandtner,miksch,rind}@ifs.tuwien.ac.at
[2] Institute of Cartography, Dresden University of Technology, Germany
alessio.bertone@tu-dresden.de

Abstract. Data Mining on time-oriented data has many real-world applications, like optimizing shift plans for shops or hospitals, or analyzing traffic or climate. As those data are often very large and multi-variate, several methods for symbolic representation of time-series have been proposed. Some of them are statistically robust, have a lower-bound distance measure, and are easy to configure, but do not consider temporal structures and domain knowledge of users. Other approaches, proposed as basis for Apriori pattern finding and similar algorithms, are strongly configurable, but the parametrization is hard to perform, resulting in ad-hoc decisions. Our contribution combines the strengths of both approaches: an interactive visual interface that helps defining event classes by applying statistical computations and domain knowledge at the same time. We are not focused on a particular application domain, but intend to make our approach useful for any kind of time-oriented data.

Keywords: Data Mining, KDD, Data Simplification, Visual Analytics.

1 Introduction

In Knowledge Discovery in Databases (KDD) [1], pattern-based Data Mining methods [2] play an important role when the user looks for specific events instead of creating a general model. Such kind of Data Mining is useful in many different application domains, like optimizing shift plans for shops or hospitals, or analyzing traffic or climate. For most methods, (e.g., [3]), an event is a tuple consisting of an integer representing the type of the event and the point in time when the event happened. A pattern (or event sequence) is a triple of an array of events with a starting and an ending time. Less formal definitions for pattern with similar meaning are given as "a local feature of data" [2] or "a local structure in the data" [4]. As time series are often very large and multi-variate, several methods for simplification have been developed (see Section 2). This simplification allows more efficient algorithms for Data Mining tasks like classification, clustering, and pattern discovery [4]. Various publications about pattern discovery [3,5,6,7,8,9,10,11] require data simplified that way. Lin et al. [12] point out

A. Holzinger and G. Pasi (Eds.): HCI-KDD 2013, LNCS 7947, pp. 400–419, 2013.

that based on the generic framework for temporal Data Mining [13], the first step for all data mining methods is "generating a simplification that fits into memory while retaining the essential features of interest". If the simplified representation of the data is symbolic (also considered discrete), methods from text processing and bioinformatics can also be used for time-oriented data [14,12]. We call the data we are dealing with time-oriented data instead of time series, as we focus on the fact that this kind of data can be multi-variate and the variables can be correlated [15]. In datasets, time usually acts as a reference domain, with each time reference pointing to one or more data values in the form $R \to C$, with R being a set of references and C a set of data values [16]. Time as a reference domain also comprises a number of important structural aspects [15], some of which are already dealt with in Data Mining [10,11]. Domain experts dealing with time-oriented data often consider the structure of time an aspect of utmost importance [17,18]. In this publication, we deal with simplification along the various data domains, while keeping the information along the reference domain (e.g., time) as unchanged as possible. Thus, other methods can be used to mine this information.

Current methods for temporal pattern discovery either assume that the data dimension already is symbolic [5,6,3,7,8,9] or include a very simple means of user-based simplification, without dealing with user interface issues [10], or doing that in a very limited manner [11]. Methods that deal with simplification of data [19,20,14,12], on the other hand, tend to have a simplistic view on time as reference domain, considering only a list of concurrent data tuples. Therefore, performing such methods can "spillage" information contained in the time domain. Moreover, they are focused on automated Data Mining methods and do not support user interaction regarding special important event classes. To bridge this gap, we present a novel visual interface that allows users to define and modify event classes either manually by giving restrictions in the data dimensions, or using a statistical method similar to [12]. Our interface gives immediate visual feedback on the effects of these modifications. Contrary to existing methods, our data simplification works on multi-variate data. It is statistically robust, and has a lower-bounding distance measure. Moreover, it preserves the (temporal) reference dimension for possible mining at later stages. Above all, our interactive user interface includes human users into the process, providing access to their domain knowledge. By doing so, we transform the existing approaches with more limited uses into a flexible tool that allows direct interaction between humans and a computer-based algorithm. The data structures and the code for calculating the event classes are already implemented. All the user interface (UI) elements have been designed and the code for calculations and visualizations has been implemented, but the connection as shown in Section 3 has to be implemented as well. So while we are showing mockups for the UI elements, the visualizations inside and the presented values are correct. Our UI is not focused on one single application domain, yet we provide an example from shift plan analysis for shops in Section 4.

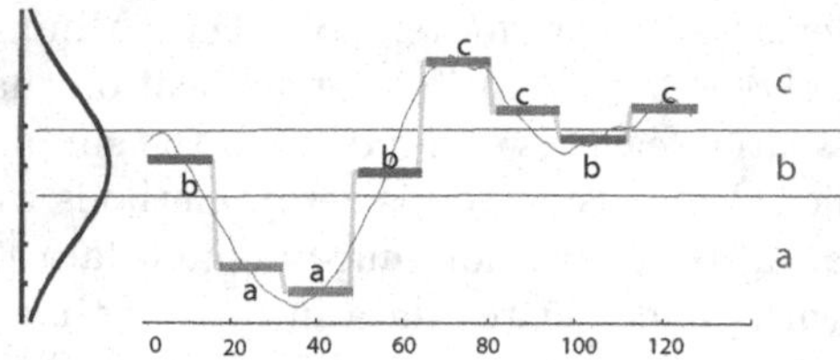

Fig. 1. A time series is discretized in SAX by using statistically determined breakpoints to map the PAA approximation into event classes [12]

2 Related Work

In this section, we focus on three questions: (1) how is data simplification currently applied, (2) which algorithms that require simplified data could be improved by better simplifications methods, and (3) how are interactive visual interfaces already successfully applied in KDD.

Methods for Data Simplification. Dealing with data locally is a newer method than developing a global model [4,21]. The first such methods [20,19] therefore were compared to global methods, like Singular Value Decomposition, Discrete Fourier Transformation, and Discrete Wavelet Transformation. The Piecewise Aggregate Approximation (PAA) by Keogh et al. [20,19] can be considered as a kind of rasterization: time-oriented data with n points in time is converted to time-oriented data with w points in time. The authors call this "dimensionality reduction" because they treat time series as vectors in high-dimensional space. The amount of data is reduced along this dimension. However, the total number of variables in the dataset is unchanged. The only parameter that can be given by users for this transformation is the number of target references w. The PAA approximation can also be considered data simplification, but the resulting data is still value-based, while important information contained in the time reference might be hidden. PAA approximation is also part of SAX [14,12]. In SAX, a time series is first normalized to a mean value of 0 and a standard deviation of 1. The authors show that most time series have Gaussian distributed data values by analyzing various time series by means described by Larsen and Marx [22]. While this may not be true for all time series, we agree with Lin et al. [12] that this kind of time series is frequent enough that it deserves primary consideration. After applying the PAA, a resulting shortened time series is transformed to a symbolic (discrete) representation by allocating each of the w time windows to one of a number of event classes. The event classes are chosen on a statistical basis: Each class contains the same number of values (see Fig. 1). Lin et al. claim that the equiprobability of classes is important for several further analysis methods that can be performed after SAX, giving some examples [23,24]. For the same reason, they provide an Euclidean distance measure for their output. To preserve temporal information, we do not include the PAA-based rasterization in our work. The discretization step can be performed without rasterization. Another approach for data simplification is the rough set approach [25,26], used to include ontologies and/or domain knowledge. In the context of Data Mining, these comprise various methods to give event classes and their characteristics names and meaning.

Time-Centered Algorithms for Pattern Finding. Lin et al. [14,12] focus on methods from text processing and bioinformatics as target of their work. When time is more than a simple counter of steps, however, different methods are capable of detecting more information. Most of those methods go back to the work by Agrawal et al. [5] who only consider patterns of events happening simultaneously (in their case, an event is the purchase of a product). They also consider time, but only as a method of separating different pattern candidates. However, they already perform planning towards further steps with more complex patterns that can overstretch time steps [6]. The events in those publications, like the purchase of a product, are inherently discrete data. Mannila et al. [3] focus further on sequences of events at different points in time. They also explicitly consider the time step at which one event occurs. They consider events to be determined by some external algorithm. Magnusson [7] is among the first to consider the time intervals between events, which is an important step to increase the consideration of time. His T-patterns are tree-shaped and therefore differ from the patterns in most other publications that are sequences. The work is focused on behavior analyses, so the events are found by some algorithm or even manually. Chen et al. [8] introduce the I-Apriori algorithm which extends the Apriori algorithm for pattern finding [4] by the consideration of intervals between events. Hu et al. [9] provide a similar approach where the focus is on patterns with events that do not need to be consecutive, as long as the time intervals are kept. Both approaches deal with events that either result from purchase and are inherently discrete, or assume that the data domain has been simplified to discrete events. Bertone et al. [10,11] provide a similar approach with user configurable time intervals that can have variable length and consider calendar aspects. They also mention the definition of events as multi-variate value combinations given by users. Therefore, Bertone et al. are also considering multi-variate data, but give limited explanation how users should deal with the complex event definition step. Summarizing those approaches, events are either considered as given, or the authors include a rather simple event finding process. Such simple event finding methods result in ad-hoc decisions that are not guaranteed to work, even if performed by very experienced users with domain knowledge. Therefore, we see a great need to find a better method of data simplification that results in more applicable event classes.

Samples for Application of Interactive Visual Interfaces in KDD. The Data Mining methods presented in the last subsection conduct pattern discovery as an automated task. Laxman et al. [4] discuss the advantages of this approach. However, interactive visual interfaces can greatly improve the applicability of the pattern finding process. Tominski [27] gives an overview how interactive visual interfaces are already used to find and display events according to the requirements and domain knowledge of users. He also provides an overview and formal model how events can be found with an interactive visual interface. Our interactive visual interface is developed with the guidelines from this publication in mind. The VISITORS system [28,29] has an interactive visual interface to explore and query patient data over time. It supports multi-variate data and combines both actual values and events. The events are determined by knowledge-based temporal abstraction methods. Similarly, Lifelines2 [30] is a system to explore events in time-oriented patient data. It provides specific interaction techniques such as alignment and temporal summaries. It expects data to be either of a categorical scale or simplified in advance. The simplified data resulting from application of our contribution are possible events that can be further analyzed by these systems. Activitree [31] provides a powerful interactive visual interface that users can employ to choose which patterns are important while performing algorithms, like those based on the Apriori approach [4]. As these examples show how interactive visual interfaces can improve various steps of Data Mining, we are heeding the call and present an interactive visual interface for the data simplification step of value discretization which has received insufficient attention in most existing work.

3 Interactive Visual Data Simplification

To present our interface, we need to (1) lay out the requirements, (2) describe the sample dataset we have used to design it, (3) show the interface itself, and (4) pay special attention to multi-variate data.

3.1 Requirements

We deal with multi-variate data as this kind of data is common among time-oriented data [15]. The number of variables can also be considered the number of dimensions in the multi-variate data space. The output should consist of a time reference and a single discrete data value. Each output element represents one single "event". Events can occur many times in any temporal order. However, the total number of different events is limited. All events that share the same characteristics and are grouped together by the user or by some kind of automated similarity computation are considered an "event class". In the current state of development, our user interface does not support giving events classes names or putting them into an ontology, however we are aware of the possible advantages of such expansions. We want users to be able to apply their domain knowledge and freely configure event classes. At the same time, we want to ensure equiprobable event classes, so that our approach can keep up with SAX [14,12] for Data Mining methods that require this kind of data discretization as input. Compared to SAX, we have to perform more complex statistical calculations in order to deal with multi-variate data. To combine the statistical and the manual approach, we need to provide three alternatives of defining the event borders between event classes:

Giving a Target Number of Event Classes. This results in classes with borders provided by the algorithm, similar to SAX, [14,12], but considerably more complex when there is more than one data variable.

Completely Free Parametrization. Similar to the one proposed by Bertone et al. [10,11]. Event classes resulting from such kind of parametrization are not equiprobable. However, it is possible to have the set of event classes partitioned into groups, with each merged group being as probable as any other group.

Giving Target Value Ranges That Are Supposed to Be Allocated into Separate Classes. The algorithm uses these ranges to calculate a number of classes (which cannot be chosen by the user in this case) with class borders at the given range borders. Below we show how these borders can be set.

All three alternatives should be performable at the same time. Making changes in one of those three alternatives has to be reflected immediately in the parameters for the other alternatives. At the same time, we need an interactive visualization for the data as well as the event classes.

3.2 Our Example Dataset

We used a synthetic dataset that is similar to the dataset analyzed by Bertone et al. [10,11]. The original dataset from that publication is not open to the public, therefore we use this alternative. The dataset covers data of a shop over one year. It is measured on a one-hour-raster a three data variables: *employees*, *customers*, and *turnover*. The dataset has the following characteristics:

- All values are higher on weekdays and lower on weekends.
- When *customers* and *employees* are high at the same time, *turnover* is high.
- When *customers* or *employees* are high while the other of those two values is low, *turnover* is average.
- When *customers* and *employees* are low, *turnover* is also low.

We will present our user interface using this dataset as exemplary input-data.

3.3 The User Interface

The user interface consists of three main parts which in turn contain different parts themselves (see Fig. 2, the parts described below are marked by red text):

1. At the top, some basic commands are presented. We include an undo button that reverts the last action.
2. At the left, we have an area where users can edit the parameters of event classes: (a) At the top, the user can specify if given variables are statistically linked or independent from each other. The user does so by clicking on the variable, shifting it between the "statistically linked" and "independent" area. We will explain the meaning of this classification below. (b) In the middle, event classes can be configured using statistical properties. (c) At the bottom, manual event classes can be defined using specific values.

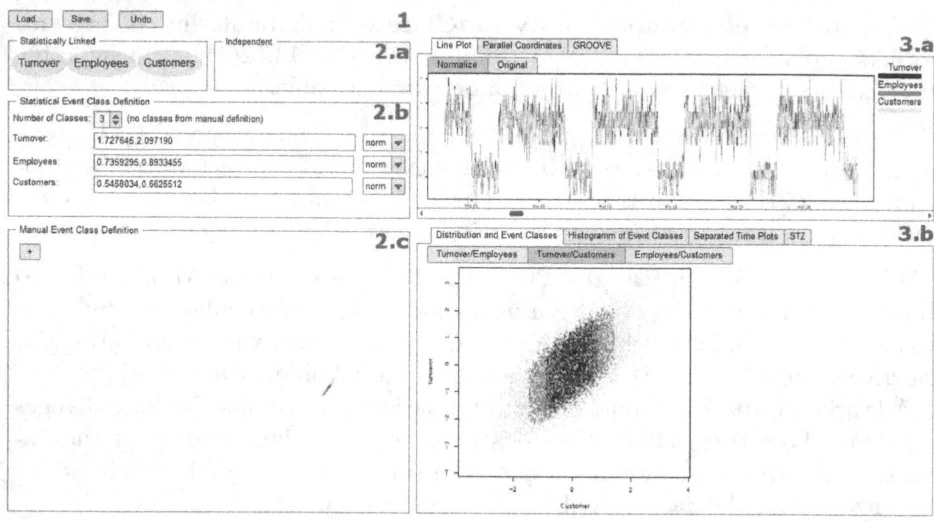

Fig. 2. Our user interface with (1) the basic commands (2.a/b) automated statistical calculation of three event classes, (3.a) a line plot, and (3.b) a value and class distribution scatter plot

3. At the right, the raw data and events are visualized. The interactive visualization allows for modification by users. (a) At the top, the dataset is shown (a line plot in Fig. 2). (b) At the bottom, the event classes are shown (a scatter plot in Fig. 2).

Event classes are calculated in two stages. The first stage is based on statistical calculations. Fig. 1 already showed how equiprobable event classes are calculated in SAX. We extended this method to n value dimensions. Due to correlation effects, the classes cannot be related to low or high values for multiple dimensions. Instead, we use distance from the mean as criterion. Fig. 3.a) shows *customers* and *turnover* of our example dataset in a 2D scatter plot with three classes. Fig. 3.c) shows all variables of our example dataset (i.e., *customers*, *employees*, and *turnover* data) in a 3D scatter plot with three classes. Interpreting the 3D view is not always straightforward and easy (see below). Therefore, we project the 3D scatter plot visualization on a plane spanned by two variables that can be switched by the user. In our user interface, we support two variants of configuring equiprobable event classes: (1) A target number can be given by the user. The system calculates equiprobable classes as done by SAX, but it works for any number of variables instead of only one. (2) Class border values can be given for any variable. The data value space can be set in normalized form (with mean of zero and standard deviation of one), or the absolute values can be used. With respect to the correlation between variables, changing the borders along one variable changes the number of classes as well as the borders along the other variables. The borders are forming hyperellipsoids (ellipsoids for three variables, ellipses for two variables). For normalized values, the border values are given on the coordinate axis of one variable, with the other variables having values of zero. Otherwise, the respective

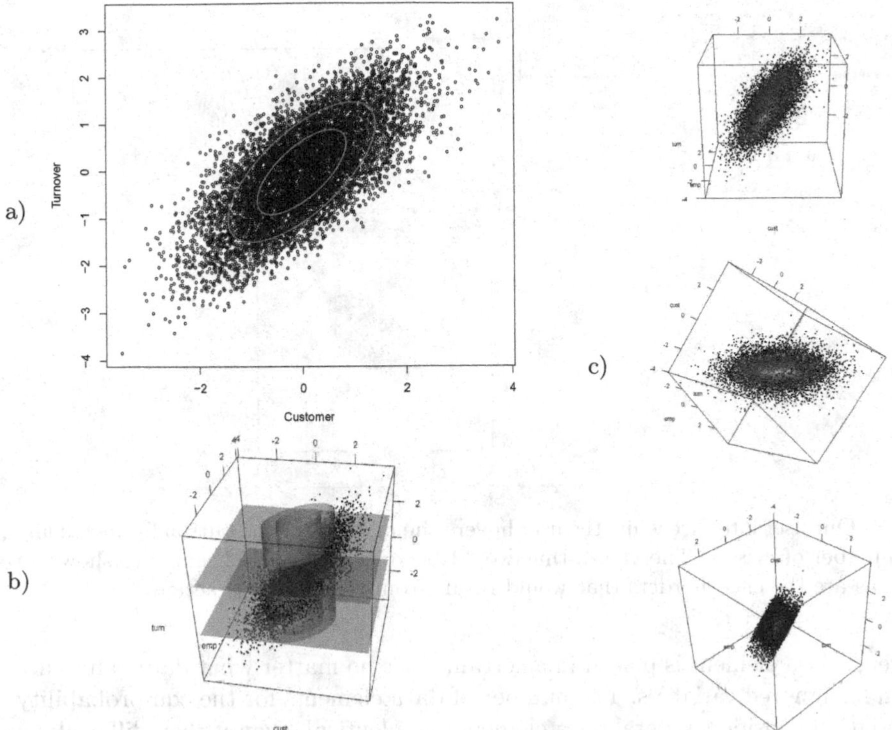

Fig. 3. a) The *customers* and *turnover* part of or example data in a 2D scatterplot. The green ellipses are the class borders. b) A combination of manual and calculated event classes in a 3D-view. c) Our example data in a 3D scatterplot from several directions. The green ellipsoids are the class borders.

mean values are used. For each value a hyperellipsoid representing a class border is formed, with its actual shape given by the correlation of the variables. Additional borders are added to the given borders in order to keep the number of events per class equiprobable, so it is likely that the algorithm will add further border position values. The various settings can all be modified; the system automatically updates the dependent values accordingly.

When the user is about to change a value, this change is reflected in the whole user interface: In case the user modifies the parameters in the text boxes on the left, the text is entered in blue as long as *Enter* has not been pressed. The newly calculated class-borders are indicated by blue ellipses in addition to the current class-borders which are shown in green (see Fig. 4). Two situations result in the system showing "possible future" values: Hovering the mouse over an arrow (see Fig. 4) or editing a value in a text box. By pressing the *Enter* key, the change is permanent (unless Undo is used). By pressing *Esc*, the text box and all values are resetted.

Variables that are not included in statistical event class definition (because the user has shifted them to manual event class definition) are ignored at this

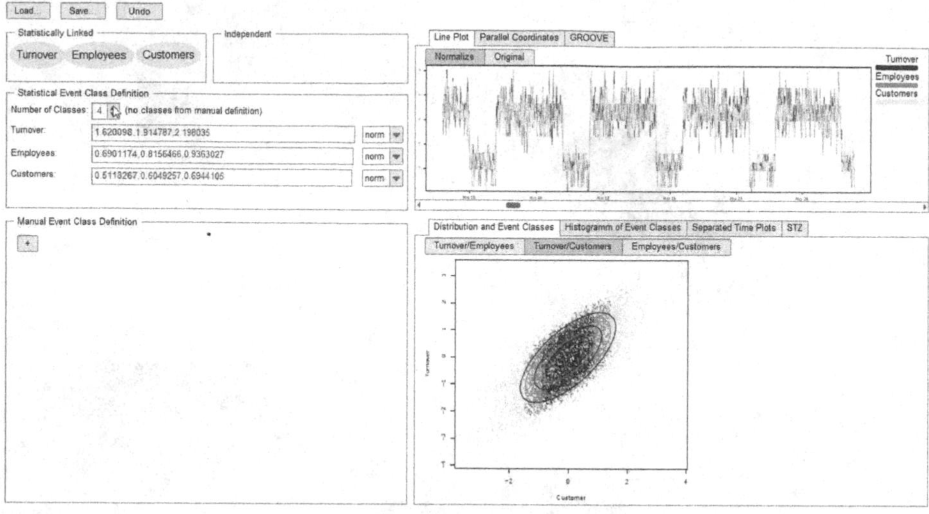

Fig. 4. Our user interface while the user hovers the mouse over the button for increasing the number of classes. The values that would become actual by clicking are shown in blue, as are the class borders that would result from this user interaction.

stage. A data element is placed in a certain class no matter what data values are given for ignored variables. The number of data elements for the equiprobability calculation considers several data elements as identical, even if they differ along the ignored variables.

For variables that are set to be independent, manual event class definition can be performed. This is done in a second stage after statistical calculations. The manual event class definition works as described by Bertone et al. [10,11]. The vertical "+" button adds new event classes. The horizontal "+" button to the right of an existing event class adds new constraints for this class (see Fig. 5). All data elements that do not fit to the constraints are placed in an "other" event class. The data value space can be set to normalized values or as absolute values. The class borders resulting from this stage are added to the class borders from the first stage. If classes from the first stage are further intersected in the second stage, equiprobability for the resulting classes cannot be imparted. However, the combinations of classes that emerge from the second stage are still equiprobable. If no manual event class definition is performed, only the classes from the first stage exist. When changes are about to be made to manual event class definition, they are reflected at all parts of the user interface in blue color. This works the same way as described above for statistical event class definition.

Fig. 2, 4, 5, and 8 show how the dataset is visualized (upper right). We provide three different visualizations of the dataset:

Line Plot is a widely known method for visualizing time-oriented data [32] (see Fig. 2 and 4). Time is plotted over the horizontal axis, while the value of a variable determines the position at the vertical axis. Multiple variables are shown in various shades of grey. Blue and green shades are not used because

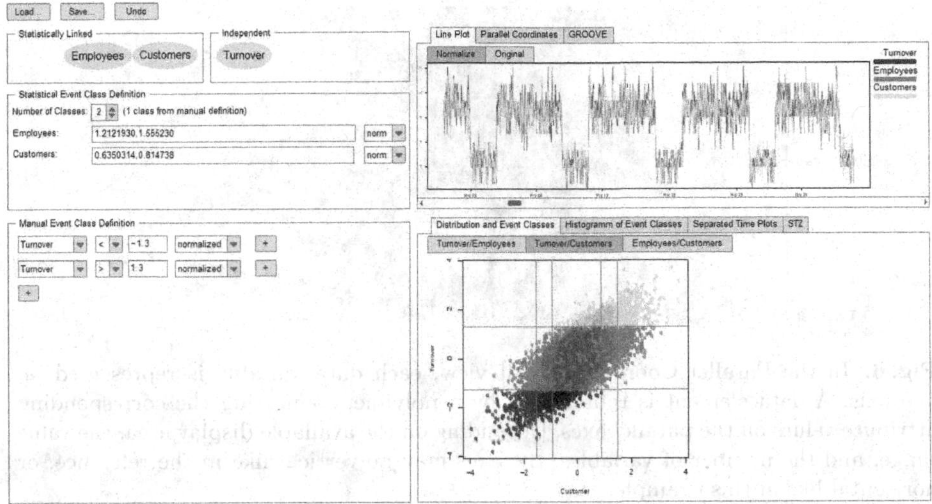

Fig. 5. Our user interface can define event classes by a combination of statistical equiprobability present our user interface using this dataset as exemplary input-data and manual borders. The resulting class-borders are indicated by green lines in the projected scatter plot on the right.

they are needed for other user interface elements. This visualization can be actived when users need to have a better view on data values.

Parallel Coordinates (see Fig. 6) by Inselberg and Dimsdale [33] do not show the flow of time. Instead, each of the axes represents one variable. The position is determined by the data values. The lines connecting the axes show the correlations between values. Therefore, this visualization can be activated when users need to investigate these correlations.

GROOVE (see Fig. 7) is a pixel-based visualization technique developed specifically for time-oriented data [34]. It is based on recursive patterns [35]. Both axes are used for time, while the data value is mapped to the color of pixels. Instead of adapting GROOVE to show multiple data variables at once, we use small multiples [36]. This visualization can be activated when users need to understand temporal structures.

Fig. 2, 4, 5, and 8 also show how event classes are visualized (lower right). Four kinds of visualization are possible:

Data Distribution and Event Classes are similar to the scatter plot [33] that has already been shown in Fig. 3.a). The points are data elements projected on a plane which can be selected by a tab bar over the visualization. The borders of event classes are shown as green lines. These borders are also projected. They represent the maximum circumference of the actual border frames. Elliptical borders result from statistical event class definition. Intersecting straight lines (as seen in Fig. 5) result either from manual event class definition or are projected elliptical cylinders. Fig. 3.b) explains this fact and shows how the classes could be represented in three dimensions.

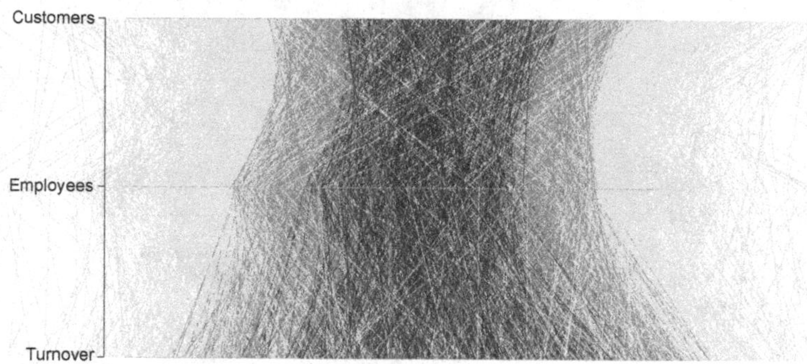

Fig. 6. In the Parallel Coordinates [33] view, each data variable is represented on one axis. A data element is represented by a polyline, connecting the corresponding attribute values on the parallel axes. Depending on the available display area, the value range, and the number of variables, the axes may be vertical like in the reference, or horizontal like in this example.

When users are about to make changes to the event class configuration, the new class borders that would result are shown as blue lines in addition to the green lines showing the current borders. It is also possible to directly interact with the class borders: They can be grabbed with the mouse and dragged. This interaction results in the creation of new borders. This border is placed directly at the mouse position (assuming a value of zero for the variables that are not part of the projection plain). When the dragged border was resulting from statistical calculation, other borders are calculated in a way so that the classes remain equiprobable (see Fig. 8). For manually defined classes, the definition of the class is simply changed. When pressing the *Esc* key before releasing the mouse button, the dragging is canceled.

Histogram of Event Classes is a visualization that shows how many data elements fall in the various event classes [37]. If only statistical event class definition is used, the classes are all the same size (and the histogram is not needed). If manual event class definition is used, the histogram shows how the probability of these classes are distributed (we aim for equal probabilities). The current state is shown in black, blue lines show predicted changes when users are currently making changes to classes. In Fig. 9, we show an example histogram for the classes defined in Fig. 5. As some of the class borders are defined in an equiprobable way, and some are defined manually, summing up these bars in the right combination would result in three equal bars.

Separated Time Plots with class borders (see Fig. 10) are a mixture between the distribution, as the points are colored according to classes, and the line plots, as the horizontal axis shows time. They are focused on displaying the event borders. Therefore, this visualization is similar to the visualization used to explain the event borders of SAX [12] (see Fig. 1).

STZ is a visualization method for qualitative abstractions and the associated quantitative time-oriented data. This interactive visualization technique, which is referred to as SemanticTimeZoom (STZ), is adopted from the

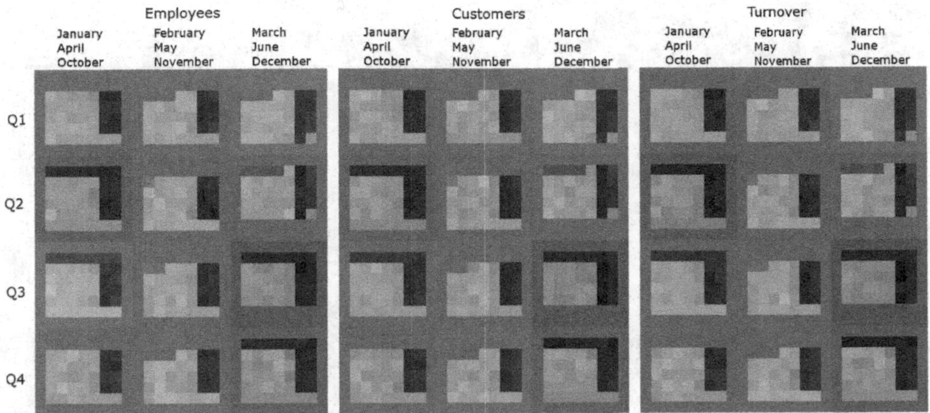

Fig. 7. GROOVE visualizations [34] of the three variables. Each part with uniform hue represents one month, red means high average, blue low average. Each pixel represents one day, bright means high value, dark means low value.

Midgaard system [38,39]. While, in previous publications the qualitative abstractions were defined on basis of a single variable, typically based on severity ranges from domain knowledge, we can apply here the event data from our classification as common abstraction for all variables. Thus, the sequence of events is shown in combination with the development of numerical raw data along the time axis (see Fig. 11). STZ allows the user to dynamically switch between different levels of detail depending on vertical space. At a low detail level it shows only events, which are represented by colored boxes. Fig. 11 demonstrates the medium level, where raw data is shown as line plot and the area is colored by event class. At high detail level the raw data is again represented in a line plot with marks at the time points when event classes change.

3.4 Dealing with More Than Three Variables

Our user interface can deal with as many variables as the computer performance and the screen space permit. For visualizations that place the variables along the dimensions, a projection on two dimensions is necessary. While a projection on a pseudo-3D-view is possible, we focus on a full 2D projection because in a pseudo 3D-view (1) the occlusion is too severe (2) the actual position of points is too hard to grasp (3) user interactions are too complicated. An example for these problems can be perceived in Fig. 3.b). For a straight projection on two dimensions, the values of the projected dimensions can be just ignored. This results in exact projections for the points, for class borders the maximum values have to be taken. So for three dimensions, the silhouette is shown. For four dimensions, when mentally picturing an animation over three dimensions, the ellipse is growing and shrinking again. Here, the projection shows the maximum circumference. For five or more dimensions, there is no real imaginable model,

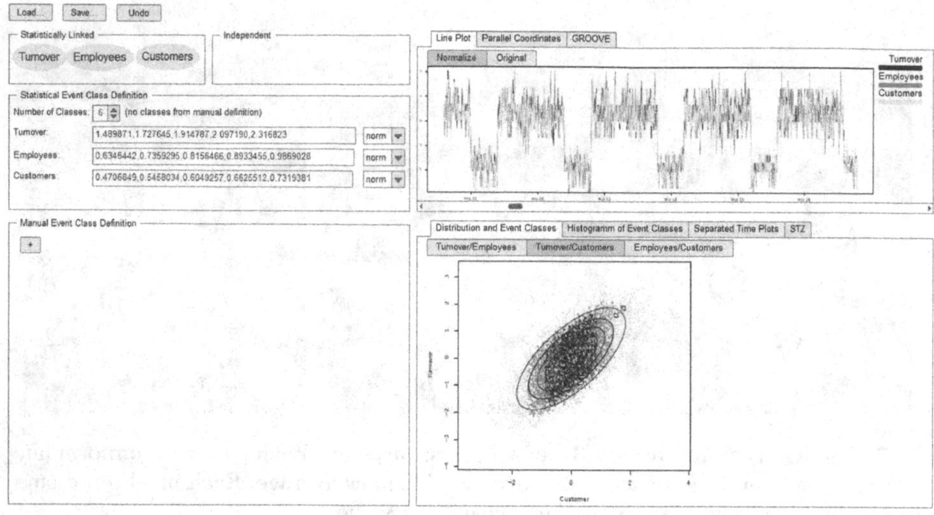

Fig. 8. Our user interface while the user drags the class borders at the right with the mouse. The potential new class borders are shown in blue. On the left side, the numerical values that represent these borders are shown, releasing the mouse sets these values, which makes them become black.

but the rule can still be applied. For values normalized to a mean of zero, the maximum circumference also exists where the other dimensions have values of zero, so dragging the class borders results in a clearly defined user interaction. Due to the fact that the class borders are showing the maximum circumference, several values that are outside one of the borders are projected inside the border. Here, the different gray levels of the data points can help to some degree.

4 Usage Scenario

We perform our usage scenario on the synthetic dataset as described in Section 3. A user wants to find important patterns in the data from her shop. This is a realistic scenario based on real-world application [10,17,18]. There are several important tasks to solve like "how many employees should be in the shop at a given time to maximize profit?", with profit being driven by *turnover*, but diminished by labor costs. By viewing the data in our user interface, it is likely that specific characteristics of the dataset given in Section 3 can already be seen. For example, the difference between weekdays and weekends becomes obvious in the GROOVE visualizations (see Fig. 7). In our example, the GROOVE visualizations are very similar among the variables. This is due to the fact that the correlation is rather high. This is an advantage though, as a high correlation simplifies finding suitable event classes. In the example case, it means that the shift plan is already not bad. The user tries several numbers of classes (see Fig. 2, 4, and 8). In the end, she finds that for her business, it is important to separate the cases of high *turnover* and low *turnover* from average *turnover*.

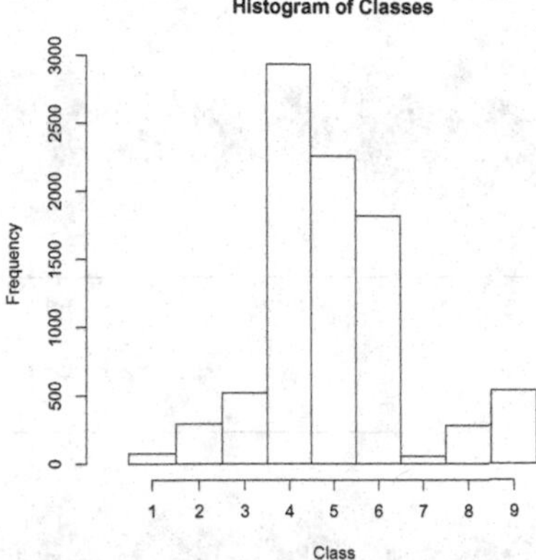

Fig. 9. A histogram [37] showing the distribution of events classes in the resulting event data (in this example, the classes are taken from Fig. 5)

The reason is that she needs to find the definite causes for the high *turnover* (hoping to reproduce them), and for low *turnover* (hoping to prevent them). Sacrificing equiprobability regarding *turnover* might be necessary. Based on her domain-knowledge, she deems a high *turnover* of about one third higher than the standard deviation and a low *turnover* of about one third lower then the standard deviation as the most important cases. Therefore, she sets manual event classes for *turnover* (see Fig. 5). This results in the categorization of nine different types of events. These classes are represented by the different gray levels of points in the lower right of Fig. 5. Of course, this view is not the solution for the whole task, but a crucial step that enables the application of Data Mining methods relying on those classes. In those future steps, sample questions that could be answered are, which situations can lead to low turnover even if there are average customers, and how high turnover can be achieved even with average customers. The classes for these situations are provided.

5 Expert Review

We have placed our user interface under review by two different experts who both have been working in a university environment with close ties to industry. They received a preliminary version of the paper and information about the focus of the review by E-Mail and replied the same way. We analyzed the textual reports and implemented as much as possible in the actual paper, while scheduling the rest for future work. As there were only two expert reviews, more complex analysis was not necessary. Goal of the reviews was to get insights about the general applicability of the approach and of possible user interaction pitfalls prior to further implementation work.

Review 1. focused on the applicability of our results for KDD. This review was done by an expert for Visual Data Mining, Temporal Data Mining, Information

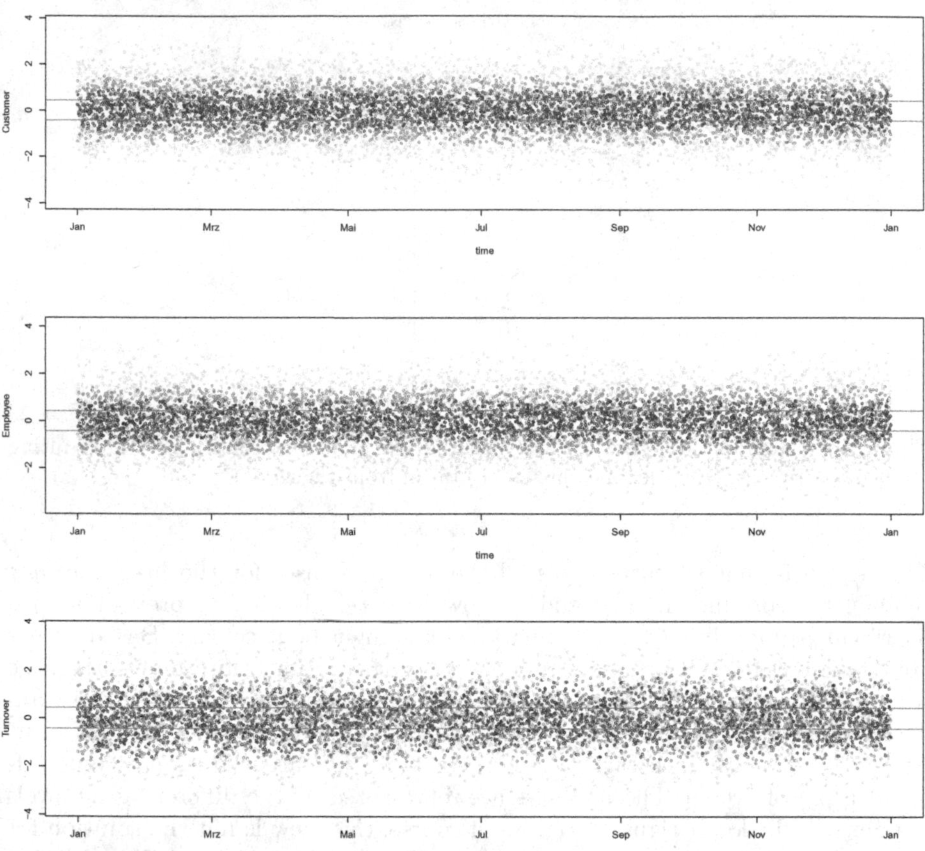

Fig. 10. Separated Time Plots: the vertical axes gives the normalized data value, the horizontal axis shows time. The points are shown in three different gray levels: each level means affiliation with one specific event class. The projected class borders are shown in green. As more than one data dimension influences the borders, the affiliation is not fully conferred by these borders.

Visualization and Visual Analytics who took about two hours to get familiar with the topic and three hours for the evaluation:

The expert considers the method "definitely of high interest, as the selection of the most suitable event classes for a given task is usually not obvious and a bad choice may negatively influence the further steps of the analysis". The description is rated as "clear and well supported by references" and the expert sees "a clear advantage for the user performing the first stages of his/her analysis, or refining not satisfactory results". Two problems are mentioned:

1. The expert is concerned about the number of attributes that can be accounted for in the visual interface and demands a better explanation for such situations. Furthermore, he proposes a Scatterplot Matrix instead of a single projection. *We improved our explanation on how to deal with more than three variables as well as our explanation of user interactions in the data distribution views. Furthermore,*

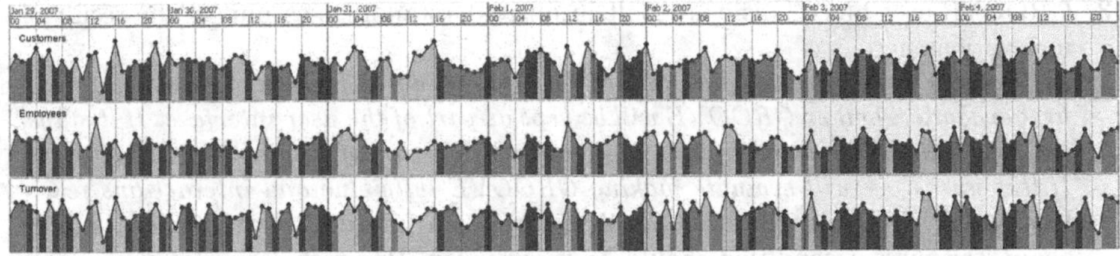

Fig. 11. The SemanticTimeZoom (STZ) adopted from Midgaard [38,39] can interactively combine numerical raw data with qualitative events

we included a discussion of the limitations of two dimensions as well as possible future developments in Section 6.

2. The expert considers the choice of the event classes "already a part of the visual analysis". Therefore, in his opinion, our technique provides a bit more than only data simplification. *We pick up the topics of visual analysis, time, and steps of the KDD process in Section 6.*

In sum, it can be said that the expert was very positive about the applicability of the method for KDD while seeing some user interface issues that we intend to deal with in the future.

Review 2. focused on the usability of our user interactions. This review was done by an expert for Interactive Visual Interfaces, Visual Decision Support, Temporal Representations, HCI, Information Visualization and Visual Analytics who took one hour to get familiar with the topic and one hour for the evaluation. We provided a list of tasks. The expert solved these tasks by hypothetically applying the user interface. Based on the answers, we made some changes to the user interface. The labels of several visualizations had to be clarified. In total, the expert had no problem dealing with the interface, but the statistical background needed to fully employ the interface seems to be rather high. Still, we think that a decent class definition result can be achieved by domain experts. The result becomes better with more statistics skills and domain knowledge—which both are even harder requirements for other methods. After answering the questions, the reviewer gave an assessment of the usability:

1. Inconsistent: when I edit textboxes or drag ellipses, the new borders are shown in blue. For the +/− buttons, this already happens when I hover them with the mouse. It would be better if that happened when the buttons are pressed.
   There actually is a small inconsistency here. However, in the current state of development, textbox content is not actually changed before pressing Enter. *The same is true for the dragging of class borders: They are not changed till releasing the mouse. When changing the number of classes with the arrow button, the click already makes the change fixed, so there is no room for a blue preview after the hovering. We think that this has to be reevaluated on an actual implementation.*
2. If I change the borders manually using some rules (with the lower left view), the new borders are shown immediately as green lines. These lines should be blue first and only become green after pressing *Enter*, removing the blue ones (see Fig. 5).
   This is a good input that will be included in the implementation of the interface.

3. GROOVE uses blue. This can result in misunderstandings as blue is the result of interactions here.
   At the same time, green is always used for class borders. To solve this issue, we will first evaluate whether GROOVE will be kept as part of the user interface. If it stays, we will evaluate whether it can be changed to an overlay of saturation and lightness. If this works, we will evaluate making GROOVE fully blue and interactions red.
4. Histogram: if the distributions resulting from an interaction are shown in blue again, the current condition should be green again. However, I am not sure if this visual metaphor should be used here, as no borders are shown, but the number of elements in the particular classes.
   We will analyze several color schemes using the working prototype.
5. There are too many possibilities to show the data, resulting in a high learning curve and confusion while switching between two different visualizations. Why are that many visualizations necessary? What is the advantage of GROOVE and simple line charts over separated time plots? Why the parallel coordinates? The correlations are hard to see in parallel coordinates and there are scatterplots anyway?
   A (smaller) choice of visualizations has to be made. However, we first have to test various visualizations and find out which ones are best, hoping that this is not too dependent on different datasets.
6. There is much overplotting in the line charts and the legend is hard to read. Why are the boxes not filled? They look like checkboxes.
   Currently, we cannot reduce the overplotting. We will look into solutions for line plot overplotting for the working prototype. The legend has already improved in the new version of the mockups shown in this paper.
7. The various time visualizations show different time spans. The time spans should be the same for all visualizations.
   The visualizations are intended to be zoomed and panned. As the mockups are static, they show states that we consider most illustrative.

6 Conclusion, Limitations, and Future Work

We have presented an interactive visual user interface that enables users to define event classes among a set of time-oriented data. These event classes can be (1) equiprobable and work similar to SAX [12], (2) freely configurable as described by Bertone et al. [10], or a combination of automated and user-based methods. The visual interface enables users to interactively develop suitable event classes for their needs. Our approach is in particular oriented to support multi-variate data. The data visualization area, and, to some degree, the event class visualization area, help users to gain an understanding of the data by means of interactive visualization. This understanding goes beyond the data variables—it also extends into understanding time-oriented effects. We consider this understanding advantageous for performing the event class configuration and, therefore, have included it into the user interface even though it surpasses the topic of data simplification. Furthermore, we can conceive a more extensive user interface that also includes other steps off KDD. In such an interface, the visualizations would be absolutely necessary. They would also help possible expansions of the user interface that let users provide names for event classes and their characteristics or even place them into an ontology. For many data variables, the two-dimensional view is a strong abstraction of the actual distribution. As a 2D-display and traditional user input devices make showing more dimensions hard to impossible, this limitation cannot be overcome directly. However, the application of linked views like a Scatterplot Matrix that show projections on different dimensions might help. Still, the number of variables that our approach can deal with in a meaningful way might be limited. To explore that actual limits of the method, a working prototype will be used. We are in the planning stage of such a prototype that will also include the proposed changes explained in Section 5. We can currently apply statistical classification under the assumption that the data has a Gaussian distribution and the Euclidian distance is an adequate measurement. Lin et al. show that this is sufficient for many important use cases [12] (see Section 2). On the other hand, Vlachos et al. [40] show that there are different tasks which require other classification and distance methods. Our approach can be extended to incorporate such methods. Furthermore, our approach is developed with time-oriented data in mind, as time is a complex data reference [15] and its complex structure is important for users [17,18]. However, other reference dimensions, like space, are also very important and show similar structures. As we focus on leaving time untouched, for other methods to deal with its structure, it is easy to extend our method to other reference domains. Many methods, like SAX, apply rasterization of the time dimension prior to further methods, because time-oriented data can be huge. However, rasterization can hide important information. Therefore, finding an optimal raster size is an important issue that can be dealt with by applying interactive visual interfaces. Others problems that are closely related to rasterization are dealing with missing values and outliers. Time-oriented data and the information contained in it is heavily influenced by certain aspects of social life and other phenomena [15]. To perform KDD on time-oriented data, special methods have been developed [10,11], but many more are needed to fully deal with all those aspects. Inventing such methods seems to be the next important step after successfully performing data simplification. Approaches like Activitree [31] show that interactive visual interfaces are an outstanding way to make such methods more accessible and effective.

Acknowledgements. This work was supported by the FWF Austrian Science Fund via the HypoVis project (#P22883).

References

1. Piateski, G., Frawley, W.: Knowledge discovery in databases. MIT Press (1991)
2. Hand, D., Mannila, H., Smyth, P.: Principles of data mining. MIT Press (2001)
3. Mannila, H., Toivonen, H., Inkeri Verkamo, A.: Discovery of frequent episodes in event sequences. Data Mining and Knowledge Discovery 1(3), 259–289 (1997)
4. Laxman, S., Sastry, P.: A Survey of Temporal Data Mining. Sadhana 31(2), 173–198 (2006)
5. Agrawal, R., Imieliński, T., Swami, A.: Mining association rules between sets of items in large databases. ACM SIGMOD Record 22(2), 207–216 (1993)

6. Srikant, R., Agrawal, R.: Mining sequential patterns: Generalizations and performance improvements. In: Apers, P.M.G., Bouzeghoub, M., Gardarin, G. (eds.) EDBT 1996. LNCS, vol. 1057, pp. 1–17. Springer, Heidelberg (1996)
7. Magnusson, M.: Discovering hidden time patterns in behavior: T-patterns and their detection. Behavior Research Methods 32(1), 93–110 (2000)
8. Chen, Y., Chiang, M., Ko, M.: Discovering time-interval sequential patterns in sequence databases. Expert Systems with Applications 25(3), 343–354 (2003)
9. Hu, Y., Huang, T., Yang, H., Chen, Y.: On mining multi-time-interval sequential patterns. Data & Knowledge Engineering 68(10), 1112–1127 (2009)
10. Bertone, A., Lammarsch, T., Turic, T., Aigner, W., Miksch, S., Gaertner, J.: MuTIny: a multi-time interval pattern discovery approach to preserve the temporal information in between. In: Proc. of ECDM 2010, pp. 101–106 (2010)
11. Bertone, A., Lammarsch, T., Turic, T., Aigner, W., Miksch, S.: Does Jason Bourne need Visual Analytics to catch the Jackal? In: Kohlhammer, J., Keim, D. (eds.) Proc. First International Symposium on Visual Analytics Science and Technology held in Europe (EuroVAST 2010). Eurographics, pp. 61–67 (2010)
12. Lin, J., Keogh, E., Wei, L., Lonardi, S.: Experiencing sax: a novel symbolic representation of time series. Data Mining and Knowledge Discovery 15(2), 107–144 (2007)
13. Faloutsos, C., Ranganathan, M., Manolopoulos, Y.: Fast subsequence matching in time-series databases. ACM SIGMOD Record 23(2), 419–429 (1994)
14. Lin, J., Keogh, E., Lonardi, S., Chiu, B.: A symbolic representation of time series, with implications for streaming algorithms. In: Proc. 8th ACM SIGMOD Workshop on Research Issues in Data Mining and Knowledge Discovery, pp. 2–11. ACM (2003)
15. Aigner, W., Miksch, S., Schumann, H., Tominski, C.: Visualization of Time-Oriented Data. Springer (2011)
16. Andrienko, N., Andrienko, G.: Exploratory analysis of spatial and temporal data: a systematic approach. Springer (2006)
17. Smuc, M., Mayr, E., Lammarsch, T., Bertone, A., Aigner, W., Risku, H., Miksch, S.: Visualizations at First Sight: Do Insights Require Training? In: Holzinger, A. (ed.) USAB 2008. LNCS, vol. 5298, pp. 261–280. Springer, Heidelberg (2008)
18. Smuc, M., Mayr, E., Lammarsch, T., Aigner, W., Miksch, S., Gärtner, J.: To Score or Not to Score? Tripling Insights for Participatory Design. IEEE Computer Graphics and Applications 29(3), 29–38 (2009)
19. Keogh, E., Chakrabarti, K., Pazzani, M., Mehrotra, S.: Dimensionality reduction for fast similarity search in large time series databases. Knowledge and information Systems 3(3), 263–286 (2001)
20. Keogh, E., Chakrabarti, K., Pazzani, M., Mehrotra, S.: Locally adaptive dimensionality reduction for indexing large time series databases. ACM SIGMOD Record 30(2), 151–162 (2001)
21. Yule, G.: On a method of investigating periodicities in disturbed series, with special reference to wolfer's sunspot numbers. Philosophical Transactions of the Royal Society of London. Series A, Containing Papers of a Mathematical or Physical Character 226, 267–298 (1927)
22. Marx, M., Larsen, R.: Introduction to mathematical statistics and its applications. Pearson/Prentice Hall (2006)
23. Apostolico, A., Bock, M., Lonardi, S.: Monotony of surprise and large-scale quest for unusual words. Journal of Computational Biology 10(3-4), 283–311 (2003)
24. Lonardi, S.: Global detectors of unusual words: design, implementation, and applications to pattern discovery in biosequences. PhD thesis, Purdue University (2001)

25. Nguyen, T.T., Skowron, A.: Rough set approach to domain knowledge approximation. In: Wang, G., Liu, Q., Yao, Y., Skowron, A. (eds.) RSFDGrC 2003. LNCS (LNAI), vol. 2639, pp. 221–228. Springer, Heidelberg (2003)
26. Chen, H., Lv, S.: Study on ontology model based on rough set. In: Int. Symp. on Intelligent Information Technology and Security Informatics, pp. 105–108. IEEE (2010)
27. Tominski, C.: Event-based concepts for user-driven visualization. Information Visualization 10(1), 65–81 (2011)
28. Klimov, D., Shahar, Y., Taieb-Maimon, M.: Intelligent selection and retrieval of multiple time-oriented records. Journal of Intelligent Information Systems 35(2), 261–300 (2010)
29. Klimov, D., Shahar, Y., Taieb-Maimon, M.: Intelligent visualization and exploration of time-oriented data of multiple patients. Artificial Intelligence in Medicine 49(1), 11–31 (2010)
30. Wang, T., Plaisant, C., Shneiderman, B., Spring, N., Roseman, D., Marchand, G., Mukherjee, V., Smith, M.: Temporal summaries: Supporting temporal categorical searching, aggregation and comparison. IEEE Trans. Visualization and Computer Graphics 15(6), 1049–1056 (2009)
31. Vrotsou, K., Johansson, J., Cooper, M.: Activitree: interactive visual exploration of sequences in event-based data using graph similarity. IEEE Trans. Visualization and Computer Graphics 15, 945–952 (2009)
32. Funkhouser, H.: A note on a tenth century graph. Osiris, 260–262 (1936)
33. Inselberg, A., Dimsdale, B.: Parallel coordinates: A tool for visualizing multidimensional geometry. In: Proc. 1st Conf. Visualization 1990, pp. 361–378. IEEE (1990)
34. Lammarsch, T., Aigner, W., Bertone, A., Gärtner, J., Mayr, E., Miksch, S., Smuc, M.: Hierarchical Temporal Patterns and Interactive Aggregated Views for Pixel-based Visualizations. In: Proc. of IV 2009, pp. 44–49. IEEE (2009)
35. Keim, D., Kriegel, H.P., Ankerst, M.: Recursive pattern: A technique for visualizing very large amounts of data. In: Proc. IEEE Visualization (Vis 1995), pp. 279–286 (1995)
36. Tufte, E.R.: The Visual Display of Quantitative Information. Graphics Press, Cheshire (1983)
37. Pearson, K.: Contributions to the mathematical theory of evolution. ii. skew variation in homogeneous material. Philosophical Transactions of the Royal Society of London. A 186, 343–414 (1895)
38. Bade, R., Schlechtweg, S., Miksch, S.: Connecting time-oriented data and information to a coherent interactive visualization. In: Proc. SIGCHI Conf. Human Factors in Computing Systems, pp. 105–112. ACM, Vienna (2004)
39. Aigner, W., Rind, A., Hoffmann, S.: Comparative evaluation of an interactive time-series visualization that combines quantitative data with qualitative abstractions. Computer Graphics Forum 31(3), 995–1004 (2012)
40. Vlachos, M., Kollios, G., Gunopulos, D.: Discovering similar multidimensional trajectories. In: Proc. 18th Int. Conf. Data Engineering, pp. 673–684. IEEE (2002)

Organizing Documents to Support Activities

Anna Zacchi and Frank M. Shipman III

Texas A&M University, College Station, TX, USA
AnnaxZacchi@gmail.com, shipman@cse.tamu.edu

Abstract. This paper describes a study in the wild of software for document organization aimed at supporting activities. A preliminary study of current practices with traditional tools was followed by the design and development of a program called Docksy. Docksy introduces workspaces explicitly zoned by movable panels and features document descriptors augmented with tags, comments, and checkboxes. Docksy was deployed for at least two weeks and users' practices with the new tool were studied. The aim of the study was to see how the new tool was appropriated and how people used the new features. The workspace structured in panels was shown to support users in clustering and separating documents, in having a holistic view of the document space, in locating files inside a workspace, and in managing temporary files. The study also shows how tags, comments, and checkboxes afforded the use of documents as explicit items in a workflow. The study suggests Docksy supports users in a variety of information and activity management tasks, including new practices for emerging activities.

Keywords: Personal Information Management, activity management, user evaluation.

1 Introduction

The computer is the primary work environment for many people today. The traditional systems offer the file system and the desktop as the only tools to manage files. The file system affords storing and retrieving files, but people today use the computer as a place to carry out activities. People are not only dealing with an increasing number of documents but they are also using the computer for many simultaneous work projects and activities. This research explores the management of people's personal environment and in particular focuses on ways people use documents and how they structure the environment to work on short-term and long-term activities and projects.

The area of research on document organization is vast. Studies found in the literature have analyzed organization of both paper and electronic documents. Prior research focuses on various aspects of organizations; among them on the organizational strategies, such as pile vs. file [1, 2], on folder structure [3], on the management based on the classification of documents, such as ephemeral vs. archival [4-6], on paper and electronic archiving strategies, on retrieval strategies [7], on relationships among job and type of organization, such as managerial vs. clerical [1, 8], and on the influence of context [9, 10].

A. Holzinger and G. Pasi (Eds.): HCI-KDD 2013, LNCS 7947, pp. 420–439, 2013.

Other studies focused on electronic document organization aimed at understanding the relation to task accomplishment or project management. Some studies focused on the way users decompose projects in folders [11], on the relationship between documents and organization and activity workflow [12], or on typical problems people have in creating and using their virtual workspaces [13]. Similarly to Kaptelinin, this research focused on the practices relative to activities or projects and workspace management, but, besides studying people working on more recent computer systems, it focused on different aspects of the personalization of the workspace; among them the use of the desktop, file names and start menu.

New technologies have been developed in support of document organization, with the goals of facilitating archival and retrieval tasks [14], of integrating different types of data files [15], and with several other objectives [11, 13, 16-23]. Colletta [24] and co-Activity Manager [25] are activity-based desktop manager that augments MS Windows with sharable virtual desktops, each one of virtual desktop collecting resources for an activity. co-Activity shares activities among collaborators. The system presented in this paper, Docksy, introduces features, such as panels, not previously used in the management of documents, and features, such as comments or checkboxes, not previously available in document widgets.

The research includes two user studies. Initially a field study is conducted to learn about current practices. Then a new tool is designed and developed, and it is studied in the wild. Rogers explains that "Designing in the wild differs from previous ethnographic approaches to interaction design by focusing on creating and evaluating new technologies in situ, rather than observing existing practices and then suggesting general design implications or system requirements. [...] Instead of developing solutions that fit with existing practices, there is a move toward experimenting with new technological possibilities that can change and even disrupt behavior" [26]. The purpose of the final study was to put in the hands of the user a tool with new features, and see how users could use them, which new uses or strategies they could come up with. The study uncovers interesting new practices afforded by the tools, and generates ideas for developing tools with some of the features introduced in Docksy. At the same time the study highlights the limitation of a study in the wild with a system that is not at production level yet. When is a tool good enough to collect information? How to have insight when the tool is in a prototype stage?

2 Preliminary Study

In the summer of 2006, a preliminary study [27] was performed to observe how people use documents on their computer desktop or in their folders structure to organize activities and tasks. The study informed the design of the system Docksy. The focus of the preliminary study was organization of documents in relation to task accomplishment or project management. The research questions were:

1. How do users integrate document management with their particular tasks?
2. Which kind of organization do users use?
3. Why did users choose a particular way to organize files?

Thirty faculty, students, and staff at Texas A&M University participated in the study. The study consisted of video recorded semi-structured interviews concerning computer use and observations of strategies for organizing documents and other resources. Each study session lasted between twenty minutes and one hour. Participants were using a variety of operating systems: four participants used Mac OSX, three Linux and 23 Windows XP.

2.1 Results

Portal. One of the things analyzed during the study was how participants structured their work environment, where they built it, which tools they used to characterize it, and which strategies they used to manage files in this environment. Each participant would typically start his activities in either the desktop or in a special folder. This starting point will be referred to as the portal or home. Participants would set up the portal in such a way to have an overview of the projects or tasks at hand. They would store the resources or the links to those resources for the projects that they are working on. 50% of the participants used the desktop as portal, 50% used a folder.

The desktop was used by the majority of participants, 87%, even the ones using a folder as portal. Participants placed documents and shortcuts on the desktop for variable time length and some gave it a structure. Most of the participants (80%) used the desktop as a temporary place before moving documents into folder in the file system, before transferring them to another device (for example to a PDA), before sending them as an attachment by email, or simply before using and deleting a document. Only one participant used the desktop exclusively for transient files, all others used it also to organize more long term resources.

Workspace Personalization. Participants arranged resources in order to have those currently used at hand. Resources included documents, folders, applications and their shortcuts. Besides arranging them in the portal of choice, they also utilized the XP quick launch bar, the XP start menu, or the Mac OS X dock. Participants customized the start menu (13%) and the launch bar (33%) by placing also shortcuts to most frequently used folders, besides applications. The main rational for the customization was to find more frequently used resources more easily; in one case one participant arranged his resources to also optimize mouse movements on the screen. Customization in the start menu consisted in grouping similar programs, changing the names of shortcuts or adding new ones. For example one participant grouped all graphical applications. Another professor used the icon of a tree for the folder that contained files for his research on fruit trees and placed this icon to the taskbar.

Organization Inside a Folder. Locating files on the desktop or inside a folder my not be an easy task when the number of files is high. To locate files inside folders participants frequently used sorting. Approximately half of the participants left the default alphabetical sorting and half occasionally switched the sorting attribute. They sorted by date, by name, or by type, with sorting by name and by date among the most used. The choice of the sorting attribute changed with participants' needs or circumstances, with attribute switching occurring even in the course of the same search. Some switched for example and sort by name. One participant used "by type" on the desktop in order to have all the shortcuts grouped in the columns on the extreme right. Another participant employed "by type" in folders containing her papers so that she could visually isolate the pictures from the text.

Participants also forced special orders by modifying the names of the files. 13% of users employed a chronological order based on the name of the files, instead of the file time stamps: they added for example a date in the filename. A professor prepended the year to the name of folders relative to classes taught: "2012_EE101 Circuits". The year he used in the filename was not the actual date of the file, the created or last modified date. He also prepared folders for classes that he will teach in the future. This forced chronological order was how he wanted his folders to appear. The strategies used to force certain sorting and the switching between one sort attribute and the other attest to the importance for users of the organization of documents on the desktop or inside folders.

File Location and Identification. Besides sorting, participants used strategies to make the files they are interested in stand out and to make their identification inside folders easier. There are two connected aspects: location and identification. Location refers to the pinpointing of a file inside a folder, and identification to recognizing the right document among several documents with similar names.

Strategies for location included sorting and icon personalization. Participants used color and modified icons to make a file stand up inside a folder or on the desktop. 13% of participants (one Mac and three XP users), changed the icons of the most frequently accessed folders and one Mac user used colors. She used red and green for urgent and important documents. One user tried to find a significant icon among the few provided by his system. He said "*I used it to distinguish between folders that I tend to use a lot. It gives me a better idea of what is going on with them*" [27].

Strategies for identification included filename coding schemes and colors. Participants extended file names with comments about the content of file, such as version, name of collaborators, status or date.

Temporary Files. Participants had files that neither were archived in the folder structure nor did they belong to current activities. These files can be divided in three categories:

- To-Process. Files that participants intend to look at and then throw away. They may eventually throw them away after some time, if they realize they do not have time or if the documents became outdated in the meanwhile.
- To-Keep. Files participants plan to file away in an archival structure.
- To-Throw. Files participants plan to delete but they have not done it yet. They may postpone it indefinitely.

Participants employed different strategies for dealing with temporary documents. Some scattered To-Process files on the desktop surface or hided them in a folder. Others used the system temporary directory or created one or several in the folder structure. One user created a "temp" directory on every network disk she was working on "*I am a big fan of the temp directory. I use it a lot but nothing there is really important. Everything in the temp directory can be deleted*" [27]. To-Keep files often were kept around because participants did not know where to put them or did not file them due to time constraints. To-Throw files were often kept in "My Documents" or in the root of the hard disk or home directory. One participant kept 30 folders under "My Documents" and hundreds of loose files: "*The folders are what is really important, but the documents under the root directory are not. They end up there and I don't use them anymore*" [27]. Participant did not always have a clear separation of those temporary files, and mixed together the different types of temporary and not temporary documents. Sometimes among the tens of files to throw away, there were files they intended to keep, but time constraints kept them from sifting them out.

Notes. While working on tasks or projects it is common to maintain notes or to-do lists. During the preliminary study participants' practices regarding notes were observed with the purpose of understanding the roles of notes in the organization of project resources. Notes were either in paper (33%) or in electronic form. Two thirds of the participants used an electronic form of notes at some time. Participants wrote notes specific to projects, similar to "readme.txt", and/or general notes. The general notes usually contained to-do lists, and were maintained in the main directory, or often on the desktop. Project specific notes were relative to projects and were maintained in the folders relatives to the project. One user kept a file called "notes" in every project directory. Another user kept a sort of diary that he called "log" on the desktop in MS WinWord format. In it he wrote all his daily activities.

2.2 Conclusion

This preliminary study showed a range of practices employed by participants to structure their activities using files. The design of the system proposed in the next section aims at providing software that better supports the current user practices regarding their work environment, while at the same time giving them the freedom to explore new practices.

3 Design

The design of the new system combined three different objectives. First, to explore the idea of expanding the concept of a computer file system and its interface to a work environment as opposed to a storage and retrieval system. Second, to support the practices related to the organization of documents uncovered during the preliminary study. Third to develop a system that is malleable enough to allow users to appropriate it for new uses or new practices.

Among the practices observed in the preliminary study, the focus is on those that allow users to integrate the organization of documents into their daily work. The following characteristics are considered important. Users should be able to:

- easily create project environments
- use files to more explicitly structure activities
- add project-specific metadata information (textual annotation) to files.

The design aims at providing infrastructure that enables the expression of meta-information about documents and the use of document representation to express information about activities. The document is traditionally represented at the interface level by the document icon, the file name, or a combination of the two. The new system augments the document representation with checkboxes and text areas that allows the insertion of tags and comment. In this way the representation of the file is transformed into something more than a handle to the file. The file representation can store information about the file; moreover activities and work flows can be structured using the new file representation.

The new system is also designed to provide an explicit portal or workspace for each project. Each workspace will provide support for the current user practice of zoning the work area. In the preliminary study users clustered documents on the desktop to circumscribe and separate different projects, or used different areas of the screen as reminder for operations to do [2].

4 Implementation

Docksy is implemented as a standalone program in the Java programming language. It presents the user with an alternative desktop composed of different workspaces, where each workspace is overlaid with panels. Fig. 1 shows one workspace with four colorful panels. Panels are areas of the desktop where users can gather together documents. The user can name panels and swap their position by dragging them. He can also resize them or change the number of the rows and columns. Fig. 2 shows one workspace with six panels and the effect of resizing.

The user can drag and drop documents from the traditional file system into panels. Once the files are in Docksy, they are displayed in a widget consisting of an icon, a title, a comment text field, a tag text field, and a series of checkboxes or flags (Fig. 3). The document widgets can be dragged from one panel to the other inside the same workspace. The widgets can be resized or deleted. Figure 3 shows five documents inside a panel named "TO DO". The document titled "Capture-Copy.JPG" has five flags or check boxes, three of which have colors, the tag area, and the comment area. Each checkbox has a label that can be shown as a tooltip.

5 Evaluation

"A central part of designing a system in the wild is evaluating the system in situ" [26]. Evaluation in the wild has many challenges. How to observe users? What to observe? For how long to observe? How to collect data? The evaluation of a system such as Docksy is made more challenging by the fact that Docksy is not an application that people use directly to accomplish a task, such as it could be a text editor for writing a text or a phone to make phone calls. Docksy is a somewhat background application that users use on their way to use other applications. Paradoxically, if a user has a system structured in such a way that he can quickly find or use the information he needs, spending few time on it is a positive aspect. At the same time structuring your information could be a meaningful task. Suppose you dump in a folder photos relative to different animals. When you have several of them, you can start seeing some patterns, and split your collection in useful subgroups. In this example spending time in the system is positive.

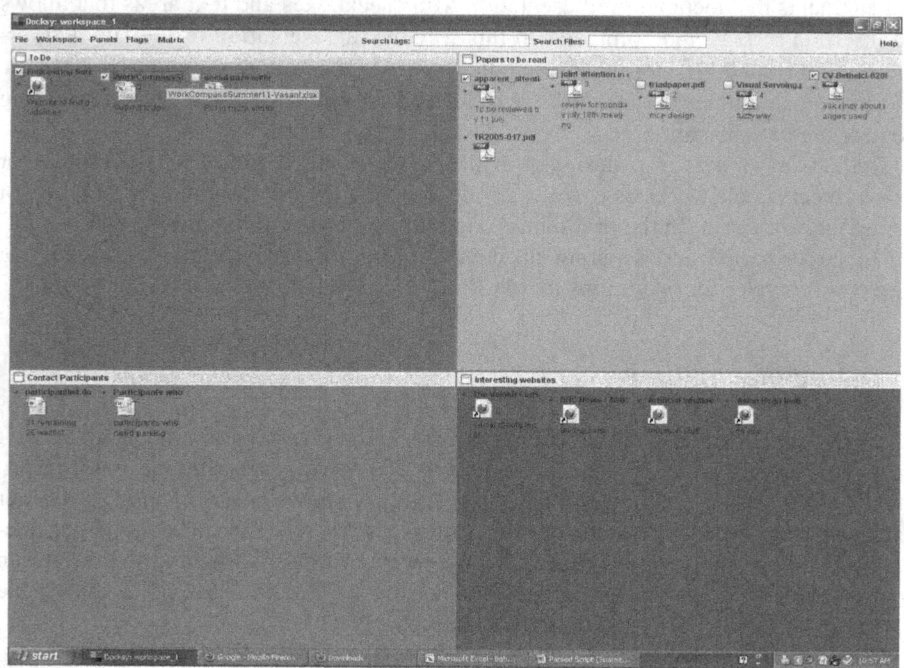

Fig. 1. Docksy's screenshot from a study participant showing one workspace with four panels

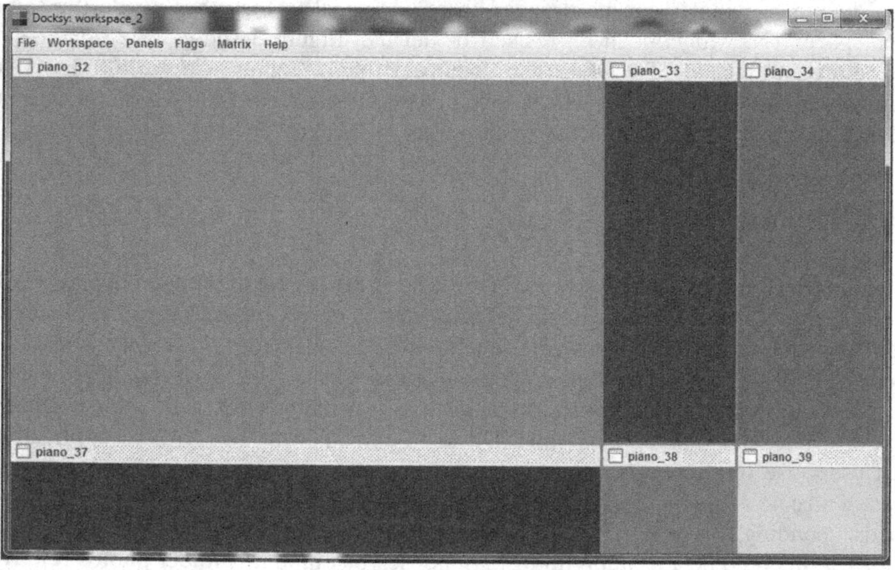

Fig. 2. Docksy: resizing panels

The solution adopted was to evaluate Docksy by looking at the possibility of the system to be adopted by users, by observing the effective support of current user practices, and by looking at its potential for changing or introducing new user practices. To do this Docksy was installed on the users' computers and participants were let to use it freely for a couple of weeks. After this period the study was run. The study consisted in a structured interview during which participants showed what they did with Docksy, which kind of structure or organizations they accomplished with the system and which rational was behind the organization. They also filled an online survey to evaluate the single features individually, such as the checkboxes or the comment areas. In short, the main research questions for Docksy's evaluation were:

- Which kinds of organization do users use in Docksy?
- Does Docksy support organization of documents for projects?
- What is Docksy's potential for changing user practices?

In the summer of 2011, 20 participants, faculty, staff, and students used Docksy for at least two weeks; Docksy was installed on their MS Windows XP, Vista, or Windows 7 systems. They were then subjected to a semi-structured interviewed about their use of Docksy and were given an online survey. The interview started by asking participants to show what they did with Docksy, and then it continued according to the particular situation at hand. The answers to open-ended questions in the survey were collected and the interviews were transcribed. Both were analyzed and keywords and concepts were grouped into thematic clusters. The next subsection reports the answer to the kind of use people did of Docksy. The title of the subsections corresponds to the thematic extracted. Fig. 4 shows a screenshot from a study participant.

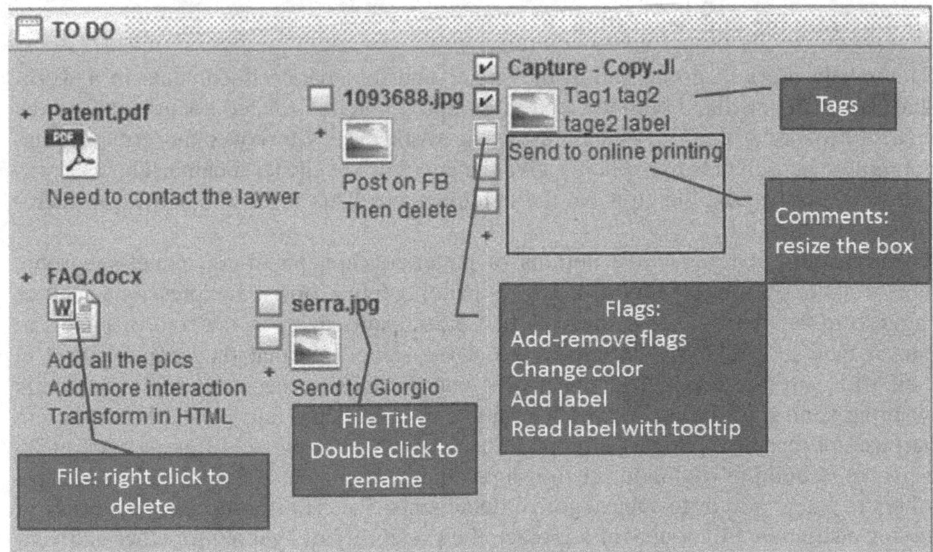

Fig. 3. Docksy's documents widgets: comments, flags, tags

5.1 Uses

Participants used Docksy for general purpose or specific projects. Some participants said that they used Docksy for general day to day activities, some for the same activities they would normally use the desktop for, and others for a specific project. A sample of the answers is: *"organizing files that had stacked up on the desktop"; "general day to day work"; "my thesis research"; "reshaping and integrating my archives"; "organizing reference material"; "organizing my pictures"; "multidimensional sorting"; "organizing documents for the project I am working on".*

5.2 Holistic View

The three aspects that participants found most useful are related to the panels: first the possibility of seeing all the documents a person is working on in the same place, to have a holistic view. Second, the fact that users do not have to open multiple folders and browse the contents to find the files they are looking for. Participants often complained of having to do too many "clicks" in the folder directories to locate files, and search for the right folder in the traditional system. Third, the possibility of easily locating a file inside a workspace, because even though all the files are present in the same workspace, they are categorized or divided by panels, making them easier to locate.

Here I get a holistic view of the documents I use on a day to day base. [...] I think it is better than the traditional file browser; I don't have to use a lot of clicks and go through every folder and see all the files. I can see all of them in one place. [Participant 10]

5.3 Grouping and Separation

Panels help users to group and at the same time to separate documents in a workspace. Participants liked both the idea of grouping together related documents and the idea of separating them in panels inside the workspace. Moreover the grouping did not require participants to explicitly give the panel or the cluster a name. The user was free to name the panel or to leave the default name. This feature is helpful in incremental formalization.

Participants juxtaposed the notions of panel-concept, panel-tag, panel-subgroup, panel-relationship, panel-logical-network, panel-activity, or used expressions such as "aggregate the spaces". This suggests that participants mapped the role of panels to one or more of the above concepts, and at the same time that they appropriated in their own way the concept of panel. For one participant the panel had a semantic meaning similar to that of a label. Grouping of documents into panels had different purposes or meanings. Some participants used panels to organize different work phases, some to build or highlight relationships among documents, similar to mind-maps, others to categorize or to subcategorize documents. Panels afforded the possibility of having a structured overview of a project, the possibility of keeping together and visible all documents relative to one subject, but at the same time to separate them. The separation also helped them in restricting the search or the browsing of documents to a subset of files, improving the efficiency of locating files.

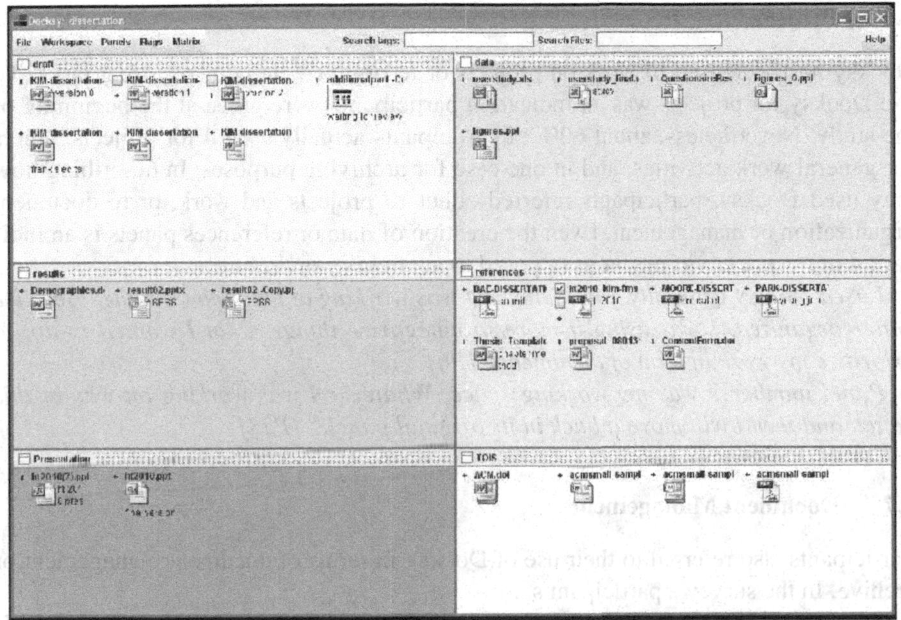

Fig. 4. Docksy's screenshot from a study participant

5.4 Focus

Some participants stated that Docksy and its panels feature helped them better focus on the current work, and relax their mind from the clutter.

It is nice to have them organized in different concepts or panels. It is uncluttering the mind. You can just look at it, where it is, and focus only on these things. I don't have to browse up and down and search anywhere. And if I am in the work mood I can just go in the work panel or workspace and my mind is just here, I can just look at what I need to see. [P5]

5.5 Comments, Tags, or Flags

The features comments, tags, and flags or checkboxes, while they were considered useful, they were used by few participants during the study. Participants that used them used tags and comments for several purposes. One use was for prioritizing among documents. Another use was for adding keywords in order to be able to find documents quicker later, or to classify them. Another was to add notes about the content, to avoid opening them again. Another use was for adding extensive comments to the files, to supplement additional information about the content, or to add summaries. Yet another use was for adding information about work or operations to do or already done about that document. Some participants said that they didn't make use of tags because they worked on Docksy only for a limited time, and they would have used those features if they had the chance to use it for a longer period.

If you have a project you can add checkbox and check off whenever you completed something. And add comments, to better remember what that file is. [P2]

5.6 Project

Docksy has been used to work on projects or to organize daily or recurrent tasks. To use Docksy for projects was an indication participants were given at the beginning of the study. Nevertheless, about 60% of participants actually used it for projects, others for general work activities, and in one case for archiving purposes. In describing how they used Docksy, participants referred either to projects and work, or to document organization or management. Even the creation of data or references panels is an indication of a work environment as opposed to an archive environment.

I used Docksy generally, for whatever I was working at that point in time. I felt like better organized. I was using it more to categorize things, what I wanted to do, to prioritize my work in term of documents. [P6]

Panel number 7 was my working space. Whatever I was working on was in this panel, and then I will move it back in its original panel." [P23]

5.7 Document Management

Participants also referred to their use of Docksy in terms of document management or archive. In the survey a participant said:

It allows me to build a meta archive relative to contingent necessities. For example with files for exams to be used in the current month, leaving the original archive unchanged.

5.8 Color

Participants liked the use of panels' background colors, but at the same time they did not like the colors proposed, that were random. Colors were also used as borders on checkboxes, but they were barely noticeable. Overall participants liked colors, but they wished for better choices. Beside the look, colors were useful in subcategorizing the different document groups, and in better segmenting the screen.

"I like being able to choose my colors. That was very helpful, because it makes me feel good with colors, I know it is silly, but ..." [P20]

5.9 Control

A couple of participants experienced a sense of control while using Docksy.

Docksy gives you the impression of control, because you can change the color of the panels, or swap the files around. [P1]

While some participants felt in control, another felt that Docksy guided him to organize the documents.

If the desktop would look like Docksy, they basically have to organize it. It will be cool to see people finding their way into organizing either they recognize it or not. Even if it looks like this, things would still be organized. [P4]

It is interesting that the above participant felt like Docksy, by forcing him to split documents in the various panels, induced him to organize his documents, and that he found this experience positive.

6 Problems

One of the problems of evaluating a prototype in the wild is that the system is not at production level and doesn't contains all the nice features users are used to. For example, users wanted to create files directly in Docksy by right clicking the mouse and using the menu item "New". That was not implemented. A user wanted to have the thumbnails for every file. That also was not implemented. The list of requests was long, but their implementation meant to create a system with most of the interface features available on current Windows OS or Apple OS. Besides the fact that that required having the resources available to a big company, it was also beyond the purpose of testing a prototype. Having a production level product will forego the intermediate stages of testing a tool during the development. Some of the requested features were implemented during the study, such the possibility to use Docksy on a network disk or on a different storage area as opposed to the local hard disk. Also a couple of bugs were fixed during the study. Nonetheless some users struggled to comprehend that a prototype doesn't have to implement everything and they were proactive in suggesting features common to other systems.

Another issue related to the prototype status was that for safety reasons all documents dragged into Docksy were duplicated. In a production level Docksy would not require a sand box but should be fully integrated into the OS. Related to this issue was the request for the possibility to drag files out of Docksy. In a production system there would not be an "out of Docksy".

Another problem was that the prototype status meant that the study would come to an end after a certain period, and this prevented some participants to invest too much into the system. They could keep the system, but since it was not a full-fledged system it meant that its support and usability where somewhat limited. By using Docksy they created a parallel system for some projects. As a user put it "*I will put a lot of effort in adding information to my files and projects, and all of this will be wasted at the end.*"

Notwithstanding the prototype status and the study period with all their consequences, problems were observed related to the usability of the features under testing and issues that raise higher-level questions about the design of Docksy.

On the interface level the choice of colors, which was random, was uncomfortable for users. Participants liked colors, but not the strong colors that came out randomly. Changing the number of panels was a task that some found not very intuitive. Another problem was the use of the comment and tag text areas. Some user didn't distinguish between the two, and actually there was nothing to distinguish them besides the position and the description in the help file.

On the design front, the main problem was the lack of scrolling, or of an equivalent feature to browse large quantity of files inside a panel. A missing layout inside panels also was a serious issue. Both these problems are in part related to the prototype status; in the design and development the focus was on some aspects while others were foregone, still these aspects had a big impact on the usability of the prototype.

One interesting problem was the conceptual difficulty to move from a system based on hierarchical folders to a system based on workspaces and panels. This is not necessary a problem, but it shows that adapting to a system with a radical different philosophy is challenging, and this influences its adoption.

7 Discussion

The following subsections discuss how the different features of Docksy supported users in their daily working practices and if they met the expectations set forth by the design.

7.1 Panels

During the preliminary study, participants used the desktop as a workspace or as a dashboard, clustered documents on the screen, or used different areas of the screen for different purposes, zoning the screen. For example one user used the center of the screen for the current work, the top left corner for the things scheduled for the week, and the top right for references, such as dictionaries. To support these practices Docksy provided panels that tile the desktop, to make explicit the clustering or the screen zoning.

The results of the Docksy study support the idea that panels were useful in supporting the user practices regarding clustering or screen zoning. Panels were the feature that participants liked the most. They liked the possibility of having background colors, the possibility to drag and drop documents from one panel to the other, and the possibility to title the panels. Besides commenting on interface characteristics, participants reported that panels helped them to better focus on the work at hand, helped them to organize files, and gave them a sense of control.

The feature that users found most important for panels was the possibility to have an overview of all their documents and at the same time to categorize or cluster the files without the need to open and browse the folders.

Participants in the study also reported that panels helped them to better locate papers inside a workspace. Search versus organization and browse is a debated issue in literature [5, 6, 14, 19, 21, 22]. While both approaches have their merits in different circumstances, there is still another related issue: what can help the user find a file when he knows the folder but the folder contains many files and the person doesn't know the title or the title and content search is not helpful. One participant highlighted the problem:

Right now I have a pdf folder and I have all of my pdfs in. But sometimes I need to categorize them, like High pressure freezing, R6, R3, R5, so that way I know this will allow me to go to the exact panel, and not have to scroll to and try to remember which author wrote which paper. So I can categorize all this as R6, which would be my title, and put all of the articles underneath that are related to it. Now I have author and date in the name of the file. And that is very difficult to remember, because I get mixed up, and I don't remember who did what, and I have go click and open the file and that takes a long time, so this will make me more efficient. [P20]

The problem is not only to locate a file in a folder, but also to remember or reconstruct the title of the file, or the content. Participants in both studies often mentioned the problem of locating documents inside a folder. They knew the folder or the parent folder containing the file they were looking for, but they complained that it took too long to locate the document inside it. Either there were too many files in the folder they were looking for, or they didn't know in which subfolder they placed it, and they spent too much time clicking and browsing the different subfolders. The preliminary study shows that participants tackled the problem by using a variety of methods. Docksy provided another way to deal with this problem: panels. Participants in the Docksy study reported that they used panels to subdivide documents inside a workspace that was equated to a big directory. Subgrouping documents inside panels helped reduce their number in a particular group being examined, and therefore speed up the location, while at the same time leaving them all visible instead of obscuring them inside a closed folder. Moreover panel color and spatial position of panels inside the workspace added contextual information that helped to speed up the location of files. Participants reported that the tag feature also helped them in locating files. A participant said that to put a document into a panel was equivalent for him to tag it with a label. Panels turned out to be a simple solution for helping users employ multiple strategies to locate documents inside a workspace.

Panels also offered support for incremental formalization. Each panel was created with a default name that the participant was free to change. The titling of the panels was not mandatory, giving users the possibility of deferring the formalization of their task or work. Not all users named the panels. One participant said:

I didn't put the names on the panels because I just forgot, I got so comfortable with it, I could look right at piano 2 and I immediately know what it is, it doesn't even need a name. It is so visible that I knew it. It never even crossed my mind to put a name. [P18]

Forcing formalization adds overhead that may discourage people or force them to overcommit too soon [28]. Panels provide a way for users to structure a workspace without imposing users to formalize it. Placing documents in different panels without forcing the selection of a title helps users create a structure without requiring too much from them.

The preliminary study described participants using a variety of different temporary files. The Docksy study showed that participants used some panels to place categories of temporary files. For example, some users delegated a panel for documents which they were planning to read, or they created a panel for references that they needed for a current task, or they used a panel for activities to do at a scheduled time. Once the task relative to those documents was accomplished, the file was either deleted or moved to another panel. This suggests that panels were also appropriated as support for managing temporary files.

7.2 Comments, Labels, and Flags

The preliminary study showed that people used a variety of strategies to organize their work. Besides positioning documents in particular places on the screen or in the folder structure, they also used colors to distinguish particular icons, they used note files inside folders, and they used the file title to add information regarding the document. To support those practices Docksy featured comments, tags, and checkboxes or flags to add to the representations or icons of documents.

The Docksy study suggests that those features were useful in structuring users' activities. Only a limited number of participants used tags, comments and checkboxes, but many participants who did not use them saw themselves as using those features in the future, provided that they would have more time to work with Docksy instead of just two weeks, and provided they would feel that their effort will not be wasted at the end of the study.

Participants who used tags, comments, and checkboxes used them to prioritize their work, to add information about the status of the documents, to add information about operations to do with the documents, to add information about the content of the file in order to avoid opening it every time they considered it, and to help locate the file. One participant added a number in the tag to prioritize files. Another user said:

I like the comment thing, if there is a volume of files, I like the fact that you can write stuff down for each file, and what you did, and what you edited, and things of that nature. So I think in terms of organization that is important, because sometimes when I edit, what I do I create a new folder for things that have been edited and I will put the edited ones in the edited folders, where here I just put a comment that it has been edited, or use a checkbox. I just find writing it is easier. [P21]

This suggests that Docksy was useful in supporting users to structure their work, and in using the documents as explicit items in the workflow description.

8 Research Questions

8.1 Which Organization?

The first research question was "Which kind of organization is used in Docksy?" The names of the panels, the interviews and surveys indicate that participants used panels to group documents belonging to subcategories, to build or highlight relationships, or to structure activities. The former can be seen in the case of the user using panels with names such as "data", "draft", "results", "references", "presentation", all related to subsections of a dissertation writing project [P9]. The second can be seen in the user that said that he would use panels to create logical networks among papers, and that documents in each panel are related to each other [P14]. The third can be seen in the case of the participant that used one panel for the work scheduled for the week, one for the work assigned to Monday, one for the work done, and others for other days of the week [P7]. Participant 23's [P23] use of one panel for current work is another example of panels' use to organize activities.

Docksy's workspace was either seen as a working space, a project space, or as an archive. A user said *"I don't see Docksy as a desktop, but as a library where I put documents that I gradually refer to"*. This also shows that Docksy has the flexibility to be used in different ways.

Besides panels, not many participants made use of the additional features in a significant way, i.e. the checkboxes, the flags, or the comments. One reason could be the length of the study period. Those features are likely to be more useful when working on a project for a longer period of time. Another reason could be that some users did not understand how they worked. Visually there was no difference between the tag and the comment text fields, and some participants did not understand the difference. A third reason, that a couple of participants mentioned, is that they did not want to invest energy into building a structure that they could not transfer to their traditional environment once the study was terminated.

8.2 Support for Projects?

The second research question was "Does Docksy support organization of documents for projects?" 59% of users used it to organize projects. It is worth to remember that at the beginning of the study participants were explicitly asked to choose at least one project to work on. While some participants actually used Docksy for projects, others used it to organize general day to day activities, and one person used it as an archive. The organization of day to day activities was still one of the desired outcomes of Docksy. Panels were the feature most used in support of projects. Panels were used in support of projects in three ways. First, participants separated activities to be performed at different times in different panels, and therefore panels had a temporal connotation or they were used as a scheduling mechanism. For example, they assigned to different panels activities scheduled for the week or for the day. Second, they partitioned in different panels documents that referred to subcomponents of a project; for example, one panel assigned to documents for the dissertation, one for the presentation, and one for the data collection. Third, they clustered in different panels documents related to each other. For example, one panel was for papers relative to fifteenth century writers, and one for sixteenth century writers.

8.3 Changing User Practices?

The third research question was: "What is Docksy's potential for changing user practice?" Participants adopted very quickly the panel organization, and some claimed that MS Windows had been lacking this feature for long time. Some participants said that they see themselves as using the comment and checkbox features in the future, assuming that they would have the chance of using Docksy for a longer time.

One participant during the interview reasoned on the way she organized files for the study and how she would do for the future, showing a change in progress between her folder structure in the traditional system, her initial use of Docksy, and her envisioned way of using it in the future.

I will divide the screen into the stages of my research project. I can put panel 1 for all material of my literature review, including my pdf documents and stuff. Panel 2 will be a Word doc plus other scholarly articles involving my method section, and I could store all my data and my interviews in Panel 3. Versus putting all of them in a single folder which is what I have right now. And that is how I would use it. I would use it for each project , as opposed to what I did right now as more organizational tool. [P23]

In Windows XP she was used to put all files in a single folder. In Docksy she initially put different project and activities in the same workspace but she said that in the future she would use one workspace for each project. She was still experimenting on which organization would better fit her work.

Other participants employed organizations of documents not possible in other systems. For example, one user created one panel for the work scheduled for the week, one for Monday, one for "done" and so one. One participant used comments and checkboxes instead of putting documents in folders. Yet another participant used tags to add priorities to documents, to indicate the sequence in which he wanted to read the documents.

One participant expressed very clearly one important aspects of the transition to Docksy:

Before, you need to change your mindset. Docksy helps you change your mental model, the way in which you are able to organize your files. You change the way in which you think to organize your files; before, you need to change. Docksy helps you to change your reasoning, in such a way that by changing your reasoning you can change the way in which you organize. [P26]

The participant used the colloquial meaning of "mental model". Docksy substitutes folders with spaces, folder "boxes" with spatial panels. This requires a different way of thinking about the system. The mental models and the practices associated with the traditional folder system change when transitioning to Docksy change. In general people appropriate new environments and use them according to their new mental models and they develop new practices.

One of the advantages of Docksy is that, although it is different from the current systems, i.e. the current desktop and the file system, it introduces a new environment that is fairly easy to transfer to and that is familiar. In this way, it does not impose a big sudden change; it affords a smooth transition. Participants can gradually move from one environment to the other, and Docksy provides scaffolding. A longer study would be able to give more insight into users' changing practices; however the above cases suggest a direction in which the practices may be changing, and overall they suggest that Docksy has the potential for changing practices.

9 Conclusion

Electronic documents are not just items to be stored and retrieved on the computer. Documents and their organization are part of a bigger work environment. Users view them as components of activities and workflows and make use of documents to structure activities. However current systems offer limited support for these practices. This raised two main research questions. First, what are users' practices in structuring their work environment on the computer? Second, which kind of environment or features will support those practices? The first part of the research addressed the first question drawing on results from a study of current practices. The second question is addressed by developing a new tool with features designed to support the practices emerging from the first study, and by studying the use of the new tool.

Overall, the studies, both the preliminary study and the Docksy study, suggested that an important aspect of organizing files for current tasks or projects is the possibility to have an overview of the workspace. The studies also suggest that participants value the possibility to categorize documents or to segment them in a variety of ways, including explicit spatial arrangement and by adding colors. While a traditional folder structure affords the segmentation of files into different groups, it doesn't provide an overview and force people to open and close folders to locate files, or to regroup files by eliminating subfolders. Participants in the Docksy study expressed their appreciation for the possibility to have an overview and a way to classify their documents different than the one offered by the desktop and folders. They found the mechanism offered by Docksy useful in providing an overview and segmentation at the same time. Participants also valued the possibility of adding information to documents, both in textual form and in other forms such as checkboxes or colors.

The final study suggests that workspaces structured in panels provide support to users' practices, and in particular for zoning the screen or clustering documents, for locating files inside a workspace, and for managing temporary files; moreover Docksy afforded incremental formalization. The study also suggests that tags, comments, and checkboxes are useful in supporting users to structure their work, and in using the documents as explicit items in a workflow description.

The introduction of a new tool in an environment for study purposes is always a challenging effort because tools and practices coevolve. Once the new tool is introduced in the environment, the user may appropriate it by adopting it and by adapting it to his needs. It is part of human nature to use things beyond their intended limits. Therefore the study of a new tool, when it is immersed in a real environment, could change the practices. Designing in the wild takes advantage of these effects, developing tools that change with the practices that both influence and are influenced by. The aim behind Docksy's design was to create a lightweight system, easy to use and flexible enough to be adapted to users' needs; easy and useful enough to be incorporated into users' daily practice, old or new. Such a system could be used to learn about practices and their evolution. Docksy showed the potential for changing user practice and the potential for the system to be adopted by users. Docksy was also a study tool that deployed in the wild helped to identify features useful in the organization of documents for accomplishing tasks, and showed some possible new strategies that users could employ when those features are provided to the users.

Acknowledgements. We would like to thank all participants in the studies, the reviewers for their helpful comments, and the editors.

References

1. Malone, T.W.: How do people organize their desks?: Implications for the design of office information systems. ACM Transaction on Information Systems (TOIS) 1, 99–112 (1983)
2. Whittaker, S., Hirschberg, J.: The character, value, and management of personal paper archives. TOCHI 8, 150–170 (2001)
3. Henderson, S.: How do people organize their desktops? In: CHI 2004 Extended Abstracts on Human Factors in Computing Systems, pp. 1047–1048. ACM Press, Vienna (2004)
4. Barreau, D., Nardi, B.A.: Finding and reminding: file organization from the desktop. SIGCHI Bulletin 27, 39–43 (1995)

5. Nardi, B., Barreau, D.: "Finding and reminding" revisited: appropriate metaphors for file organization at the desktop. SIGCHI Bull. 29, 76–78 (1997)
6. Boardman, R., Sasse, M.A.: Stuff goes into the computer and doesn't come out: a cross-tool study of personal information management. In: Proceedings of the SIGCHI Conference on Human Factors in Computing Systems (CHI 2004), pp. 583–590. ACM Press, Vienna (2004)
7. Ravasio, P., Schär, S.G., Krueger, H.: In pursuit of desktop evolution: User problems and practices with modern desktop systems. TOCHI 11, 156–180 (2004)
8. Bondarenko, O., Janssen, R.: Documents at Hand: Learning from Paper to Improve Digital Technologies. In: Proceedings of the SIGCHI Conference on Human Factors in Computing Systems (CHI 2005), pp. 121–130. ACM Press, Portland (2005)
9. Barreau, D.K.: Context as a factor in personal information management systems. Journal of the American Society for Information Science (JASIS) 46, 327–339 (1995)
10. Kelly, D.: Evaluating personal information management behaviors and tools. Commun. ACM 49, 84–86 (2006)
11. Jones, W., Munat, C., Bruce, H.: The Universal Labeler: Plan the project and let your information follow. In: 68th Annual Meeting of the American Society for Information Science and Technology (ASIST 2005), vol. 42, Wiley Subscription Services, Inc., A Wiley Company, Charlotte, NC, USA (2005)
12. Dourish, P.: The appropriation of interactive technologies: some lessons from placeless documents. CSCW 12, 465–490 (2003)
13. Kaptelinin, V.: Creating computer-based work environments: an empirical study of Macintosh users. In: Proc. SIGCPR/SIGMIS 1996, pp. 360–366. ACM Press, Denver (1996)
14. Dumais, S., Cutrell, E., Cadiz, J.J., Jancke, G., Sarin, R., Robbins, D.C.: Stuff I've seen: a system for personal information retrieval and re-use. In: Proceedings of the 26th Annual International ACM SIGIR Conference on Research and Development in Information Retrieval, pp. 72–79. ACM Press, Toronto (2003)
15. Karger, D.R., Jones, W.: Data unification in personal information management. Commun. ACM 49, 77–82 (2006)
16. Bardram, J., Bunde-Pedersen, J., Soegaard, M.: Support for activity-based computing in a personal computing operating system. In: Proceedings of the SIGCHI Conference on Human Factors in Computing Systems (CHI 2006), pp. 211–220. ACM Press, Montreal (2006)
17. Dourish, P., Edwards, W.K., LaMarca, A., Lamping, J., Petersen, K., Salisbury, M., Terry, D.B., Thornton, J.: Extending document management systems with user-specific active properties. TOIS 18, 140–170 (2000)
18. Henderson Jr., D.A., Card, S.: Rooms: the use of multiple virtual workspaces to reduce space contention in a window-based graphical user interface. ACM Transactions on Graphics (TOG) 5, 211–243 (1986)
19. Voida, S., Mynatt, E.D.: It feels better than filing: everyday work experiences in an activity-based computing system. In: Proceedings of the 27th International Conference on Human Factors in Computing Systems (CHI 2009), pp. 259–268. ACM Press, Boston (2009)
20. Kaptelinin, V.: UMEA: translating interaction histories into project contexts. In: Proceedings of the SIGCHI Conference on Human Factors in Computing Systems (CHI 2003), pp. 353–360. ACM Press, Ft. Lauderdale (2003)
21. Fertig, S., Freeman, E., Gelernter, D.: Lifestreams: an alternative to the desktop metaphor. In: Tauber, M.J. (ed.) Conference Companion on Human Factors in Computing Systems: Common Ground (CHI 1996), pp. 410–411. ACM Press, Vancouver (1996)

22. Cutrell, E., Dumais, S.T., Teevan, J.: Searching to eliminate personal information management. Communications of ACM 49, 58–64 (2006)
23. Moran, T.P., Zhai, S.: Beyond the Desktop Metaphor in Seven Dimensions. In: Kaptelinin, V., Czerwinski, M. (eds.) Designing Integrated Digital Work Environments: Beyond the Desktop Metaphor, pp. 335–354. The MIT Press, Cambridge (2007)
24. Oleksik, G., Wilson, M.L., Tashman, C., Rodrigues, E.M., Kazai, G., Smyth, G., Milic-Frayling, N., Jones, R.: Lightweight tagging expands information and activity management practices. In: Proc. CHI, pp. 279–288. ACM, Boston (2009)
25. Houben, S., Vermeulen, J., Luyten, K., Coninx, K.: Co-activity manager: integrating activity-based collaboration into the desktop interface. In: Proceedings of the International Working Conference on Advanced Visual Interfaces, pp. 398–401. ACM, Capri Island (2012)
26. Rogers, Y.: Interaction Design Gone Wild: Striving for Wild Theory. Interactions, 58–62 (July-August 2011)
27. Zacchi, A., Shipman, F.: Personal Environment Management. In: Kovács, L., Fuhr, N., Meghini, C. (eds.) ECDL 2007. LNCS, vol. 4675, pp. 345–356. Springer, Heidelberg (2007)
28. Shipman, F., Marshall, C.: Formality considered harmful: experiences, emerging themes, and directions on the use of formal representations in interactive systems. The Journal of Collaborative Computing (CSCW) 8, 333–352 (1999)

Author Index